G 26

DUMBARTON OAKS STUDIES

XXIII

The Fortifications of Armenian Cilicia

The Fortifications of Armenian Cilicia

Robert W. Edwards

Dumbarton Oaks Research Library and Collection
Washington, D.C.

©1987 DUMBARTON OAKS
TRUSTEES FOR HARVARD UNIVERSITY
WASHINGTON, D.C.

Published with the assistance of the J. Paul Getty Trust

UG
432
.C64
E38
1987

Library of Congress Cataloging-in-Publication Data

Edwards, Robert W., 1949–
The fortifications of Armenian Cilicia.

(Dumbarton Oaks studies ; 23)
Bibliography: p. xxiii
1. Fortification—Cilicia—History. 2. Cilicia—
History—Armenian Kingdom, 1080–1375. 3. Turkey—
Antiquities. 4. Cilicia—Antiquities. I. Title.
II. Series.
UG432.C64E38 1987 623′.12′09564 86-32948
ISBN 0-88402-163-7

To Peter Kasavan and Eric Palson,

for their dedication to the
principle that the Armenian
monuments of Anatolia should
be studied and preserved

Contents

Preface to the Text

This study is the first attempt to define the military architecture of the Armenian kingdom in Cilicia. Hitherto what we knew of these medieval fortifications was confined to a selection of articles and a single monograph, which described, at times superficially and with some repetition, about twenty-five forts in the easily accessible regions of Cilicia Pedias and the surrounding Highlands. No one had undertaken a detailed survey of the Taurus Mountains encircling the plain, although these barriers were the heart and soul of the Armenian kingdom. Nor had any commentator attempted to identify the peculiarities of Armenian architecture in order to distinguish Armenian construction from the contributions of the other medieval builders in Cilicia. This was the task that I began in the spring of 1973. After many years and dozens of exploratory missions I have nearly trebled the number of medieval forts for which we have accurate plans and descriptions. It is now possible to establish reliable paradigms for the Armenian military architecture in Cilicia. The seventy-five sites that I have studied probably constitute the majority of forts within the confines of the Armenian kingdom, but many still remain to be discovered. Since my surveys covered all the major regions in and around Cilicia Pedias and the Highlands, as well as explorations in the Tracheia west of the Calycadnus, a good geographical cross section is represented.

Prior to the 1930s we have only occasional comments on medieval architecture in the travelogues of professional explorers.[1] The two works that mention more than just a few forts are V. Langlois's *Voyage dans la Cilicie* (Paris, 1861) and Father Alishan's *Sissouan ou l'Arméno-Cilicie* (Venice, 1899). Both authors give summary histories for selected sites, but there is no effort to describe and compare them as architectural entities. The first attempt at a partial survey of Armenian monuments in Cilicia was undertaken by J. Gottwald in the 1930s and published in *Byzantinische Zeitschrift*. Unfortunately his descriptions are uneven, and his plans were executed without the aid of a transit or other measuring devices. In the last fifty years the more general architectural surveys of the northern Levant have included a few "representative examples of fortress construction" in Cilicia; among these are the works of E. Herzfeld and S. Guyer, W. Müller-Wiener, and R. Fedden and J. Thomson. When M. Gough completed his excellent survey of the late antique city of Anazarbus, he included only a few pages of comments on the adjacent Armenian castle, which is one of the largest fortifications in the Middle East. The best studies of Cilician forts were published during the 1960s in the

[1] The most complete bibliography of 19th- and 20th-century travelogues on the region of Cilicia is in Boase, 191–93.

Journal of Anatolian Studies by J. G. Dunbar and W. W. M. Boal,[2] G. R. Youngs, and F. C. R. Robinson and P. C. Hughes. These surveys describe with great precision and sensitivity five of the principal monuments of the Armenian kingdom. In 1976 H. Hellenkemper published our only monograph on the Cilician forts. In it he performed two functions: he collected and summarized what had already been written on the fortified sites in Edessa and Cilicia, adding supplementary comments and a few new plans, and he attempted to study the military architecture in its geographical context. These commentators as well as this writer have been handicapped by the present prohibition on the excavation of Armenian sites. Despite this limitation, I believe that my formal analysis of the surface remains will provide for the first time a reliable assessment of the architectural contributions of the Armenians in the northern Levant.

However, there are still some uncertainties with regard to the military architecture of Cilicia, especially in matters involving the chronology, the medieval toponyms, and the specific identification of non-Armenian forts. That is why this study is also a prelude to a second stage of investigation which will involve the formal and systematic excavation of *all* of these sites. The careful analysis of now buried artifacts will lead to more precise definitions. Considering the Armenian penchant for executing dedicatory inscriptions and the handful of epigraphs that still survive today, many inscriptions, either whole or in part, must await the excavator's shovel. It is quite possible or even probable that some of my conclusions will be altered by new discoveries. In my descriptions of these sites I have been careful to include directions for those wishing to travel to each fortification. I am painfully aware of my own limitations as an archeologist, and I would heartily welcome criticism of my assessments as well as new explorations in Cilicia. As an ever expanding Turkish population settles in the Highlands, the ancient and medieval sites are plundered for their stones with increasing frequency. This vandalism is not due to malice toward the previous occupants of Cilicia but is the result of necessity and ignorance. It is imperative to save these forts in photographs and plans.

The text of this study is organized into two Parts, the Introduction and the Catalogue, followed by three Appendices. The latter deal with subjects that are not large enough to constitute sections of the Introduction and yet are too general to be integrated into any single unit of the Catalogue. The Catalogue contains the description of every fortified complex that I have surveyed in Cilicia. These seventy-five sites are arranged in alphabetical order and can readily be located on my map (fig. 2) by coordinating the degrees and minutes that follow the name of each fort at the beginning of Part II. When a fort appears on any of the other charts in the List of Maps, it is noted in the Catalogue. This list is not a complete glossary of all the charts on Cilicia but only a selection of maps that are readily available in the libraries of North America. In the Catalogue the architectural analysis of each site, which involves a systematic and critical assessment of the masonry, plan, and separate buildings, is prefaced by introductory remarks on the location, adjacent roads, sources of water, precise topography, and specific history (if it is known). Whenever relevant, intervisibility with neighboring sites is noted. For

[2]More recently Dunbar and Boal have published a survey of Azgıt; see Boase, 85–91.

the most part, the sections of the Introduction are a synthesis of the large corpus of material in the Catalogue. Section 1 is designed to give only a historical overview of the period by relating the salient facts of Armenian political history. It is intended as a touchstone to allow the reader of the Catalogue to place the dates and inhabitants of a particular fort into a larger framework. The reasons for intentionally limiting the historical scope are discussed in the opening paragraphs of Section 1. Sections 2 through 5 define the architectural features that distinguish Armenian from non-Armenian construction in Cilicia. Forts are divided into types, and their topographical and structural components are isolated and compared. The reader will discover that the Armenians employ, with an almost uncanny consistency, types of masonry and architectural elements that are peculiar to their culture. Although I have already published a large corpus of Armenian churches and chapels in Cilicia, their distinctive features are summarized in Section 2. In addition, I include herein descriptions of the Armenian chapel at Babaoğlan and the Greek churches at Milvan, Vahga, and Çem. The Templar fortress of Bağras (near Cilicia) was published separately; the other Crusader sites within or near the confines of the Armenian kingdom are mentioned in Section 5. In Section 6 I discuss the problems of dating the Armenian forts. A reassessment of the boundaries of the Armenian kingdom and a few general conclusions on how the strategy of defense shaped the civilian settlements as well as the social fabric of Armenian society are contained in Section 7. I have tried to establish paradigms and methods by which the military architecture in other regions of Anatolia and the Levant can be analyzed. However, there is no attempt here to compare and evaluate the Armenian contribution in Cilicia with contemporary developments in the Arab, Greek, and Crusader regions. I expect to undertake this task in separate publications.[3]

Certain peculiarities of my treatment of this material should be noted. I have catalogued all seventy-five sites (except the Pillar of Jonah) by their modern Turkish names. These designations will allow researchers to locate the sites on modern maps. In those cases where the medieval name of the fort is *securely* known, it will be mentioned in the description of each site. Occasionally a few of the variant spellings will also be listed, but there is no attempt at a critical study of the etymology of the toponyms. I have been very conservative in associating medieval names in the texts with the surviving sites. The great danger in assigning place names from inadequate evidence is that the concomitant reinterpretation of the local history will carry an undeserved credibility. As some of the fortifications have not even been named by the locals, I assigned to them the names of the nearest village. Because the rendering of the topography and the insertion of the settlement toponyms have left my map (fig. 2) rather crowded, the names of appropriate rivers, streams, and mountains have been placed only with the description of each site in Part II. I have made every attempt to follow modern Turkish usage with regard to spelling, which sometimes results in inconsistencies, such as Bucak Köy, Bucakköy, and Bucak Köyü. The reader will note a definite bias for "Kalesi" rather than "Kale"; as most of the sites are named after a region or town, the possessive suffix seems more appropriate. No distinction is made between the dotted and undotted "i" in upper case. For the transliteration of Armenian I have tried to follow the Hübschmann-Meillet-Benveniste

[3]The first to appear is Edwards, "Pontos."

system; a few very familiar latinized forms of Armenian proper nouns do appear (for example, Constantine and Ruben as opposed to Kostandin and Ṙubēn). The rendering of most of the Arabic is derived from F. Pareja, *Islamologia* (Rome, 1951), 8.

Over the last decade many people from all walks of life have unselfishly extended their help to my Cilician project. The real heroes are the denizens of modern Cilicia. Never have I felt more protected and never have I received so much from people who have so little as when hiking through the Taurus Mountains. The generosity of the Turkish peasants was not limited to food and lodgings, which they gave without solicitation, but to the free use of pack animals and jeeps as well as their service as guides. The encroachment of paved roads and modern civilization into the mountains is bringing a change to the traditional values of this Moslem society. I was fortunate to see this region before it was completely spoiled.

From 1973 to 1979 the directors of the Archeological Museum in Adana, Orhan Aytuğ Taşyürek and his successor, Metin Pehlivaner, gave their unhesitating support to my work at a time and place where lesser men would have closed the door.

My greatest debt is to the University of California at Berkeley and its faculty. The abridged and unrevised version of this manuscript, presently distributed by University Microfilms International under the title *The Fortifications of Medieval Cilicia,* was submitted on March 17, 1983, to complete the requirements for a Ph.D. in Ancient History and Mediterranean Archeology at U.C. Berkeley. To the members of my dissertation committee, Professors John K. Anderson, Raphael Sealey, and Guitty Azarpay, I am most grateful for their unfailing judgment and steadfast support. Professor Thomas F. Mathews (Institute of Fine Arts, New York University) not only provided valuable criticism of the draft of this text but the kind of encouragement that makes a career in academia so gratifying. To others I owe many thanks, especially: Jean Bony (University of California, Berkeley), Nina Garsoïan (Columbia University), Dickran Kouymjian (Fresno State University, California), Stephen Miller (American School of Classical Studies, Athens), James Russell (University of British Columbia), and Avedis Sanjian (University of California, Los Angeles). Dr. Hansgerd Hellenkemper and Professor Friedrich Hild reviewed the text and plans of my manuscript in September 1983 and May 1984 respectively, and I have introduced a few changes based on their comments. Unfortunately the draft of their forthcoming gazetteer on Cilicia (in the *Tabula Imperii Byzantini*) was not made available to me. Permission was given to Dr. V. L. Parsegian, project director of the microfiche series *Armenian Architecture,* to consult my dissertation for his forthcoming volume (No. 5) on Cilicia. I am *not* responsible for the text in that volume nor the selection of its illustrations.

For their financial and moral support I would like to thank: V. L. Parsegian (Armenian Educational Council), the Armenian General Benevolent Union, Alex Manoogian, the Calouste Gulbenkian Foundation, Richard and Beatrice Hagopian, Jack and Mary Aslanian, Allen Odion, Carlo Uomini, and many friends in the greater San Francisco Bay Area.

The revisions to this manuscript were made during a very productive Fellowship at Dumbarton Oaks. I am indebted to Dr. Gianfranco Fiaccadori for his careful translation from the Arabic of the passages that I selected on Cilician geography.

I would like to express my sincere appreciation to Frances Kianka, Manuscript Editor, and to Glenn Ruby, Publishing Manager, for so sedulously guiding this manuscript through publication. To Professor Robert Thomson, Director of Dumbarton Oaks, and a number of unspecified readers, I am grateful for final adjustments to the text.

RWE
February 1986
Dumbarton Oaks

Preface to the Plans

From March 1973 to November 1974 I was a permanent resident of Cilicia and was periodically assisted in my surveys by Messrs. Guy Bocian and Mason Cox. From June to September 1979 Messrs. Peter J. Kasavan and Eric Palson were my cosurveyors during another Cilician field survey. Mr. Ross MacLeod acted as my cosurveyor in June and July 1981. I owe a debt of gratitude to these five assistants, who so often had to endure limited rations, an inadequate supply of clean water, and the total absence of sanitary facilities. Since many of the sites in this study are isolated, we had to subsist for as long as ten days on what we could carry on our backs. More than anything else this study is a testimony of their patience and high professional standards.

The instruments used during the surveys were the following: two Brunton transits; two 60-m fiberglass tapes; two 3-m metal tapes; three optical range finders; and a pocket calculator. I am responsible for all black & white and color photographs. Two Nikon FM (SLR) camera bodies were used in conjunction with either 50 mm or 35 mm Nikkor lenses. The black & white prints are made from Kodak Pan-X and Tri-X film; the color plates are produced from Kodachrome and Ektachrome transparencies.

With regard to the plans, which appear with the descriptions of most of the sites in the Catalogue, certain peculiarities should be noted. Of the seventy-five fortifications considered in this study I was unable to execute plans for six (Anacık, Mansurlu, Misis, Pillar of Jonah, Sarı Seki, and Trapesak) because of restrictions imposed by the Turkish military or the near total destruction of some sites. Eleven of the plans from the remaining sixty-nine sites are "based" on earlier surveys. This means that I have not remeasured the site but merely added corrections from my own observations to an earlier, published plan. These sites include: Amuda, Anavarza, Ayas (two parts), Azgit, Bodrum, Gökvelioğlu, Korykos (two parts), Kum, Toprak, Tumlu, and Vahga. Since the eleven plans appear to be accurate I could not justify the time and expense required to conduct entirely new surveys. If the legend indicates that my plan is "adapted" from another survey, this implies that I have resurveyed the site, borrowing some information from earlier studies. The "adapted" plans are: the church at Anavarza (included on the main plan of the site) and Yılan. My assistants and I executed original plans for the fifty-seven remaining sites between 1973 and 1981. Six of the fifty-seven—Anahşa, Çalan, Gülek, Lampron, Silifke, and Sis—had plans of varying quality published prior to 1973, but they were not consulted during my surveys. Nine other sites that I surveyed—Babaoğlan, Çandır, Çem, Gösne, Kız (near Dorak), Payas, Savranda, Saimbeyli, and Sinap (near Çandır)—were sighted and described in print (all rather briefly) by modern explorers before 1973, but no plans were published.

On my plans, when a particular element in the fort is enlarged in detail, it is always framed by double borders with its own scale. When depicting different levels of a building, normally the ground floor is shown with the rocks and contour lines, and the upper level of a particular building is seen to the side of the main plan and always at the same scale. "Level" implies a unit or group of units on the same plane which are defined by solid walls. Rarely are structures in garrison forts more than two stories. Sometimes an open fortified terrace stands atop the upper level. Terraces are not shown on the plans, but they are described in the text and photos. Battlements (that is, merlons and crenellations) are seldom extant. They are often the first stone elements in the fort to collapse. If they do exist, I describe their nature and location in the text. The embrasured loopholes with casemates are normally below the battlements, and I try to show them on the plans when possible (for example, they are depicted at Meydan Kalesi but not on all portions of Yılan).

The "legend" sheet appended to this Preface (fig. 1) is designed to define in graphic terms the symbols that appear on the plans. Any peculiarities of an individual plan are noted in the description and the first or prefatory footnote on that site in the Catalogue. In order to avoid confusion the following is intended as an explanatory supplement to the legend sheet and to the analysis in Part I.2.

Alien masonry (shown by cross-hatching): For want of a better term, this simply means that one type of masonry has been refaced by a *completely* different type. The alien masonry represents the earlier phase of construction (for example, Sis). On the plans no distinction is made symbolically in cases where one type of masonry *surmounts* another at the same floor level. Such distinctions are noted in the narrative. For a variety of reasons, I decided not to depict the masonry of separate building periods on the plans with an elaborate system of symbols. The majority of sites show only one period of construction. Also, it is impossible to portray accurately on a single drawing the overlapping of different types of masonry. At a fort with a complex building history extremely precise elevations and multiple ground plans, which should properly be published as transparent overlays, would have to be completed. The time and instruments required for such undertakings were beyond the scope of my regional field surveys.

Contour lines: For various reasons the contour lines were not always surveyed for the plans but occasionally approximated. On a particular plan the interval-of-distance between contour lines is noted in the prefatory footnote on each site. The ascent of contour lines is marked by elevation arrows. Since most sites are located on outcrops, elevation arrows define the ambiguous terrain.

Doors and windows: Doors and windows that lack jambs will be shown merely as simple straight-sided holes. When jambs are present their size is rendered by contracting the width of the door/window to the appropriate degree. On the plan the jambs are framed with lines parallel to the course of the wall. This indicates that the arch or lintel over the jambs is at a different level than the covering over the rest of the door/window. Because of the limitations of scale, vault lines, which depict the shape of the arched covering over the door/window, are shown only when space permits (for example, Belen Keşlik). Windows and doors that have been blocked with either debris or masonry are depicted with a light stippling.

Pivot hole (above): This simply means that a circular hole has been bored into the soffit of a lintel or arch over a door/window. Based on present remains, it appears that the hole was simply a socket for the vertical post which attached to the end of a swinging door (or shutter).

Pivot housing: The pivot housing is a corbeled block in which a pivot hole has been bored. Sometimes a similar corbeled element projects from the upper course of the jamb into the space of the threshold (for example, Gülek and Meydan). These blocks probably protected the vulnerable corners of the wooden doors.

Post hole: This hole is *always* at ground level and frequently appears in the sill of a door/window. In the latter case it can function as a pivot hole at the lower level. Occasionally, the pivot hole above appears without post holes, and the reverse is true. Obviously the medieval builders of Cilicia could choose various methods to secure the rotating end of a door or shutter. In the narratives of the Catalogue I will mention frequently "double wooden doors." This means that the door consists of two wooden panels that open in the center and are anchored at the ends to vertical pivot-rods. Riveted sheets of metal covered the exterior of the panels.

Rock cut in section: This applies only to surveys where strict horizontal plans were rendered or where an overhanging cliff was cut away to expose the ground-level plan of the site (for example, Gediği).

Rubble: Rubble is used to indicate the area that is covered by the masonry of collapsed structures. No rubble is shown under a standing vault, but occasionally a rock mass that protrudes into the body of a vaulted chamber is shown (for example, Geben).

Slot machicolation: This is simply a hole between two architectural elements. Normally it appears between an arrière-voussure (that is, adjoining arches) over a door; of the two arches, the innermost rests on the jambs, while the outermost forms the exterior arch for the door. Symbolically the slot machicolation is difficult to represent on the scale of most plans, and is is shown in this study as a tripartite door. When a slot machicolation is represented in detail (for example, Anahşa) the width of the slot is defined by dashed lines.

Superimposed windows: When a window is placed above a door in the *same* floor level and its dimensions are smaller than the door below, it is depicted with dashed lines.

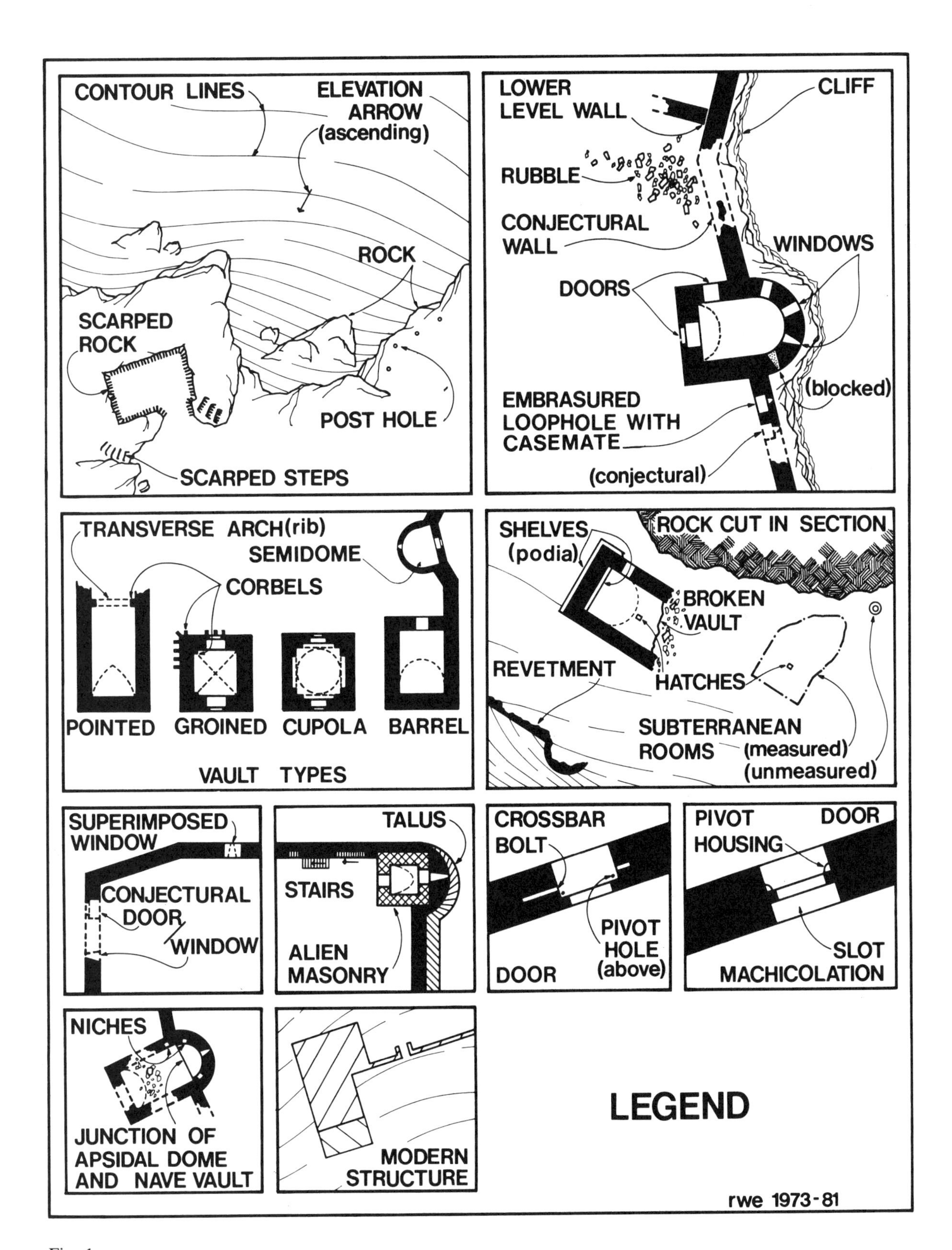

CONTOUR LINES
ELEVATION ARROW (ascending)
ROCK
SCARPED ROCK
POST HOLE
SCARPED STEPS
LOWER LEVEL WALL
CLIFF
RUBBLE
CONJECTURAL WALL
WINDOWS
DOORS
(blocked)
EMBRASURED LOOPHOLE WITH CASEMATE
(conjectural)
TRANSVERSE ARCH(rib)
SEMIDOME
CORBELS
POINTED
GROINED
CUPOLA
BARREL
VAULT TYPES
SHELVES (podia)
ROCK CUT IN SECTION
BROKEN VAULT
REVETMENT
HATCHES
SUBTERRANEAN ROOMS
(measured)
(unmeasured)
SUPERIMPOSED WINDOW
CONJECTURAL DOOR
WINDOW
TALUS
STAIRS
ALIEN MASONRY
CROSSBAR BOLT
PIVOT HOLE (above)
DOOR
PIVOT HOUSING
DOOR
SLOT
MACHICOLATION
NICHES
JUNCTION OF APSIDAL DOME AND NAVE VAULT
MODERN STRUCTURE
LEGEND
rwe 1973-81

Fig. 1

List of Maps

Abbreviations[1]

Adana (1)	Türkei, 1:200,000, Deutsche Heereskarte (Berlin), Blatt Nr. I–X (1941).
Adana (2)	Türkei, 1:400,000, R. Kiepert, Karte von Kleinasien (Berlin), Blatt Adana D IV (1916).
Central Cilicia	Syria & Turkey, 1:500,000, Tactical Pilotage Chart (St. Louis, Missouri), Sheet G–4A (1966).
Cilician Gates	Turkey, 1:250,000, War Office of Great Britain (London), Cilician Gates (1919).
Cilicie	Turquie, 1:800,000, K. J. Basmadjian, Carte de Cilicie (Paris, 1918).
Elbistan (1)	Türkei, 1:200,000, Deutsche Heereskarte (Berlin), Blatt Nr. G–XI (1941).
Elbistan (2)	Turkey, 1:200,000, Army Map Service (Washington, D.C.), Elbistan (1941).
Everek	Türkei, 1:200,000, Deutsche Heereskarte (Berlin), Blatt Nr. G–X (1941).
Figure 2	Turkey, the Fortifications of Armenian Cilicia, R. W. Edwards (1981).
Gaziantep	Türkei, 1:200,000, Deutsche Heereskarte (Berlin), Blatt Nr. H–XI (1941).
Kozan	Türkei, 1:200,000, Deutsche Heereskarte (Berlin), Blatt Nr. H–X (1941).
Malatya	Turkey, 1:800,000, Army Map Service (Washington, D.C.), Malatya (1941).
Maras	Turkey, 1:200,000, Army Map Service (Washington, D.C.), Maras (1942).
Marash	Turkey, 1:250,000, War Office of Great Britain (London), Marash (1915).
Mersin (1)	Türkei, 1:200,000, Deutsche Heereskarte (Berlin), Blatt Nr. I–IX (1941).
Mersin (2)	Turkey, 1:200,000, Army Map Service (Washington, D.C.), Mersin (1943).
Ulukişla	Türkei, 1:200,000, Deutsche Heereskarte (Berlin), Blatt Nr. H–IX (1941).

[1]These abbreviations are derived from the actual title of each map. *Maras* and *Ulukişla* do not reflect modern Turkish pronunciation and orthography which render the toponyms as Maraş and Ulukışla.

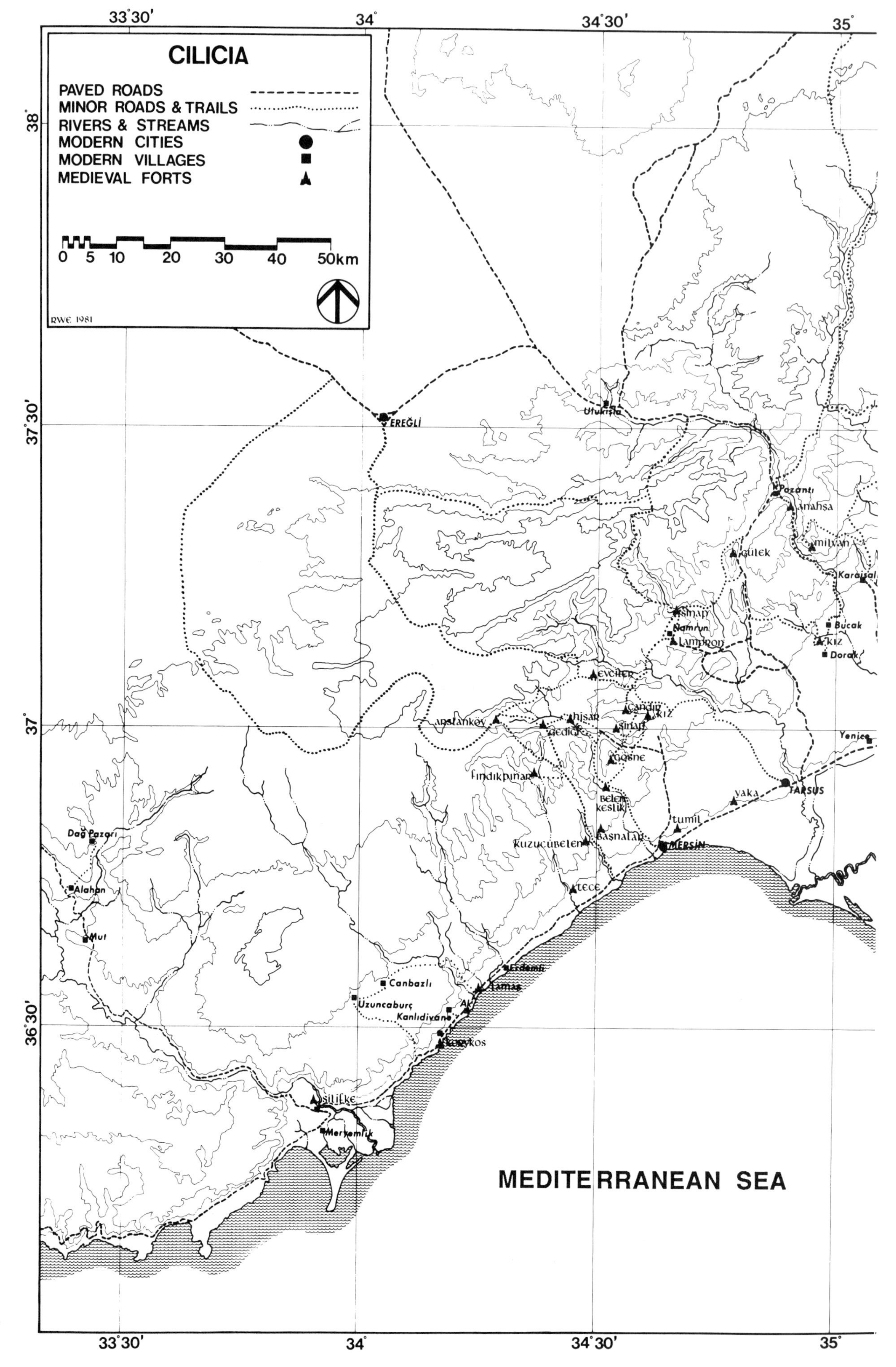

CILICIA
PAVED ROADS
MINOR ROADS & TRAILS
RIVERS & STREAMS
MODERN CITIES
MODERN VILLAGES
MEDIEVAL FORTS
0 5 10 20 30 40 50km
RWE 1981
33°30′
34°
34°30′
35°
38°
37°30′
37°
36°30′
EREĞLİ
Ulukışla
Pozantı
Anahşa
Gülek
Milvan
Karaisalı
Sinap
Namrun
Lampron
Bucak
Kız
Dorak
Evciler
Çandır
Kız
Hisar
Sinap
Arslanköy
Gedikli
Ağoşne
Fındıkpınar
Belen keslik
Tumil
Vaka
TARSUS
Yenice
Kuzucubelen
Başnalar
MERSİN
Tece
Dağ Pazarı
Alahan
Mut
Erdemli
Lamas
Canbazlı
Uzuncaburç
Kanlıdivane
Ak
Korykos
Silifke
Meryemlik
MEDITERRANEAN SEA

Fig. 2

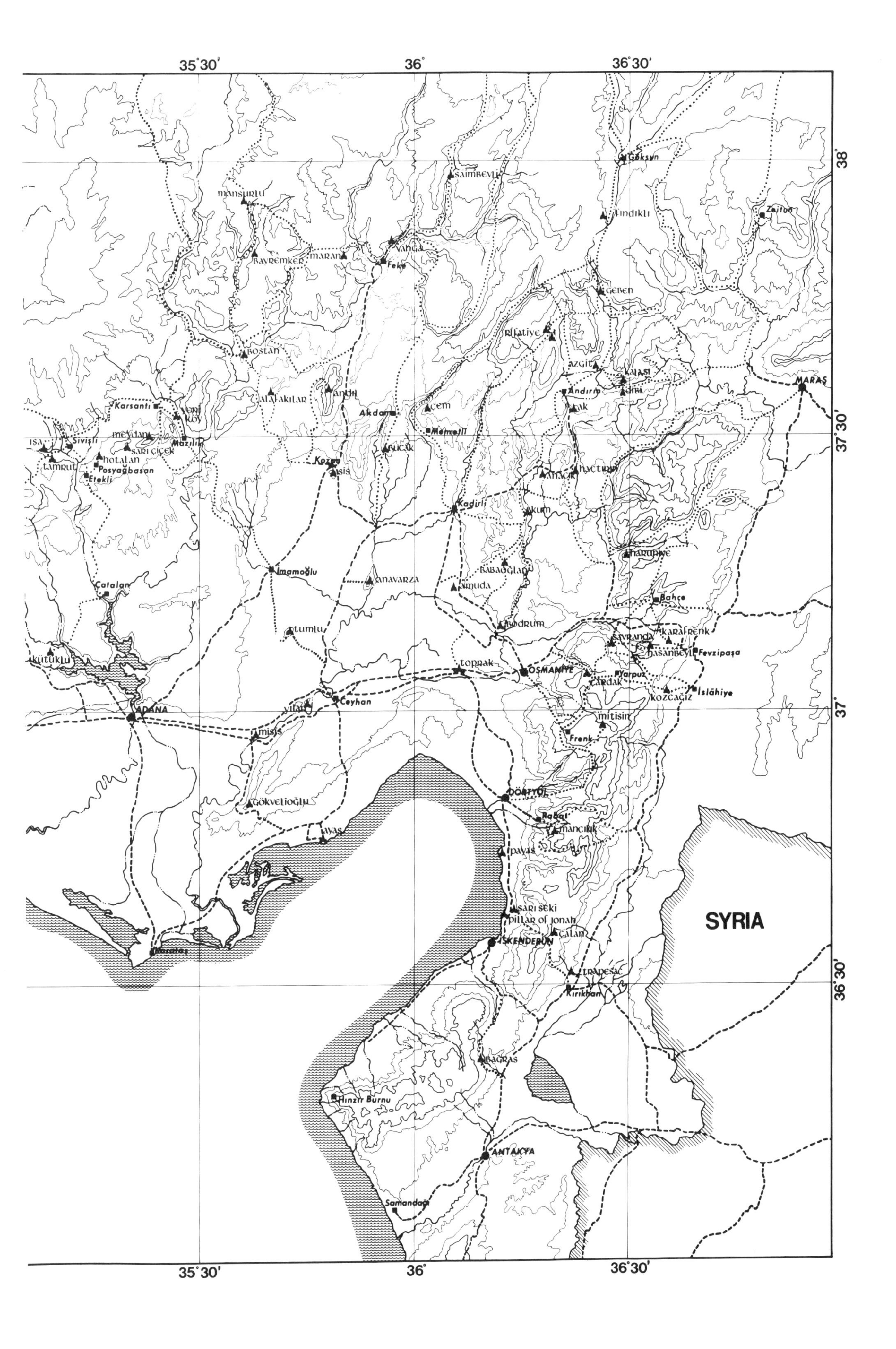

35°30′
36°
36°30′
38°
37°30′
37°
36°30′
Göksun
SAIMBEYLI
Mansurlu
Fındıklı
Zeitun
Vanga
Bayremker
Maran
Feke
Geben
Rifatiye
Bostan
Azgit
Kalası
Andırın
MARAŞ
Alafakılar
Anahşa
Karsantı
Yeni Köy
Akdam
Cem
Memetli
Meydan
Sarı Çiçek
Mazılık
Sivişli
Hotalan
Posyağbasan
Tamrut
Etekli
Kozan
Sis
Vahka
Kadirli
Anacık
Haçtırın
Akum
Haruniye
Babaoğlan
İmamoğlu
Anavarza
Amuda
Bahçe
Catalan
Tumlu
Bodrum
Savranda
Karafrenk
Hasanbeyli
Fevzipaşa
Kütüklü
Toprak
OSMANIYE
Yarpuz
Çardak
Kozcağız
İslâhiye
ADANA
Yılan
Ceyhan
Misis
Mitisin
Frenk
Gökvelioğlu
DÖRTYOL
Rabat
Mancılık
Ayas
Payas
Sarı Seki
Pillar of Jonah
Çalan
İSKENDERUN
Trapesac
Kırıkhan
SYRIA
Bağras
Hınzır Burnu
ANTAKYA
Samandağ
Karataş

Abbreviations and Selected Bibliography

All references in the footnotes are to page or column numbers unless otherwise indicated.

Abūʾl-Fidāʾ: Aboulféda, *Géographie d'Aboulféda,* trans. S. Guyard, vol. 2.2, Paris, 1883.
Aghassi: Aghassi, *Zeïtoun,* trans. A. Tchobanian, Paris, 1897.
Ainsworth: W. Ainsworth, *Travels and Researches in Asia Minor, Mesopotamia, Chaldea, and Armenia,* vol. 2, London, 1842.
Al-Balāḏurī: Al-Balâdhuri, *Kitâb futûh al-buldân,* vol. 1, trans. P. Hitti, New York, 1916.
Alishan, *L'Armeno-Veneto:* G. Alishan, *L'Armeno-Veneto compendio storico e documenti delle relazioni degli Armeni con Veneziani,* 2 parts, Venice, 1893.
Alishan, *Léon:* G. Alishan, *Léon le magnifique, premier roi de Sissouan ou de l'Arménocilicie,* Venice, 1888.
Alishan, *Sissouan:* G. Alishan, *Sissouan ou l'Arméno-Cilicie. Description géographique et historique,* Venice, 1899.
Alkım: U. B. Alkım, "Ein altes Wegenetz im südwestlichen Antitaurus-Gebiet," *Anadolu Araştırmaları* 1.2, 1959.
Al-Maqrīzī: Al-Maḳrīzī, *Histoire des sultans mamlouks, de l'Egypte,* trans. E. Quatremère, 2 vols., Paris, 1837–42.
Anna Comnena: Anna Comnena, *Alexiad,* vol. 3, ed. B. Leib, Paris, 1945.
AOL: Archives de l'Orient latin.
AS: Anatolian Studies.

Bar Hebraeus: *The Chronography of Gregory Abûʾl Faraj . . . Bar Hebraeus,* trans. E. A. W. Budge, vol. 1, Oxford, 1932.
Barker: W. Barker, *The Birthland of St. Paul. Cilicia: Its Former History and Present State,* ed. W. Ainsworth, London, 1855.
Beaufort: F. Beaufort, *Karamania, or a Brief Description of the South Coast of Asia Minor,* 2nd ed., London, 1818.
Bedoukian: P. Bedoukian, *Coinage of Cilician Armenia,* American Numismatic Society, Numismatic Notes and Monographs, No. 147, New York, 1962.
Berkian: A. Berkian, *Armenischer Wehrbau im Mittelalter,* Diss., Darmstadt, 1976.
BGA: Bibliotheca geographorum arabicorum.
Boase: *The Cilician Kingdom of Armenia,* ed. T. S. R. Boase, Edinburgh, 1978.
Bossert and Alkım: H. Bossert and U. Alkım, *Karatepe,* Istanbul, 1947.

Brocquière: Bertrandon de la Brocquière, *Le voyage d'outremer,* ed. Ch. Schefer, Paris, 1892.

Budde: L. Budde, *Antike Mosaiken in Kilikien,* 2 vols., I. *Frühchristliche Mosaiken in Misis-Mopsuhestia,* II. *Die heidnischen Mosaiken,* Recklinghausen, 1969–72.

Byz: Byzantion.

BZ: Byzantinische Zeitschrift.

Cahen: C. Cahen, *La Syrie du nord à l'époque des croisades et la principauté franque d'Antioche,* Paris, 1940.

Canard: M. Canard, *Histoire de la dynastie des H'amdanides de Jazîra et de Syrie,* Algiers, 1951.

Canard, "Le royaume": M. Canard, "Le royaume d'Arménie-Cilicie et les Mamelouks jusqu'au traité de 1285," *REArm,* n.s. 4, 1967.

Canard, *Sayf al Daula:* M. Canard, *Sayf al Daula,* Algiers, 1934.

Canard and Grégoire: M. Canard (trans.) and H. Grégoire, *Extraits des sources arabes,* in *La dynastie macédonienne,* A Vasiliev, ed., *Byzance et les Arabes,* CBHB 2.2, Brussels, 1950.

Carne: J. Carne, *Syria, the Holy Land, Asia Minor, &c,* 3 vols., London, 1838.

CBHB: Corpus bruxellense historiae byzantinae.

CFHB: Corpus fontium historiae byzantinae.

Chauvet and Isambert: C. Favre and B. Mandrot, "Cilicie," in A. Chauvet and E. Isambert, *Itinéraire descriptif, historique et archéologique de l'Orient,* vol. 3, Paris, 1882.

Choniatēs: Nicetas Choniatēs, *Historia,* ed. J. van Dieten, CFHB 11.1, Berlin, 1975.

Christomanos: C. Christomanos, "Feudal-Organisation des Königreiches Klein-Armenien," *Jahrbuch der kaiserlichen heraldischen Gesellschaft Adler,* Vienna, 16–17, 1890.

Cinnamus: John Cinnamus, *Epitome rerum ab Ioanne et Alexio Comnenis gestarum,* CSHB, ed. A. Meineke, Bonn, 1836.

CSCO: Corpus scriptorum christianorum orientalium.

CSHB: Corpus scriptorum historiae byzantinae.

Cuinet: V. Cuinet, *La Turquie d'Asie. Géographie administrative, statistique, descriptive et raisonnée de chaque province de l'Asie-Mineure,* vol. 2, Paris, 1892.

Dardel: Jean Dardel, *RHC, DocArm,* vol. 2.

Davis: E. J. Davis, *Life in Asiatic Turkey,* London, 1879.

Delaville Le Roulx: J. Delaville Le Roulx, ed., *Cartulaire général de l'ordre des Hospitaliers de St.-Jean de Jérusalem,* vol. 2, Paris, 1897.

Dennis: *Three Byzantine Military Treatises,* ed., trans., and notes G. T. Dennis, Dumbarton Oaks Texts 9 (CFHB 25), Washington, D.C., 1985.

Der Nersessian: S. Der Nersessian, "The Kingdom of Cilician Armenia," *A History of the Crusades,* ed. K. M. Setton, vol. 2, Madison, 1969.

Deschamps, I: P. Deschamps, *Les châteaux des croisés en Terre Sainte,* vol. 1, *Le Crac des Chevaliers,* Paris, 1934.

Deschamps, III: P. Deschamps, *Les châteaux des croisés en Terre Sainte,* vol. 3, *La défense du comté de Tripoli et de la principauté d'Antioche,* Paris, 1973.

Deschamps, "Servantikar": P. Deschamps, "Le château de Servantikar en Cilicie," *Syria* 18, 1937.

DOP: Dumbarton Oaks Papers.

DOS: Dumbarton Oaks Studies.

Dulaurier: E. Dulaurier, "Etude sur l'organisation politique, religieuse et administrative du royaume de la Petite-Arménie," *JA,* October 1861.

Dunbar and Boal: J. G. Dunbar and W. W. M. Boal, "The Castle of Vahga," *AS* 14, 1964.

Dussaud: R. Dussaud, *Topographie historique de la Syrie antique et médiévale,* Paris, 1927.

Edwards, "Artvin": R. W. Edwards, "The Fortifications of Artvin: A Second Preliminary Report on the Marchlands of Northeast Turkey," *DOP* 40, 1986.

Edwards, "Bağras": R. W. Edwards, "Bağras and Armenian Cilicia: A Reassessment," *REArm,* n.s. 17, 1983.

Edwards, "Doğubeyazit": R. W. Edwards, "The Fortress at Doğubeyazit (Daroynk^c)," *REArm,* n.s. 18, 1984.

Edwards, "Donjon": R. W. Edwards, "The Crusader Donjon at Anavarza in Cilicia," *Abstracts of the Tenth Annual Byzantine Studies Conference,* Cincinnati, 1984.

Edwards, "First Report": R. W. Edwards, "Ecclesiastical Architecture in the Fortifications of Armenian Cilicia," *DOP* 36, 1982.

Edwards, "Pontos": R. W. Edwards, "The Garrison Forts of the Pontos: A Case for the Diffusion of the Armenian Paradigm," *REArm,* n.s. 19, 1985.

Edwards, "Şebinkarahisar": R. W. Edwards, "The Fortress of Şebinkarahisar (Koloneia)," *Corso di cultura sull'arte ravennate e bizantina* 32, 1985.

Edwards, "Second Report": R. W. Edwards, "Ecclesiastical Architecture in the Fortifications of Armenian Cilicia: Second Report," *DOP* 37, 1983.

Edwards, "Yılan": R. W. Edwards, "On the Supposed Date of Yılan Kalesi," *Journal of the Society for Armenian Studies* 1, 1984.

EI: The Encyclopaedia of Islam, Leiden, 1913–34; new edition (cited as *EI*^{2}), Leiden, 1960–85.

Engels: D. Engels, *Alexander the Great and the Logistics of the Macedonian Army,* Berkeley, 1978.

Ēp^{c}rikean: H. Ēp^{c}rikean, *Patkerazard Bnašxarhik Bararan,* 2 vols., Venice, 1903–5.

Favre and Mandrot: C. Favre and B. Mandrot, "Voyage en Cilicie, 1874," *Bulletin de la Société de Géographie de Paris,* 1878.

Fedden and Thomson: R. Fedden and J. Thomson, *Crusader Castles,* London, 1957.

Flemming: B. Flemming, *Landschaftsgeschichte von Pamphylien, Pisidien und Lykien im Spätmittelalter,* Wiesbaden, 1964.

Forstreuter: K. Forstreuter, *Der deutsche Orden am Mittelmeer,* Bonn, 1967.

Frech: F. Frech, "Die armenischen Burgen," *Zeitschrift der Gesellschaft für Erdkunde zu Berlin* 9, 1915.

Gaudefroy-Demombynes: M. Gaudefroy-Demombynes, *La Syrie à l'époque des Mamelouks d'après les auteurs arabes,* Paris, 1923.

Gottwald, "Burgen": J. Gottwald, "Burgen und Kirchen im mittleren Kilikien," *BZ* 41, 1941.

Gottwald, "Kirche": J. Gottwald, "Die Kirche und das Schloss Paperan in Kilikisch-Armenien," *BZ* 36, 1936.

Gottwald, "Til": J. Gottwald, "Die Burg Til im südöstlichen Kilikien," *BZ* 40, 1940.

Gough: M. Gough, "Anazarbus," *AS* 2, 1952.

Gregory: Gregory the Priest, *RHC, DocArm,* vol. 1.

Grigor Aknercʿi: Grigor of Akancʿ, *History of the Nation of the Archers (The Mongols),* ed. and trans. R. Blake and R. Frye, Cambridge, Mass., 1954.

HA: Handes Amsorya.

Haldon and Kennedy: J. Haldon and H. Kennedy, "The Arab-Byzantine Frontier in the Eighth and Ninth Centuries: Military Organisation and Society in the Borderlands," *Zbornik Radova Vizantološkog Instituta* 19, 1980.

Handbook: A Handbook of Asia Minor, vol. 4, part 2, *Cilicia, Antitaurus, and North Syria,* Naval Staff Intelligence Department (U.K.), May 1919.

Haykakan: Haykakan Sovetakan Hanragitaran, vols. 1–10, Erevan, 1974–84.

Heberdey and Wilhelm: R. Heberdey and A. Wilhelm, "Reisen in Kilikien," *Denkschriften der kaiserlichen Akademie der Wissenschaften in Wien, Philosophisch-historische Classe* 44.6, 1896.

Heffening: W. Heffening, "Eine Burgruine im Taurus," *Repertorium für Kunstwissenschaft* 45, 1925.

Hellenkemper: H. Hellenkemper, *Burgen der Kreuzritterzeit in der Grafschaft Edessa und im Königreich Kleinarmenien,* Bonn, 1976.

Herzfeld and Guyer: E. Herzfeld and S. Guyer, *Meriamlik und Korykos,* vol. 2, Monumenta Asiae Minoris Antiqua, Manchester, 1930.

Hetʿum: Hetʿum of Korykos, *RHC, DocArm,* vol. 1.

Hetʿum II: King Hetʿum II, *RHC, DocArm,* vol. 1.

Heyd: W. Heyd, *Histoire du commerce du Levant au moyen âge,* 2nd ed., trans. F. Raynaud, 2 vols., Leipzig, 1923.

Hild: F. Hild, *Das byzantinische Strassensystem in Kappadokien,* Vienna, 1977.

Hild and Hellenkemper: F. Hild, G. and H. Hellenkemper, "Kommagene-Kilikien-Isaurien," *Reallexikon zur byzantinischen Kunst,* vol. 4, Stuttgart, 1984.

Hild and Restle: F. Hild and M. Restle, *Kappadokien,* Tabula Imperii Byzantini 2, Vienna, 1981.

Hogarth and Munro: D. G. Hogarth and J. A. R. Munro, *Modern and Ancient Roads in Eastern Asia Minor,* Royal Geographical Society, Supplementary Papers, vol. 3, London, 1893.

Honigmann: E. Honigmann, *Die Ostgrenze des byzantinischen Reiches von 363 bis 1071,* A. Vasiliev, ed., *Byzance et les Arabes,* CBHB 3, Brussels, 1935.

Humann and Puchstein: K. Humann and O. Puchstein, *Reisen in Kleinasien und Nordsyrien,* Berlin, 1890.

IA: Islâm Ansiklopedisi, Istanbul, 1940–85.

Ibn al-Furāt: Ibn al-Furāt, *Ayyubids, Mamlukes and Crusaders, Selections from the Tārīkh al-duwal wa'l-Mulūk,* vol. 2, trans. U. and M. Lyons, notes J. Riley-Smith, Cambridge, 1971.

Ibn Bībī: Ibn Bībī, *Die Seltschukengeschichte,* trans. H. Duda, Copenhagen, 1959.

Ibn Ḥawqal: Ibn Hauqal, *Configuration de la terre (Kitab surat al-ard),* intro. and trans. J. Kramers and G. Wiet, vol. 1, Beirut-Paris, 1964.

Imhoof-Blumer: F. Imhoof-Blumer, *Kleinasiatische Münzen,* vol. 2, Vienna, 1902.

JA: Journal asiatique.

Jacquot: P. Jacquot, *Antioche,* vol. 1, Antakya, 1931.

Janke: A. Janke, *Auf Alexanders des Grossen Pfaden. Eine Reise durch Kleinasien,* Berlin, 1904.

JHS: Journal of Hellenic Studies.

JÖB: Jahrbuch der österreichischen Byzantinistik.

Jones: A. H. M. Jones, *The Cities of the Eastern Roman Provinces,* 2nd ed., Oxford, 1971.

Karst: Smbat Sparapet, *Sempadscher Kodex aus dem 13. Jahrhundert oder mittelarmenisches Rechtsbuch,* ed., trans., and comm. J. Karst, 2 vols., Straßburg, 1905.

Keil and Wilhelm: J. Keil and A. Wilhelm, *Denkmäler aus dem Rauhen Kilikien,* vol. 3, Monumenta Asiae Minoris Antiqua, Manchester, 1931.

Kᶜēlēšean: M. Kᶜēlēšean, *Sis-Matean,* Beirut, 1949.

King: E. King, "A Journey through Armenian Cilicia," *Asian Affairs* 24, 1937.

Kirakos: Kirakos Ganjakecᶜi, *Patmutᶜyun Hayocᶜ,* ed. K. Melikᶜ-Ohanǰanyan, Erevan, 1961.

Kirakos, M. Brosset: Kirakos of Gantzac, *Histoire d'Arménie,* in *Deux historiens arméniens,* trans. M. Brosset, St. Petersburg, 1870.

Kotschy: T. Kotschy, *Reise in den cilicischen Taurus, über Tarsus,* Gotha, 1858.

Langendorf and Zimmermann: J. Langendorf and G. Zimmermann, "La forteresse de Séléfké (Turquie)," *Genava* 12, 1964.

Langlois, *Cartulaire:* V. Langlois, *Le trésor des chartes d'Arménie ou cartulaire de la chancellerie royale des Roupéniens,* Venice, 1863.

Langlois, *Inscriptions:* V. Langlois, *Inscriptions grecques, romaines, byzantines et arméniennes de la Cilicie,* Paris, 1854.

Langlois, "Lampron": V. Langlois, "Les ruines de Lampron en Cilicie," *Revue de l'orient, de l'Algérie et des colonies* (Paris) 12, 1860.

Langlois, *Rapport:* V. Langlois, *Rapport sur l'exploration archéologique de la Cilicie et de la Petite-Arménie, pendant les années 1852–1853,* Paris, 1854.

Langlois, *Voyage:* V. Langlois, *Voyage dans la Cilicie et dans les montagnes du Taurus, exécuté pendant les années 1852–1853,* Paris, 1861.

Le Strange, *Caliphate:* G. Le Strange, *The Lands of the Eastern Caliphate,* Cambridge, 1905.

Le Strange, *Palestine:* G. Le Strange, *Palestine under the Moslems,* London, 1890.

Lilie: R. J. Lilie, *Byzanz und die Kreuzfahrerstaaten,* Munich, 1981.

Lüders: A. Lüders, *Die Kreuzzüge im Urteil syrischer und armenischer Quellen,* Berlin, 1964.

Magie: D. Magie, *Roman Rule in Asia Minor,* 2 vols., Princeton, 1950.

Makhairas: Leontios Makhairas, *Recital Concerning the Sweet Land of Cyprus, entitled 'Chronicle',* ed., trans., and notes R. Dawkins, 2 vols., Oxford, 1932.

Manandian: H. A. Manandian, *The Trade and Cities of Armenia in Relation to Ancient World Trade,* 2nd ed., trans. N. Garsoïan, Lisbon, 1965.

Mas Latrie: R. de Mas Latrie, ed., *Chroniques d'Amadi et de Strambaldi,* 2 vols., Paris, 1891–93.

Matthew of Edessa: Matthew of Edessa, *RHC, DocArm,* vol. 1.

Michael Italikos: Michel Italikos, *Lettres et discours,* ed. P. Gautier, Archives de l'Orient chrétien 14, Paris, 1972.

Michael the Syrian: Michael the Syrian, *RHC, DocArm,* vol. 1.

Michael the Syrian, J. B. Chabot: Michael the Syrian, *Chronique de Michel le Syrien, patriarche jacobite d'Antioche,* 3 vols., trans. J. B. Chabot, Paris, 1899–1924.

Mikaeljan: G. Mikaeljan, *Istorija Kilikiĭskogo armjanskogo gosudarstva,* Erevan, 1952.

Müller-Wiener: W. Müller-Wiener, *Castles of the Crusaders,* trans. J. Brownjohn, London, 1966.

Nersēs of Lampron: Nersēs of Lampron, *RHC, DocArm,* vol. 1.

Oikonomidès: *Les listes de préséance byzantines des IXe et Xe siècles,* intro., text, trans., and comm. N. Oikonomidès, Paris, 1972.

Oskean: H. Oskean, *Kilikiayi vankᶜerə,* Vienna, 1957.

PECS: The Princeton Encyclopedia of Classical Sites, ed. R. Stillwell, Princeton, 1976.

Pōłosean: Y. Pōłosean, *Hačəni əndhanur patmutᶜiwnə ew šrǰakay Gōzan-tałi Hay giwłerə,* Los Angeles, 1942.

PPTS: Palestine Pilgrims' Text Society.

Popper: W. Popper, *Egypt and Syria under the Circassian Sultans 1382–1468 A.D.: Systematic Notes to Ibn Taghrî Birdî's Chronicle of Egypt,* Berkeley, 1955.

RA: Revue archéologique.

Ramsay: W. Ramsay, *The Historical Geography of Asia Minor,* Royal Geographical Society, Supplementary Papers, vol. 4, London, 1890; rpr. 1972.

RE: Paulys Realencyclopädie der classischen Altertumswissenschaft, Stuttgart-Munich, 1894–1978.

REArm: Revue des études arméniennes.

Rey: E. Rey, "Les périples des côtes de Syrie et de la Petite Arménie," *AOL* 2, 1884.

RHC, DocArm: Recueil des historiens des croisades: Documents arméniens, 2 vols., Paris, 1869–1906; rpr. Farnborough, 1967.

RHC, HistOcc: Recueil des historiens des croisades: Historiens occidentaux, 5 vols., Paris, 1844–95.

RHC, HistOrien: Recueil des historiens des croisades: Historiens orientaux, 5 vols., Paris, 1872–1906.

Ritter: C. Ritter, *Die Erdkunde von Asien,* vol. 9.2, *Klein-Asien,* Berlin, 1859.

Robinson and Hughes: F. C. R. Robinson and P. C. Hughes, "Lampron—Castle of Armenian Cilicia," *AS* 19, 1969.

Röhricht: R. Röhricht, comp., *Regesta regni Hierosolymitani, 1097–1291,* Innsbruck, 1893.

Rüdt-Collenberg: W. H. Rüdt-Collenberg, *The Rupenides, Hethumides and Lusignans,* Paris, 1963.

Samuel of Ani: Samuel of Ani, *RHC, DocArm,* vol. 1.

Sanjian: A. Sanjian, *Colophons of Armenian Manuscripts, 1301–1480, A Source for Middle Eastern History,* Cambridge, Mass., 1969.

Schaffer: F. Schaffer, "Cilicia," *Petermanns Mitteilungen,* Ergänzungsheft 141, 1903.

Schultze: V. Schultze, *Altchristliche Städte und Landschaften,* vol. 2.2, *Kleinasien,* Gütersloh, 1926.

Seton-Williams: M. Seton-Williams, "Cilician Survey," *AS* 4, 1954.

Sevgen: N. Sevgen, *Anadolu Kaleleri,* Ankara, 1959.

Smbat: The Constable Smbat (or Smbat Sparapet), *RHC, DocArm,* vol. 1.

Smbat, G. Dédéyan: *La chronique attribuée au Connétable Smbat,* intro., trans., and notes G. Dédéyan, Paris, 1980.

Sterrett: J. Sterrett, *An Epigraphical Journey in Asia Minor,* Papers of the American School of Classical Studies at Athens, vol. 2, Boston, 1888.

Sümer: F. Sümer, "Çukur-ova Tarihine Dâir Araştırmalar," *Tarih Araştırmaları Dergisi* 1.1, 1963.

Taeschner: F. Taeschner, *Das anatolische Wegenetz nach osmanischen Quellen,* 2 vols., Leipzig, 1924–26.

Tēr Łazarean: H. Tēr Łazarean, *Haykakan Kilikia,* Antilias-Beirut, 1966.

Texier: C. Texier, *Asie Mineure, description géographique, historique et archéologique,* Paris, 1862.

Tomaschek: W. Tomaschek, *Zur historischen Topographie von Kleinasien im Mittelalter,* Sitzungsberichte der kaiserlichen Akademie der Wissenschaften, Philosophisch-historische Classe, 124, Vienna, 1891.

Toumanoff: C. Toumanoff, *Manuel de généalogie et de chronologie pour l'histoire de la Caucasie chrétienne (Arménie-Géorgie-Albanie),* Rome, 1976.

Tritton and Gibb: A. S. Tritton and H. A. R. Gibb, "The First and Second Crusades from an Anonymous Syriac Chronicle," *Journal of the Royal Asiatic Society,* 1933.

Vahram of Edessa: Vahram of Edessa, *RHC, DocArm,* vol. 1.

Vasiliev: A. Vasiliev, *Les relations politiques de Byzance et des Arabes à l'époque de la dynastie macédonienne, 867–959,* A. Vasiliev, ed., *Byzance et les Arabes,* CBHB 2.1, Brussels, 1968.

Wilbrand von Oldenburg: Wilbrand von Oldenburg, *Peregrinatio,* ed. and trans. J. Laurent, Hamburg, 1859.

Youngs: G. R. Youngs, "Three Cilician Castles," *AS* 15, 1965.

Yovhannēsean: M. Yovhannēsean, *Kilikean Berder,* Venice, 1985.

List of Figures and Plates

Figures

The plans (figures 4–78) appear in an alphabetical sequence with the description of each fortification in the Catalogue (Part II).

Plates

The letter-designations in the captions are derived from the plan(s) of each site. The numbers for color plates are in boldface type.

Part One

Introduction

1. Historical Background

Considering the fragmentary and all too often contradictory nature of the primary sources for the period of the Armenian kingdom in Cilicia, it is difficult to write a definitive political and military history of the region. Historians find it impossible to locate many Russian and Soviet publications and must rely principally on the two volumes of Armenian documents in the late nineteenth-century edition of the *Recueil des historiens des croisades*. Neither volume contains a real apparatus criticus to provide the researcher with variant readings and the location of possible scribal errors or lacunae. Nor is there any consistent attempt to deal with the chronological problems in the texts of Samuel of Ani or Het'um of Korykos.[1] Only one chronicle relating the military history of Cilicia, that written by the Constable Smbat, has received serious attention in the past thirty-one years.[2] The Armenian inscriptions from the period of the kingdom are less valuable as historical sources. A handful of epigraphs has survived[3]—most so badly mutilated that it is impossible to confirm the nineteenth-century transcriptions made from them[4] or deal critically with the translations.[5] The Armenian colophons provide valuable evidence for the chronology of the kingdom, but the editions are incomplete.[6]

Greek, Arabic, and Latin sources contribute important bits of information on Armenian history and toponyms.[7] However, the only non-Armenian source that offers a detailed account of life in Cilicia, the thirteenth-century chronicle of the Teutonic monk Wilbrand von Oldenburg, is not available in a serious critical edition.[8] The printed text, published in 1859 by

[1]A first step in this direction was undertaken by V. A. Hakobyan in his edition of *Manr žamanakagrut'yunner, XIII–XVIII dd.*, 2 vols. (Erevan, 1951–56). See also A. Abrahamyan, "Samuel Anec'u," *Ējmiatzin* (1952), no. 1, 30–37, no. 2, 34–43.

[2]Smbat Sparapet, *Taregirk'*, ed. S. Akelean (Venice, 1956); idem, *Leotpis*, trans. A. Galstyan (Erevan, 1974); S. Der Nersessian, "The Armenian Chronicle of the Constable Smbat," *DOP* 13 (1959); Smbat, G. Dédéyan. Refer to the bibliography of Armenian sources in K. Bardakjian and R. Thomson, *A Reference Guide to Armenian Literature* (forthcoming).

[3]These inscriptions are discussed below in the Catalogue.

[4]Langlois, *Inscriptions*, 12 ff; Alishan, *Sissouan*, passim.

[5]For the sake of completeness and uniformity I have cited Langlois's translations throughout the text in the Catalogue.

[6]L. Xač'ikyan, ed., *XV dari hayeren jeṙagreri hišatakaranner*, parts 1 and 2 (Erevan, 1955–58); idem, *XIV dari hayeren jeṙagreri hišatakaranner* (Erevan, 1950); Yovsēp'ean, Garegin I Kat'ołikos, *Yišatakarank' jeṙagrac'*, I (Antilias, 1951); A. Mat'evosyan, ed., *Hayeren jeṙagreri hišatakaranner, XIII dar* (Erevan, 1984); Sanjian.

[7]The most recent publications on Armeno-Crusader relations are the very competent studies by J. Riley-Smith ("The Templars and the Teutonic Knights in Cilician Armenia") and A. Luttrell ("The Hospitallers' Interventions in Cilician Armenia: 1291–1375") in Boase, 92 ff and 118 ff.

[8]Wilbrand von Oldenburg, 6–30. Cf. idem, *Peregrinatores medii aevi quatuor*, ed. J. Laurent (Leipzig, 1864). 162–90.

J. Laurent, is accompanied by a rather fanciful translation and notes that reflect just how little was known of Cilician topography in the nineteenth century. To complicate matters, one recent commentator[9] simply republished Laurent's edition of the text under the assumption that the modern reader is familiar enough with medieval Latin to understand the grammatical inconsistencies of a German monk who was writing in a foreign language. What follows in this section is a *brief* historical survey of the Armenian kingdom in Cilicia. Because this study is primarily architectural in nature, I will not attempt to reinterpret here any of the questions concerning chronology or political events, but the footnotes will refer the reader to the most recent scholarship on the kingdom.[10]

The history of Armenian Cilicia[11] begins with the ill-conceived policies of Constantinople. The tenth century was a period of relative prosperity for the Armenians under the Bagratid dynasty. However, in the first two decades of the eleventh century the Greeks succeeded in occupying much of Greater Armenia. By 1045 they forced King Gagik II of Ani to relinquish his throne. Soon Emperor Constantine Monomachus ordered the forced resettlement of Armenians into the regions of Caesarea, Sebastea, Tarsus, Antioch, Edessa, and Maraş. The latter three cities quickly became prominent Armenian principalities that endured until they were absorbed into various Crusader states. The reduced numbers of indigenous Armenians in the regions of Van and Ani left the Byzantines poorly prepared to halt the Seljuk invasions. The fateful battle occurred in 1071 on the hilly plain of Malazgirt (Manzikert), where the Turkish forces decisively defeated the Greeks and captured Emperor Romanus Diogenes.[12] As a result, all of Anatolia was opened to nomadic invasion.

Although Cilicia was to become the most successful and enduring of the Armenian settlements, the Greeks seem to have held political suzerainty there throughout most of the eleventh century. The earliest known Armenian migration into Cilicia occurred in the first quarter of the tenth century when fifty petty nobles from the house of Sasun moved there together with their families.[13] By the mid-eleventh century the trickle had turned into a flood. The Byzantines

[9] Hellenkemper, 17–26.

[10] The information in Part I.1 is in part derived from the primary and secondary sources cited in the Bibliography and from the following: N. Iorga, *Brève histoire de la Petite Arménie* (Paris, 1930); P. Charanis, *The Armenians in the Byzantine Empire* (Lisbon, 1964); S. Der Nersessian, *Armenia and the Byzantine Empire* (Cambridge, Mass., 1945); R. Grousset, *Histoire des croisades et du royaume franc de Jérusalem,* 3 vols. (Paris, 1934–36); S. Runciman, *A History of the Crusades,* 3 vols. (London, 1951–54); A. Sukiasjan, *Istorija Kilikiĭskogo armjanskogo gosudarstva i prava (XI–XIV vv.)* (Erevan, 1969).

[11] Throughout the ancient and medieval periods the region of Cilicia was defined by the Taurus and Anti-Taurus Mountains to the north and east and by the Mediterranean to the south. Only the western border of Cilicia seems to have fluctuated constantly; see T. Mitford, "Roman Rough Cilicia," *Aufstieg und Niedergang der römischen Welt,* 2.7.2 (1980), 1230 ff. Just a small portion of what is now called Cilicia Tracheia was part of the Armenian kingdom of Cilicia; see below, Part I.7.

[12] C. Toumanoff, "The Background of Manzikert," *Proceedings of the XIII Congress of Byzantine Studies* (Oxford, 1967), 411 ff; Matthew of Edessa, 158.

[13] Although this first migration was small, the Armenians made their presence felt by the late 10th century. Al-Maqdisī (*Aḫsan at-Taqāsīm fī maʿrifat al-aqālīm,* ed. M. J. de Goeje, *Descriptio Imperii Moslemici,* BGA III [Leiden, 1877], 189, line 1) in his discussion of the Ǧabal Lukkām expresses annoyance that in his day (ca. A.D. 985) this region is controlled by the Armenians. See also G. Dédéyan, "L'immigration arménienne en Cappadoce au XIe siècle," *Byz* 45 (1975), 14 ff.

were repopulating the garrison forts of the Taurus with these displaced Armenians. As trusted vassals the Armenians also assumed administrative roles in the province. For at least four years between 1072 and 1085 we know that the Byzantine governor of Tarsus was the Armenian noble Aplłarip. His father, Hasan, was a lesser prince in the province of Vaspurakan and had served in the army of Emperor Michael V. Aplłarip is said to have given the fort of Lambrōn (Lampron) to one of his generals, Ōšin.[14] Ōšin settled his family and retainers at Lambrōn; his heirs founded the Het῾umid dynasty, one of the two great families in Armenian Cilicia. Not only did the Het῾umids support Greek policy through the first half of the twelfth century, but many even adopted the calendar and liturgy of the orthodox Greek Church.

The history of the founding of the Rubenids, the rivals of the Het῾umids, is not so easily confirmed in our extant sources.[15] It seems that the Bagratid King Gagik II, who had resettled in the district of Caesarea, was openly hostile to his Greek neighbors. Aplłarip, perhaps at the request of his Greek masters, sought an alliance with Gagik by offering his daughter to Gagik's elder son, David. Soon after the marriage David was imprisoned in the castle of Lambrōn (or in Papeṙōn/Çandır). Gagik made the long journey to Tarsus in order to ransom his son. On his return trip Gagik was captured by Greeks who were waiting in ambush, and taken to the fortress of Cybistra, where he was killed. His impaled body was chained to the ramparts of the fort. Ruben, one of Gagik's semilegendary chieftains (and perhaps relative), eventually fled with his retinue to the fort of Kopitaṙ[16] in Cilicia. We know that around 1097 Constantine, Ruben's son, captured the fortress of Vahga, a strategic link on the road connecting Sis to Cappadocia. Both Constantine and Ōšin of Lambrōn provided supplies to the First Crusade in 1097.[17] Constantine avoided any involvement in the struggle between Baldwin, Tancred, the Greeks, and the Turks for control of Cilicia Pedias. Eventually the Crusaders moved to the south and east, retaining control of Tarsus, Adana, and Misis as well as the fortresses at Anavarza, Savranda, and Toprak.

At Constantine's death (between 1100 and 1102) his son, T῾oros I, assumed the leadership of the Rubenid clan and framed a policy for expansion that was eventually to lead to the unification of Armenian Cilicia under one king. T῾oros built a number of mountain fortresses and, perhaps by 1111, had seized Anavarza from its Frankish (or Byzantine) occupants. T῾oros was the first Armenian leader to establish permanent settlements in the plain. The event in T῾oros' reign to which the Armenian chronicles devote considerable attention is his capture of Cybistra. In killing the sons of Mandale, the lords of Cybistra and murderers of King

---

In the 960s Cilicia was rapidly depopulated of its Arab inhabitants, making it (along with Cappadocia) an attractive site for Armenian settlement. See G. Dagron, "Minorités ethniques et religieuses dans l'orient byzantin à la fin du Xe et au XIe siècle: L'immigration syrienne," *Travaux et mémoires* 6 (1976), 176–89, 208 ff.

[14] J. Laurent, "Arméniens de Cilicie: Aspiétès, Oschin, Ursinus," *Mélanges offerts à G. Schlumberger,* I (Paris, 1924), 159 ff; Lüders, 18.

[15] N. Adontz, "Notes arméno-byzantines, IV, l'aïeul des Roubéniens," *Byz* 10 (1935), 185 ff. Genealogies of the Het῾umids and Rubenids can be found in Toumanoff, 275 ff, 439 ff; Rüdt-Collenberg, passim. See also Dardel, 5 note 3.

[16] Smbat, G. Dédéyan, 71 note 98.

[17] G. Ter Grigorian Iskenderian, *Die Kreuzfahrer und ihre Beziehungen zu den armenischen Nachbarfürsten* (Weida-Leipzig, 1915), 26 ff.

Gagik, he avenged his clan and preserved the honor of Armenia. Tᶜoros also brought back a considerable amount of booty, including an icon that he placed in his newly constructed baronial church at Anavarza.[18] Tᶜoros was successful in maintaining good relations with the Turks, Crusaders, and Byzantines by following a policy of non-involvement. The only exception was in 1118 when he dispatched a company of Armenian troops under his brother Levon to assist Roger of Antioch in his capture of ᶜAzāz.

After the death of Baron Tᶜoros in 1129 his brother Levon I continued his policies.[19] Having persevered against Turkish and Frankish opposition in the early 1130s, Levon temporarily occupied the major cities of the plain and in 1135 captured Savranda, the Frankish possession that guarded the Amanus pass. Because of his new alliance with the Danişmendids, Levon felt confident enough to counter the Frankish threat. However, Baldwin of Maraş and his allies reacted quickly, and, despite the help provided by Joscelin of Edessa, Levon was captured and detained for two months in Antioch. The Latins changed their opinion of Levon when they heard that Emperor John Comnenus was leading an army southeast into the Levant. The Crusaders quickly made an alliance with the Rubenid baron. From 1137 to 1138 the Greeks, backed by their faithful Hetᶜumid allies, systematically captured all of the Frankish and Rubenid possessions in the plain and successfully carried their campaign into the Taurus Mountains. Baron Levon I and two of his sons, Ruben and Tᶜoros II, were captured and sent to Constantinople in chains. Only Tᶜoros would survive and return to Cilicia. Mleh and Stephen, the two other sons of Levon I, took refuge with their cousin Joscelin of Edessa.

Between 1142 and 1144 Tᶜoros II escaped from Constantinople and began the reconquest of Cilicia Pedias, which he completed in 1151. In 1152 Emperor Manuel sent his cousin Andronicus Comnenus to subdue Tᶜoros, but the latter routed the Greek army and killed a number of their Hetᶜumid allies, including Smbat of Papeṙōn. Among the captured Hetᶜumid nobles were Ōšin II of Lambrōn, Vasil of Barjrberd, and Tigran of Prakana. Emperor Manuel next tried to persuade the Seljuks to attack Tᶜoros; when that failed he turned to Reginald of Antioch. However, Manuel refused to send the money he had promised to Reginald, and the latter made an alliance with Tᶜoros. The two allies jointly plundered the Byzantine possessions on Cyprus. In 1158 the Byzantine armies quickly moved across Cilicia in a surprise attack. Eventually Tᶜoros and Manuel were reconciled at Misis.[20] The Armenian baron assumed the role of a penitent and received from the emperor the title of sebastos. He was left in control of the mountain forts and even joined the Greeks and Crusaders in 1164 during the campaign against Nūr ad-Dīn. Although technically a vassal of the Byzantine state, he maintained a remarkable degree of independence in his foreign policy. Before his death in 1168 Tᶜoros expelled his younger brother Mleh, who had become troublesome to the Armenian barons.

[18]T. Boase, "The History of the Kingdom," in Boase, 10. Tᶜoros I appears to have collected icons of the Virgin, for in 1104 he purchased one from Tᶜatᶜul, the prince of Maraş. See Matthew of Edessa, 75.

[19]Cinnamus, 16–19; Der Nersessian, 635–41; C. Cahen, *Pre-Ottoman Turkey,* trans. J. Jones-Williams (London, 1968), 94 ff; Ibn al-Qalānisī, *The Damascus Chronicle of the Crusades,* ed. and trans. H. Gibb (London, 1932), 241, 246, 349; Tritton and Gibb, 99, 275 ff; Michael Italikos, 252 f.

[20]F. Chalandon, *Jean II Comnène (1118–1143) et Manuel I Comnène (1143–1180)* (Paris, 1912), 418 ff; Kirakos, M. Brosset, 63 ff. One year before his ascension to the Katᶜołikate (1166), Nerses IV met in Misis with Alexius, the brother-in law of Manuel, to begin a reconciliation between the Greek and Armenian Churches.

Mleh's banishment was to have far-reaching consequences for Armenian Cilicia.[21] He took refuge with Nūr ad-Dīn, who supplied him with a sufficient number of Moslem troops to depose Tᶜoros' son, Ruben II, and the regent Thomas. Ruben II was later murdered by his uncle's agents in the castle of Hṙovmklay. With the support of his Turkish and Arab allies Mleh drove the Crusaders from the castles of the Amanus and captured the Byzantine governor of Tarsus, whom he handed over to Nūr ad-Dīn in exchange for the district of Maraş. When Nūr ad-Dīn died in 1174 the Armenian barons, long fearful of the Moslem alliance, assassinated Mleh. Since Stephen, the brother of Tᶜoros II, had been murdered ten years earlier by the Byzantines in the castle of Hamus, the Armenian nobles turned to his two sons, Ruben III and Levon II. During the reign of Mleh they had both been safely hidden at Papeṙōn and were now ready to assume the leadership of the Rubenid dynasty. The elder, Ruben, took firm control over the administration of his lands. In 1181 he arranged to marry the daughter of Humphrey of Toron in Jerusalem, thus forging a bond with the leading Frankish states in the Levant. After a series of skirmishes with the Seljuks on his northern border, Ruben made an alliance with Kilij Arslan. Ruben succeeded in capturing the few remaining sites in Cilicia that were still under Byzantine control.[22] Despite his many years of association with the Hetᶜumids at Papeṙōn, he was not successful in preventing them from periodically raiding the Rubenid settlements in the plain. Following the example of his ancestors, Ruben laid siege to Lambrōn. In response, the Hetᶜumids, who had been abandoned by their Byzantine allies, turned to Bohemond III of Antioch. Bohemond was fearful of Ruben's influence and seized him at a banquet in Antioch. Levon II carried on the siege of Lambrōn (unsuccessfully), and eventually Bohemond released Ruben for a sizable ransom, which included the fortresses of the Amanus. On his release Ruben recaptured the Amanus and retired to the monastery of Drazark in 1187.

His brother, Baron Levon II, at once showed himself to be an able leader and succeeded in driving the bothersome Turkish nomads from his domains.[23] Because of political intrigues in Constantinople and Iconium as well as the resounding defeat of the Crusaders at Hattin (1188), Levon was in a position to enlarge and consolidate his barony. He not only secured all of the forts from the Calycadnus (Göksu) to the Anti-Taurus Mountains, but he occupied La Roche de Roissol, La Roche Guillaume, and Bağras, three Templar sites near the plain of Antioch. Levon even led raids as far north as Caesarea and briefly occupied (until 1216) the Byzantine fort of Loulon. Perhaps Levon's greatest accomplishments were internal. By tricking the Hetᶜumid barons into attending a festival in Tarsus, Levon's troops were able to capture Lambrōn (1201).[24] Levon cemented an alliance with the Hetᶜumids by marrying his niece to Ōšin of Lambrōn and by giving the fortress of Lambrōn to his own mother, Ritᶜa, who was Hetᶜumid by birth. The political chaos in the Crusader Levant induced Levon to grant com-

[21] Der Nersessian, 642 f.

[22] W. Hecht, "Byzanz und die Armenier nach dem Tode Kaiser Manuels I 1180–96," *Byz* 37 (1967), 60 ff.

[23] Alishan, *Léon,* 105 ff; T. Rohde, *König Leon II von Kleinarmenien,* Diss. (Göttingen, 1869), 3–44; A. Savvides, *Byzantium in the Near East: Its Relations with the Seljuk Sultanate of Rum in Asia Minor, the Armenians of Cilicia and the Mongols, A.D. c 1192–1237* (Thessaloniki, 1981), 94 f, 116–20, 130, 145–47; Röhricht, 201, 208 f, 212 f, 218 f. This success, however, was of short duration; see Ibn Bībī, 23, 55, 70–75.

[24] N. Akinean, "Hetᶜum Heṫi Tēr Lambroni," *HA* 59 (1955), 397–405.

mercial privileges to the Venetians and Genoese, who were eager to enlarge the safe port at Ayas. Transportation duties on the caravans from Erzurum and Trabzon, as well as Cilician timber, goat hides, and wheat, brought substantial revenues to the Rubenids.

In 1190 the unfortunate drowning of Frederick Barbarossa in the Calycadnus (near Silifke) temporarily dashed Levon's hope of formal recognition from European princes and a crown of his own. Levon periodically assisted the Crusaders, and he even silenced the annoying Bohemond III of Antioch. The latter was captured by Levon at Bağras (repeating the ruse that Bohemond had earlier performed on Ruben III) and was soon released (ca. 1193/94) when he promised to recognize Levon's acquisitions on the east flank of the Nur Dağları and to wed his son, Raymond, to Alice, Levon's niece. After agreeing (at least nominally) to certain papal demands regarding changes in the Armenian liturgy,[25] Levon finally received recognition of his royal status and independence from the Europeans. On 6 January 1198/99[26] in the presence of the Greek metropolitan of Tarsus, the Syrian Jacobite patriarch, the Armenian katʿołikos, and the papal legate, Conrad of Mainz, Levon was crowned king of Armenia. However, Levon's kingly title did not insure the success of his policies. His attempt to install Raymond-Ruben (the issue of his niece Alice) on the throne of Antioch had at first mixed results and eventually led to the Armenian abandonment of Bağras and severely strained relations with his European allies.[27] While Levon remained hostile toward the Templars, he granted certain forts in Cilicia to the Hospitalers and Teutonic Knights.[28] Before his death in May 1219 he

[25] Regarding the relationship between Armenian Cilicia and the Greek, Latin, and Syrian Churches see: E. Ter-Minassiantz, *Die armenische Kirche in ihren Beziehungen zu den syrischen Kirchen* (Leipzig, 1904), 130 ff; C. Frazee, "The Christian Church in Cilician Armenia: Its Relations with Rome and Constantinople to 1198," *Church History* 45 (1976), 166–84; C. Kohler, "Lettres pontificales concernant l'histoire de la Petite Arménie au XIVᵉ siècle," *Florilegium ou recueil de travaux d'érudition dédiés à M. le Marquis Melchior de Vogüé* (Paris, 1909), 303–27; P. Tekeyan, "Controverses christologiques en Arméno-Cilicie dans la seconde moitié du XIIᵉ siècle (1165–1198)," *Orientalia christiana analecta* 124 (1939), 5–121; M. Oudenrijn, "Uniteurs et Dominicains d'Arménie," *Oriens christianus* 40 (1956), 94–112, 42 (1958), 110–33, 43 (1959), 110–19, 45 (1961), 95–108, 46 (1962), 99–116; A. Balgy, *Historia doctrinae catholicae inter Armenos unionisque eorum cum ecclesia Romano in Concilio Florentino* (Vienna, 1878); A. Ter-Mikelian, *Die armenische Kirche in ihren Beziehungen zur byzantinischen (vom IV. bis zum XIII. Jahrhundert)* (Leipzig, 1892); J. Prawer, "The Armenians in Jerusalem under the Crusaders," *Armenian and Biblical Studies*, ed. M. Stone (Jerusalem, 1976), 223–36; H. F. Tournebize, *Histoire politique et religieuse de l'Arménie* (Paris, 1910), 235–388, 644–753; idem, "Les cent dix-sept accusations présentées à Benoît XII contre les Arméniens," *Revue de l'Orient chrétien* 11 (1906), 163–81, 274–300, 352–70; idem, "Les Frères-Uniteurs ou Dominicains arméniens (1330–1794)," ibid., 22 (1920–21), 145–61, 251–79; Nersēs of Lampron, 569 ff; A. Atiya, *A History of Eastern Christianity* (London, 1968), 332–34; M. Baldwin, "Missions to the East in the Thirteenth and Fourteenth Centuries," *A History of the Crusades*, ed. K. Setton, V (Madison, 1985), 463, 469 f, 478, 485 f, 489–93, 506, 510; A. Heisenberg, "Zu den armenisch-byzantinischen Beziehungen am Anfang des 13. Jahrhunderts," *Sitzungsberichte der Bayerischen Akademie der Wissenschaften, Philosophisch-philologische und historische Klasse. Jahrgang 1929*, 6, 3–20; Grigor Aknercʿi, 378–80; D. Robert, "Négociations ecclésiastiques arméno-byzantines au XIIIᵉ siècle," *Studi bizantini e neoellenici* 5 (1939), 146–51; B. Hamilton, "The Armenian Church and the Papacy at the Time of the Crusades," *Eastern Churches Review* 10 (1978), 61–87; J. Gay, *Le Pape Clément VI et les affaires d'Orient (1342–1352)* (Paris, 1904; rpr. New York, 1972).

[26] Der Nersessian, 648 note 23; A. Atamian, "The Data of the Coronation of Levon I," *Armenian Review* 32 (1979), 280 ff; Kirakos, M. Brosset, 78 f. For the purpose of this study, I shall refer to Baron Levon II as King Levon I after this date.

[27] Cahen, 526 ff; Edwards, "Bağras," 431 f.

[28] Forstreuter, 59–67.

had succeeded not only in unifying the Armenians of Cilicia, but in creating the most powerful Christian state in the northern Levant and eastern Anatolia.

Since Levon left no direct male heirs, his daughter Zapēl became heiress to his estates and title.[29] Zapēl's first marriage to Philip, the son of Bohemond IV of Antioch, ended in disaster. It seemed that Philip openly favored his Latin barons for court appointments and refused to accept the teachings of the Armenian Church. After only a brief period as king of Armenia, Philip was arrested, imprisoned, and eventually forced to drink poison. In June 1226 Zapēl married Hetʿum, the son of Constantine of Papeṙōn.

Most of the reign of Hetʿum I was marked by relative peace and prosperity. Two Seljuk invasions, in 1233 and 1245, did little damage, but the Turks did manage to extract some tribute from the Armenian crown.[30] The coastal sites west of Silifke, which had been briefly occupied by Armenians during the period of Levon's reign, were captured by the Seljuks. The stunning defeat of the Seljuks in 1243 at the hands of the Mongols moved Hetʿum to seek some sort of rapprochement with these new protagonists. In 1247 he dispatched his brother the Constable Smbat on an embassy to the Mongols. Three years later Smbat returned with a treaty that guaranteed the protection of all Armenian settlements in Anatolia as well as the promise of Mongol help in recapturing the Armenian possessions in Seljuk hands. In 1253 Hetʿum traveled to the court of the Great Khan at Karakorum.[31] The Armenian king formally acknowledged Mongol supremacy in the region; in return the Khan promised that all Armenian monasteries in the Mongol dominions would be freed from taxation. Hetʿum also committed large contingents of Armenian troops to fight in Mongolian campaigns. King Hetʿum returned home through Greater Armenia and received proper recognition from the clergy and *naxarars*. In 1254 he married his daughter to Bohemond VI and succeeded in briefly extending Armenian influence over Antioch. With Mongolian help the Armenians regained the district of Maraş and even captured Behesni on the Euphrates frontier. Hetʿum also rode into Aleppo and Damascus in company with the triumphant Mongolian forces. In a series of pitched battles on the western frontier of Cilicia Hetʿum repulsed the Karamanids and killed their leader Karaman. But the savor of success was short-lived, for the battle of ʿAin Jālūt (1260) ended Mongolian invincibility and opened Cilicia to Mamluk penetration. Hetʿum, convinced of the soundness of the Christian-Mongol alliance, was slow to realize the Egyptian threat.[32] In 1266, after he refused to placate Baybars by surrendering a border town, the Mamluks invaded his kingdom with full force. The few cities of Cilicia, including the capital at Sis, were plundered and burnt. In 1269 Hetʿum retired and relinquished his throne to King Levon II.

[29] Der Nersessian, 651 f.

[30] P Žavoronkov, "Nikeiskaja imperija i vostok," *Vizantiĭskiĭ Vremennik* 39 (1978), 93–101; Ibn Bībī, 140–42.

[31] E. Bretschneider, "Notices of the Medieval Geography and History of Central and Western Asia," *Journal of the North-China Branch of the Asiatic Society,* n.s. 10 (Shanghai, 1876), 297–302; R. Hennig, *Terrae Incognitae (A.D. 1200–1415),* III (Leiden, 1953), 61–64. G. Bezzola, *Die Mongolen in abendländischer Sicht* [*1220–1270*] (Bern, 1974), 151–54, 179, 182, 190–92; J. Boyle, "The Journey of Hetʿum I, King of Little Armenia, to the Court of the Great Khan Möngke," *Central Asiatic Journal* 9.3 (Sept. 1964), 175–89; Kirakos, M. Brosset, 176 ff; Grigor Aknercʿi, 312–14, 340–42; J. Klaproth, "Aperçu des entreprises des Mongols en Géorgie et en Arménie, dans le XIIIᵉ siècle," *JA* 12 (1833), 206–14.

[32] Grigor Aknercʿi, 352–72; Canard, "Le royaume," 217 ff.

The collapse of the Mongolian alliance and repeated attacks by the Mamluks forced Levon to accept a treaty with Cairo in 1285. As a result, the Armenians were prohibited from building forts and Levon had to pay a staggering tribute of one million dirhams per annum. After Levon's death in 1289 his son, Het῾um II, sought to revive the Mongolian alliance and to forge treaties with Cyprus and Constantinople. Yet dissension among the Armenian barons forced Het῾um from his throne on three different occasions. Any hope of maintaining the Christian-Mongol alliance ended in 1304 when the Mongols officially adopted Islam. On a visit to Anavarza in 1307 the Mongol Emir Bilarghu treacherously murdered the new king, Levon III, and his predecessor Het῾um.[33] What followed in the nearly seventy years that the kingdom survived was a continuous series of disputes over succession and blood feuds. On three separate occasions the Lusignans ascended the Armenian throne.[34] The last of these Cyprio-Armenian kings was Levon V, who was crowned at Sis in September 1374. Within six months he was forced to flee from his capital because of the treachery of his own nobles and the besieging Egyptian army. His freedom was short-lived for he and his family were captured at the fortress of Geben and taken to Cairo in chains. He eventually won his release and died in an obscure suburb of Paris in 1393.[35] As the last king of Armenia, he was accorded the final honor of a royal sepulcher at St. Denis.

2. *The Characteristics of Armenian Fortifications: Twenty Distinctive Features*

Throughout Cilicia the Armenian fortifications display certain peculiar characteristics with a high degree of consistency. This does not imply any lack of variation within expected norms nor the complete absence of anomalies. However, by carefully comparing the following table of twenty characteristic features in Armenian military architecture with the individual forts in the Catalogue, the reader should become aware of the reasons for separating Armenian from

[33] Dardel, 16 f.

[34] J. Richard, ed., *Chypre sous les Lusignans: Documents chypriotes des Archives du Vatican (XIVe et XVe siècles)* (Paris, 1962); V. Langlois, "Documents pour servir à l'histoire des Lusignans de la Petite Arménie," *RA* 16.1 (1859), 109–16, 143–69, 216–34; idem, "Documents pour servir à une sigillographie des rois d'Arménie," *RA* 11.2 (1855), 630–34; K. Basmajean, *Lewon V Lusinean* (Paris, 1908); A. Galstyan, "Kilikyan Hayoc῾ t῾agavorut῾yan korcanman harc῾i surǰə," *Lraber* 9 (1971), 50–63; L. de Mas Latrie, *Histoire d'isle de Chypre*, II, *L'isle de Chypre sous le règne des princes de la maison de Lusignan, 1192–1489* (Paris, 1853); G. Hill, *A History of Cyprus*, I, *The Frankish Period 1192–1432* (Cambridge, 1948); G. Raynaud, *Les Gestes des Chiprois* (Geneva, 1887); Mas Latrie, I, 88–90, 276–80, 299 f, 313–15, 324 ff, 377 f, 390–92, II, 25 ff; U. Robert, "La chronique d'Arménie de Jean Dardel, évêque de Toriboli," *AOL* 2 (1884), 1–15.

[35] See A. Luttrell in Boase, 132 f. One equally valid tradition places Levon V within the fortress of Sis at the time of his capture in 1375.

non-Armenian fortifications in Cilicia (see below, Part I.5). On a number of occasions a small fort may lack a few of the architectural features that are common to the larger sites, but what survives will clearly mark it as Armenian. From the outset the reader ought to be aware that only two of the twenty characteristics (slot machicolations and ecclesiastical construction) are unique to Armenian architecture; the other features appear in ancient and late antique fortresses of the Mediterranean and the East.[1] The significance of the Armenian contribution is that for the first time all twenty characteristics are used simultaneously and they are brought to a degree of perfection that had not been seen before the tenth century. This implies that the Armenians had highly developed theories on military architecture. The almost total lack of experimentation with new designs in Cilicia is a reflection on the relative backwardness of their enemies with regard to tactics. Military architecture is always a response to the nature of siege warfare.[2] If the designs that the Armenians brought from their homeland in Greater Armenia served them well in Cilicia, then there was no impetus to improve them. The Armenians were not influenced by the Arab and Byzantine forts that they found in Cilicia. Many of these non-Armenian forts were dismantled or covered over with new constructions (Çardak is one of the few exceptions). In the twelfth century the Crusaders had no effect on Armenian architecture, but the reverse may be true.

Apart from architectural paradigms, there are two methods used to determine if Armenians are responsible for building a fort in Cilicia. The most reliable way is to find an Armenian dedicatory inscription in the circuit wall or one of the towers. Three fortresses—Tamrut, Sis, and Mancılık—have such inscriptions and share most of the twenty characteristic features of Armenian military architecture. However, this inscriptional material must be used with caution. At three other fortresses—Korykos, Silifke, and Anavarza—the Armenians placed their

[1] I plan to publish a study which in part deals with the nature of military architecture in Armenia Major, how it developed and later influenced the west through Armenian settlements in Cilicia, Jerusalem, Edessa, and Cairo. Let me note in passing that the Armenians were the successors of the Urartians in the region of Van and Ayrarat. The Urartians had evolved a highly sophisticated system of fortresses that resisted for two centuries the aggressive siege tactics of the Assyrians. Later, as a Christian nation in the East, the Armenians maintained wide-ranging contacts throughout much of the world. The Armenian fortresses of Cilicia are the result of a long process of evolution. Cf. Edwards, "Doğubeyazit," 435–45; idem, "Artvin," 178.

[2] For this study I have decided not to describe siege tactics in the medieval Levant and Anatolia since this subject is so fully discussed in a number of modern publications. For example: F. Lot, *L'art militaire et les armées au moyen âge en Europe et dans le proche orient,* 2 vols. (Paris, 1946); R. Smail, *Crusading Warfare (1097–1193)* (Cambridge, 1956); R. Contamine, *La guerre au moyen âge* (Paris, 1980); H. Eydoux, *Les châteaux du soleil* (Paris, 1982); C. Oman, *The Art of War in the Middle Ages,* rev. H. Beeler (Ithaca, 1960); Maurice, *Das Strategikon des Maurikios,* ed. G. Dennis, CFHB 17 (Vienna, 1981); Dennis, 37 ff; A. Bivar, "Cavalry Equipment and Tactics on the Euphrates Frontier," *DOP* 26 (1972); J. Haldon, "Some Aspects of Byzantine Military Technology from the Sixth to the Tenth Century," *Byzantine and Modern Greek Studies* 1 (1975); M. Benvenisti, *The Crusaders in the Holy Land* (Jerusalem, 1970), 284–86; C. Cahen, "Un traité d'armurerie composé pour Saladin," *Bulletin d'études orientales* 12 (1947–48), 141–63; "The Crusades and the Technological Thrust of the West," in L. White, Jr., *Medieval Religion and Technology, Collected Essays* (Berkeley, 1978), 283–89; T. Wise, *The Wars of the Crusaders 1096–1291* (London, 1978), 208–18; D. Hill, "Trebuchets," *Viator* 4 (1973), 99–114; J. Partington, *A History of Greek Fire and Gunpowder* (Cambridge, 1960), 21–41; R. Payne-Gallwey, *The Crossbow,* 2nd ed. (London, 1958), 261 ff.

Medieval sources that recount in detail the sieges of Armenian forts: Choniatēs, 25–27 ff; Matthew of Edessa, trans. and comm. A. Dostourian, Diss. (Rutgers University, 1972), 137–42, 587 note 14.

inscriptions on non-Armenian constructions after they completed repairs. At Korykos, which I class as a non-Armenian site, the Armenians reconstructed only a part of the sea castle using their characteristic masonry and towers. Likewise, Silifke Kalesi has an Armenian inscription that was added *after* the Hospitalers built the present castle. Onto the donjon of Anavarza, one of the few non-Armenian structures in the massive fortress complex, the Armenians affixed an inscription to record a period of reconstruction. Of the seventy-five sites in this study only these six have Armenian inscriptions in situ; at many of the other fortresses a shallow niche survives in the circuit wall to indicate where an inscription once stood. Translations of the six inscriptions can be found in the Catalogue.

The second approach is to study carefully the medieval Armenian chronicles for references to prolonged periods of Armenian occupation and construction at forts. Sites that can thus be securely identified today are: Anavarza, Çandır, Geben, Gülek, Lampron, Savranda, Sis, and Vahga.[3] Along with Tamrut and Mancılık, these eight sites share most of the twenty characteristic features of Armenian military architecture. Thirty-four forts that cannot be identified by one of the two methods, but that share most of the twenty features, are *probably* Armenian. These sites are: Ak, Anacık, Anahşa, Andıl, Arslanköy, Azgit, Babaoğlan, Bayremker, Belen Keşlik, Bodrum, Bostan, Bucak, Çem, Fındıklı, Fındıkpınar, Gediği, Gökvelioğlu, Gösne, Hisar, Hotalan, Işa, Kız (near Dorak), Kuzucubelen, Maran, Meydan, Rifatiye I, Saimbeyli, Sarı Çiçek, Sinap (near Çandır), Sinap (near Lampron), Tece, Tumlu, Yeni Köy, and Yılan. Thus from my total of seventy-five fortifications in Cilicia, forty-four are the result of Armenian construction.[4]

The Twenty Characteristic Features

1. The plan of an Armenian fortification is as spontaneous as the ground on which it lies. The circuit walls follow carefully the sinuosities of the rock and conform to the demands of the local topography. Since the resulting outline of the circuit is never standard or symmetrical, the plans of no two Armenian forts are alike. For example, had the Armenians constructed the fort of Çardak, we should expect to see a more oval plan in response to the shape of the mountain. This rule of consistent irregularity holds true for all garrison forts and watchtowers since these are confined to the irregular heights of an outcrop. Not only is it difficult for an

[3] Ayas, which can be identified with the modern Yumurtalık, was certainly part of the Armenian kingdom. But the extent to which it was inhabited by Armenians is unknown. Milvan, Korykos, Silifke, and Toprak can be securely identified, but we know from the chronicles and physical evidence that they were not built, but only repaired, by Armenians.

[4] Fourteen of the forty-four sites (Anahşa, Anavarza, Babaoğlan, Çandır, Çem, Geben, Gökvelioğlu, Lampron, Rifatiye I, Savranda, Sis, Tumlu, Vahga, and Yılan) have the remains of earlier, non-Armenian periods of construction, but they are so extensively rebuilt that they are classed as "Armenian" forts. Because of extensive damage the percentage of the Armenian construction at Babaoğlan is difficult to determine. Only six of the sites classed as "non-Armenian" show minor traces of Armenian construction; see below, Part 1.5. Tece and Kız (near Dorak) were built by Armenians using Latin plans. I have classed the "site" of Ayas as non-Armenian because the largest fortified unit there, the land castle, is a Turkish construction. Its small sea castle was probably built by the Armenians.

enemy to advance against such a fort, but it rules out the possibility of mining.[5] The fortified estate houses, which are constructed on relatively flat land, do have regular plans.

2. Natural rock walls, cliffs, promontories, and spurs are exploited to the degree that they become integral parts of the defensive system. Walls of masonry are constructed into the rock mass as if they were natural extensions of the outcrop. Wherever the point of access to the summit becomes less severe walls are built; they increase in thickness and height to the degree that the path of approach becomes easier. This is quite apparent at forts like Bodrum and Gökvelioğlu. By exploiting the natural defenses the builders can reduce the expenses involved with construction and concentrate their engineering skills on one or two weak points.

3. Protruding angles and corners are avoided on the exterior of circuit walls. There is a tendency to show always a rounded, curving face to an advancing enemy. The reason is obvious. Since a corner usually consists of a series of horizontally placed keystones that anchor the adjacent walls, a sapper simply has to remove the lower corner stones to weaken or even collapse both walls. With a rounded face, all of the facing stones bear an equal weight and no part of the exterior facing is especially susceptible to sappers. There is also a tendency, when the Armenians build rectangular structures on the interior of a fort (for examples, chapels), to avoid the use of quoins (that is, oversized corner stones). In all but one of the Armenian estate houses the exterior angles are covered with rounded salients.

4. The circuit walls are always constructed with an inner and outer facing of cut stone. Both units are anchored to a central poured core. The shape and quality of facing stones vary greatly, but certain types are *always* used in certain areas of the fort. Because this complex subject involves many types of facing stones, I will defer further discussion until Part I.3.

5. All circuit walls in Armenian forts are built with a slight batter. A batter, which is simply the inward tilt of a wall, offers two advantages. The principal benefit is that an attack with a ram at the base of the wall does not cause the top of the wall to recoil outward from the shock so violently. Secondarily, since the top of the circuit has a slightly smaller circumference than the base, the weight of the upper half of the wall is displaced through the sides as well as through the base. A cupola works on a similar principle. In fact, if the circuit wall of an Armenian fort was extended to its maximum height, it would form an irregular cone.

6. The circuit wall of an Armenian fort does not rise from a plinth but simply from a shallow trench cut in the natural rock. As one would expect, the base of the wall is substantially thicker than the top. Since mining is impractical, a special foundation was deemed unnecessary. This is why taluses are so rare in Armenian fortifications and where they do appear they normally cover small, irregular clefts in the rock foundation (cf. Gökvelioğlu and Meydan with Toprak).

7. There is a tendency in the large garrison forts to employ a series of ascending circuit walls to create separate, defensible baileys. Normally two baileys are created, as we see at Çem, Savranda, and Geben.[6] Yılan and Anavarza are exceptional with three baileys. Fre-

[5]Mining can be accomplished when fortresses stand on level ground. An enemy simply digs underneath a wall or tower and props up the foundation with wooden beams as the soil is removed. Once a sufficient section is undermined, the wood is fired and the wall collapses.

[6]Dibi Kalesi is the only small garrison fort that has two baileys.

quently, the largest bailey is at the lowest level, blocking the most direct line of access into the fort. The smaller upper bailey girdles the summit of the outcrop. The theory behind the bailey system is simple. If an enemy breaches the lower bailey, the defenders can retreat to the separate upper ward and fire down on an attacker, who is now confined in the enclosed lower bailey and who still must make a difficult ascent to dislodge the defenders.[7] In most cases the upper bailey is fitted with cisterns, storage rooms, and other buildings that would help the defenders to withstand a siege.

8. Armenian garrison forts never employ donjons, keeps, or large isolated towers.[8] Such structures are passive in nature and thus contrary to the Armenian concept of an offensive fortification. Men who retreat into a donjon expect that its massive walls will be their primary defense; a keep cannot offer the broad lines of attack that is possible from the upper ward of an Armenian fort.

9. All types of Armenian fortifications consistently employ rounded towers and eschew the use of square or polygonal bastions. Like the curving face of the circuit (cf. no. 3 above), the rounded front of the tower can more efficiently deflect projectiles and avoid the problem of exposed corners. The Armenians employed two types of rounded towers: the semicircular and the horseshoe-shaped. The tactical advantage of the semicircular tower is that the archers can fire at a distant enemy from any position on the half circle. Atop square towers only the archers at the extreme end of the salient have a clear view outward. The big advantage of the rectangular tower is that the two flanking sides provide an excellent platform from which to attack an enemy near the circuit wall. This may explain why the horseshoe-shaped tower became so popular in Armenian Cilicia; it has the extended sides of a rectangular tower and the broad, rounded apex of the semicircular tower. Most of the Armenian towers contain rooms that are covered by apsidal vaults. The large towers usually have two vaulted stories. Depending on a tower's location in the fort, the lower floor could be used for a cistern, storage, or a barracks. Sometimes the lower floor (for example, Tamrut) and normally the upper floor are opened by embrasured loopholes. The three-quarter rounded towers only appear on Byzantine fortifications (for example, Evciler and Çardak).

10. Battlements on the tops of the circuit and towers are the rule rather than the exception. Battlements are normally flanked on the interior side by a stone wall walk. The battlements and wall walk are not continuous for the entire length of the circuit but periodically are cut off by towers that rise significantly above the height of the circuit (for example, the west side of the circuit in the upper bailey of Yılan Kalesi). Armenian merlons are frequently rectangular with slightly rounded tops. The exceptions are at Yılan Kalesi where a series of merlons have pyramidal crowns and are pierced by embrasured loopholes. Because merlons have so little support they are the first elements to collapse. Consequently most forts show only fragments of these indentations. On most of the plans in the Catalogue I have chosen not to represent battlements and wall walks; any exceptions will be noted in the commentaries.

[7] Gökvelioğlu is unusual in that the higher bailey is the first line of defense.

[8] The donjon at Anavarza is a Crusader construction, which was only repaired by the Armenians; see the discussion of Anavarza in the Catalogue, and Edwards, "Anavarza," 53–55.

11. In Armenian garrison forts the path of approach to the main gate of the lower bailey always flanks the base of the circuit, thus exposing any potential enemy to prolonged fire from above. A fine example of this is at Savranda Kalesi where all the towers on the east flank have a clear line of fire at the path of ascent. In order to prevent an advance on a fort from a direction other than the intended approach, an outwork can be attached to the exterior of the gatehouse. Outworks are rare in Armenian military architecture because the site for the fort is so carefully chosen that the topography allows only one convenient line of approach.

12. Moats and ditches are never constructed because the steep and rocky nature of these fortified sites is an adequate deterrent against assault with mobile siege towers.

13. All entrances, including posterns, are either in the lee of a tower or next to a bulge in the circuit wall. Although most of the doors are protected by jambs and secured by crossbar bolts, an adjacent salient allows some degree of camouflage and provides flanking fire for those entering and leaving.

14. The main opening into a castle or large garrison fort usually is a complex entrance that is incorporated into a gatehouse.[9] The most common type is the bent entrance, which requires the approaching party to make a 90-degree turn in a confined, box-like space. Examples of these entrances can be seen at Savranda, Anavarza, Yılan, Tamrut, Lampron, and Gökvelioğlu. At Çandır and Vahga a stepped, snake-like vaulted passage is attached to one end of the bent entrance to create an even more restrictive access. Another less common type of gatehouse is the vaulted corridor, which consists of a narrow rectangular room with a single door for entrance and exit at each end (for example, Meydan). In this type of gatehouse the approaching party is not required to turn on the interior but is exposed to fire through machicolations (ports) in the vaulted ceiling.[10] Usually the outer door of a gatehouse is covered by a compound arch. Towers are either adjacent to these gatehouses or built around them. In most of the smaller garrison forts the entrance appears to be simply a compound arch in the lee of a tower. But on closer examination this arched door is always positioned in a natural jog of the outcrop which forces the approaching party to make a turn (usually 60–90 degrees) just before entering (for example, Gülek, Geben, and Ak). In other words, this is a natural, unvaulted bent entrance. All of the entrances efficiently limit the speed and movement of an advancing party.

15. The vast majority of compound arches in the entrances of Armenian garrison forts incorporate a slot machicolation. The slot machicolation appears to be an Armenian invention.[11] Unlike the machicolations that are simply ports in the floors of second-level rooms or overhanging chambers, the slot machicolations can only be fitted between the apices of the lower outer arch and the higher inner arch of the exterior door of a gate. The gap is usually no more than 40 cm in width and allows the defenders to fire down on someone at the threshold.

[9]My views on Armenian gatehouses were first presented at the 1980 Conference of the American Oriental Society. A summary appears in the *Abstracts of the 119th Meeting of the American Oriental Society,* San Francisco (U.S.A.), April 15–17 (1980), 39 f.

[10]Some of the bent entrances also have machicolations on the interior (e.g., Vahga).

[11]To my knowledge, the slot machicolation first makes its appearance at Van Kalesi in the 7th century.

16. The most common type of window that opens the circuit wall and towers is the embrasured loophole.[12] This opening can simply be a splay running through the entire thickness of the wall, or the interior half can be widened with straight sides to form a small vaulted chamber known as a casemate. The casemate is designed to accommodate the body of an archer in both a kneeling and standing position. The inner side of the casemate is always open. On the exterior of the fort the loopholes appear simply as narrow slits. Many of the loopholes have a stirrup base that allows the archer greater dexterity in firing at an enemy near the wall. The top of the loophole is rounded on the exterior because the embrasure is normally round-headed.[13] In the small embrasured loopholes and windows (for example, the apsidal opening in a chapel) the rounded top is *always* a monolith in which a half cone has been cut. This is quite different from the technique used by the Mamluks, who employ many narrow voussoirs in the cover (see Toprak in the Catalogue). In their large embrasured openings, where a monolith is not practical, the Armenians use only a few voussoirs (normally no more than six) to form the rounded splayed cover.

Broad, straight-sided windows open the interior rooms of the fort; they are extremely rare in circuit walls and towers.

17. There is a proclivity, but not an absolute preference, for using pointed arches and pointed vaults in Armenian military architecture. Barrel vaults and simple half-round arches are sometimes used side by side with their slightly pointed counterparts. Semicircular arches are frequently used over postern gates and small doors, while pointed arches are more often confined to the large doors in gatehouses. Sometimes when two pointed arches abut to form the exterior of a door (creating a compound arch) one of the arches will be less pointed and the inner arch will always have a lintel or transom below the springing level.[14]

Normally, rectangular rooms are covered by simple longitudinal vaults. Square rooms or bays (especially gatehouses and other structures that must sustain the stress of a siege) are covered by groined vaults. Cupolas are very rare in Armenian fortifications, appearing only in a cistern at Yılan Kalesi and a bathhouse at Lampron.

18. Armenian undercrofts are never a continuous series of interconnecting galleries studded with embrasured loopholes. Usually they are the vaulted structures below the battlements and loopholes (for example, Yılan and Gökvelioğlu). In contrast, both of the levels of an Armenian tower may be opened by embrasured loopholes. Generally, undercrofts are not as numerous in Armenian forts as they are in Crusader architecture. The big advantage of undercrofts is that they act as internal buttresses along the interior of the circuit walls. Like towers, undercrofts were used for storage, kitchens, and barracks.

19. Cisterns are ubiquitous in Armenian fortifications. Normally they are so cleverly designed that the runoff is collected by drains scarped in the natural rock. In many cases the

[12] The residential complex at Çandır has tall broad windows on the exterior because the altitude of the site removed the apartments from harm's way.

[13] With only a few exceptions, embrasured openings are covered with rounded tops.

[14] Lintels and transoms are necessary since they form a secure, square head for the top of the swinging door. The resulting tympanum was filled in with a stone wall.

wall walks are designed with a slope to channel the rain along the battlements. Most cisterns are adapted into natural sinks and supplemented with masonry.

20. Armenian ecclesiastical architecture in Cilicia always has a unique, standard, and very predictable plan that is readily distinguishable from its Byzantine counterpart.[15] Not every fort has a chapel or church, since the garrisons at many sites could worship in a neighboring village. It is significant that all of the sites with Armenian dedicatory inscriptions (except Silifke) and the majority of Armenian-occupied sites mentioned in the chronicles have these peculiar churches and chapels. Eight forts, which have no inscriptions and which cannot be positively identified with sites mentioned in the texts, have Armenian chapels. These are: Ak, Çem, Maran, Yılan, Meydan, Saimbeyli, Sarı Çiçek, and Işa. The majority of these chapels were physically incorporated into the circuit walls at the time these sites were built. When the Armenians occupied the preexisting land castle at Korykos and Çardak Kalesi they built freestanding chapels on the interior of the bailey without reconstructing the forts.

The Armenian chapel consists internally of a simple rectangular nave that terminates at the east with an apse. The center of the apse is normally pierced by one low-level embrasured window. As with the Armenian church in Cilicia, the north and south walls of the apse have low niches. Frequently a niche will appear in the nave. One of the apsidal niches may have functioned as the prothesis.[16] The recess in the nave may have held votive offerings, such as pictures of saints and candles; in a few cases it functioned as a piscina. The south and west walls of the nave are often opened by a door. The chapel is topped by a barrel vault and gabled roof. The masonry of the chapel is generally the same well-coursed ashlar that is used to construct the circuit wall.

Like the unattached chapel, the Armenian church has a rectangular facade. The chevet consists of a central apse and flanking apsidioles. Internally two types of plans can be employed; the nave can be flanked by aisles and covered by longitudinal barrel vaults, or four piers can support a central cupola. Inscriptions on the fortress churches at Anavarza (ca. 1111) and Çandır (1251) confirm their origin.

[15] See Edwards, "First Report" and "Second Report" for a complete description of the ecclesiastical buildings. The peculiar plan of Armenian tower chapels is repeated in Armenia Major; see A. Altun, *Mardin'de Türk Devri Mimarisi* (Istanbul, 1971), 20.

[16] *Burchard of Mount Sion,* trans. A. Stewart, PPTS (London, 1896), 110 f.

3. *The Masonry of Cilician Fortifications*

It is with hesitation that I catalogue the masonry of medieval Cilicia.[1] Too often in the past archeologists have been tempted to use such catalogues to determine the relative chronology of buildings when other evidence is lacking. Many of these attempts have been criticized when later excavations and systematic regional surveys uncovered contradictory evidence.[2] Medieval builders and their masons, pressed by the limitations of time and money, may be inconsistent in their use of masonry. Byzantine and Crusader fort builders readily mix the architectural spoils from the long-abandoned cities of the Levant with their own masonry. If work crews from different regions were commissioned to construct a fort, then minor variations in the appearance of facing stones should be expected. Architects can be anachronistic and recommend the use of archaic types of facing stones for purely aesthetic reasons. One should assume multiple periods of construction only when there is an abrupt change in the style of masonry and the consistency of the poured core.[3]

I have arranged the masonry of Cilicia into types because the principal builders in the medieval period, the Armenians, possessed a theory of masonry and employed it with a remarkable degree of uniformity in their fortifications. In both military and ecclesiastical constructions the Armenians always use a poured-wall technique.[4] This consists simply of an inner and an outer facing of cut stone and a poured core. The core, which appears to be layered at each course level,[5] is made up of the same limestone mortar that seals the interstices of the facing stones as well as an abundance of small fieldstones and potsherds. On the exterior of the circuit walls, towers, and gatehouses, and in all places where an enemy could inflict

[1]The forts of the Byzantines and Armenians are considered here. Only the military churches of the Armenians are included in fig. 3. Mamluk reconstruction, as well as the peculiar masonry of Arab and Crusader forts, is excluded from this study, because too few examples from those periods survive to establish reliable paradigms.

[2]One example is R. L. Scranton's *Greek Walls* (Cambridge, Mass., 1941). While this study still offers valuable insights into the technical aspects of masonry and construction, his system of chronology based on the types of masonry has been revised, because in part the types that he labels as early Greek appear in late Hellenistic buildings. See F. Winter, *Greek Fortifications* (Toronto, 1971), 80–91.

[3]By "consistency" I mean the homogeneous appearance of the core and a certain uniformity in the chemical components of the mortar (see below, Appendix 2).

[4]Most recently A. Berkian (42–47) has discussed the use of the poured-wall technique in Greater Armenia. This technique, which was employed by the Byzantines in Cilicia before the arrival of the Armenians, was developed in the pre-classical and classical world. See J. Ward-Perkins, "Notes on the Structure and Building Methods of Early Byzantine Architecture," *Great Palaces of the Byzantine Emperors, Second Report,* ed. D. Rice (Edinburgh, 1958), 62–77, 82 ff; G. Wright, *Ancient Building in South Syria and Palestine* (Leiden-Köln, 1958), I, 332–513, II, 290–367.

For information on the tooling of facing stones in the Crusader Levant, see Deschamps, I, 227 ff.

[5]Unfortunately we have no specific information on actual construction practices. Contemporary techniques for stone construction in western Europe have been discussed in some detail; see G. Binding, "Baumeister und Handwerker im Baubetrieb," *Ornamenta ecclesiae, Kunst und Kunstler der Romanik,* I, ed. A Legner (Cologne, 1985), 171–83, especially the bibliography on p. 183.

Because support holes are completely absent in the facing stones, we can assume that tongs were not used for lifting. It is likely that a winch or block and tackle hoisted the cut stone in a rope sling. See D. Leistikow, "Aufzugsvorrichtungen für Werksteine im mittelalterlichen Baubetrieb: Wolf und Zange," *Architectura* 12 (1982), 20–33.

damage with siege weapons, the Armenians consistently use large, well-coursed, rusticated facing stones (either types V, VI, or VII) whose interior sides are pointed to bind firmly with the poured core.[6] Stones with narrow exterior margins are less vulnerable to the assaults of sappers. The firm anchorage in the core allows the exterior face to absorb and transfer the shock of battering rams more efficiently. It is significant that the Armenians prefer types V and VII masonries; they are the exterior facing stones in about 80 percent of the circuit walls. The principal advantage of types V and VII is that their rusticated centers protrude from the margins, protecting further the delicate junctions from the rams. The Armenians almost never employ types I through IV or type IX[7] as an *exterior* facing in areas subject to direct attack. Also, they never use brick to construct the walls of fortifications. On a few occasions they seem to have recycled Byzantine brick for use in churches and minor fixtures in the fort.[8] On the interior the Armenians have been known to use all nine types of masonry. But normally the interior facing of an Armenian wall will be type III or IV because the inner side is not critical for defense. There is a definite logic to the placement of specific types of masonry in an Armenian fortress.

It is interesting that the Byzantines mainly employ brick and types III and IV masonries as an exterior facing stone. In two forts, Korykos and Lamas, the Byzantines recycled the fine ashlar from neighboring ancient settlements to construct the defenses. At both sites headers are used in the walls. But in other forts with known periods of Byzantine construction headers are less frequent and other types of masonry are employed. What characterizes Byzantine construction is the inconsistent use of masonry types. The original Byzantine wall at the far south end of Anavarza Kalesi (pl. 10a) is built with an *opus listatum* (that is, alternating courses of brick and ashlar). In a later period of Byzantine construction type IV is used for the exterior facing of the east circuit of Anavarza (pl. 289a).[9] At Gökvelioğlu the Greeks constructed the northeast circuit with type III masonry (pl. 82a).[10] It may be significant that in other forts where Byzantine periods of occupation are suspected, types III and IV masonries appear as exterior facing stones in conjunction with the peculiar elements of Byzantine design (for example, a symmetrical plan, square and/or three-quarter rounded towers at regular intervals, donjons, exposed corners, etc.; cf. Evciler and Çardak). Types V, VI, and VII masonries are *never* associated with features peculiar to Greek construction.[11]

[6]This is somewhat analogous to the anchorage of a tooth in the mandible and is remotely related to the *anathyrosis* of classical architecture. See A. W. Lawrence, *Greek Architecture* (Baltimore, 1957), 225; A. Orlandos, *Les matériaux de construction et la technique architecturale des anciens Grecs,* trans. V. Hadjimichali, II (Paris, 1968), 71–184; R. Martin, *Manuel d'architecture grecque,* I, *Matériaux et techniques* (Paris, 1965), 190 ff.

[7]Occasionally, type IX will be used for the frames of doors and windows.

[8]These examples include the vaults over the Church of T'oros I at Anavarza, the apsidal dome in the chapel of Meydan Kalesi, and cisterns at Anavarza, Yılan, Tumlu, and Gökvelioğlu. See: Youngs, 115 ff; Edwards, "First Report" and "Second Report."

[9]Gough, 121. The use of type IV by the Byzantines may represent a later expansion of the original brick fort (probably under Nicephorus Phocas).

[10]Youngs, pl. 21a.

[11]There is no known example in any period of Byzantine military architecture where type VII masonry is used.

It should be noted that the nine types of Cilician masonry are not absolute. Often within a given type the size of the individual stone may vary as well as the thickness of the core. Represented in figure 3 are the somewhat ideal types that are relatively homogeneous in appearance. The specific criteria for determining a homogeneous type are: average shape and size of stones, the nature of the interstices, the extent to which each block has been tooled, the regularity and nature of the courses, and the thickness of the core in relation to the stones. Because these are average types I will often qualify them in the course of my description of each site. For example, the exterior facing of Çardak Kalesi consists of a large well-cut type IV. Type IV is already a "large crude," which means that the type IV stones at Çardak are slightly larger than those in figure 3 and that the margins and facing of each stone are also finer. If a type of masonry does not fall into one of the categories, then I will discuss it separately as an anomaly. Also, some types of facing stones are frequently mixed. In the same wall types III and IV, VII and VIII, V and VI, or V and VII can appear side by side. I will try to point out in the Catalogue where these combinations occur. Most of the masonry described herein consists of limestone. Occasionally basalt, serpentine, and slate are used for construction. The physical reasons for dividing the masonries into separate categories are made evident in the following paradigms.

Type I. This dry rubble masonry or "trockenmauerwerk" occurs only on the interior of forts (pl. 46a). Today extant examples of this masonry are extremely rare since it is subject to rapid decomposition from weathering and earthquakes. It consists of uncut fieldstones that are shaped into a very crude wall. No mortar or fill is used, and often the large stones are anchored in place by smaller ones. The type I wall never supports vaults and is often used to build pens for animals.

Type II. This masonry is a mixture of small unhewn fieldstones, potsherds, and massive amounts of mortar. This type is essentially a heavily mortared rubble and is analogous to the Roman *opus caementicium*.

In those areas of Armenian citadels that have no direct strategic importance type II masonry is evident. Armenians use type II extensively for the vaulted ceilings of structures on the interior (pl. 197b); in gatehouses and other frontline defenses designers would employ a stronger type of masonry. For example, the undercrofts behind the southwest curtain wall at Tumlu Kalesi are barrel-vaulted structures that have only two ribs of neatly faced voussoirs (type IX) to support the entire ceiling of type II masonry.[12] Here the ceiling thickness varies from 30 cm at the apex to 70 cm at the springing. There are two obvious reasons for the use of type II in the vaulted ceilings: first, it can readily be moulded to the demands of particular buildings due to its initial plastic nature while on the centering; and second, it is much cheaper and easier to use irregular stones and mortar than to face and angle large rectangular blocks neatly. Aside from its use in undercrofts, type II masonry is often employed in the ceilings and walls of cisterns. Certain minor buildings, such as storehouses, are made entirely of type

[12] Youngs, 116.

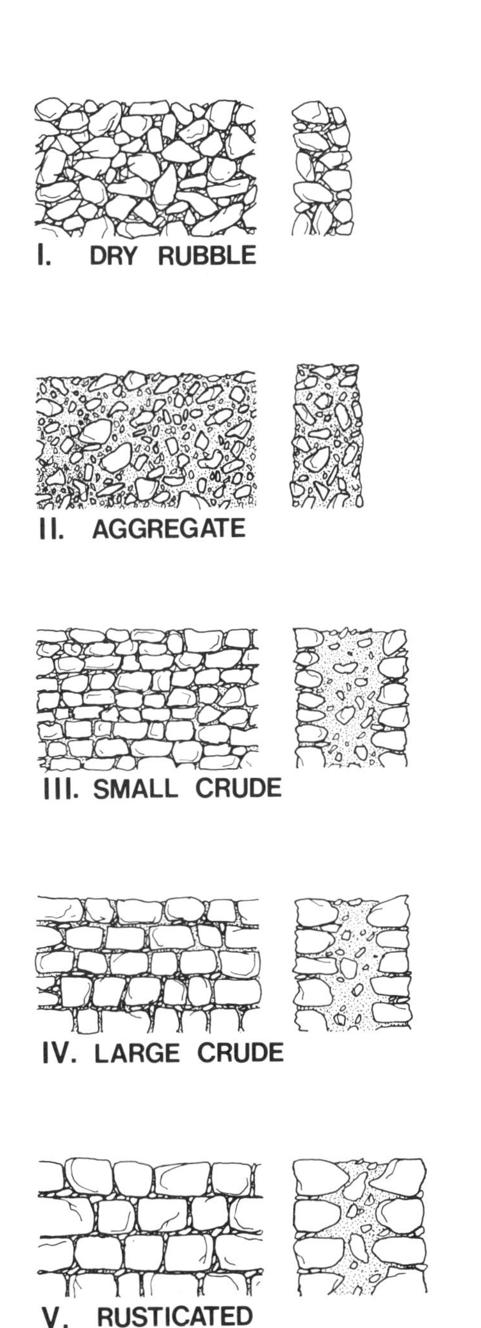

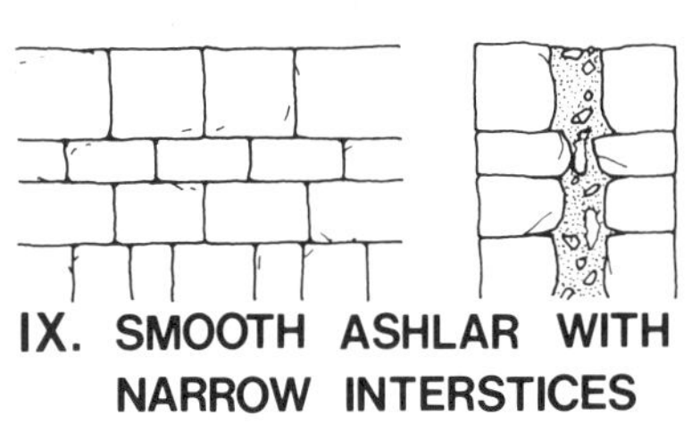

THE MASONRY OF MEDIEVAL CILICIA

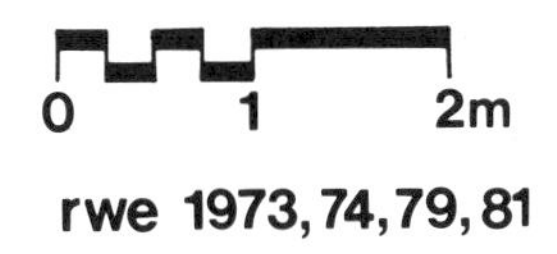

Fig. 3

II. The denizens of Armenian forts occasionally made repairs to other types of masonries with type II.

Type III. This class consists of small, often poorly cut rectangular and square stones whose interstices are filled with an abundance of rock chips and mortar (pl. 219a). The thickness of the core is always greater than the combined depth of the exterior and interior facing. This class is actually a combination of two types, one that has its courses in regular layers and the other with irregular courses. In figure 3 the latter is shown on the top and bottom of the paradigm. The core consists of mortar, uncut fieldstones, and potsherds.[13]

In the Armenian forts of Cilicia type III is used only on the interior of the fort. It is used in a variety of ways, most often for the walls of freestanding cisterns (for example, Gökvelioğlu) and small storage rooms. Sometimes it is the interior facing for the circuit walls as well as the masonry of non-strategic vaults.

Type IV. This masonry is used extensively on all interior constructions in Armenian fortifications (pl. 1a). The large size of type III is distinguishable from the smaller size of type IV in three ways: first, the stones of type IV are generally cut with more precision, having a very symmetrical appearance (rarely polygonal); second, the courses of type IV are always regular; and third, the thickness of the core is smaller than the combined depth of an exterior and interior facing. Like type III, the interstices of type IV are filled with rock chips and mortar. Type IV is readily distinguishable from types V and VI by observing the core. In type IV there is no *consistent* attempt to make the interior side of each block pointed. A problem can arise when the core is not visible in a sample that appears to be type IV and when the facing was either stuccoed at one time or has become covered with minerals oozing from the inner core. In some instances a large type IV can resemble (on the exterior) a small type V or VI.

In Byzantine forts that employ type IV as an exterior facing stone an identical masonry is often used on the interior side (cf. Çardak and Evciler).

Type V. This masonry consists of large rusticated blocks with neatly trimmed edges (pls. 32a, 152a). The interstices are filled with occasional rock chips and mortar. These stones are tapered on the interior; a rather thin core anchors the blocks. Type V masonry is slightly less common as an exterior facing stone than type VII.

The critical (and sometimes the only tangible) difference between a large type IV and the small type V is the pointed interior side of the latter. Type VII, which is always associated with Armenian architecture in Cilicia, frequently appears in proximity to type V. The stones of type V have narrower interstices than those of type IV. At times the distinction between types III, IV, and V can become blurred, and these designations must be applied with caution. To credit a fort to the Byzantines simply because type IV seems to be used as an exterior facing stone is risky. The elements of design must be the determining factor, and the masonry should only reinforce the conclusions.

[13]The cores of types III through IX are identical in this respect.

Type VI. The stones of this masonry are quite similar to those of type V (pl. 220a). It is used as an exterior facing stone and at times for the repair of structures. Type VI is distinguishable from type V in that the edges of the exterior face of the latter are still visible on the outside. In type VI the margins have been stuccoed with a broad band of mortar. Occasionally the center of the band is studded with a row of pebbles that are frequently different in color from the limestone blocks. The pebbles are a purely decorative motif and have no functional value.

Type VI is occasionally mixed with other types (for example, type V at Tumlu). The problem with identifying this masonry is that the stuccoed margins can break off, leaving what appears to be a type V masonry. Also, the medieval occupants of a fort could have stuccoed the edges of a preexisting type V. It is possible that stones that I identify as types V and VI at the same site are from the same building period.

Type VII. This masonry consists of well-cut rectangular blocks that have a projecting boss on the outer face (pl. 19b). The edges of the outer face have neatly drafted margins that facilitate the alignment of the blocks during construction. Normally this masonry is pseudo-isodomic. Usually the exterior margins show a thin line of mortar and occasionally rock chips. This masonry differs from type V in that it has the distinctive drafted margins.

As mentioned earlier, this and type V are the most common masonries for the exterior of Armenian circuit walls. On a few rare occasions it is actually employed as the interior facing stone. There is a tendency in type VII, as with some other types, for the larger stones to be at the base of the wall and the rest of the stones to become uniformly smaller toward the top. Obviously the larger stones can more efficiently carry the weight of the wall and offer better protection against battering rams.

Type VIII. Type VIII has blocks that have a smooth squared face on the exterior and are tapered on the interior (pl. 263a). Mortar and rock chips are occasionally visible on the exterior.

The Armenians rarely use this masonry as an exterior facing stone (preferring V, VI, and VII), but they frequently employ it for interior structures (for example, Geben). It is not used as frequently as types III, IV, and VII, and it is avoided in ecclesiastical architecture.

Type IX. This masonry consists of perfectly coursed rectangular and square stones whose sharp exterior edges and flat faces form extremely tight margins where almost no traces of mortar are visible (pls. 7a, 141a). The flush interstices are continued through the interior where *no* attempt is made to taper the inner faces of these stones.

This masonry is normally for internal structures of great importance (for example, the churches at Çandır, Anavarza, and Kız [near Gösne] and the baronial apartments at Lampron and Çandır).[14] This masonry, which is expensive to execute (due to the smooth symmetrical nature of all sides), is not suitable against a frontal attack since the stones are not anchored firmly in the extremely thin core.

[14]Cf. Berkian's analysis of the identical masonry in Greater Armenia; see above, note 4.

Type IX is used without a core in situations where its flat symmetrical surfaces can be beveled for abutments and joints. For example, it is used in door frames, vaults, ribs, and arrière-voussures; it is readily adaptable to the precise and sharp contours of embrasured loopholes and the delicate jambs of windows. Often where it is employed for the exterior frames of gates and posterns (pl. 1b), a protruding boss is left in the center of the outer face of the block to protect the interstices from rams.

In a few surviving non-Armenian Crusader constructions in Cilicia type IX is used as an exterior facing. At Yaka, Kütüklu, and Tumil (three possible Crusader sites), the type IX masonry has a thin rusticated center, broad drafted margins, and an unusually thick core. At Silifke a uniformly smooth type IX facing is also used with a thick core.

4. A Typology for Armenian Military Architecture

Within the general context of being an integral part of a large network of defense, an Armenian fortification[1] can have as many as three specific functions. A fortification can be a watch post, the fortified home of a baron, and a residence for a permanent garrison. Many of the fortified structures in Cilicia embody two of these functions, and a few sites only one. All fortifications in a major or minor way function as watch posts; yet some are built exclusively for this purpose. The introduction to each Armenian site in the Catalogue mentions the specific typology of the defenses.

Watch Posts. A watch post is a very small fortified unit that can permanently house as many as five men. It is positioned in a mountainous area or in the rocky foothills so that it has a commanding view of at least one major road into the plain and can communicate, either by signal fire or horseman, with a neighboring garrison fort.[2] Among the structures that fall into this category are: Haçtırın, Hotalan, Kuzucubelen, Alafakılar, Mitisin, and perhaps Hasanbeyli and the Pillar of Jonah.[3]

As architectural entities, watch posts are passive in nature. They do not have towers or other structural devices that permit an aggressive defense. If these posts were attacked, the

[1] In this monograph the words "fortification" and "fort" refer to any type of fortified unit. A fourth category of military architecture, that of city walls, will not be discussed. We have no evidence for this type of construction during the period of the Armenian kingdom. What little is known of the urban defenses of Misis is summarized in Part II (the Catalogue).

[2] Dennis, 150 ff.

[3] It is quite probable that the latter four sites were built by non-Armenians, but at some period after their construction they were occupied by Armenians. Hasanbeyli and the Pillar of Jonah may have also functioned as toll stations.

topography of the sites and their thick walls were expected to protect the few defenders. Only three of these sites (Haçtırın, Kuzucubelen, and Hasanbeyli) seem to have had embrasured loopholes or other openings to accommodate archers.[4] Because of their size, watch posts could never secure roads from marauding bands but only warn nearby garrisons.

Aside from the above-mentioned observation posts, there are a small number of isolated cloisters that could serve as watch posts. These sites do not have towers, bent entrances, or many features associated with military architecture. Occasionally, simple gates and precinct walls are visible. The four sites included in this category are: Frenk, Gediği, Kız (near Gösne), and Milvan.[5] All but Kız have commanding views over important valleys and roads. Gediği probably had intervisibility with three garrison forts. If the residents in these locales spotted suspicious activities, they could certainly communicate this information.

Estate Houses. For want of a better term, these structures are simply the fortified estates of petty barons. While they show many of the peculiar architectural features of the Armenian fortification (see Part I.2), they differ in one respect—their plan. These stone structures are simply rectangular houses that are found over the broad expanse of Armenian Cilicia. These sites include Anacık, Belen Keşlik, Gösne, Sinap near Lampron, and Sinap near Çandır. The simplicity of their design is due to the simplicity of their function. They are the most European in appearance of all the Armenian architecture in Cilicia in that they resemble donjons. They are not like the gigantic, twelfth-century donjons at Richmond and Rochester (England), which functioned as major administrative centers and housed a large number of retainers and troops. The fortified estates in Cilicia normally have only two vaulted stories and rarely cover more than 200 sq m of surface area. They are similar to the thirteenth-century residence at Kolossi (Cyprus)[6] and were intended to house a single family and a few retainers. However, there is no evidence that the Armenians derived this plan from the west. In fact, in the early tenth century large fortified estate houses were being constructed in the regions of Greater Armenia controlled by the Bagratids. There is a particularly fine example of one of these structures at Tignis, in the vicinity of Ani.[7] It seems likely that the Armenians and Europeans were simultaneously drawing on an older architectural tradition.[8] The estate houses in Cilicia are located in or near the agricultural districts that the resident barons controlled.

Three of the five estate houses listed above show only minor variations in their plans. Anacık and the two Sinaps are almost identical in size; they have a very small tower in each

[4]Extensive damage at some sites may have erased evidence of other openings.

[5]Because Frenk and Kız have the remains of *Armenian* churches (and not much else), both are published in Edwards, "First Report" and "Second Report." Full descriptions of Gediği and Milvan appear below in the Catalogue. Of the four cloisters, only Milvan seems to have also had an active military function.

[6]A. Megaw, *Kolossi Castle, A Guide* (Nicosia, 1963).

[7]Berkian, 120 f.

[8]Similar fortified estates are seen in early medieval Georgia; see R. Mepisashvili and V. Tsintsadze, *The Arts of Georgia* (Leipzig, 1979), 47 ff. The estate houses may have been derived from the Hellenistic tower residences in south Anatolia; see A. W. Lawrence, *Greek Aims in Fortification* (Oxford, 1979), 187–92; cf. H. Butler, *Syria*, II B, *Architecture of Northern Syria*, Publications of the Princeton University Archaeological Expeditions to Syria in 1904–5 and 1909 (Leiden, 1920), 137 f, 223–35.

of the corners. The towers did not function as fighting platforms but served merely as salient buttresses to protect the delicate corners.[9] The three sites have a single lower-level entrance, usually in the long south wall. The second story is always supported by a stone vault over the lower-level room and normally fitted with embrasured loopholes. Belen Keşlik differs in that its four exterior corners are not protected by salient buttresses. These four sites show no evidence of ever having been surrounded by a circuit; they lie exposed on fairly flat ground, which is contrary to the normal practice of protecting fortifications atop outcrops. Gösne has a unique feature in that a polygonal building is attached to the donjon by a short narrow wall.

If these estate houses seem especially vulnerable, it should be remembered that they had intervisibility with or immediate proximity to neighboring garrison forts in the mountains and were consequently protected.

There are two sites that functioned principally as residences but differ architecturally from the other fortifications in this category. One is Sarı Çiçek, which may be a summer palace or a monastery. This complex is isolated in the Taurus Mountains and has a number of buildings, including a chapel. The other exception is the so-called fort of the sparrow hawk, known today as Andıl Kalesi. It crowns the top of a mountain and once served as a retreat for the Armenian kings during times of invasion. Both Andıl and Sarı Çiçek are not designed specifically to administer a district.

The seven sites mentioned above are readily identifiable as small residences. There are two sites in Armenian Cilicia, however, that have characteristics of both estate houses and garrison forts: Tece and Kız (near Dorak). Both are four-story, keep-like structures surrounded by a large, independent circuit wall with towers. Undoubtedly the additional space and defenses indicate the presence of a garrison. Neither Tece nor Kız (near Dorak) are like the Byzantine garrison forts of Evciler and Kozcağız. The latter have relatively crude masonries and small donjons that are integral parts of the circuit walls. One of the important principles in Cilician military architecture (see Part I.2) is that the Armenians do not employ a donjon in conjunction with a circuit wall. Since the masonry of Tece and Kız (near Dorak) resembles the Armenian paradigm for fortifications in Cilicia, both sites may have been fiefs granted to Crusaders who in turn employed Armenian masons to build their forts to Latin specifications. It is quite interesting that of all the structures in this category only Tece and Kız are located neither in the mountains nor in the Highlands.[10] Tece is on the coast, like the Hospitaler fort at Silifke, and Kız crowns an outcrop beside a major road in the plain, just like the Teutonic fort at Amuda. The Armenians *never* granted the Crusaders a site in the mountains or Highlands.

Garrison Forts. The distinction between a watch post and a small garrison fort is not arbitrary. For example, the small forts at Bucak, Hisar, and Işa are in excellent locales to observe troop movements along roads and are large enough to house at least ten to thirty

[9]The similarity between the plan of the estate house and the symmetrical garrison forts at Karafrenk, Kütüklu, Yaka, Tumil, and Kum is only superficial. The estate houses are not castra or open enclosures.

[10]Tece and Kız are not in the proximity of any Armenian forts.

men.[11] They are protected by towers, elaborate entrances, and battlements. In other words, they are offensive structures. Not only can troops defend themselves vigorously inside the fort, but they can also sally forth to protect the roads. Azgit, Maran, and other forts of their size even have postern gates. Garrison forts are located in all regions of Armenian Cilicia.

The large garrison forts like Savranda have even more sophisticated defenses, such as outworks and machicolations over the entrances. Many of the large garrison forts were the permanent homes of resident barons. It is sometimes difficult to determine which of these large forts were also baronial residences. Normally an abundance of buildings (suitable for habitation) on the interior of the enceinte as well as references in the medieval chronicles distinguish some sites as the homes and administrative centers of district lords. This is certainly the case for Vahga, Anavarza, Lampron, and Çandır. In this study I will refer to fortifications that appear to be both troop depots and baronial residences as "fortresses" or "castles," while the smaller, non-baronial sites will be labeled simply as "garrison forts." In a few cases the king of Armenian Cilicia was also the chief lord of a mountain castle,[12] thus creating the additional distinction of a baronial and a royal castle. It should be noted that even if a particular baron is designated as the lord of a certain fortress on the Coronation List (below, Appendix 3), this is no indication that he resided permanently therein.

5. The Arab, Byzantine, and Crusader Military Construction in Cilicia

The majority of fortified sites in medieval Cilicia are clearly recognizable as Armenian constructions. In those forts that are intimately associated with Armenian history and bear Armenian dedicatory inscriptions there is a surprising degree of similarity in architectural features and masonry. These peculiar features are discussed in detail in Part I.2–3. When identical features appear in a fort that has neither inscriptions nor a specific mention in the medieval chronicles, that site is credited to the Armenians. Of the seventy-five sites discussed in the Catalogue, forty-four conform to the paradigms for Armenian fortifications in Cilicia. Fourteen of these Armenian sites have some remains from earlier non-Armenian periods of construction.[1] Two of the thirty-one fortifications that are not included in the category of Armenian

[11] Forts like Bayremker, Arslanköy, and Kalası are difficult to categorize because they are so heavily damaged. Since two have towers and one has two baileys, they appear to be small garrison forts.

[12] See below, Appendix 3.

[1] See above, Part I.2, note 4.

sites cannot be evaluated architecturally. One of these, Misis, was dismantled by the early 1970s, and the other, Sarı Seki, is presently inside a high-security military compound and is off limits to all visitors. Both of these sites played an important role in the history of Armenian Cilicia. The Arabs, Byzantines, and Crusaders probably constructed the remaining twenty-nine fortified sites: Alafakılar, Amuda, Ayas,[2] Başnalar, Çalan, Çardak, Dibi, Evciler, Haçtırın, Haruniye, Hasanbeyli, Kalası, Karafrenk, Kozcağız, Korykos, Kum, Kütüklu, Lamas, Mansurlu, Milvan, Mitisin, Payas,[3] Pillar of Jonah, Rifatiye II, Silifke, Toprak, Trapesak, Tumil, and Yaka. After their construction all of these twenty-nine sites were either occupied by the Armenians when they assumed control over Cilicia or were under their suzerainty. A number of these forts (Çardak, Kozcağız, Korykos, Milvan, Silifke, Toprak, and Trapesak) were repaired or supplemented slightly with Armenian-type construction. Many of the twenty-nine sites share a few of the features of Armenian military architecture, but their plans and masonry are normally quite alien. Only one of these forts, Dibi Kalesi, has a number of features that are common in Armenian architecture, but its crude exterior facing stones are quite atypical. There is the remote possibility that Dibi, which has no recorded history or inscriptions, may be an Armenian construction.[4] The reader must be aware that the paradigms established by this writer for the Armenian forts of Cilicia are generalizations that allow for variations within certain norms but not entirely new forms. Fortunately there are specific historical references that associate fourteen of the twenty-nine forts with prolonged periods of occupation by Arabs, Byzantines, and/or Crusaders. Some references credit construction to non-Armenian occupants. Fifteen forts have no recorded history to confirm their origins, but they are sufficiently different from Armenian sites and share characteristics with the known examples of Arab, Byzantine, and Crusader architecture to be reliably be classed as "non-Armenian." Unfortunately there is not enough specific evidence to establish reliable paradigms for the Arab, Byzantine, and Crusader military architecture in Cilicia. Thus I will simply refer to these fifteen undocumented forts as "medieval." When the Mamluks assumed control of Armenian Cilicia in the fourteenth century there was no need to construct new forts, but they did repair a number of sites. What is known of the Arab, Byzantine, and Crusader military occupation will be summarized in the following pages.

Many of the pre-Arab urban centers in Cilicia Pedias and the coastal cities of eastern Cilicia Tracheia had protective circuit walls.[5] Two cities, Anazarbus and Sis, each had an adjoining fortified acropolis. These urban defenses date from the four centuries of the Roman Empire. However, there is no mention in ancient and early Byzantine sources of garrison forts

[2] Both Ayas (land castle) and Payas are Ottoman constructions; see the appropriate entries in the Catalogue.

[3] See above, note 2.

[4] One could hypothesize that for financial reasons Armenian masons used a unique (crude) type of masonry that never appears again on their Cilician forts. Dibi could simply be the result of Crusader builders copying Armenian plans with their own masonry.

[5] H. Hellenkemper, "Zur Entwicklung des Stadtbildes in Kilikien," *Aufstieg und Niedergang der römischen Welt*, 2.7.2 (1980), 1275 ff. The techniques used for the construction of the pre-Arab period city walls differ from those seen in fortifications built after the 7th century.

In the regions near Silifke some of the large towns had protective circuits as well. See J. Bent, "A Journey in Cilicia Tracheia," *JHS* 12 (1891), 222 f.

prior to the invasion of the Arabs. Such garrison forts would be located along the major roads and in the rural areas of the borderlands. A string of interdependent forts would constitute a *limes*. While dispersed garrisons were necessary to maintain order in the troubled provinces farther east, Cilicia was always regarded as a safe place of passage. When the Isaurian raiders moved across the Calycadnus in the fourth and fifth centuries they were not countered by resident Cilician garrisons but by armies sent from Constantinople or Antioch.[6] In the second quarter of the seventh century, when the Arabs captured most of northern Syria and began to make punitive raids into Cilicia, Emperor Heraclius ordered the urban militia to evacuate all the towns east of Adana and to lay waste to the abandoned district because the Byzantines had no way of defending Cilicia Pedias.[7] The Arabs easily captured the rest of Cilicia.

The Arab occupation of Cilicia (650–963) began a new period of settlement in which the circuit walls around the cities were repaired and garrison forts were constructed for the first time. In the second half of the seventh century the Omayyads merely used Cilicia as a convenient thoroughfare where they occasionally stopped to plunder the defenseless Greek settlements. By the early eighth century Cilicia became a frontier and a staging area for the annual invasions into Cappadocia and western Asia Minor. Entire Arab settlements were moved from Syria en masse to the newly rebuilt cities of Cilicia Pedias. In the last two decades of the eighth century the Abbasid Caliph Hārūn ar-Rašīd repaired the walls of Tarsus and stationed there a permanent garrison of eight thousand men.[8] The city was said to have double walls, eighty-seven towers, six gates, and a dry moat. Tarsus was one of the largest staging areas for Arab invasions. Ibn Ḥawqal reported seeing a hundred thousand mounted warriors within the walls.[9] The defenses of both Adana and Misis were likewise repaired under Hārūn ar-Rašīd. Adana had actually experienced at least two prior periods of Arab construction under the Omayyad Caliph Walīd (743) and the Abbasid Caliph al-Manṣūr (758). Part of the fortified city, which was reported to be on the west bank of the Seyhan River, may date from the late

[6]G. Dagron, ed. and trans., *Vie et miracles de Sainte Thècle,* Subsidia Hagiographica, 62 (1978), 113 ff; R. Edwards, "The Garrison Forts of Byzantine Cilicia," *Eighth Annual Byzantine Studies Conference* (Chicago, 1982), 45 f.

[7]M. Canard, "Cilicia," *EI*[2], 35 ff. The Greeks seem to have had military depots in or near civilian settlements where weapons were stored and where troops could be sent in emergencies. But there is no evidence for permanent garrison forts, only fortified cities and towns which Arab chroniclers often call "fortresses." See: E. Brooks, "The Arabs in Asia Minor (641–750), from Arabic Sources," *JHS* 18 (1898), 182–208, esp. 203–5; Haldon and Kennedy, 83 f, 96 note 60, 97; al-Balāḏurī, 253; A. Stratos, *Byzantium in the Seventh Century,* III, 642–668, trans. H. Hionides (Amsterdam, 1975), 33 f; A. Santoro, *Byzantium and the Arabs during the Isaurian Period, 717–802 A.D.*, Diss. (Rutgers University, 1978), 212 ff.

[8]Le Strange, *Caliphate,* 132; H. Kennedy, "Arab Settlement on the Byzantine Frontier in the Eighth and Ninth Centuries," *Yayla* 2 (1979), 22–24; cf. E. Brooks, "Byzantines and Arabs in the Time of the Early Abbasids," *The English Historical Review* 15 (1900), 733–40, 16 (1901), 84–87; Haldon and Kennedy, 108 ff; B. Darkot, "Tarsus," *IA*, 18–24; Ritter, 197 ff; al-Balāḏurī, 260–63; R. J. Lilie, *Die byzantinische Reaktion auf die Ausbreitung der Araber* (Munich, 1976), 60–65 (notes 3, 16, 19, 22), 75, 85 ff, 112, 114 f, 119, 125, 127 ff, 139, 161 (note 60), 165, 168 (note 25), 171 (note 36), 322.

[9]The number of troops is probably an exaggeration. Even in the 10th century Tarsus was the principal military depot; see al-Ṭabarī, *History,* XXXVIII, *The Return of the Caliphate to Baghdad,* trans. F. Rosenthal (Albany, 1985), 14, 41, 71, 74, 83, 90 f, 97, 119 f, 151, 155, 180 f, 192, 196, 204; Vasiliev, 79–81, 87, 101 f, 121–23, 125, 160 ff; Canard and Grégoire, 45, 57 f, 127, 135 f, 152 f, 156, 167, 201, 240–42; S. Stern, "The Coins of Thamal and of Other Governors of Tarsus," *Journal of the American Oriental Society* 80 (1960), 217–25; G. Miles, "Islamic Coins from the Tarsus Excavations of 1935–1937," *The Aegean and the Near East, Studies Presented to Hetty Goldman* (New York, 1957), 297–312.

Omayyad period.[10] Unlike Tarsus and Adana, Misis was abandoned by order of Emperor Heraclius; throughout the first fifty years of the Omayyad period that city and its fortifications were inhabited by only a handful of Greeks. It was not until 703 that ᶜAbd Allāh, the son of the Omayyad Caliph ᶜAbd al-Malik, occupied the town and rebuilt the fortified circuit on its original foundations.[11] Two of the inland cities in Cilicia Pedias, Sis and Anazarbus, were occupied by Moslem settlers. During the reign of the Abbasid Caliph al-Mutawakkil, the grandson of Hārūn ar-Rašīd, the small fortified acropolis adjacent to the town of Sis was repaired.[12] Anazarbus, whose city circuit and citadel were repaired and incorporated into the frontier defense by Hārūn ar-Rašīd, seems to have suffered severely from the retaliatory raids carried out by the Byzantines in the early and mid-ninth century.[13] It was not until the successful campaign of Nicephorus Phocas in 962 that these five cities again came under Greek control.[14]

Unfortunately it is impossible to determine the peculiarities of Arab military architecture in Cilicia from an examination of these city defenses because so little remains. Only at two sites, Tarsus and Anazarbus, have fragments of Arab construction survived.[15] In both cases the Arabs simply recycled masonry from surviving buildings to reconstruct the original plans of the city walls.

If any determination is to be made about the nature of Arab fortifications, then it will have to come from the three Arab garrison forts of Cilicia that are mentioned frequently in the chronicles: al-Kanīsah, al-Muṯaqqab, and al-Hārūniyyah. Unfortunately only the latter has been securely located. Al-Kanīsah as-Sawdāʾ is a fort built of black stone in the vicinity of Haruniye and Misis. I have located only two forts in this area that are constructed with black basalt: Toprak and Karafrenk. There is no certain evidence to determine which of the two dissimilar sites is al-Kanīsah. Al-Muṯaqqab is located near al-Kanīsah, but the specifics of its location are contradictory. It is said to stand not only on the sea but in the mountains of the Ǧabel Lukkām (Nur Dağları) and in the plain.[16] This description could fit any number of sites

[10] Le Strange, *Caliphate,* 131; Fr. Taeschner, "Adana," *EI²*, 182 f; N. Maggiore, *Adana città dell'Asia Minore* (Palermo, 1842), 22 ff. Unfortunately all traces of the medieval walls of Adana have disappeared. Before their destruction they were never the subject of a scholarly investigation. See also: Budde, II, 13–30, 219 f; Carne, I, opp. 33; Sevgen, 17 ff; Brocquière, 95 note 1, 95–98; Maṭrākçī Naṣūḥ, *Beyān-ı Menāzil-i Sefer-i ᶜIraḳeyn-i Sulṭān Süleymān Ḫān, ed. and comm. H. Yurdayın (Ankara, 1976), pl. 108b; V. Langlois, "Voyage dans la Cilicie, Adana," RA* 11.2 (1855), 641–51; Haldon and Kennedy, 107 ff; B. Darkot, "Adana," *IA,* 127–29; Ritter, 165 ff.

[11] E. Honigmann, "Miṣṣīṣ," *EI,* 521; Le Strange, *Caliphate,* 130; Haldon and Kennedy, 107 note 6, 108 ff; refer to Misis in the Catalogue.

[12] V. Büchner, "Sis," *EI,* 453.

[13] M. Canard, "ᶜAyn Zarba," *EI²*, 789 f; idem, "La prise d'Héraclée et les relations entre Hārūn ar-Rashīd et l'empereur Nicéphore 1er," *Byz* 32 (1962), 372 f; Gough, 98. Caliph al-Mutawakkil carried out repairs to the circuit in the second half of the 9th century; see Anavarza in the Catalogue.

[14] Canard, 278–86, 650 ff, 805–38. The occupation of Misis was delayed by several years.

[15] Gough, 109, 119; H. Goldman, ed., *Excavations at Gözlü Kule, Tarsus,* I (Princeton, 1950), 29, 37; F. Day, "The Islamic Finds at Tarsus," *Asia* 41 (1941), 143–46; Honigmann, 43; Budde, II, 111–26, 222 f. The Arab strategy was to link the urban defenses with a network of garrison forts that stretched from the Euphrates to Tarsus. See: E. Honigmann, "al-Thughūr," *EI,* 738 f; Le Strange, *Palestine,* 37 ff; Dennis, 138, 221.

[16] Le Strange (*Palestine,* 447 and 510) has carefully collected some of the references in Arab texts to al-Kanīsah and al-Muṯaqqab. These references are examined in the Catalogue under Gökvelioğlu and Karafrenk; see also Çardak and Toprak.

from Amuda and Babaoğlan in the north to Savranda in the southeast and Gökvelioğlu in the southwest. Only al-Hārūniyyah, modern Haruniye, is clearly identified by its location and name as an Arab garrison fort. In general this fort is characterized by an oblong symmetrical plan, an exterior facing of smooth ashlar masonry (type IX masonry in the first major period of construction), a single stripe of light-colored stone across the salient north end, and a simple gate preceded by an open corridor. However, this one example of an Arab fort is not sufficient to establish a paradigm for Arab military architecture in Cilicia.

Historically we know slightly more about the Byzantine garrison forts of Cilicia than those of the Arab occupation.[17] The earliest mention of a Byzantine station and a front line of Greek defense is at the Lamas River. Al-Maṣᶜūdī reports that in the eighth century Lamas served as a place to exchange Greek and Moslem prisoners.[18] Since there is no mention of a fort at Lamas in the late antique period the present site is probably a late seventh- or eighth-century Byzantine construction. To the west of Lamas there is specific mention in the history of Anna Comnena that Alexius I ordered the construction of garrison forts at Silifke and Korykos after 1097.[19] Unfortunately at the former site only minor traces of what may be the Byzantine period survive. Today almost all of the land fortress at Korykos is from the Byzantine period. Like Lamas, it is built entirely with masonry plundered from nearby abandoned cities. This recycling of material tells us little about Byzantine masonry techniques, except that headers are used with some frequency (more so at Korykos than at Lamas).[20] Both sites have certain characteristic features: they have a symmetrical (or even geometrical) plan, square and/or rounded towers at regular intervals, a double trace (at Lamas only in the west end), and there is some sort of donjon or isolated area on the inner circuit wall. After 964, when Nicephorus Phocas succeeded in driving out the Arabs from Cilicia, the Byzantines are credited with building and/or repairing a number of fortified sites. Among these forts are Anavarza, Savranda, Lampron, and Vahga.[21] Unfortunately Armenian reconstruction at these sites was so extensive that most traces of the Byzantine plans have been covered or destroyed. What does remain are sections of Byzantine circuit walls. In all four cases the exterior facing stones consist of large, crudely cut blocks (type IV masonry).[22] If this type of masonry occurs in a fort that is undocumented and bears many of the characteristic features of Lamas and Korykos, then it is possible that the site is a Byzantine construction. This is certainly the case for Evciler and Kozcağız.

With the establishment of the Armenian kingdom various orders of Crusaders were given quantities of land and permission to build fortifications by the Armenian kings. The Crusader sites seem to be confined to the border areas at the south, west, and east, leaving the central and northern forts under Armenian control. From the available evidence it appears that the massive castle at Silifke was constructed almost entirely by the Hospitalers and that the keep

[17]On the tactics of the Byzantine reconquest of Cilicia see J. Howard-Jonhston, *Studies in the Organization of the Byzantine Army in the Tenth and Eleventh Centuries,* Diss. (Christ Church, Oxford, 1971), 189 ff.

[18]Le Strange, *Caliphate,* 133; "Cilicia," *EI*², 36. See also Lamas in the Catalogue.

[19]Anna Comnena, 45 f.

[20]Headers are not employed in any of the known examples of Armenian, Arab, or Crusader architecture in Cilicia.

[21]See the descriptions of these four sites as well as Silifke and Gökvelioğlu in the Catalogue.

[22]Cf. the Byzantine construction in the fortress of Bağras; see R. Edwards, "Bağras," 419 ff.

at Amuda was built during its occupation by the Teutonic Knights. The Armenians also gave the Arab complex at Haruniye to the Teutonic Knights, who may be responsible for repairs at the site. Most of the garrison fort at Trapesak was the undertaking of the Templars.[23] With regard to Silifke, Amuda, and Trapesak, no three structures could be more dissimilar in their masonry and architectural features.[24] This dissimilarity is due to the fact that the three forts were built by three distinct orders of Crusaders at different times. The military theories that shaped these sites, as well as the crews of masons, differed.[25] It is surprising that the Crusaders did not rely on Armenian masons and architects to build these three forts.[26] There are at least two fortified sites, Tece and Kız (near Dorak), that bear the imprint of Armenian masons but have plans that resemble Crusader forts in the Levant. Three other fortifications—Hasanbeyli, Savranda, and Toprak—had periods of Crusader occupation, but the extent to which the Crusaders contributed to their construction (if at all) cannot be determined.[27] This lack of uniformity in the three known forts of Crusader construction in Cilicia means that it is impossible to establish a "Crusader" paradigm. However, there is one rather exciting correlation. The Crusader fortress at Silifke exclusively has type IX masonry as an exterior facing. Many of these blocks are cut for the fortress, and some stones have been recycled from the city below. As noted above (Part I.3), type IX masonry is *never* used as an exterior facing stone in the circuit walls of an Armenian fort. It is quite interesting that three forts on the south coast between Silifke and Tarsus (Kütüklu, Tumil, and Yaka) have *only* newly cut type IX as an exterior facing. Many of these facing stones have a small, thin, rough boss in the center. These three coastal forts are not only in an area that has no record of an Armenian military settlement, but they have similar plans that consist of a square or rectangular circuit with four corner towers. The Armenians *never* use this standard castrum design in their military architecture. These forts may be Crusader constructions.[28]

There are four sites where Mamluk repairs were carried out on Cilician forts in the fourteenth and fifteenth centuries: Anavarza, Sis, Toprak, and Tumlu. Mamluk construction

[23] It should be noted that Trapesak (like Bağras) briefly became part of the Cilician kingdom during the reign of King Levon I. Cf. Çalan, Sarı Seki, and the Pillar of Jonah in the Catalogue.

[24] The plans of the three forts are similar only in that their circuit walls have a general symmetrical shape; see also Kum in the Catalogue.

[25] Armenian influence on Crusader constructions is quite evident at Silifke Kalesi with its horseshoe-shaped towers. But non-Armenian motifs, such as the ditch and continuous talus, still persist.

[26] There is evidence that shows that the Crusaders had their own master architects and that they even transported prefabricated architectural elements to sites under construction. A. Luttrell was kind enough to let me see the draft of his forthcoming work on the Crusader castle at Bodrum (Turkey), where evidence indicates the prefabrication of material on Rhodes.

[27] Savranda was only captured and briefly held by the Crusaders; most of the fortress is an Armenian construction. See also Edwards, "Donjon," 53–55.

[28] It is also possible that the three forts date from the Arab occupation. However, the link with the Crusader period seems to be stronger in that much of the coastline west of Ayas (including the estuaries of the Ceyhan) was granted to the Order of the Hospital in 1214. Later the grant may have been extended inland to include the unidentified fort of Govara (near Yılan?). See: Delaville Le Roulx, 115 f; Hellenkemper, 260 note 1. The masonry of these three coastal forts bears a remarkable resemblance to the Roman *opus caementicium* with an ashlar facing. Because their plans are similar to the Roman castrum, it is within the realm of possibility that these sites are late Roman. See: F. Deichmann, "Westliche Bautechnik im römischen und rhomäischen Osten," *Mitteilungen des Deutschen Archäologischen Instituts, Römische Abteilung* 86 (1979), 473–527; F. Rakob, "Opus Caementicium—und die Folgen," ibid., 90 (1983), 359 ff. The reader must remember that until excavations are completed we can do no more than speculate on the origins of these sites. Cf. Kum and Karafrenk in the Catalogue.

at Tumlu and Toprak has distinctive embrasured loopholes in that numerous voussoirs form the rounded tops.

It would be a disservice to try to assign all the non-Armenian forts to Arab, Byzantines, and Crusaders. In discussing each of these forts in the Catalogue I will do no more than note the possibility that a site may belong to one of the three builders because of features that are similar to those in the known forts. Formal excavations at these sites would greatly enhance our ability to assign these forts to their rightful builders.

6. A Chronology for Armenian Fortifications

At the outset the reader should be forewarned that I have not established a specific and verifiable scheme for the dating of Armenian architecture in Cilicia. In fact, many of the comments in this section are somewhat negative in that they are directed at disproving the previously published hypotheses on the chronology of Armenian forts. The greatest hurdle facing the modern researcher of this topic is the small number of extant dedicatory inscriptions and the incredible paucity of *specific* references to construction in the medieval Armenian chronicles. Not until the majority of Armenian sites are systematically excavated and the pottery, coins, and other artifacts are carefully analyzed will anyone be able to propose a chronology that is more than theoretical.[1]

The nineteenth- and twentieth-century commentators on the kingdom of Armenian Cilicia have proposed and accepted the apparently "logical" maxim that Levon I, the first king of a united Cilicia, was the only man who had the economic resources and the administrative apparatus to construct most of the magnificent forts in the mountains and plain.[2] They cite the inscriptions on the donjon at Anavarza (1187/88) and from the sea castle at Korykos (1206) that credit Levon I as the builder.[3] In this accepted view it is held that Levon's policy of fortifying Cilicia was pursued by his successors until the end of the thirteenth century when the kingdom began to decline.[4] Frequently the comment of Michael the Syrian, that Levon I

[1] At present, modern science can provide no final solutions to these problems of chronology. Because the Armenians seldom used wood in the cores of their walls, dendrochronology has only a limited application in medieval Cilicia. When the techniques for applying mortar analysis to dating are perfected, then a resolution will be at hand.

[2] Alishan, *Léon,* 67 ff; idem, *Sissouan,* 64 ff; V. Kurkjian, *The Armenian Kingdom of Cilicia* (New York, 1919), 6 f; Fedden and Thomson, 35–39, 96–101; Hellenkemper, 262, 269 f.

[3] Hellenkemper, 291; Langlois, *Inscriptions,* 16 f, 48; idem, *Voyage,* 215; see Anavarza in the Catalogue.

[4] The inscription at Mancılık Kalesi dates to 1290; see Hellenkemper, 105, 292. Mancılık has the latest dedicatory inscription still in situ. In the mid-19th century Langlois (*Inscriptions,* 28) recorded a dedicatory inscription (now lost) in Tarsus that had been removed from its castle. The inscription, which dates from 1319, seems to record the completion of an unattested castle. Also see Boase, 183, "Teghenkar."

had sixty-two forts under his suzerainty,[5] is cited as support. In his discussion of the donjon at Anavarza Hellenkemper says: "Leon I. wollte ein Symbol setzen; Überlegenheit und Anspruch dokumentiert er in der Inschrift. Den Donjon nimmt er als Träger seiner Absicht. Der Bau spricht eine eigene Sprache: Weder vorher noch nachher erreichte in Kleinarmenien ein Burgenbau eine so dichte Formulierung der Bauausführung. Der Hinweis auf die kurze Bauzeit bezeugt zugleich, dass Leon sofort nach der Thronbesteigung auch ein Zeichen setzen wollte, um seine Stellung als primus inter pares zu unterstreichen."[6]

There are inherent problems in this somewhat romanticized view of Levon and his importance for Armenian architecture. The inscription at Anavarza is from a period when Levon was a baron and not the king of a united Cilicia. Anavarza is one of the largest garrison forts in the Middle East, and the repair of its Crusader-built donjon in 1187/88 marks the *last* Armenian building period at the castle. The question then arises: who did build the massive complex of rooms and the nearly 4 km of circuit walls of Anavarza Kalesi? On the other side of Cilicia the Armenian construction at Korykos is essentially confined to repairs carried out on the small Byzantine island castle by both Levon I and his successor, Het῾um I. In order to understand better the rationale and problems with the accepted system of chronology, specific excerpts from the most recent commentaries on Armenian architecture will be cited along with analyses in the following pages. The excerpts are from very competent scholars who over the last twenty years were the first to bring some sort of architectural and historical justification to this system of dating.

One of the best surveys of an Armenian fortress in Cilicia was undertaken by J. G. Dunbar and W. W. M. Boal at the Rubenid site of Vahga. In their conclusion the two authors tried to date this castle by comparing it to Yılan and Anavarza.

> In particular, attention may be drawn to the fact that the groined vault with distinctive, cross-shaped keystone, seen in the Outer Gatehouse at Vahga, finds an almost exact parallel in the main gatehouse at Yılan, while a similar but not identical construction appears in one of the vaults of the main gatehouse at Anavarza.
>
> This latter building is dated by an inscription to the year 1188, while the gatehouse at Yılan bears a relief carving depicting between two lions rampant, an enthroned king, whom it is tempting to identify as Leo II, the first king of Little Armenia, who was crowned in 1198. The identification is not proven, but in any case the gatehouse can hardly have been erected before the inauguration of the Armenian kingdom, nor, by reason of the subsequent decline in the political fortunes of Leo's successors, is the structure likely to be much later in date than the third quarter of the thirteenth century.[7]

The use of cross-shaped keystones is not peculiar to the three sites mentioned above but occurs elsewhere in Armenian Cilicia, even at castles like Çandır. The latter was taken over by the

[5] Michael the Syrian, 405.
[6] Hellenkemper, 292.
[7] Dunbar and Boal, 183 f.

Armenians in the 1080s. Also in Greater Armenia, monolithic, cross-shaped stones were frequently employed in intersecting vaults long before the donjon at Anavarza was built.[8] Such a common motif in Armenian architecture cannot be used to date any fort, even to a particular century. The relief above the door of the upper bailey gatehouse at Yılan Kalesi does not represent Levon I. It shows a badly weathered figure seated with his legs crossed and his arms outstretched. This depiction is similar only to the numismatic portraits of Kings Hetʿum I (1226–70), Hetʿum II (1289–1306), and Levon III (1301–7).[9] Since these three monarchs were vassals of both the Turks and the Mongols, an Oriental posture is not unusual. However, during the independent and highly successful reign of Levon I such a depiction would be ludicrous. Also, this relief is associated with the reconstruction of the gatehouse and even if it did represent Levon I, the original builder of the fortress is still unknown.

In their study of the great Hetʿumid castle at Lampron F. C. R. Robinson and P. C. Hughes speculate on the date of construction of the baronial apartments. These apartments are built exclusively with a uniformly smooth ashlar (type IX).[10] The method used by the two authors is simply to establish the dates when a similar smooth ashlar is supposedly used by other Armenians in Cilicia and then within those chronological limits to find corresponding periods in the history of Lampron when construction seems likely. The critical part of their assessment is as follow: "Mason's marks and similar masonry have also been seen at Anazarbus and Silifke while there is a resemblance at the island castle of Corycos (today Kız Kalesi). Part of the Armenian construction at Anazarbus is dated to 1188, Silifke was completely rebuilt between 1210 and the early 1220's while Corycos was constructed between 1206 and 1251. Similar masonry therefore was being employed between the 1180's and 1250's."[11] As already noted, the inscription at Anavarza is affixed only to the Crusader donjon and was posted when the Armenians repaired the complex. Most of the original donjon is *not* constructed with smooth ashlar but with a curious type of cyclopean masonry that is seen nowhere else in Cilicia. Silifke Kalesi does have smooth ashlar as an exterior facing, but it is a fortress built by the Hospitalers, not the Armenians. The smooth ashlar in the *Byzantine* fortresses at Korykos consists entirely of materials recycled from the neighboring late antique city. The military construction at Korykos can be dated specifically to the first years of the twelfth century.[12]

It is important to remember that the use of a smooth ashlar facing with a poured core is not a technique invented by the Armenians in Cilicia. It has antecedents in the classical world and was consistently used from the fifth century onward in Greater Armenia for most eccle-

[8]Particularly fine examples are in the mid-12th-century vaulted library at the monastery of Hałbat; see *Haghbat,* Documenti di architettura Armena, I (1974). As discussed in the Catalogue, I believe the donjon at Anavarza to be a Crusader construction.

[9]Edwards, "Yılan," 23 ff; cf. Youngs, "Castles," 130; Gottwald, "Burgen," 87 ff. See also Yılan in the Catalogue.

[10]In my discussion of the castle at Lampron I speculate on why the Hetʿumids used this rather costly masonry; see Lampron in the Catalogue.

[11]Robinson and Hughes, 202; cf. Hellenkemper, 235 f.

[12]For a detailed discussion of the masonry and plan of the donjon at Anavarza as well as the sites of Silifke and Korykos, see the Catalogue.

siastical architecture and some secular buildings.[13] The first datable use of smooth ashlar by the Armenians in Cilicia is at the baronial church of Tᶜoros I. According to textual references and the inscription on the church, it was dedicated about A.D. 1111.[14]

This church has additional significance in that it is located in the south bailey of Anavarza. There are traces of Arab, Byzantine, and Mamluk construction in the circuit wall of the south bailey, but most of it is the result of *one* period of Armenian construction.[15] The chronicles specifically state that Tᶜoros I descended from his mountain stronghold at Vahga in the first two decades of the twelfth century and repaired a number of forts he had captured in the plain. He is also credited with the construction of many monasteries on the fringes of the plain.[16] If Tᶜoros took Anavarza by siege, then it may have resulted in damage to the south bailey (the most accessible point in the fortress).[17] A hundred and fifty years earlier this site had suffered severe damage during the raids conducted by Nicephorus Phocas on the Arab garrison.[18] Because of its central location Tᶜoros chose Anavarza as the administrative center for his new barony. The conspicuous church he built in the south bailey was meant to be a symbol of a new Armenian hegemony in east Cilicia. In the dedicatory inscription on that church Tᶜoros carefully mentions his heroic ancestors as well as his parents and children. We should not expect him to place his dynastic church inside a dilapidated or war-ravaged circuit. The Armenian walls and horseshoe towers that make up most of the south bailey must come from the reign of Tᶜoros I. More than any other Armenian leader Tᶜoros I is associated with a prolonged residence at Anavarza. His successors temporarily lost the site; Baron Levon II probably transferred the Rubenid capital from Anavarza to Sis by the mid-1190s. It is significant that the Armenian construction at Anavarza dates from the first quarter of the twelfth century; it shows that the rusticated facing stones (types V and VII) of the circuit as well as the distinctive towers were in use at a fairly early period in the Armenian occupation.[19] This rusticated masonry made its appearance in Greater Armenia at least four hundred years before the battle of Manzikert.[20] Because of the rich architectural traditions that the Armenians brought into Cilicia, it is impossible to associate one type of masonry with any period in the Armenian kingdom.[21] Thus at Lampron there is no reason why the Armenian construction with smooth ashlar cannot date to the earliest period of occupation. From the 1080s to the

[13] F. Gandolfo, *Aisleless Churches and Chapels in Armenia from the IV to the VII Century,* Studies on Medieval Armenian Architecture, II (Rome, 1973), 143 ff; T. Tᶜoramanyan, *Nyowtᶜer haykakan čartarapetowtᶜyan patmowtᶜyan,* I (Erevan, 1942), 22 ff.

[14] Samuel of Ani, 448 f.

[15] Gough, 119–21. It was the walls of Tᶜoros I that so impressed Choniatēs (25).

[16] Vahram of Edessa, 499; Samuel of Ani, 448 f.

[17] It is significant that during the Crusader occupation of Anavarza (ca. 1098–1110) the Franks ignored the south bailey and constructed the keep to block access into the central bailey. Perhaps when the Franks evicted the Byzantines the circuit of the south bailey was damaged.

[18] Parts of the east wall of the south bailey may actually be the result of repairs carried out by Nicephorus Phocas. Gough believes that this Byzantine construction is rather poor and inadequate for the defense of the south bailey.

[19] Savranda and Vahga are two other sites that have type VII as an exterior facing and an early period of Armenian occupation.

[20] Berkian, 46; Edwards, "Doğubeyazit," 437, 440.

[21] Cf. Hellenkemper, 269–71.

1140s the Het'umids were officially recognized in Constantinople as the administrators of the province and those responsible for protecting Byzantine interests and collecting taxes. To assume that Lampron, dynastic seat of the Het'umids, was refortified and beautified during the period when its owners had hegemony over the part of Cilicia west of the Pedias is not unwarranted.[22]

In my opinion, there are no historical reasons to conclude that the vast majority of the architectural remains from Armenian Cilicia date to the reign of King Levon I and his immediate successors. By the last quarter of the eleventh century the Armenian migration into Cilicia was almost complete. To assume that the Armenians patiently waited in the ruined forts of the previous Arab and Byzantine tenants until a king suddenly appeared is to ignore the forces that shape history. A kingdom, no matter how small, can only arise once it has established the kind of political and economic stability brought by a strong system of fortifications. Undoubtedly most of the sixty-two forts said to be under the suzerainty of Levon I were built before his reign and before the final expulsion of the Byzantines. Construction continued during the reign of Levon I and throughout the thirteenth century.[23]

7. The Role of Military Architecture in Medieval Cilicia: The Triumph of a Non-Urban Strategy

During the medieval period the Armenians established permanent settlements in that part of Cilicia which the topography has shaped into a natural defensive unit.[1] The western and northern borders are formed by the curving range of the Taurus Mountains. The Anti-Taurus Mountains (or Nur Dağları), which abruptly merge into the Taurus near Göksun (fig. 2), constitute the eastern flank of Armenian Cilicia. Together they create a semicircular ring around the alluvial plain. The Mediterranean Sea forms the southern terminus of the Pedias.

The Armenian settlers did not simply inhabit this region as spectators in a theater who calmly sat back and witnessed for almost two hundred years the clash of the two great titans on their borders—the Saracens and the Crusaders—but from the inception of their suzerainty

[22]The widespread use of the expensive smooth ashlar (type IX) at Lampron would logically correspond to a period of prosperity.

[23]The only official restriction on the construction of fortifications in Armenian Cilicia was imposed by the Mamluks in 1285; see M. Canard, "Le royaume," 216 ff.

[1]This region constitutes roughly the eastern half of the Roman province of Cilicia. See: G. Bean and T. Mitford, *Journeys in Rough Cilicia in 1962 and 1963* (Vienna, 1964); K. Hopwood, "Policing the Hinterland: Rough Cilicia and Isauria," *Armies and Frontiers in Roman and Byzantine Anatolia,* ed. S. Mitchell (Oxford, 1983); above, Part I.1, note 11.

in Cilicia they were drawn into the conflict and eventually consumed by it. The reason that they did succeed as a viable political and military entity for the period of the Crusades is due to the mountainous borders of Cilicia and the integrated network of defenses that they built into them. The Taurus and Anti-Taurus Mountains do not form a solid vertical barrier but are punctured and torn by river valleys and deep gorges. Since such openings provide passage into Cilicia, the Armenians constructed chains of forts to guard these routes. In most cases it is not one site on a single road but a series of forts that prevents an enemy from advancing on the major roads as well as the auxiliary routes. Most of the forts have intervisibility which allows for rapid communication and the efficient mustering and dispatch of troops. A surprise attack through the mountains was all but impossible. In many cases the Armenian regiments encountered the enemy long before he could descend into the plain. This gave the defenders a tactical advantage in that they could set an ambush and choose the time of battle in familiar territory. Also, if an enemy failed to capture a mountain fort, he would be reluctant to advance farther with his rear undefended. The fortifications became a deterrent against invasion through the passes of Cilicia.[2]

The southern border, the Mediterranean Sea, would seem to be the most vulnerable flank. Yet historically we know that not one of the major attacks, which eventually brought about the downfall of the Armenian kingdom, came by sea. The reason for this is twofold. First, the shoreline of Armenian Cilicia, which stretches roughly from Iskenderun to Erdemli, provides few good ports. Because of the constant and heavy siltation of the coast, sandbars, just below the surface of the water, frequently extend into the sea as far as 500 m. Thus a ship, even with a moderate draft, may have difficulty approaching land. A knowledge of the location and approaches into the deepwater coves made the Cilician pirates especially troublesome to the Roman navy. The second reason that the Mediterranean remained a safe flank is simply that the Armenians never reoccupied the abandoned coastal municipalities (except Ayas) that were the centers of Cilician civilization during the Roman and early Byzantine periods. The Armenian-occupied cities, such as Misis, Adana, Sis, and Tarsus, were located inland. Any invasion by sea would require that an army traverse the marshy coastlands (the areas east of Mersin) with horses and carts. The logistics of such an undertaking would eliminate the element of surprise.

The Armenians merely maintained a presence at two coastal ports: Ayas and Korykos.[3] Only Ayas seems to have had a civilian population. Korykos has a sea and a land castle which once guarded a harbor. Ayas was especially valuable to the Armenian kingdom because there

[2]Wilbrand von Oldenburg (15 f) was the first to discuss the Armenian strategy of blocking the passes with a network of forts. In the 10th-century work on skirmishing the value of occupying the mountain passes and auxiliary routes is stressed (see Dennis, 154 ff). This Byzantine source describes the tactics of mountain warfare where few fortresses have been constructed.

In Cilicia at least three of the mountain forts were originally founded by the Arabs (see above, Part I.5). The Armenians were the first settlers who succeeded in securing all the mountains around the plain.

[3]The Armenian military presence in Ayas is sporadic. The ancient port of Pompeiopolis is occasionally mentioned in the medieval chronicles (Smbat, 666, 674 f). This small port was not an emporium for European merchants; it was the seat of a bishop.

the trade from the Orient was loaded on Italian ships. Much of the merchandise that was shipped from Ayas was stored in the inland port of Misis. From Misis it was either floated on flat-bottom barges down the Ceyhan River to the sea (a distance of about 60 km)[4] or it was transported on the road that linked Misis and Ayas (fig. 2). The large number of garrison forts in the immediate vicinity of Misis testifies to the importance of Venetian and Genoese warehouses there. That the Armenians neglected their coastline and were often too willing to hand over the administration of the vital eastern trade to the Italians is a reflection on their origins. The Armenians are a mountain-dwelling people, and they readily adapted to life in the Highlands of the Taurus and Anti-Taurus Mountains of Cilicia.

The two principal approaches through the east flank of the Armenian kingdom are the Belen and Amanus passes.[5] The Belen pass connects Kırıkhan and Antakya to Iskenderun. The guardian on the east flank of the pass in the foothills of the Anti-Taurus is Bağras. Prior to Iskenderun there is no fort on the west flank of the pass. Since this small port town is not mentioned in the chronicles as having an Armenian settlement and since Bağras was a Crusader possession for most of its history, it is probably safe to assume that medieval Alexandretta (Iskenderun) was a Crusader port.[6] This conclusion is supported by the presence of a smaller and more difficult pass 30 km north of Belen. The pass is guarded by the Crusader fort at Çalan. This Çalan pass terminates at a point north of Iskenderun where the Pillar of Jonah (the "Portella") is located. The Pillar was a toll station that the Crusaders possessed for most of the twelfth and thirteenth centuries. About one kilometer to the northeast of the Pillar is the fort of Sarı Seki; both sites seem to mark the southeastern extent of the Armenian kingdom.[7] Sarı Seki could easily control traffic moving north from Kırıkhan (via the Çalan pass) or Iskenderun. Between Sarı Seki and the second major opening in the Anti-Taurus Mountains, the Amanus pass, are five minor passes that cut across in an east-west direction. They are

[4]*Handbook,* 59. In the medieval period the river must have been broad, for the Byzantine fleet customarily anchored in the Pyramos (Ceyhan); see Choniatēs, 50.

[5]The specific topography of the major passes leading into Cilicia has been discussed extensively. See: Ainsworth, 71 ff; U. B. Alkım, "Ein altes Wegenetz im südwestlichen Antitaurus-Gebeit," *Anadolu Araştırmaları 1.2 (1959), 207–22; J. G. C. Anderson, "The Road-System of Eastern Asia Minor with the Evidence of Byzantine Campaigns," Journal of Roman Studies* 17 (1897); Chauvet and Isambert, 754 ff; R. P. Harper, "Podandus and the Via Tauri," *AS* 20 (1970), 149–53; Hild, 51 ff; Hild and Restle, 41 ff; Humann and Puchstein, 105 ff; Ramsay, 349 ff; idem, "Cilicia, Tarsus and the Great Taurus Pass," *Geographical Journal* 22 (1903), 357–413; Hogarth and Munro, 643 ff; W. Brice, "The Roman Roads through the Anti-Taurus and the Tigris Bridge at Hasan Keyf," *Serta Indogermanica. Festschrift für Gunter Neumann* (Innsbruck, 1982), 19–33; K. Miller, *Itineraria romana* (Stuttgart, 1919), 749–54; Taeschner, I, 136–51, II, 30 f; *Handbook,* 66 ff; D. Van Berchem, "Le port de Séleucie de Piérie et l'infrastructure logistique des guerres parthiques," *Bonner Jahrbücher* 185 (1985), 47–87; D. French, *Roman Roads and Milestones of Asia Minor,* I, *The Pilgrim's Road* (Oxford, 1981), 92–95, 98–100, maps 6 and 7; Janke, 31–49.

[6]Wilbrand von Oldenburg (16) crossed the Belen pass when Bağras was briefly held by the Armenians. For Wilbrand the contemporary Alexandretta was a town of little consequence surrounded by a ruined circuit wall. However, by the late 16th century John Sanderson (*The Travels of John Sanderson in the Levant 1584–1602,* ed. W. Foster [London, 1931], 63, 167 f, 176, 199, 213, 222) found that Alexandretta had become one of the principal ports of the Ottoman north Levant. Cf. Brocquière, 94.

[7]Mas Latrie, I, 89; Rey, 332 f; Michael the Syrian, 349 note 1; Gregory, 171 note 1, 171 f note 2; Deschamps, III, 70. In the 14th century this line of demarcation probably shifted north to Canamella/Caramela; see K. Kretschmer, *Die italienischen Portolane des Mittelalters* (Berlin, 1909; rpr. Hildesheim, 1962), 237.

guarded directly or indirectly by Armenian forts (fig. 2). The Amanus pass is protected at its east end by Hasanbeyli and at the west by Savranda. The Amanus route is sometimes confused with the Bahçe pass which lies 6 km to the north. Part of the modern paved road that links Adana to Gaziantep actually cuts a new route between both passes. From Maraş there are at least four trails that traverse the northeast boundary of Cilicia. Kalası, Dibi, Geben, and Haruniye are four forts on these routes. Undoubtedly further exploration along these Maraş trails will uncover other garrison forts.

Between Saimbeyli and Süleymanlı (Zeitun) there are two major trails that move from north to south. The most important one is the Göksun-Andırın trail that leads to Kadirli and the plain. This route is guarded by six forts. To the west the Saimbeyli-Kozan road is protected by three forts. Between this route and the Tarsus-Pozantı road through the Cilician Gates there are three other trails which today are passable only by jeep. However, in medieval times these trails were thought worthy enough to be guarded by forts like Bostan, Meydan, Tamrut, and Milvan. The fortress at Gülek has a commanding view of the Cilician Gates and intervisibility with Tarsus. Three other routes to the west of Gülek also enter Armenian Cilicia and are guarded by a series of forts.

That the Armenians chose to defend and inhabit the valleys in the mountains is not surprising. Their traditional homelands, such as Vaspurakan and Ayrarat, were too often the site of violent confrontations between the Latin and Greek west and the Persian and Arab east. The mountains have always provided a secure sanctuary. However, the Armenians in Cilicia also settled part of the broad, almost sea-level plain by constructing garrison forts and even occupying a few of the late antique cities.[8] Misis, Sis, Adana, and Tarsus were protected by the enveloping ring of mountain fortresses, castles in the plain, and the barren coastline. Thus Armenian Cilicia was a bipartite kingdom consisting of the mountainous settlements, some of which were the great baronial seats like Lampron, Çandır, and Vahga, and the *inland* (or northern) plain. If an invading force penetrated and overran the Armenian defenses of the latter, then the Armenian barons who resided in the plain could take refuge in the mountains.

A variety of factors necessitated the Armenian habitation of at least the northern half of the plain. The fertile soil, a constant supply of water, and a temperate climate make the Cilician plain the most fertile agricultural area in all of Anatolia. By controlling the plain the Armenians also licensed and taxed (as middlemen) the storage of the goods brought through the Taurus Mountains from the Orient.[9] To have control of the mountains alone would simply allow them to charge a toll on passing caravans.

We are fortunate to have a brief description of one of the oriental trade routes that ran from Iran to Armenian Cilicia. In the early fourteenth century Francesco Balducci Pegolotti, a Florentine scholar and entrepreneur, describes the road that leads from Ayas to Tabriz.[10] The largest section of this route, which runs east from Sivas to Tabriz (via Erzincan, Erzurum, and

[8] Most of the antique cities in Cilicia were founded by Hellenistic kings. See Jones, 191 ff.

[9] Alishan, *L'Armeno-Veneto,* 13 ff; J. Richard, "Agricultural Conditions in the Crusader States," *A History of the Crusades,* ed. K. Setton, V (Madison, 1985), 265.

[10] Francesco Balducci Pegolotti, *La pratica della mercatura,* ed. A. Evans (Cambridge, Mass, 1936), 28 f, 389.

Xoy), has been securely located and described in a number of modern commentaries.[11] Pegolotti mentions that the caravans pass from Ayas through Colidara, Gandon, and Casena before reaching Sivas. Of these three sites between Ayas and Sivas, Casena has been identified with Caesarea in Cappadocia.[12] The locations of Colidara and Gandon are uncertain. In both cases Pegolotti is trying to render Armenian place names so that they can be understood by his Italian readers. This rendition probably resulted in a slight change of spelling and pronunciation. H. A. Manandian accepts W. Heyd's identification of Colidara with the Armenian fortress of Copitar (Kopitař).[13] However, this fortress has never been securely identified; G. Alishan hypothesized that the district of Copitar lies between Partzerpert (Barjrberd) and Molevon.[14] These regions are northwest of Ayas and nearer to the area of the Cilician Gates. For most of the twelfth and thirteenth centuries the district immediately north of the Cilician Gates (near modern Ulukışla and the medieval Loulon) was under the control of the Rum Seljuks. Thus the Tarsus-Pozantı road through the Cilician Gates would not have been a suitable route for Armenian trade. I believe that the site of Gandon can be identified with Geben (Kapan/Gabon) in northeast Cilicia. Geben is a logical choice for we know that in 1201 King Levon I signed a trade agreement with the Genoese that specifically stipulated that the Italians had to pay an extra toll when passing the region controlled by Geben's lord.[15] The route from Ayas to Geben initially follows the course of the Ceyhan River before turning into the valley cut by the Andırın Suyu. In the Armenian chronicles there is mention of a fortress by the name of Govara near the Ceyhan River.[16] If this fortress is Colidara, then the route described by Pegolotti predictably follows the most heavily fortified road in Cilicia. M. H. Kiepert also discusses another trade route that runs north from Sis through Saimbeyli and eventually leads to Caesarea.[17] From the account of Marco Polo it appears that Ayas was the traditional point of departure for travel to the Far East.[18]

A strong Armenian presence in the plain resulted in improved communications. Normally, contact between the forts in the Taurus Mountains (that is, the areas to the north and west of the plain) was on a north-south axis, following the alignment of the river valleys. This meant that Azgit, which could communicate quickly (either by horseman or signal fire) with the forts at Geben, Kalası, or Ak, was able to send rapid messages only along the valley cut by the Andırın Suyu. The fortresses at Çem and Vahga or even sites farther to the west like Çandır might not learn of a major invasion in the Andırın area until days later. The great advantage of having fortifications in the plain is that they could communicate on an east-west axis and

[11] Manandian, 187 ff; Heyd, II, 112 ff; M. Yule, *Cathay and the Way Thither,* II (London, 1866), 299 ff.

[12] Manandian, 191 f; Heyd, II, 114; Boase, 165.

[13] Ibid.

[14] Alishan, *Sissouan,* 167; cf. Meydan (note 2) in the Catalogue; below, Appendix 3, note 7.

[15] See above, note 11. Manandian believes the Geben area to be an important route for trade, but he has confused some of the place names in his text. The most direct route from Geben to Caesarea is through Göksun (fig. 2).

[16] Alishan, *Sissouan,* 240; Hellenkemper, 169, 186. Unfortunately it is impossible to identify Govara with a specific fort in the plain. The likely candidates are Bodrum, Gökvelioğlu, and Tumlu.

[17] M. H. Kiepert, "Über Pegolotti's vorderasiatisches Itinerar," *Sitzungsberichte der philosophisch-historischen Classe der Berliner Akademie* (1881), 901 ff.

[18] Marco Polo, *The Travels,* trans. R. Latham (New York, 1958), 37, 39, 46.

quickly dispatch troops into troubled areas. Every garrison fort in Cilicia Pedias has intervisibility with at least two other forts in the plain and most can communicate directly with the Highland valleys.[19] Since the garrison forts of the plain are built on the occasional outcrops, they do not differ architecturally from the mountain forts. Generally the garrison forts in the plain are slightly larger than those in the mountains, and because there are fewer obstructions in the plain they are spaced widely apart. It should be stressed that their exact location in the plain always depends on the proximity of an outcrop. The close relationship between the forts in the mountains and in the plain was also symbiotic in that the fortified port of Ayas (the only significant Armenian site on the coastal edge of the plain) brought in various taxes for the Armenian kingdom. Those taxes, in turn, could finance the maintenance of existing garrisons and even the construction of new forts in the mountains. This increased security made Ayas the safest harbor in southern Anatolia and the Levant, insuring a steady and increasing flow of revenues.

The control of the plain and the wealth that resulted from it had an important political consequence for Armenian Cilicia. The rulers of the Armenian kingdom took an active and often offensive role in the political affairs of their neighbors. However, they could only maintain such activities as long as they had a secure kingdom at home. For example, when King

[19] In my opinion, the ability of men in fortifications to communicate with fire and smoke to their counterparts has been exaggerated. Over the last century many of those who have studied Cilician forts have published maps that claim to show the lines of visual communication (Favre and Mandrot, 145–51; Müller-Wiener, 108 f; Fedden and Thomson, 12, 54; Hellenkemper, 262, pl. 92). In many instances my own observations have confirmed their conclusions, and I have listed in my discussion of each site (Catalogue) the neighboring fortifications that are clearly visible. However, in some cases it is geographically impossible for forts in very close proximity to communicate visually. For example, Yılan cannot communicate directly with Gökvelioğlu at the southwest. The Cebelinur Dağı, which rises to a height of 840 m, blocks a direct line of sight. Since both forts have intervisibility with Misis, it was possible to communicate through a second party. It has been claimed that Azgit (Fedden and Thomson, 54) and Ak (Hellenkemper, pl. 92) have intervisibility with Anavarza. But in both cases the Taurus Mountains in the region of Andırın block any view into the plain. Azgit Kalesi is actually on a small outcrop in the bottom of a valley. Ak Kale is situated on an ascending outcrop above a river, but it is not on the summit of the gorge. The greatest flexibility with respect to intervisibility is with the forts in the plain. Even here there are limitations in that a centrally placed fortress like Anavarza cannot communicate directly with Adana or Bodrum.

The chief impediment to visual communication among forts that have intervisibility is the weather. By the nature of its topography and location, Andıl must have intervisibility with Sis. Yet for all the weeks I spent at Sis and Andıl, never once was I able to see the other site because of a thick cloud cover that seemed to be anchored to the 1,500-m peak of Andıl Dağı. During the late spring, summer, and fall there is an almost constant thick haze that is created naturally when the intense heat evaporates the water in the wet alluvial plain. The recent expansion of agriculture and modern industrial pollution certainly contributes to the present loss of visibility. In the 1970s visibility in the summer from most of the plain forts rarely exceeds 10 km. It is very difficult to imagine that even in the medieval period a site like Mancılık could communicate with Ayas or Sarı Seki and that Gülek had constant intervisibility with Tarsus. Also, it should be remembered that communication by signal in the medieval period was not accomplished with a beam of high-intensity light but with fire and smoke. In the bright Cilician sun this might not be readily visible; night may have been a more efficient time for such communication. Although intervisibility had its limitations, it is apparent that fire signals as well as mounted couriers played an integral part in the security network around Cilicia.

See: Ramsay, 352 f; P. Pattenden, "The Byzantine Early Warning System," *Byz* 53 (1983), 258–99, esp. notes 1 and 10; S. Parker, "Preliminary Report on the 1982 Season of the Central *Limes Arabicus* Project," *Preliminary Reports of ASOR-Sponsored Excavations 1981–83, BASOR Supplement No. 23* (Winona Lake, Indiana), 18 f.

Levon I occupied the Templar fortress of Bağras on the eastern flank of the Anti-Taurus Mountains, his intent was merely to interfere in the internal affairs of Antioch.[20] When the Crusaders from the south and the Turkmen from the north began to penetrate the borders of the Armenian kingdom and harass the population, Levon permanently withdrew from Bağras.

When speaking about the mutual dependence of the fortifications in the plain and mountains, one should not assume that the urban centers played an important role in the defense of the kingdom. All of the late antique cities of Cilicia, which are within the confines of the Armenian kingdom and inhabited (that is, Tarsus, Adana, Ayas, Misis, Hieropolis/Castabala, Sis, and Anazarbus), are located in the plain and described as having neighboring garrison forts, but there is *no* evidence that city walls were ever constructed by the Armenians. Nor is there any mention that the Roman-Byzantine-Arab walls around cities like Tarsus and Adana were ever repaired.[21] A close inspection of Ayas reveals that the surviving land fort is an Ottoman construction; if that plan follows the original medieval circuit, then it was adequate to protect only the Venetian and Genoese warehouses in one corner of the port. In the first decade of the thirteenth century Wilbrand von Oldenburg expresses wonder at the magnificent fortifications of the Armenian kingdom, but he cannot hide his disappointment when he first sees the Armenian capital at Sis. He describes it as almost an unwalled village.[22] He reserves his praise for the fortress and walled palace at Sis. His description of Anazarbus leaves the clear impression that the late antique city is abandoned and that only the adjacent fortress is occupied.

> . . . we came to Naversa, which is a very strong castle positioned on the top of an outcrop (*quod est castrum optimum in alto monte situm*), which nature has centrally placed in the plain of that land to the best advantage of the king. For this reason the king himself usually calls his own banner "Naversa."
>
> At the base of the outcrop was once located a city (*in pede huius montis sita fuit quedam civitas*), and that its influence was great is today attested also by a wondrous aqueduct stretching there on tall columns for a distance of two miles.[23]

[20] Edwards, "Bağras," 431 f.

[21] H. Goldman, ed., *Excavation at Gözlü Kule, Tarsus,* I (Princeton, 1950), 29, 37; F. Day, "The Islamic Finds at Tarsus," *Asia* 41 (1941), 143–46; see below, note 27.

[22] Wilbrand von Oldenburg, 18.

[23] Ibid., 20. In concluding that the late antique city of Anazarbus was uninhabited during the Armenian period, I am at variance with the opinions of M. Gough. Gough (103) accepts the earlier suggestions of R. Heberdey and A. Wilhelm that the foundation of the city circuit was Byzantine and the reconstruction was Armenian. While the repair to the circuit does have a poured core, the facing stones consist of relatively smooth ashlar. Only the Arabs are known to have used a comparable ashlar as an exterior facing stone (cf. Haruniye). The fact that the Cufic inscription of the Caliph al-Mutawakkil was found in a *square* tower of the city circuit would indicate that the repair is from the Arab period. Square towers are anathema in Armenian military architecture. Choniatēs (25) specifies that in his day the garrison fort atop the outcrop is the well-populated city (πόλις) of "Anazarba." There is no evidence in our texts of a civilian Armenian population in the late antique city. My views on urbanism in Cilicia during both the Arab and Armenian periods differ from those of L. Rother, *Gedanken zur Stadtentwicklung in der Çukurova (Türkei)* (Wiesbaden, 1972), 31–49.

It is significant that the Armenian archbishop of Anavarza resided permanently at the monastery of Kastałōn, over 115 km from the castle itself. Kastałōn (Kastałoni Vankᶜ), a presently unidentified site, was located near Vahga. The monastery may have been established by Greeks and later reoccupied by Armenians. The

Canamella (Hieropolis/Castabala?) was also a ghost town, for Wilbrand twice passes the site and only refers to it briefly as a *castellum* (that is, a citadel, fort, or stronghold).[24] When Wilbrand does encounter an inhabited city he always comments on the population. Of Misis he says that the city is encircled by a turreted wall that had collapsed through old age (*sed antiquitate corrosum*) and that the population is small (*paucos . . . habens inhabitatores*).[25] Wilbrand characterizes Adana as *non divites habens cives*.[26] The only city that Wilbrand labels as populous (*multos habens inhabitatores*) is Tarsus.[27] It also happens to be the only city with

founders of the Rubenid dynasty were buried there as well as the ascetic Markos. By the fourth quarter of the 12th century the head of the monastery was also the archbishop of Anavarza. Of the two positions, the latter was by far the more prestigious (a selection made for political reasons by the crown) and probably brought with it the appointment at Kastałōn. At the coronation of Levon I, Tēr Kostandin held both positions. Considering that Sis is situated between Anavarza and Vahga, the ecclesiastical jurisdiction of this archbishop could not have extended from the plain to Kastałōn. After 1198 the chief cleric at Sis certainly administered the regions adjacent to the capital. The archbishop of Anavarza resided in the mountainous Kastałōn primarily because it was safer than the vulnerable, underpopulated plain and, secondarily, because his largest area of ecclesiastical jurisdiction was north of Sis. By the second half of the 13th century the decline of Anavarza's importance and the frequent invasions in the plain resulted in the permanent separation of this joint position. See: Smbat, 610; Smbat, G. Dédéyan, 74 note 18; Matthew of Edessa, 48, 79; Vahram of Edessa, 498; Samuel of Ani, 448; Alishan, *Sissouan,* 173, 280; Ēpʿrikean, II, 281 f; below, note 34. There is a similar experience in the Greek world, where Byzantine bishops who feared Turkish raids seldom saw their sees; see S. Vryonis, *Decline of Medieval Hellenism in Asia Minor and the Process of Islamization from the Eleventh through the Fifteenth Century* (Berkeley, 1971), 288–350.

[24] Wilbrand von Oldenburg, 17, 20, 72 note 98. J. Laurent's association of Canamella with Hieropolis/Castabala is uncertain.

[25] Ibid, 17.

[26] Ibid., 18. Adana may not have been rebuilt after its destruction by the Turks in 1137; see Tritton and Gibb, 276 f.

Marco Polo (*Travels* [above, note 18], 46), who makes a point of referring to only "villages" and "towns" in Cilicia (in contrast to the "cities" of the Far East), provides one explanation for the general neglect of urban centers when he describes Cilicia's humid climate as unhealthy and extremely enervating. Cf. Brocquière, 99 note 1. Both Tarsus and Adana played a minor role in Cilician trade.

[27] Wilbrand von Oldenburg, 17 f. To what extent King Hetʿum I repaired the *city walls* in 1228 cannot be determined; he may have simply refortified the citadel within Tarsus. See: Langlois, *Voyage,* 315–28; idem, *Rapport,* 34; idem, "Notes sur trois inscriptions arméniennes de l'église de la vierge à Tarse (Cilicie)," *RA* 10.2 (1854), 745. Most of the Armenian inscriptions from Tarsus are religious in nature and date from the 1260s to 1319.

The one fully extant ecclesiastical construction of medieval Tarsus, which some scholars mistakenly label as the Church of St. Peter (cf. C. Enlart, *Les monuments des Croisés dans le royaume de Jérusalem,* II [Paris, 1928], 379 f, fig. 487; Hild and Hellenkemper, 281; Alishan, *Sissouan,* 319 f) despite evidence associating it with a structure dedicated to St. Paul or the Virgin, is known today as Kilise Cami/Eski Cami. This church is not included in my catalogue of Armenian ecclesiastical architecture (Edwards, "First Report" and "Second Report"), because the architectural features indicate that it was built by the Franks during Baldwin's suzerainty there in the first decades of the 12th century. While the single nave and two flanking aisles of the church's hall plan, as well as the apsidal niches, gabled roof over three longitudinal barrel vaults, pointed arches, and limited use of the Muqarnas, all have precedents in the Armenian religious architecture of both Cilicia and Armenia Major, there are other features whose origin can be traced to the traditions of European Romanesque and Crusader architecture. For example, the numerous and closely spaced transverse arches of the nave vault rise on foliated capitals with attached pilasters that stretch from the arcade spandrels (cf. S. Langè, *Architettura della Crociate in Palestina* [Como, 1965], 162–65; Enlart, *Monuments,* figs. 29 f, 30–41, 216 f, 241, 498, 582). In Armenian construction the transverse arches normally spring directly from the capitals of an arcade, turning the spandrels into abutments for diaphragm walls and eliminating any horizontal molding (as we see here) between the spandrel wall and the springing of the barrel vault (one of the few exceptions can be found in J. M. Thierry, "L'église Surb-Yovhannēs de Biwrakan," *REArm,* n.s. 13 [1978–79], 210, fig. 2). Also, the unarticulated, freestanding, massive round columns of the arcades in the Tarsus church are common in Romanesque structures, completely

a predominantly Greek population. It is defended principally by a dilapidated circuit wall, but it does have a strong citadel in the center (*muro cingitur pre antiquitate mutilato, sed castrum habet in capite sui firmum et bonum*).[28] Whether the Armenian garrison in the citadel would have given refuge to part of the Greek population of Tarsus during an invasion is open to speculation.

The vast majority of the population of Armenian Cilicia was rural, and most residents were probably confined to the relatively secure Highland valleys in the mountains. Those who occupied the northern half of the Cilician plain during the period of the kingdom lived outside of cities. The characteristic method of settlement was to have a large garrison fort with a number of scattered villages in the immediate area. The villages were agricultural settlements. This is the kind of description that Wilbrand gives us for the fort of Amuda. During a period of invasion the residents in the immediate vicinity of a fort could flee there for safety. Those who lived farther away simply had to fend for themselves.[29] The rural nature of Armenian society not only insured the survival of the feudal system but protected it from any potential middle class. All trade agreements and major economic concerns were firmly in the hands of the royal family. Considering the absence of customary amenities (for example, civic buildings, aqueducts, baths, etc.) in the few sparsely populated urban centers of Armenian Cilicia, one might aptly conclude that the late antique cities have become central market towns or even strategic hamlets. At Misis and Adana a small urban garrison protected the trading activities of the residents and neighboring farmers; church services were undoubtedly coordinated with market days.

Those scholars who are concerned with the survival of late antique cities into the Byzantine Middle Ages must surely view the course of events in Cilicia with a sense of irony.[30] For here the Arabs, who are all too often credited with the destruction of the traditional polis, have faithfully rebuilt, repopulated, and defended the urban centers until the mid-tenth-century reconquest by the Greeks. It is the brave new Hellenes under Nicephorus Phocas who evacuate and plunder the cities of Cilicia in their headlong rush to reoccupy northern Syria. For a century and a half the scattered garrisons of Greeks quietly watched the cities of the Pedias (except Tarsus) decay. By the late eleventh century such conditions permitted the Armenians to introduce their new strategy for defense and system of habitation.

What is most interesting about the Armenian experience in Cilicia is that it represents the antithesis of settled life in the Crusader Levant and even in the Bagratid capital of Ani (Armenia Major). Joshua Prawer notes: "The most characteristic trait of Crusader society is its complete urbanization . . . to assure its military striking power the minority (i.e., the Christians) gathered itself in walled cities and in castles, and from those centers it kept the countryside in

---

unknown in Armenian Cilicia, and rare in the ecclesiastical architecture of Armenia Major. When they do occur, they are not made with small ashlar blocks (as here) but with monolithic drums. The inscriptions in this Tarsus church probably record its occupation and repair by the Armenians.

[28] Wilbrand von Oldenburg, 17.

[29] Ibid., 20. Refer to the section on Amuda in the Catalogue. This policy of not defending the outlying villages was a common practice in Anatolia following the Arab invasions; see Dennis, 164, 222 ff. There is evidence of medieval villages beside the forts at Tumlu, Gökvelioğlu, and Toprak; see Youngs, 188.

[30] A. Kazhdan and A. Cutler, "Continuity and Discontinuity in Byzantine History," *Byz* 52 (1982), 429 ff.

thrall . . . the Crusader cities were, consequently, almost the only habitat of the conquerors."[31] Later the Crusaders expanded from the cities to protect their borders and their income from agricultural produce. Their strategy was simply to guard the ends of the military and commercial roads. There was no attempt to secure a region the size of Cilicia. Two centuries before the great migration into Cilicia some of the Armenian *naxarars* in the regions of Van and Ayrarat made themselves kings, and by efficiently controlling the routes of trade they began a cultural renaissance that culminated for one family in the construction of the great Bagratid city of Ani.[32] Here Armenian society became so "urbanized" that many peasants no longer paid obligations in kind but in coin. Townsmen and artisans received special class privileges from the aristocracy. However, the experiment at Ani was exceptional. The bulk of Armenian society was rural, and the partial destruction and depopulation of Ani in the eleventh century reinforced its traditional contempt for urban life. Instinctively, the Armenians migrating into Cilicia avoided the vulnerable cities in favor of dispersed and fortress-oriented settlements in the mountains and northern plain.[33] The two great dynastic families, the Het῾umids and Rubenids, and later the Armenian kings backed by a faithful retinue of barons, provided the political and cultural adhesives that formed a homogeneous society. Because of the security they provided, the mountains and not the plain became the heart of the Armenian kingdom. This explains why the royal capital was moved from Anavarza to Sis at the foot of the Taurus Mountains and why the kingdom functioned for over a hundred years after its capital was sacked by the Mamluks. The king and his court simply moved into the mountain strongholds. The allure of the mountains was so strong that even in the dynastic period, when the Het῾umids were the vassals of the Byzantine crown and the official administrators of western Cilicia, they chose not to govern from Tarsus but from their mountain strongholds at Lampron and Çandır. Likewise, it is not surprising that most of the archives, libraries, workshops for ecclesiastical silver, and schools for miniature painting were located in the monasteries and castles of the Taurus.[34]

Precisely how the forts were integrated into the feudal system of Armenian Cilicia is difficult to determine. The Armenian sources from the period of the Cilician kingdom do not supply us with a picture of the socio-political fabric of Armenian society. We do know that

[31] J. Prawer, *Crusader Institutions* (Oxford, 1980), 51 f, 102, 474 ff. See also: idem, "Crusader Cities," in *The Medieval City. Studies Presented to R. Lopez,* ed. H. Miskimin et al. (New Haven, 1977), 179 ff; J. L. LaMonte, *Feudal Monarchy in the Latin Kingdom of Jerusalem* (Cambridge, Mass., 1932); J. Richard, *Le royaume latin de Jérusalem* (Paris, 1953); J. Riley-Smith, *The Knights of St. John of Jerusalem and Cyprus, 1050–1310* (London, 1967).

[32] Manandian, 133 ff. In a forthcoming article Prof. N. Garsoïan (Columbia University) rejects the views of Manandian and supports the thesis that pre-Seljuk Armenian society was essentially non-urban, even in the Bagratid period.

[33] It is regrettable that no documentation survives from the Armenian kingdom that permits a demographic survey along the lines of Prof. A. Laiou-Thomadakis', *Peasant Society in the Late Byzantine Empire* (Princeton, 1977).

[34] H. Oskean, "Kilikiayi vank῾erə," *HA* 69 (1955), 373–96, 505–25; idem, 70 (1956), 26–45, 163–95, 320–36, 452–67; Oskean, passim; A. Kakovkin, "Pamjatniki xudožestvennogo serebra kilikiĭskoĭ Armenii," *Patma-banasirakan Handes* 69.2 (1975), 192–208; Grigor Aknerc῾i, 390 f note 63; P. Akinean, "Akanc῾ kam Akneri vank῾ə," *HA* (1948), 217–50; S. Der Nersessian, *Armenian Art* (Paris, 1977–78), 123–62, esp. 253.

King Levon I granted titles and fiefs to Armenian barons and to Crusaders,[35] but we do not know if taxes were collected by baronial or royal officials, how garrisons were trained and paid, or what traditional and oath-bound responsibilities specifically joined a baron to his king. The few summary commentaries we have on these subjects from modern writers are never supported by textual evidence and almost always are derived from paradigms of an earlier feudal system in Armenia Major. A good example is in Manandian's study of the trade and cities of Armenia. His entire assessment of the feudal system in Cilicia is as follows:

> The kingdom of the Rubenids, which had continuous relations with European states, was under the influence of western European feudalism, but at the same time it preserved the chief and specifically Armenian characteristics of the ancient feudal-*naxarar* system. The Cilician kingdom was divided into counties and baronies; it had a class of knights as did western European kingdoms, the Ancient Armenian *azats* called *jiavor,* as well as a class of bound peasants called *šinakan* or *paroikoi*. The following offices existed at the court of the Armenian kings: the chancellor, who supervised the royal chancery, the commander of the army called *connetable* or *spasalar,* the chaplain, the *seneschal,* and others.[36]

Although the *Assises d'Antioche* was translated into Armenian by the late thirteenth century, no assizes of Armenian Cilicia have yet been discovered.[37] Aside from Smbat's rather vague comment that a baron's castle passed to his heirs as inalienable property, we know little about the feudal districts. Since the territory of Armenian Cilicia has a higher concentration of military architecture than any other region in the Middle East of equivalent size, the organization required to coordinate any common activity among the barons must have been huge.

Aside from the discovery and identification of many hitherto unknown forts in Armenian

[35] Appendix 3 contains what may be a partial list of Cilician barons and their castles. See also the sections on Haruniye, Amuda, and Silifke in the Catalogue; Forstreuter, 59–67. Cf. N. Oikonomidès, "The Donations of Castles in the Last Quarter of the 11th Century (Dölger, *Regesten* n° 1012)," *Polychronion, Festschrift Franz Dölger,* ed. P. Wirth (Heidelberg, 1966), 413–17.

[36] Manandian, 175. Sirarpie Der Nersessian (*The Armenians* [London, 1969], 58–60; Der Nersessian, 651) has a similar assessment and has listed the titles of other court officials, including the bailiff (*bayl*), chancellor (*chantsler*), marshal (*marachakhd*), and the chief baron (*avak baron*). Der Nersessian has rightly noted that not only was the nomenclature of western Europe and the Latin Levant adopted by the Armenian court, but old Byzantine titles, such as *sebastos* and *proximos,* were also retained. E. Dulaurier (289 ff; *RHC, DocArm,* I, lxxiv ff) mentions similar officials and carefully lists the names of barons who held these offices. V. Hagopian ("The Relations of the Armenians and the Franks during the Reign of Leon II," *Armenia,* February [1909], 17 ff) depicts a feudal system in Cilicia that is essentially derived from that of the Franks. However, no commentaries have yet been published that specifically describe the duties of court officials. One detailed view of the Cilician court survives in *Burchard of Mount Sion,* trans. A. Stewart, PPTS (London, 1896), 106–10.

The best description of the feudal system (or more appropriately dynastic system) in Armenia Major was composed by Nicholas Adontz (*Armenia in the Period of Justinian,* trans. and rev. N. Garsoïan [Lisbon, 1970]). This view of Armenian society before the Arab conquests sheds no light on the social structure in Armenian Cilicia. Cf. C. Dowsett, "A Twelfth-century Armenian Inscription at Edessa," in *Iran and Islam,* ed. C. Bosworth (Edinburgh, 1971), 205 ff.

[37] G. Alishan, ed. and trans., *Assises d'Antioche* (Venice, 1876), vi–xxiii. Der Nersessian (above, note 36) believes that as the authority of the Latin Assizes increased in Cilicia it was necessary for Armenians to translate and adopt them. But there is no evidence in the Armenian chronicles that the Assizes of Antioch or any other

Cilicia, the most important result of my many field surveys is that now we have a better idea of the boundaries of the Armenian kingdom. It is much smaller than modern commentators had ever suspected.[38] The forts of Armenian construction follow the west flank of the Anti-Taurus Mountains until the latter merge into the Taurus Mountains just south of the latitude of Göksun.[39] The line of forts then follows the curving south flank of the Taurus Mountains until reaching the sea. The most northerly of the Armenian forts appears to be Saimbeyli. King Levon I is said to have carried out raids north of the modern village of Ulukışla, which is on the north flank of the Taurus Mountains. He is also said to have captured and briefly held the Byzantine fort of Loulon (near Ulukışla).[40] But there is no evidence of Armenian construction at Loulon. What is most surprising is that Cilicia Tracheia, which was often included with Cilicia Pedias in forming the administrative district of "Cilicia" during the late

---

Latin laws were *adhered to* in Cilicia. If an attempt was made to introduce the Assizes of Antioch, it may have been as unsuccessful as King Levon's short-lived attempt to introduce Roman Catholicism.

One frequently cited piece of evidence that seems to link the Armenian feudal system with Latin law is a short excerpt in the Assizes of Jerusalem (*RHC, DocArm,* I, 605 note 1). In this passage the Baron Constantine asks John of Ibelin to decide whether he (Constantine) has the right to bestow the castles at Korykos on his younger son (Ōšin) when the older son (Constable Smbat) claims the gift. John of Ibelin decides that Constantine does have the right to grant Korykos to Ōšin despite the protests of Smbat. However, there is an air of unreality about the whole incident. The Grand Baron Constantine was the father of King Het῾um I, and the grant of the castles probably took place around 1240. Why would a baron from the most prominent family in Cilicia seek a solution from a foreign government? The Armenians in Cilicia had been living in fortifications for nearly two centuries and undoubtedly a reliable and accepted system for the transfer of ownership had been in use. The Armenian chronicles never refer to any problem with regard to the transfer of estates. In my opinion, the story in the Assizes of Jerusalem is spurious. Cf. J. Karst, *Grundriß der Geschichte des armenischen Rechtes,* II (Stuttgart, 1907), 8 note 7.

In 1265 Smbat completed his legal Codex under a commission from King Het῾um I. The vast majority of the sections in this document deal with laws on property, sale, family, and ecclesiastical administration. It is clear that the king regarded the church as an extension of his authority. What receives only the briefest mention in the Codex are the rights and prerogatives governing the barons and their castles (Karst, I, 20–26, 103–5, II, 2–31, 94–98). When discussing the barons, Smbat does not mention the traditions, precedents, and ancient customs, which are prominent elsewhere in the Codex. I suggest that his brevity and vagueness are intentional. At this time Het῾um's position was weak due to internal dissension and external threats. He probably lost considerable power with respect to his barons, and he wanted neither to advertise nor codify this reality. In chronicles written after 1265 the Codex is never cited as a source of authority in disputes between the barons and the king.

See also Rüdt-Collenberg, 27 f, and the bibliography of modern Armenian commentaries in K. Bardakjian and R. Thomson, *A Reference Guide to Armenian Literature* (forthcoming).

[38] A. Sukiasjan, *Istorija Kilikiĭskogo armjanskogo gosudarstva i prava (XI–XIV vv.)* (Erevan, 1969), map; Alishan, *Sissouan,* 71 ff; Alishan, *Léon,* 171 ff; Langlois, *Voyage,* 3 ff; Mikaeljan, map; Tēr Łazarean, 63 ff; Boase, map (front); Smbat, G. Dédéyan, opp. 136. The aforementioned commentators hypothesized the unusually large and unjustified boundaries for the Armenian kingdom based *only* on the ambiguous guest list for Levon I's coronation (see my response below, Appendix 3) and the *brief* Armenian occupation of the regions near Anamur (see below, note 42).

My conclusion is based on two presuppositions: first, that the architectural traditions of the Armenian kingdom are fairly consistent, so that it is possible to identify unattested forts (i.e., structures that have neither inscriptions nor specific mention in the medieval chronicles) as Armenian constructions; and second, that my explorations in Cilicia and its environs were quite adequate to get at least a good sample of the buildings in a given region.

[39] My survey of the regions north and west of Maraş was adequate but not complete. Further study of these regions may yield more Armenian forts. Forts and monasteries have been identified near Fırnıs (Fawṙnaws) and Zēyt῾un.

[40] Hild, 51; see below, Appendix 3.

Roman and early Byzantine periods, was almost ignored with regard to long-term Armenian settlement. There are records of Armenian campaigns as far west as Alanya[41] and even a totally unjustified claim that the coastal fort at Anamur is an Armenian construction.[42] In reality the only site under Armenian suzerainty west of Korykos was the large castle at Silifke. This site was given to the Crusader Order of the Hospital by King Levon I, and most of the medieval construction at Silifke belongs to the Crusader period of occupation. Silifke Kalesi was so far removed from the major Armenian settlements to the east that it was considered a frontier whose defenses could safely be entrusted to others.[43] Korykos and Lamas are not Armenian constructions, but the former was occupied by an Armenian garrison because of its strategic value as Cilicia's other port.

The western border that marks the beginning of Armenian construction appears to run from Tece to Arslanköy. I believe that the reason for the limited western expansion of the Armenian kingdom is twofold: geography and climate. The line from Arslanköy to Tece is the point where the Taurus Mountains turn into the Mediterranean Sea. For the Armenians to expand farther west would have put them in the natural, mountainous buffer that separated their kingdom from the territory of the Sultan of Rum (Konya) and later the Karamanoğlu emirs.[44] The second and perhaps the most important reason why the Armenians did not expand farther west was that the regions north and west of Erdemli are hot, dry, and barren.[45] Even

[41] S. Lloyd and D. Strom Rice, *Alanya (ʿAlāʾiyya)* (London, 1958), 2 ff.

[42] D. Lang, *Armenia, Cradle of Civilization* (London, 1970), 210; cf. Boase, 151 f and Hellenkemper, 260. It is quite possible that repairs were carried out to the medieval fort above Eski Anamur during the period of Levon's brief occupation (roughly from 1196 to 1219). The neighboring fort of Softa Kalesi may in part be Armenian. But there was no attempt by subsequent rulers to recapture the littoral of Cilicia Tracheia. It was never treated as a permanent part of the medieval kingdom of Armenia. Although Rašīd ad-Dīn (*Die Frankengeschichte des Rašīd ad-Dīn,* intro., trans., and comm. K. Jahn [Vienna, 1977], 44) indicates in a general way that the Armenian kingdom ("Little Armenia") extends from Malatya to the borders of Rum in his day (the late 13th century), he adds the specific qualification that its territory is surrounded by mountains and is possessed of five major cities: Ayas, Sis, Tarsus, Adana, and Misis. The absence of any mention of Maraş, Silifke, or Anamur indicates that the kingdom has contracted into its natural mountainous defenses. His mention of twenty thousand flourishing settlements is obviously an exaggeration.

Some Armenians lived in proximity to Anamur until 1923. An Armenian church that once stood near the coastal fortress of Anamur is poorly attested in modern sources. The relief decoration on this church postdated the 15th century. See: Alishan, *Sissouan,* 379 ff; Ēpʿrikean, I, 165–67; J. Strzygowski, *Die Baukunst der Armenier und Europa,* II (Vienna, 1918), 540; Sevgen, 49 f.

No verifiable sites of Armenian construction were located in the Tracheia during the surveys of Bean and Mitford. Two presently unpublished Armenian inscriptions are said to be at Lausada. See G. Bean and T. Mitford, *Journeys in Rough Cilicia 1964–1968* (Vienna, 1970), 121, 195, and 210.

[43] No non-Armenians were given a fort in the mountains or Highlands; see above note 35. Often the grant of a fortress to a Crusader order was intended to be temporary. It was viewed as part of a short-term, politically expedient alliance and a source of free military labor. Occasionally, this ephemeral and very manipulative aspect of Armenian politics led to misunderstandings among the European "allies." In 1328 Pope John XXII (acting on behalf of the Hospitalers) demanded from the Cilician crown the return of Frankish estates on the borders of Cilicia Pedias. Since Levon IV wanted to keep the Hospitalers at a distance, he enthusiastically offered (in 1332) to give to the order in perpetuity two forts near Anamur (deep in Cilicia Tracheia), which the Armenians had been unable to garrison for over a hundred years! See *Annales ecclesiastici,* ed. C. Baronius, XXIV (Bar-le-Duc, 1872), 499 f.

[44] The castle at Silifke was a prize for which the Sultan of Konya contended; see Silifke in the Catalogue.

[45] It may be more than coincidental that this natural boundary north of Erdemli is very near the Lamas River, which cut the border between the two Cilicias in ancient times and later separated the Byzantines from

today the population in these areas is sparse, and the few inland settlements can barely earn a living from the soil. Armenian Cilicia was confined to a self-contained geographical unit. The Armenians defended themselves by simply shoring up the openings in a natural barrier to create a continuous semicircular march. While this was the first attempt to secure Cilicia in such a manner,[46] it was a strategy that the Armenians had earlier seen in the area of Lake Van. Their predecessors in the Van region, the Urartians under Rusas I, had constructed a network of garrison forts in the mountains to seal off most of the fertile regions around the shores of the lake.[47]

---

the Arabs. See C. Huart, "Lamas-Su," *EI*, 13; T. Mitford, "Roman Rough Cilicia," *Aufstieg und Niedergang der römischen Welt*, 2.7.2. (1980), 1241; Texier, 726; Strabo 14.5.6; Janke, 5; Lamas (note 3) in the Catalogue.

[46]The Neo-Hittites, Byzantines, and Arabs fortified only select areas in Cilicia.

[47]C. Burney, "Urartian Fortresses and Towns in the Van Region," *AS* 7 (1957), 38. It should be noted that the ring of defenses in Cilicia was by no means impregnable in the late 13th and 14th centuries. See: Ibn Bībī, 140–42.

Part Two

Catalogue

Seventy-Five Medieval Fortifications in Cilicia

name	*dates of study*	*latitude (N)*	*longitude (E)*
Ak	*1973,79*	*37°33′*	*36°22′*
Alafakılar	*1981*	*37°36′*	*35°40′*
Amuda	*1973,81*	*37°14′*	*36°05′*
Anacık	*1973*	*37°26′*	*36°18′*
Anahşa	*1979*	*37°22′*	*34°54′*
Anavarza	*1973,74,79,81*	*37°15′*	*35°54′*
Andıl	*1979*	*37°35′*	*35°48′*
Arslanköy	*1979*	*37°01′*	*34°17′*
Ayas	*1974,81*	*36°46′*	*35°47′*
Azgit	*1973,81*	*37°38′*	*36°25′*
Babaoğlan	*1979*	*36°16′*	*36°12′*
Başnalar	*1981*	*36°49′*	*34°30′*
Bayremker	*1981*	*37°50′*	*35°38′*
Belen Keşlik	*1979*	*36°56′*	*34°31′*
Bodrum	*1973, 81*	*37°09′*	*36°12′*
Bostan	*1981*	*37°39′*	*35°36′*
Bucak	*1973*	*37°28′*	*35°56′*
Çalan	*1981*	*36°36′*	*36°19′*
Çandır	*1979*	*37°03′*	*34°34′*
Çardak	*1974,79*	*37°03′*	*36°24′*
Çem	*1973,79*	*37°33′*	*36°02′*
Dibi	*1979*	*37°35′*	*36°29′*
Evciler	*1979*	*37°06′*	*34°29′*
Fındıklı	*1979*	*37°54′*	*36°26′*
Fındıkpınar	*1981*	*36°54′*	*34°22′*
Geben	*1974,79*	*37°46′*	*36°26′*
Gediği	*1979*	*37°00′*	*34°23′*
Gökvelioğlu	*1973,81*	*36°50′*	*35°37′*
Gösne	*1979*	*36°57′*	*34°32′*
Gülek	*1973,79*	*37°18′*	*34°47′*
Haçtırın	*1974,79*	*37°26′*	*36°22′*
Haruniye	*1973,79*	*37°17′*	*36°29′*
Hasanbeyli	*1981*	*37°07′*	*36°34′*
Hisar	*1979*	*37°01′*	*34°26′*
Hotalan	*1981*	*37°28′*	*35°17′*
Işa	*1981*	*37°29′*	*35°07′*

Kalası	*1979*	*37°36′*	*36°29′*
Karafrenk	*1981*	*37°09′*	*36°35′*
Kız (near Dorak)	*1979*	*37°09′*	*34°58′*
Korykos	*1973,74,81*	*36°28′*	*34°10′*
Kozcağız	*1981*	*37°02′*	*36°35′*
Kütüklu	*1981*	*37°07′*	*35°08′*
Kum	*1974,79*	*37°22′*	*36°16′*
Kuzucubelen	*1981*	*36°48′*	*34°29′*
Lamas	*1974,79*	*36°34′*	*34°15′*
Lampron	*1973,79*	*37°09′*	*34°39′*
Mancılık	*1974,79*	*36°47′*	*36°19′*
Mansurlu	*1981*	*37°56′*	*35°37′*
Maran	*1979*	*37°49′*	*35°51′*
Meydan	*1974,79*	*37°31′*	*35°23′*
Milvan	*1981*	*37°18′*	*34°57′*
Misis	*1973,74,81*	*36°57′*	*35°38′*
Mitisin	*1981*	*36°58′*	*36°26′*
Payas	*1973,81*	*36°45′*	*36°12′*
Pillar of Jonah	*1981*	*36°38′*	*36°13′*
Rifatiye I	*1979*	*37°41′*	*36°19′*
Rifatiye II	*1979*	*37°42′*	*36°18′*
Saimbeyli	*1981*	*37°59′*	*36°05′*
Sarı Çiçek	*1981*	*37°29′*	*35°20′*
Sarı Seki	*1981*	*36°39′*	*36°14′*
Savranda	*1973,79*	*37°07′*	*36°28′*
Silifke	*1973,79*	*36°23′*	*33°54′*
Sinap (near Çandır)	*1979*	*37°00′*	*34°33′*
Sinap (near Lampron)	*1979*	*37°12′*	*34°40′*
Sis	*1973,74,79,81*	*37°27′*	*35°48′*
Tamrut	*1981*	*37°28′*	*35°09′*
Tece	*1981*	*36°43′*	*34°27′*
Toprak	*1973,79,81*	*37°04′*	*36°06′*
Trapesak	*1979*	*36°32′*	*36°22′*
Tumil	*1981*	*36°50′*	*34°40′*
Tumlu	*1974*	*37°09′*	*35°43′*
Vahga	*1974*	*37°50′*	*35°57′*
Yaka	*1981*	*36°53′*	*34°47′*
Yeni Köy	*1979*	*37°33′*	*35°26′*
Yılan	*1973,79,81*	*37°01′*	*35°45′*

Ak

Ak Kalesi[1] guards the east flank of the strategic road that runs from Kadirli to Göksun and points north. This garrison fort crowns the top of a lofty limestone outcrop that is 2 km southeast of the town of Andırın (fig. 4). The locals report that a trail from the east merges with the Kadirli–Göksun road just south of Ak.[2] This trail may be an alternate route to Maraş and gives even more strategic importance to Ak Kalesi. Today the closest source of year-round water to the fort is a well in the small hamlet of Eski Kürtullu. In the spring and early summer a few mountain streams flow around the base of the outcrop. The medieval name of this site and its date of construction are unknown. The easiest approach to the fort is made from the north, near the base of the outcrop at Eski Kürtullu. In ascending the outcrop the path hugs the north face until it turns south and eventually reaches gate C.

The masonry of Ak Kalesi is uncomplicated. On the interior types III and IV masonries predominate (pl. 1a). Chapel E, which is located on the highest points in the north half of the fort, is built largely with type III masonry and quoins for support.[3] On the exterior of the circuit wall rough forms of types V and VI predominate, except for gates A and C where a rusticated type IX is used for the jambs and voussoirs (pl. 1b). The masoned walls and natural rock have been carefully integrated into a spontaneous plan that is so typical of the Armenian forts. Ak does not have a masoned wall to separate the fort into baileys, but the east half is divided from the west by a barrier of rocks and a sharply falling cliff (cf. the south bailey at Meydan).

At the lowest point in the fort is gate A (pl. 2a). Unlike the other gate, A is in an excellent state of preservation. It is flanked by a tower at the east and is surmounted by three high-placed corbels. No doubt the corbels supported a breastwork or machicolation that was manned from the wall walk along the length of the southwest wall. The depressed arch over the exterior of the door is anchored firmly by a large keystone. On the interior of the door the vault is pointed and set much higher than the depressed arch (pl. 1a). There are provisions for a crossbar bolt; two pivot housings with drilled holes once accommodated double doors. On the interior side of the circuit near A and to the west there are numerous joist holes which may indicate that wooden buildings were once attached. East of gate A is a two-level windowless tower (pl. 2a). The lower level consists of a round masoned pit that may have served as a cistern; at the lowest point in the fort, this tower would be sure to collect the maximum amount of runoff. There is no indication today that the pit was covered by a vault or cupola. Above the pit a second level rises until it merges into the wall walk. The interior of this level is rectangular in shape and opened at the northeast. The wall walk follows the inner contours of the rounded tower. A wooden roof for the second level probably rested on the edge of the wall walk. It seems that the northeast wall of the second floor was also made of wood. From this tower the circuit wall ascends to the southeast. There is a small stairway in the wall walk to facilitate traffic uphill. Only fragments of battlements and a few merlons are still visible along the circuit.

Tower B is simply a hollow salient with a few descending steps at the northeast. Minor traces of springing stones indicate that B may once have been covered by a barrel vault. From B to gate C the wall jogs and narrows in places. Gate C, now partially collapsed, stands at the highest point on the outcrop (pl. 1b). Like A, gate C is placed in the lee of a tower that provides flanking fire. Present remains indicate that the design of C is identical to A, including the corbeled pivot housings.

Northeast of C is the large protruding salient D. The two corbels in the side wall of D once supported a ceiling for the interior room. Unlike the southwest wall, the circuit from B to D shows no signs of battlements. Building E, the badly damaged chapel, is distinctively Armenian in its shape and in the alignment of doors and niches.

East of gate C is a series of caves that I did not have time to explore. This fort is definitely Armenian in design and masonry, and, like Azgit, it was a minor garrison fort on an important highway.

[1]The plan of this hitherto unsurveyed site was begun in March 1973 and completed on my second visit in July 1979. The contour lines are placed at intervals of 50 cm.

[2]Because of circumstances, I was unable to hike this trail to the east; the trail does not appear on my map of Cilicia. See Fedden and Thomson, 12, 22, 54; Boase, 149, 153; Alkım, map 3.

[3]Quoins are rare in the Armenian architecture of Cilicia, and when they do appear they are used with a poorer quality masonry. For a description of the chapel see Edwards, "First Report," 165 f.

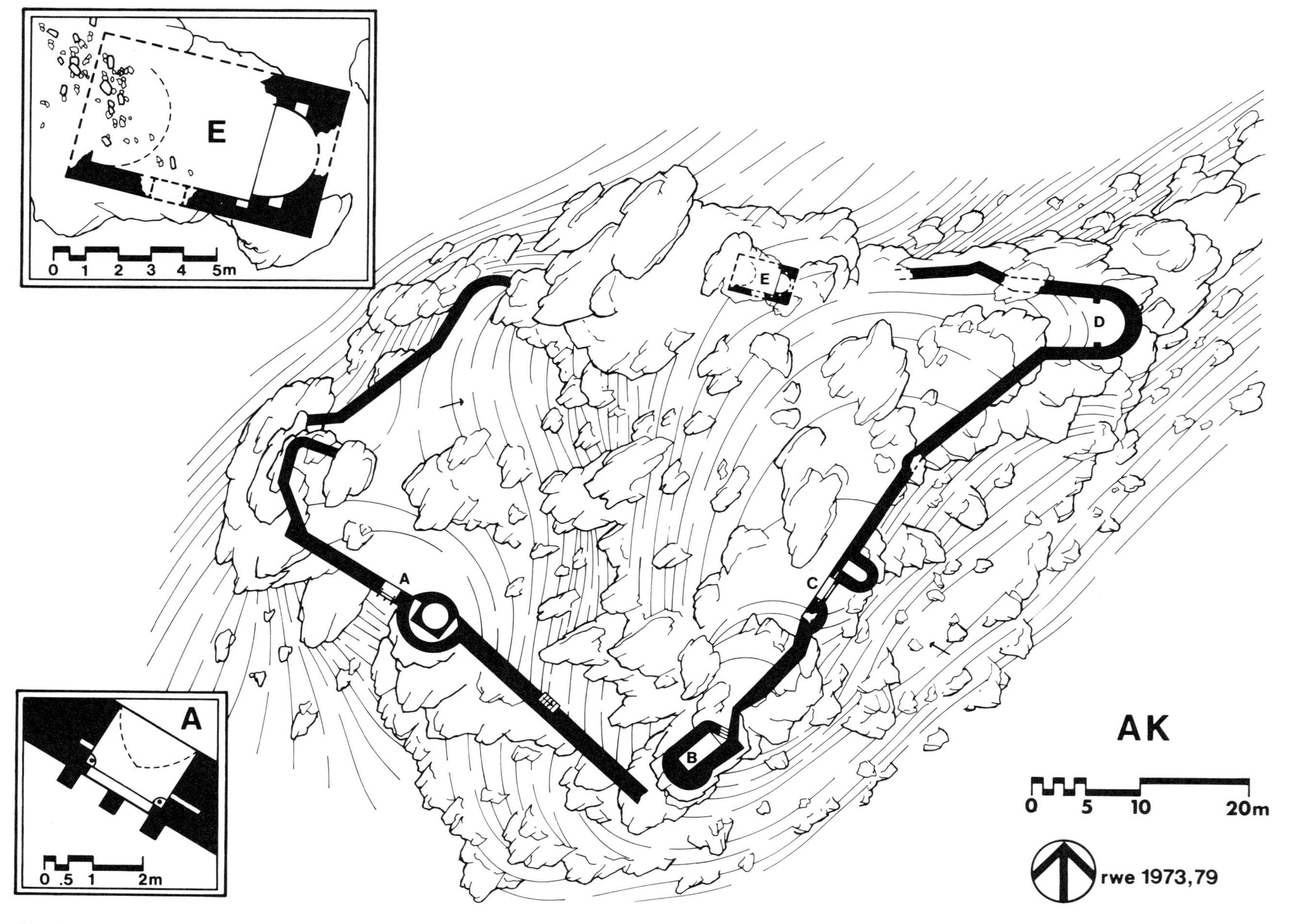

Fig. 4

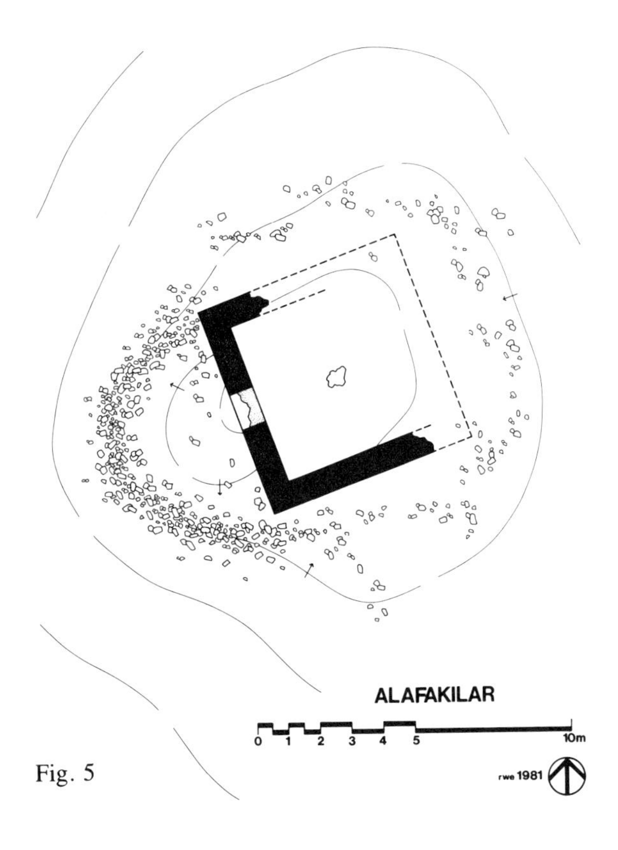

Fig. 5

Alafakılar

Alafakılar Kalesi[1] is a small fortified watch post on the east flank of a very strategic road that connects the northern centers of Mansurlu, Bostan, and Maran with Kozan (Sis). Geographically, this route is situated between two major north-south arteries: one links Kozan to Vahga and the other connects Kozan to the vale of Karsantı. In 1981 this site could be reached with a jeep by driving northwest from Kozan for about seventy minutes. The northwest road from Kozan is comfortable, consisting of alternating patches of asphalt and packed gravel. Fifteen kilometers out of Kozan there is a narrow, winding dirt trail that leads northeast off the main road to Bostan and passes the village of Kızılar. The tiny village of Alafakılar is about 6 km from the main road on this dirt trail.[2] The watch post stands atop a gently sloping hill about 2 km southeast of Alafakılar Köy. There is an easy path of access from the village to the fort. Because the rocky soil immediately around the site has been farmed periodically, this small watch post appears to rise rather abruptly from a barren landscape. Once atop the hill it becomes obvious why the fort was placed here. From Alafakılar Kalesi there is a clear view of Anavarza to the southeast and Andıl Kalesi directly east. To the north and northwest the mountainous trails and road to Mansurlu are visible, and information about hostile traffic moving south or north could be communicated quickly to the garrisons at the east and south. Unfortunately the history and even the medieval name of this site are unknown.

Alafakılar Kalesi consists simply of a square tower (fig. 5). All of the east wall and most of the north wall have collapsed and are covered by debris and shrubs. At the west the center of the wall stands over 2 m in height (pl. 2b). The other extant fragments of wall are lower and decaying rapidly. A tree near the northwest corner is shifting part of the west wall. There are no apparent divisions on the interior, only a rounded hatch-opening in the floor. Since recent digging has exposed what is probably the lintel of a door in the west wall, it is likely that this fort is at least a two-story structure. It is quite possible that this door is actually a porthole, opening the top of a cistern wall. The exposed floor level at the top of the mound is also the ceiling level of the first floor. Because my permission to explore this and other sites did not extend to excavations, I was unable to examine the first level. The wall thicknesses shown on my plan represent only the second level. Because of its size, it seems likely that this watch post held only a few men.

The masonry of Alafakılar Kalesi is highly unusual. Instead of the limestone, which is so ubiquitous in Cilicia, the walls here are constructed with sandstone. I could not determine where the stones were quarried. This sandstone was obviously taken from very symmetrical bedding planes. It cleaves in neat, parallel layers, which explains why the stones show little evidence of being cut. The inner and outer facing stones are anchored by a poured core and thick margins of mortar (pl. 2b). This masonry does not conform to any of the established paradigms for the masonry of medieval Cilicia. The stones are set in very regular courses, but their size varies so much that the average dimensions are impossible to determine. In general, quoins are not used in the corners. The white mortar, which is generously mixed with pebbles and rock fragments, is in sharp contrast to the gray sandstone.

The design and masonry of this fortified watch post are not Armenian in character. Considering that the Byzantines occupied both Andıl and Anavarza, it is quite possible that Alafakılar Kalesi is a Byzantine construction.

[1] The plan for this hitherto unsurveyed site was executed in July 1981. The contour lines are spaced at intervals of 1.5 m.

[2] Alafakılar is not noted on the *Kozan* map, but the neighboring villages of Kızılar and Kaleboğazı are listed. The "Kale" near these villages is Alafakılar Kalesi.

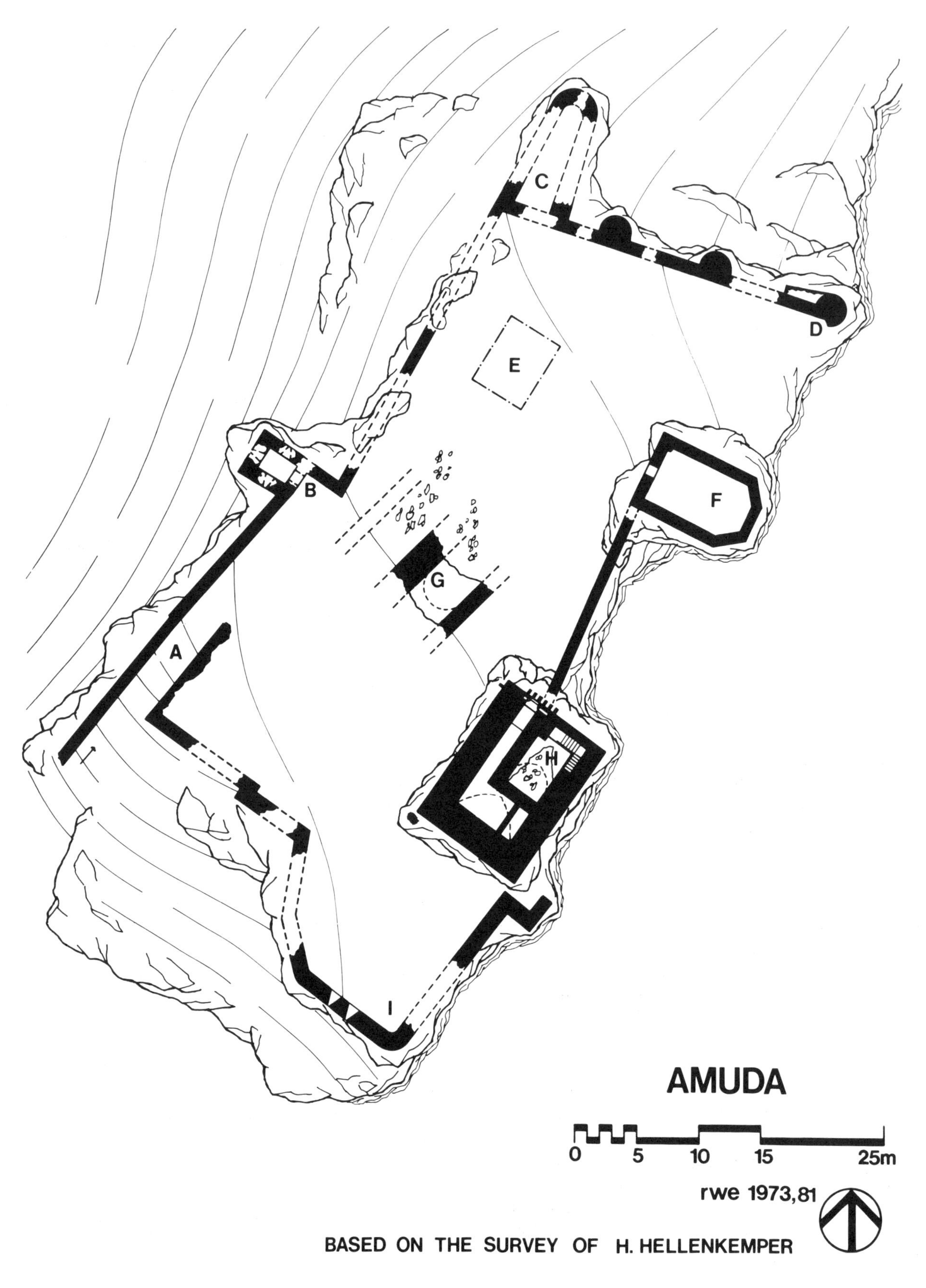
A
B
C
D
E
F
G
H
I
AMUDA
0
5
10
15
25m
rwe 1973,81
BASED ON THE SURVEY OF H. HELLENKEMPER

Fig. 6

Amuda

Atop a finger-like projection of the Gökçe Dağı, Amuda Kalesi[1] has a panoramic view into the eastern lobe of Cilicia Pedias. From this 80-m-high perch of limestone the medieval garrison of Amuda[2] had clear intervisibility with Bucak and Anavarza, as well as Tumlu, Yılan, and Toprak. As with all of the medieval sites in the plain (except Ayas and Karafrenk), Amuda is abandoned. However, the modern town of Hemite has expanded from the north flank of the fortified outcrop, which was the original site of the medieval village, along the banks of the Ceyhan Nehri. The region is lush and fertile, and the river provides fish to the village. In 1212 Wilbrand von Oldenburg mentioned that the fish were especially abundant around the feast of Palm Sunday.[3] Indeed, the presence of a broad, navigable river, which continues its southward journey to the sea, explains the position of the fort. Not only could Amuda Kalesi guard the river and relay messages to neighboring forts, but it commanded the strategic north-south road from the Amanus pass to Kadirli and Kozan (Sis). Also, a second road runs east from the outcrop through a valley of the Gökçe Dağı. This road passes just north (and within view) of Babaoğlan and joins the strategic trail that links Kum, Andırın, Geben, and Göksun. Today Hemite can be reached easily by car from either Osmaniye, Toprak, or Kadirli (via Yenikent).

Amuda Kalesi is frequently mentioned by nineteenth- and twentieth-century travelers. V. Langlois mistakenly associated Wilbrand's Adamodana (The German Amuda) with Tumlu Kalesi, a fortress to the southwest.[4] Other travelers, such as Davis, Gottwald, and J. Thomson, merely mention the location of Amuda Kalesi without discussing its architecture in detail.[5] The latter two, as well as Cahen, accept Alishan's correct identification of Amuda with the modern site of Hemite.[6] The only serious discussion of the architecture of this fort was published in 1976 by H. Hellenkemper.[7] My purpose is merely to relate what little we know of the history of this fort and to give the results of my field survey.

From our available sources it appears that the garrison fort of Amuda plays a minor role in the history of Armenian Cilicia. Sometime between 1146 and 1148 T῾oros II made this site his first conquest after escaping from Constantinople.[8] This implies that the outcrop was fortified prior to the mid-twelfth century, but we know nothing of its history. Amuda appears only in the less reliable edition of Smbat's Coronation List.[9] In 1212 King Levon I gave the fort to the Teutonic Knights during the visit of their master, Hermann von Salza.[10] Over ten years later Bohemond IV of Antioch was tricked by Baron Constantine into traveling to Amuda in search of his son Philip, whom he believed to be imprisoned there.[11] Bohemond began his march into Cilicia unaware that his son, who had briefly become king of Armenian Cilicia, had been assassinated. The next major event at Amuda occurs in 1266 when the Mamluk forces under Baybars captured the region that Amuda commanded.[12] It is reported that the 2,200 refugees who fled into the fort to escape the Egyptian army were captured.[13] Many women and children were taken to Cairo as slaves, while the men were put to death. Thereafter, it is unknown whether a Christian garrison occupied the site; in the last years of the thirteenth century it was quietly retaken by the Mamluks.[14] The date that the Teutonic Knights abandoned the fort is unknown.[15]

Today, as in medieval times, the easiest line of access from the base of the outcrop to the summit of the fort is from the southwest (fig. 6). The entire east flank is a relatively sheer cliff with some rather large caves at the base. Caves are frequently associated with the limestone formations in Cilicia. To the north and west the approach to the summit is less severe than at the east. There is no evidence of a gatehouse or door at the southwest, but merely a ramp at point A. Only the foundation of the retaining wall for the ramp is visible today.

The circuit walls and structures within the fortress consist of four distinct types of masonry. Type VI appears only as the exterior facing of keep H (pl. 3a); the interior of H is a large well-cut type IV (pl. 3b), where occasionally rough polygonal blocks are used. The exterior facing of the lower west circuit and building F consists of type IX (pl. 5a). The upper level of the west circuit and almost all of the exterior facing of the north and south circuit walls are built of type IV (pl. 4a). The interior facing of the circuit wall consists of a combination of types III and IV, with the latter predominating. Because of the high ground levels in the fort, the interior facing, which corresponds to the type IX, cannot be determined. All of the facing stones in the vaulted room G are type IV. In the opinion of this writer, the use of types IX, IV, and VI as an exterior facing represents three major periods of construction.[16]

Unfortunately most of the circuit wall southeast of A has collapsed. However, the southeast corner of the south circuit (at point I) has survived to a substantial height (approximately 2.8 m) and shows two distinct levels. In the first level are two simple, square-headed, embrasured loopholes. Just above these loopholes the wall becomes substantially thinner, creating a shelf. The square joist holes above the shelf indicate that some sort of wooden covering protected the first level. The second-level walls (not shown on the plan) are so thin that they probably were no more than a line of battlements. Considering the relatively easy line of access into the fort from the south, all of the original south circuit was probably lined with similar embrasures. The south circuit wall does not join keep H because the latter is elevated on a rock platform.

The west circuit is carefully fitted over the undulating rock mass. The south end of the west circuit, which is part of the northwest wall of the ramp, consists entirely of a well-cut type IV and does not have a lower level of smooth ashlar like the section southwest of tower B (pls. 4a, 4b). At the point where the ashlar ends, the exclusively type IV wall shifts direction slightly to the south, leaving the end of the smooth ashlar wall to protrude. Since the type IV surmounts the ashlar in an irregular fashion throughout tower B, it must represent a later period of construction. All of the south circuit has an exterior facing of type IV. Thus the ramp was not part of the first building period which employed the type IX.[17] The location of the south circuit from the first building period cannot be determined.

The plan for the present west circuit from the protruding end of the ashlar wall to the north side of the square bastion B was formed in the first period of construction. The ashlar in the lowest courses has fairly narrow interstices, but the upper courses of ashlar near the junctions with the type IV have very broad margins that are filled with numerous rock chips and mortar like type IV masonry. Obviously many of the upper-level ashlar blocks were recycled in the second building period. It was also in the second period that the upper half of the tower with its door and three windows was formed (pl. 4b). Because of extensive damage the shape of the windows and the location of the door cannot be determined precisely. The relatively flat top of the tower may indicate that it was not covered by a vault. The jog in the circuit wall immediately east of tower B, as well as the north half of the west circuit, is from the second phase of construction.

Today almost nothing is left of the rounded salient C. Hellenkemper reports that in 1939 tower C stood to 5 or 6 m in height, had a single embrasured window at the north, and was from a different building period than the rest of the north circuit.[18] None of this information can be verified today. A few facing stones from the foundation show that the masonry of tower C was probably the same as the rest of the north circuit, a small crude type IV. The south wall of C probably had a door to give access to its hollow interior. The only other towers along the shattered north circuit are three small semicircular buttresses. There is a small revetment wall below the easternmost of these solid buttresses.

Outside of keep H the only evidence of a cistern in the fort is the vaulted, subterranean chamber E, which is positioned about 8 m south of tower C.

Today the hexagonal room F is almost completely demolished (pl. 5a). A section of its exterior facing at the northwest stands to just over 2 m in height, but on the interior only the upper two courses (just over 90 cm in height) are visible (pl. 5b). The disparity in the height of the inner and outer facing is due to debris inside of F and to the fact that F is positioned atop a rock and the exterior facing fills in many natural clefts. Most of the exterior facing is type IX; however, its presence here is as a recycled material. Unlike the lower level of tower B in which the smooth ashlar shows only the smallest traces of mortar in its regular, thin margins, the ashlar in F is badly chipped and always has broad stripes of mortar or pebbles in the interstices. F is not contemporary with the first building period (that is, the type IX of tower B and its wall to the south). Because so little of room F survives, its function is impossible to determine. There is no reason to assume that F is a chapel simply because it has a triangular east end. This is not especially suited for an apse, nor is there any division between a potential apse and nave.[19] Hellenkemper is probably correct in assuming that there was a single entrance at the west because of the shape of the surrounding rock mass. However, his hypothesis that the room was covered by a groined vault cannot be supported by any extant remains.[20] A barrel vault with a faceted east end is equally possible as an alternative covering. In the northwest corner of F there is a small trefoil-shaped base for what may have been a cluster of three attached columns (pl. 5b). This plinth is about 60 cm in height.[21] It may have a smaller counterpart at the northeast. There is no evidence of plinths on the south side; they were either removed or never constructed.

In the middle of the fort are the fragmentary remains of the vaulted hall G. The specific function of G is unknown.

With respect to its exterior facing stones, keep H is unlike any other structure in the fort (pl. 3a).[22] There is no evidence of the type IX facing from the first period of construction, not even as a recycled stone. The only time the type IV appears on the exterior is in areas of repair(?) above the corbels at the north. Thus it is quite likely that the type VI of the keep belong to a third and quite separate period of construction. Like Hellenkemper, I believe that the keep is very European in design and was probably built during the occupation of the fort by the Teutonic Knights.[23]

Although this keep does have four floors, as J. Thomson mentioned,[24] only three are apparent on the exterior. The single entrance into the keep at the north is actually the door to the second level. The sill of the door is about 1.7 m above a natural rock step. The rock step is very narrow, and its location on the relatively flat rock pedestal under keep H determined the location of the door. Below the floor of the second level is one (and perhaps two) cistern(s) (pl. 3b). Part of the vault over the cistern located in the northeast corner of the keep has collapsed, exposing the crude masonry of its slightly pointed cover. The cistern was partially adapted to a natural rock cleft, and its masoned walls are not completely stuccoed. This cistern was fed by drainage ducts in the thickness of the wall. Another cistern may be hidden under the southwest corner of the second level.

The most interesting features of the second level of H are the entrance at the northeast and the staircase at the northwest. The entrance portal has jambs and a monolithic lintel. A crossbar bolt secured a double wooden door. Two circular pivot housings anchored the door in position. Since the vault of the basement (or first) level cistern in the northeast corner is slightly higher than the floor of the second level, a narrow corridor was created behind the door. It appears that an approaching party walked over the vault of the cistern from its south end to reach the bent staircase in the thickness of the northeast corner (pl. 3b). This staircase leads to the third level. A large, slightly pointed vault covers the entire second level.

All but the north wall of the third level (not shown on the plan) has collapsed. This north wall was opened by two windows. The exterior frames of both windows are now shattered, but the interior sides appear to be roundheaded.[25] Above the windows is a series of joist holes which indicates that a wooden roof covered a third level. Just over the two windows on the exterior are six corbels, which supported some sort of protective brattice. There is a fragment of another corbel at the north end of the west wall (pl. 3a). It is likely that corbels were also located along the west and south walls since these sides could also be attacked from inside the fort. The east side of the north wall and probably all of the east wall did not have corbels. The sheer cliffs on the east flank would make an enemy assault impossible. The fourth level was an open terrace that was surrounded by battlements. The two surviving merlons at the north appear to have a pyramidal top. There was also a small north portal at this terrace level to give access to the corbeled brattice. This small door was roundheaded.

What is surprising is the paucity of evidence for the reuse of type IX masonry from the first building period in the subsequent reconstructions at Amuda Kalesi. This may indicate that the initial construction on the outcrop was confined to only a small part of the summit and the ruined segments yielded little masonry. I found no evidence of type IX masonry in any of the buildings in the village below. It is evident that a very systematic plan was imposed on this irregular outcrop; there was no attempt in the periods of reconstruction to copy the plans of Armenian military architecture (for example, multiple baileys, bent entrance, etc.).

[1] The contour lines on the plan are spaced at intervals of 75 cm.

[2] Amuda, Amuta, and Amoutay are the designations most commonly used in Armenian, while ʿAmûdhâ is found in Syrian texts. The Franks also use the variant Hamuda. The German friar Wilbrand von Oldenburg refers to the village and fort as Adamodana. After 1266 it is known as Qalʿat al-ʿAmūdayn ("fortress of the two columns") in the Arab texts. Today most locals refer to it as Hemite Kalesi, and some even call the fort "Amuda." Hemite ultimately derives from the early Arab toponym Ḥamaṭīye.

On most maps the site is referred to as Hemite or Amuda; see *Adana (2)* and *Kozan*. The site appears on the maps of certain travelogues; see: Davis; Favre and Mandrot; Schaffer, 44.

[3] Wilbrand von Oldenburg, 20.

[4] Langlois, *Voyage*, 445; Youngs, 113. This error was also made by the editor of *RHC, DocArm*, I, 810. Honigmann (map III) mistakenly associates Toprak with Amuda (Ḥamaṭīye).

[5] Davis, 127, 134; Gottwald, "Til," 92 note 1; Fedden and Thomson, 36 and 44; *Handbook*, 310 f; Boase, 150 f.

[6] Alishan, *Sissouan*, 225–29; Cahen, 148.

[7] Hellenkemper, 123–31.

[8] This incident is mentioned by Bar Hebraeus (275), Michael the Syrian (341), and Smbat (618); cf. Samuel of Ani (453).

[9] Appendix 3, note 21; Hellenkemper, 129 note 1; Smbat, 637; Smbat, G. Dédéyan, 76 note 36; Christomanos, 152; Alishan, *Léon*, 174 note 1.

[10] Hellenkemper, 129; Alishan, *Sissouan*, 225 ff. A papal bull that discusses this transfer is dated 27 February 1213. When Wilbrand von Oldenburg (20) made his visit in March 1212, the fort was already in Crusader hands. See also: Forstreuter, 59–61, 234 f; Mikaeljan, 160; Cahen, 618; Langlois, *Cartulaire*, 117–20; J. Riley-Smith in Boase, 113; E. Strehlke, ed. *Tabulae ordinis Theutonici ex tabularii regii Berolinensis codice potissimum* (Berlin, 1869; rpr. Toronto, 1975), 37–40, 126; Röhricht, 229 f.

[11] Vahram of Edessa, 516 note 1; Cahen, 635; Bar Hebraeus, 381.

[12] E. Honigmann, "Miṣṣīṣ," *EI*, 525; Canard, "Le royaume," 231.

[13] Alishan, *Sissouan*, 227; Ibn al-Furāt, 99, 217.

[14] Al-Maqrīzī, II.2, 61 note 18.

[15] Hellenkemper, 130. Cf. Yovhannēsean, 193–97. It is likely that the Crusaders were still in residence in 1236; see J. Riley-Smith in Boase, 115.

[16] There is no reliable way to date these three periods. It appears that type IX was used in the first period of construction. This

masonry is never used by the Armenians as an exterior facing, but it is used by the Arabs at Haruniye. We know that the region around Amuda was heavily fortified by the Arabs in the late 8th century; see Le Strange, *Caliphate*, 127 ff. Since the Byzantines are suspected of using types III and IV as an exterior facing in Cilicia (e.g., Gökvelioğlu and Anavarza), its presence here may be due to a period of Byzantine occupation.

[17] I do not share Hellenkemper's view (131) that "Die Steinmetzarbeit [of the ashlar in tower B] ist ähnlich den Werkstücken am Donjon von Anavarza, das Material stammt vermutlich aus dem gleichen Bruch." The ashlar in the keep at Anavarza is three times the size of the type IX at Amuda. The ashlar in the former is a very unique type; see Anavarza in the Catalogue.

[18] Hellenkemper, 125, 131. Unfortunately Hellenkemper does not discuss the source of this information.

[19] Hellenkemper, 126. Excavations may determine someday the actual function of chamber F.

[20] Ibid.

[21] Ibid., pl. 28b.

[22] I found no evidence that the keep had an earlier period of construction. In what survives of the keep, the type VI masonry is fairly uniform. We have no way of knowing if Armenian masons aided the Crusaders in building the keep. The bossed centers on the quoins may not be a link to the Armenians since the latter never use quoins in military architecture. Even in Armenian ecclesiastical architecture quoins are rare; see Edwards, "First Report," 166.

[23] Hellenkemper, 130. It is interesting that type VI facing stone predominates in the forts at Kum and Toprak, where Crusader construction is also suspected.

[24] Fedden and Thomson, 44; cf. Hellenkemper, 126 note 1.

[25] Hellenkemper, pl. 27a.

Anacık

Anacık Kalesi[1] is a large fortified estate house that stands just south of the strategic road from Kadirli to Andırın on the fringe of northeast Cilicia Pedias.[2] This small fort commands a broad Highland valley about 12 km north of Kum Kalesi. The dozens of small streams that cross the valley flow south and eventually join the Ceyhan Nehri. The structure at Anacık has no recorded history. The dedicatory inscription, once in a niche over the south door, is missing (pl. 288a).

The exterior facing stones of Anacık Kalesi are a very uniform type V masonry, while the interior facing is often a mixture of types III and IV.

The site is quite similar in design to two other fortified country estates, Sinap near Lampron and Sinap near Çandır. Like the latter two, Anacık is rectangular with small, round corner towers (pl. 287a). However, Anacık is slightly longer and has an additional tower in the center of the north wall and two very thin salients flanking the door at the south (pl. 288a). Also, Anacık has embrasured arrow slits in the south and west walls of the lower level as well as in the upper level. The first level chamber has no dividing walls on the interior but is covered by an undamaged, slightly pointed vault. The only apparent damage to the first level is in the west and east walls where three low-level punctures now give access to the chamber. Because of their low position and size, none of the three breaches could have been windows or doors.

Only a few fragments of the upper-level walls are standing today. A tall, roundheaded window is still present in the west wall of the second level. A cantilevered stairway in the west wall gives access to the story above. Considering the thickness of the upper-level walls, it is likely that this story was once vaulted. Anacık, like the two Sinaps, is probably an Armenian construction.

[1] In fall 1973 I visited this unsurveyed site briefly for the first and last time. Circumstances did not allow me to execute a plan.

[2] Anacık Kalesi and its neighboring village appear on the *Gaziantep* map. See also Alkım, map 3.

Anahşa

As the Kayseri–Tarsus road descends south from the heights of the Taurus Mountains to Pozantı,[1] the huge, almost kidney-shaped Anahşa Dağı[2] blocks a direct route to the Mediterranean. The major road from Pozantı veers to the west through the Cilician Gates, passing Gülek Kalesi and eventually arriving in the Cilician plain. An alternate route from Pozantı follows the course of the Çakıt Suyu along the east flank of the mountain to Milvan, Karaisalı, and Adana. The old Baghdad railway is tunneled into the cliffs above this road. On a conspicuous spur of Anahşa Dağı at an altitude of 1,940 m is the guardian of the northern end of this highway.[3] This garrison fort, which bears the same name as the mountain, has no recorded history or inscriptions, and consequently its original medieval name is unknown. The historical names assigned to this site by Ramsay, Harper, and others are purely speculative.[4] The only architectural study of this fort was undertaken by Heffening in 1918.[5]

Anahşa Kalesi can best be reached by driving south from Pozantı (on the route to Tarsus) for a few kilometers and taking the turnoff to the left at Soğukpınar. At this point a forestry road, passable only by lumber trucks and four-wheeled vehicles, climbs the

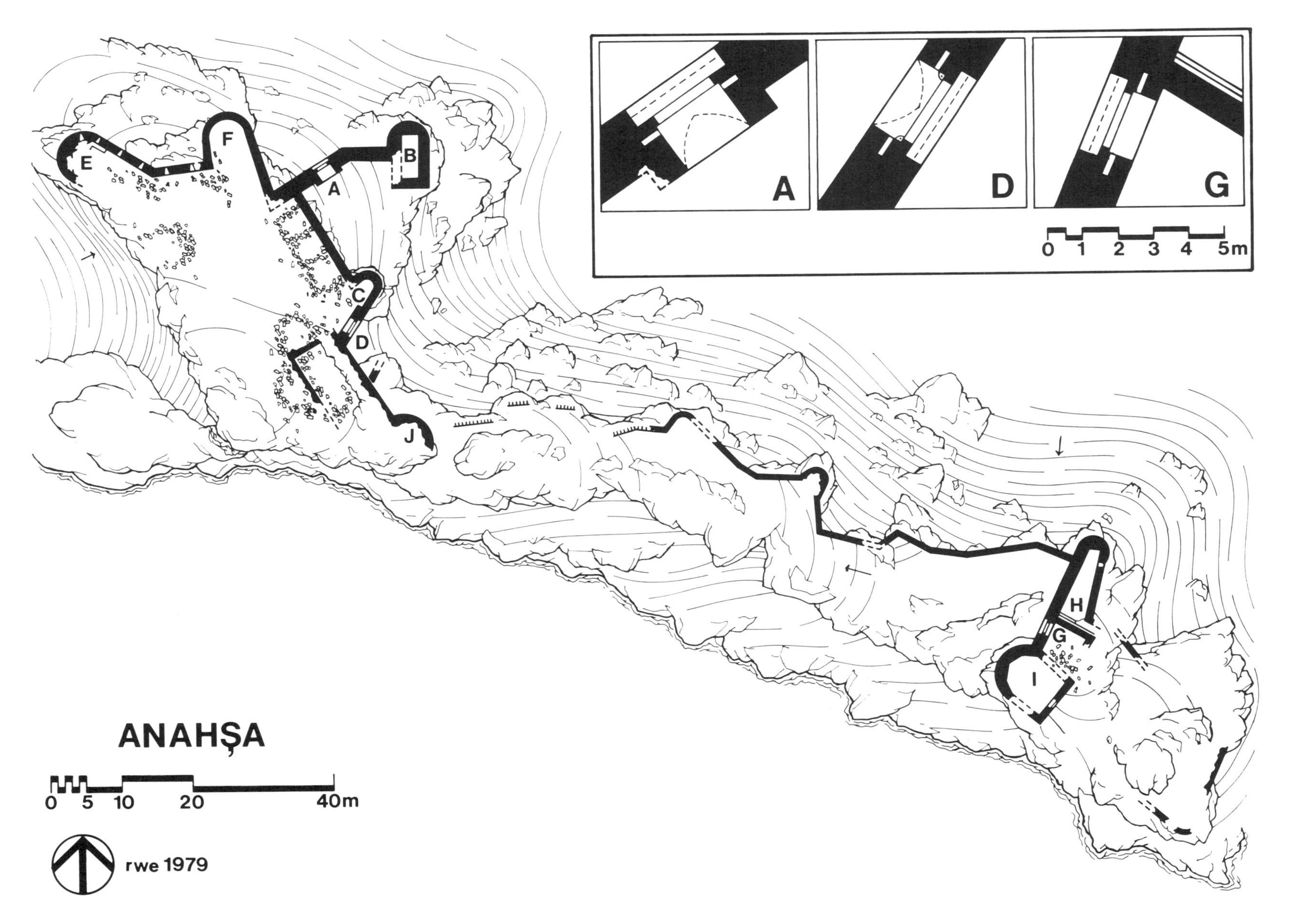

Fig. 7

mountain in a northeasterly direction for about 10 km before reaching the fort. From the right flank of the road the enclosure is about a 15-minute hike. The trail to the site, which is probably medieval in origin, leads to the northwest tip of the fort (pl. 6a). At the east and the south the outcrop is defined by steep cliffs that abruptly descend over 500 m (fig. 7). From this fort there is a commanding view of the Çakıt valley at the south. The vale of Pozantı is not entirely visible at the north because of towering outcrops of rock that hem off the gorge.

The extant remains of Anahşa Kalesi indicate that the fort is the result of at least three periods of construction, with the majority of the buildings belonging to the Armenians. Structures A through D and parts of the circuit wall west of H have type V as an exterior facing stone (pl. 7b). In some locations type VI is also visible on the exterior (for example, towers E and I). Generally the interior facing is either types III or IV and occasionally a combination of both. Type IX is used to construct gates A, D, and G (pl. 7a); all three entrances are constructed in the typical Armenian pattern.[6] In a few instances the type IX has a boss and drafted margins on the exterior face. The only major anomaly in the western half of the complex is a small parapet that surmounts tower E (pl. 6a). This wall consists of something similar to type III masonry and certainly represents a later addition. Two small windows (not shown on the plan), which are roundheaded on the interior side and squareheaded on the exterior, pierce this parapet; these are quite unlike the embrasured arrow slits in the main wall of the tower. The parapet and its two windows may have been built by Ibrahim Paşa in the 1830s to accommodate small cannons.[7]

In the eastern half of the fort the east wall of Salient H and a section of the east wall of tower I have an exterior facing of type IV masonry (pl. 8b). The exterior facing in the west wall of H and the entire west facade of tower I are built exclusively with types V and VI masonries. In the west wall of H the type VI is confined to the northern half and the type V to the southern half. The northern half, which resembles the west facade of tower I, may represent a second period of Armenian construction. This conclusion is supported by the presence of two different types of poured cores in the west wall of H. The type IV present in H, I, and sections of the circuit west of H may be the remnants of an original Byzantine fort which was built on a smaller scale. All masonry at Anahşa Kalesi is limestone, which was quarried at the site. Many sherds of glazed pottery lie among the ruins.

At the extreme west, tower E rises from the edge of the sheer cliff that forms the natural barrier along the southern perimeter of the fort (pl. 6a). Six embrasured loopholes dot the north wall of tower E up to the point of its junction with tower F. Although the interior sides of the windows are now shattered, there is evidence that they were more narrowly splayed than usual and constructed with type III masonry throughout (normally ashlar would be used to build the interior frame).

Tower F, to the east of E, forms the center of the three adjoining bastions that together defend the vulnerable northwest slope like a thrusting trident. Between towers F and B is the first entrance into the fortress, the tripartite gate A (pl. 7a). This door is constructed with a narrow slot machicolation (pl. 6b). This machicolation was manned from a thin parapet atop the door. Despite the assertions of Heffening,[8] this opening could not have accommodated a portcullis because any falling door would not be framed by projective jambs on the outside. There is no indication of wear on the sides of the machicolation nor any evidence of a slot or receptacle for a portcullis in the sill of the door. A depressed arch of seven voussoirs rests on the jambs of A; the voussoirs have joggled joints and are anchored into position by a single keystone. Behind the jambs are sockets for a crossbar bolt. On the interior of the door the pointed vault is made with relatively crude stones. This is quite unlike the pointed arch on the exterior which makes up the machicolation. Once through gate A the advancing party finds itself in a small court framed by the south wall of tower B and salient C (pl. 7b). At the upper level of salient C three large blocks are corbeled from the tower to the adjoining north wall. Their function is unknown.

A narrow pathway leads around tower C to gate D (pl. 8a). Gate D is designed like gate A, but it is not as well preserved. The machicolation is badly damaged, and the outer (or east) arch has lost its adjoining walls. Unlike gate A the twin keystones of the outer arch are oversized. The arch over the jambs is identical to the one in gate A except that in D pivot housings for a double door abut on the interior side. On the interior side of gate D, the vault, which rises to a slightly depressed point, is made of ashlar masonry. Gate D gives access to the west bailey, which is defined by towers E, F, C, and J. This area has the remains of many collapsed buildings. From tower J the circuit continues to the east; immediately to the east of J these walls have collapsed, but their scarped trenches are still visible. The area between J and complex G through I constitutes a rather rocky central bailey, which has no visible remains of stone buildings. Here, as at all Armenian forts, the walls carefully follow the sinuosities of the rock; the topography alone determines the shape of the baileys.

Gate G, which gives access to the small east bailey, has collapsed (pl. 8b). Only the jambs and part of the springing on the south side are present. The remains indicate that the apex of the interior vault stood 2.4 m from the current ground level and that this door was designed like gates A and D with a slot machicolation and a crossbar bolt. A few thin metal bars have been inserted between the stones of the jambs of gate G; their function is unknown. The only apparent entrance to room H is the gap in its southeast corner. On the interior side of the south wall of H are two descending podia. There is a series of joist holes on the interior of the east and west walls of H; these holes are about 1.3 m from the present ground level. The joist holes supported the floor of the second level. Whether the upper floor of H had a vault of stone cannot be determined. The only other distinctive feature in H is the broadly splayed window in the east wall. The window is covered by twin lintels. Tower I at the south has a polygonal interior and a single door at the southeast. There is no evidence of stairs outside the door.

Considering the expansion and almost total reconstruction of this Byzantine site by the Armenians, we can assume that Anahşa played a vital role in defending Cilicia from Turkish raids.

[1]The plan for Anahşa was completed during a single visit in August 1979. The interval of distance between the contour lines is 50 cm.

[2]This mountain is also referred to as Tekir Dağı.

[3]Anahşa Kalesi is southeast of Pozantı and northwest of Karapınar. The location given for this fort on the *Ulukişla* map is inaccurate; it should be located farther to the southeast. This site also appears on the maps of the *Cilician Gates* and *Cilicie*. Cf. *Handbook*, 275 f.

[4]W. Ramsay, "Cilicia, Tarsus and the Great Taurus Pass," *Geographical Journal* 22 (1903), 383 f; R. P. Harper, "Podandus and the Via Tauri," *AS* 20 (1970), 150; Kotschy, 242 ff, 334; Alishan, *Sissouan*, 139; Langlois, *Voyage*, 166, 377–79; Heffening, 188 f; Schaffer, 81; Boase, 151. See also: Hild, 57; Hild and Restle, 185. Hild's association of this site with the Byzantine Gypsarion is quite possible, but the evidence is circumstantial at best. Cf. Ēpʿrikean, I, 167 f; Taeschner, I, 138 f.

[5]Heffening, 179–89. Hellenkemper (224 f) did not visit this site but merely summarized the account of Heffening. See also Yovhannēsean, 103–6.

[6]Hellenkemper's assertion (225) that the design and masonry of the gates permit the dating of the fort to the first half of the 13th century is without support.

[7]M. Canard, "Cilicia," *EI*², 38. Ibrahim Paşa is known to have fortified the area near Pozantı.

[8]Heffening, 185.

Anavarza

On a large majestic outcrop of limestone in the center of Cilicia Pedias is the castle known as Anavarza.[1] From an altitude of 200 m this most strategic site has clear intervisibility with Toprak, Bodrum, Amuda, Bucak, Sis, Tumlu, and Yılan. By its proximity it commands all the north-south roads through the eastern lobe of the plain. The Sombaz Çay flows past the east flank of Anavarza before joining with the Ceyhan River. The entire west flank of the outcrop consists of almost vertical cliffs, while the broad south end and east flank slope more gently to the alluvial floor of the plain (fig. 8). The circuit walls of the late antique city of Anazarbus abut on the west flank; the entire outcrop is over 4.5 km in length and once served as the acropolis for the city. The modern village of Anavarza,[2] which is located south of the old circuit walls, can easily be reached by driving along the paved road from Ceyhan to Kadirli and taking the designated turnoff.

Much of the history of late antique and medieval Anavarza has survived.[3] The classical history of the city is less detailed, but we do know that Anazarbus prospered during the reign of Vespasian. Through the third century it remained the center for viticulture and flax; the city even called itself a "metropolis." This title took on a formal character in the fifth century during the reign of Theodosius when Anazarbus became the chief city of a newly formed Cilicia Secunda. During the sixth century the site was renamed Justinopolis and later Justinianopolis, after these emperors financed repairs of the city following successive earthquakes.[4] In the second half of the seventh century[5] Arab invasions devastated the town and left it abandoned until it was resettled en masse by the Abbasids. The first known attempt to refortify the site occurred under Hārūn ar-Rašīd in 796. In 827 the Moslem colony was greatly supplemented by a resettlement of Egyptians in the area. The first large-scale Byzantine raids on Anazarbus occurred in 804, 806, 835, and 855. The last raid was so successful that the Caliph al-Mutawakkil undertook reconstruction at this site before 861. Near a tower in the city circuit are the fragments of a Cufic inscription that bear his name.[6] In the tenth century the city still had a large population and was of such strategic importance as to warrant the expenditure of three million dirhams by the Hamdānid Sayf ad-Dawlah on its fortification.

The especially brutal campaign of Nicephorus Phocas in 962 resulted in the surrender of the city, the death of many of its inhabitants, and the felling of over fifty thousand palm trees.[7] Repairs were carried out on

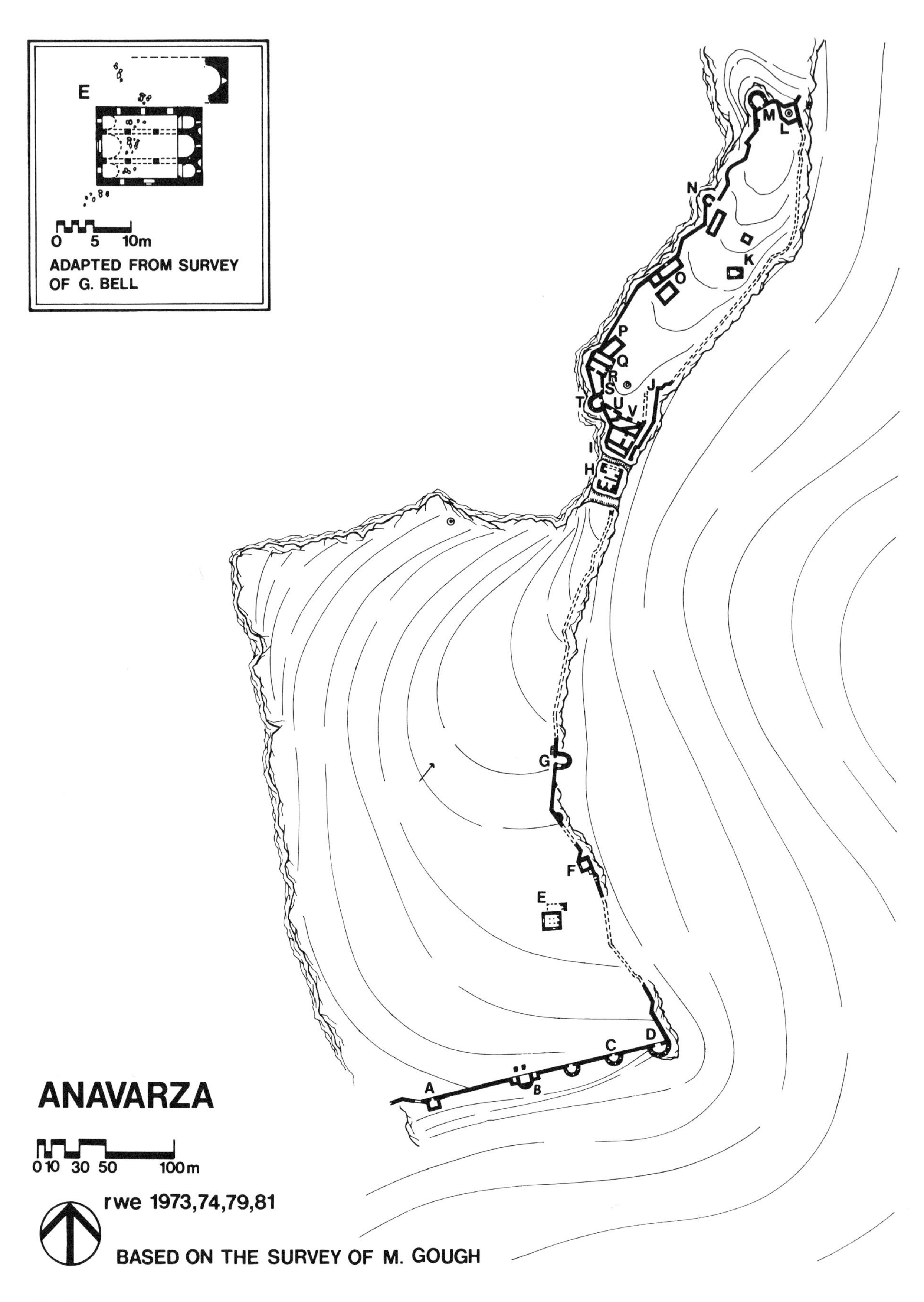
E
0 5 10m
ADAPTED FROM SURVEY
OF G. BELL
M
L
N
K
O
P
Q
R
S
T
U
V
J
I
H
G
F
E
D
C
A
B
ANAVARZA
0 10 30 50 100m
rwe 1973,74,79,81
BASED ON THE SURVEY OF M. GOUGH

Fig. 8

the defenses, and a Byzantine strategos with his garrison resided in only the fortress of Anavarza through most of the eleventh century.[8] In late 1097 or early 1098 it was captured by the armies of the first Crusade and was later incorporated into Bohemond's Principality of Antioch. The Franks allegedly recognized the suzerainty of Alexius Comnenus over Cilicia in 1108.[9] By the year 1111 the fortress was captured by the Rubenid Baron T^coros I. Vahram of Edessa seems to imply that T^coros undertook extensive construction at Anavarza and its environs.[10] No doubt rebuilding had to take place after the devastating earthquake in 1114. Anavarza remained the center of the far-flung Rubenid barony under Levon I until 1137, when Emperor John II succeeded in capturing the outcrop after a siege of thirty-seven days.[11] Baron Levon I and his immediate family were transported in chains to the Byzantine capital. As part of his booty, the emperor brought to Constantinople the icon of the Virgin, which earlier T^coros I himself had plundered from a Byzantine site in Cappadocia and installed in his newly built church at Anavarza. T^coros II, Levon's son, managed to escape from Constantinople and on his return to Cilicia recaptured Anavarza (ca. 1148). Emperor Manuel I occupied Anavarza in 1158/59 only to return this site as well as the rest of Cilicia Pedias in 1162 to T^coros. The latter offered submission to the emperor and received in turn the title of sebastos. During the reign of Baron Levon II (later to become King Levon I) the only evidence of Armenian construction on the outcrop is the repair of the donjon in 1187/88. Within a few years the administrative center was moved north from Anavarza to Sis. The first of the Mamluk raids, which were to have a disastrous effect on Anavarza, occurred in the 1270s.[12] The Armenians were permanently removed from the outcrop in 1374.[13] Through most of the fifteenth century it was maintained as a Mamluk garrison.

During the last century and a half many visitors have commented on the site, but only M. Gough has succeeded in publishing a thoroughly professional account.[14] His narrative places much more emphasis on the Roman and late antique city at the base of the outcrop than on the medieval castle. It is my intent here to supplement his architectural description. What is still required is a full-scale archeological survey of the entire outcrop.

Anavarza Kalesi can be divided into three baileys. The north bailey, which I did not adequately explore, is less distinct. It consists of a single wall and occasional towers that run for a considerable distance along the east flank of the outcrop. Like the south bailey, the western limit of the north enclosure is formed by almost vertical cliffs.

The south wall of the south bailey is an awesome barrier (pl. 287b). It consists of four horseshoe-shaped bastions[15] at the east (B through D) and a single square tower (A) at the west. The five periods of construction in the south bailey have five distinct types of masonry. The only certain remnant of the pre-Arab, Byzantine construction is the collapsed arch of the original entrance which is directly north of tower B (pl. 10a).[16] When the west wall was rebuilt during the Armenian period this gate was closed off and may have been incorporated into some sort of auxiliary building. The east flank of this gate has suffered extensive damage, but the west half clearly shows that it was built with an *opus listatum*. This type of masonry, which appears nowhere else in the castle, consists of four layers of brick tiles sandwiched between a single course of roughly cut ashlar. Occasionally the courses of ashlar extrude into the space of the bricks. The beds of mortar separating the brick tiles are often uneven; the mortar itself is mixed with an unusually large amount of rock fragments. Excavations may reveal more about the plan of this gate.

Type IV masonry was used extensively as an exterior facing in the south bailey. Except for some Armenian additions (discussed below), most of the east circuit has this Byzantine masonry.[17] Fortunately many fragments of this type IV have survived. Since the Armenian masonry overlaps this relatively crude stone, we can assume that the latter dates either from the sixth century (repairs carried out by Justin I and Justinian after a series of earthquakes) or, more likely, from the prolonged period of occupation after the tenth-century conquest by Nicephorus Phocas.[18]

Another major period of construction encompasses tower A, the two square salients flanking B, and the circuit between A and B (pl. 9b). This masonry consists of extremely smooth recycled blocks of ashlar that have been placed in somewhat irregular courses. The foundation of the circuit east of tower A and the lower 30 percent of the square bastions flanking B have relatively small crude stones that form a socle. I believe that this very prominent construction occurred during the Arab occupation. The city circuit wall below has a similar recycled ashlar, square towers, and a dedicatory inscription in Cufic.[19] When the Arabs built their wall they extended slightly to the south the original line of the south defenses. It appears that B was still the entrance since the Arabs constructed two closely spaced bastions as if to flank a gate. The Armenians later built a rounded front to join the two square bastions. All traces of the Arab circuit east of B have disappeared.

The fourth and most significant period of construction in the south bailey is marked by the consistent use of the rusticated type V and VII masonries. It appears as an exterior facing in the areas from towers B through D (pl. 288b) and in the east circuit north of D at points F and G (pl. 10b). Type V is frequently used as the interior facing; occasionally type IV is employed. As I discussed above (Part I.6), it appears that Tᶜoros I is responsible for the Armenian construction in the south bailey (ca. 1111–29).

The fifth and final period of construction in the south bailey involved repairs to the south circuit. Principally, these repairs occurred with a type III masonry in the area west of tower A (pl. 9a). Occasionally, large ashlar blocks are mixed with the more crude stones. Immediately west of tower A there is a gaping hole that once served as a postern. What also characterizes this period is the use of thin square merlons. This is quite different from the rounded tops on the thick Armenian-period merlons.[20] A few of the square merlons in type III masonry appear as repairs just north of tower D and possibly south of F (pl. 10b).[21] However, most of the Armenian wall north of D has the original battlements. Considering that the only known period of military occupation after the Armenians was under the Mamluks, it is likely that these crude repairs belong to the period after 1375. It is significant that much of the circuit wall west of A (not properly surveyed on the plan) had to be replaced. Perhaps an earthquake severely damaged this area. Because of the nature of the slope at the west, towers and a high circuit were not deemed necessary in this region. It is quite possible that the postern adjacent to A is a Mamluk addition. This could mean that the Armenians, who simply relied on the Arab circuit to protect the western half of the south wall, had no south entrance and used only the bent entrance at G.

The entire south wall east of A (pl. 10a) and what remains of the east wall appear to have had a wall walk (pl. 10b). Cantilevered stairs (not shown on the plan) at strategic places along the walls provide convenient access from ground level to the battlements. In the three Armenian towers east of B (pl. 11a), the embrasured loopholes are all flanked by casemates with slightly pointed tops. Typically, the rounded hoods of all the embrasures are constructed out of monolithic blocks. Only a few of the embrasured slits have stirrup bases on the exterior. Recently, on the interior of tower D treasure hunters have cleared the windowless, lower-level chamber of the bastion (pl. 11a). Parts of the walls of this room are constructed with brick tiles. This brick is not used in an *opus listatum* and probably represents materials recycled by the Armenians. This lower chamber in tower D appears to have been a cistern.

Along the heavily damaged east circuit the only visible constructions are towers F and G. Gough believes that the Armenians did not pay sufficient attention to the east flank of the lower bailey because this predominantly Byzantine wall is so badly damaged *today.*[22] It is quite possible that this Greek construction was well preserved in medieval times and has collapsed (along with Armenian construction) during more recent earthquakes. The square Byzantine tower and flanking circuit at F were partially refaced in the Armenian period. A small Armenian chapel is built in the parapet directly south of F. To the west of this chapel is the baronial church of Tᶜoros I; at a later period the church may have been called St. Zōravaracᶜ.[23] The tower-gatehouse G and the surviving section of the circuit to the south are Armenian constructions (pl. 289a). This short section of attached wall terminates in a small tower. This wall is interesting in that the stones in the lower half are a slightly more crude type V than in the upper half. Also, what appear to be four(?) roundels are visible near the center.[24] Tower G is a typical bent entrance with dual portals (pl. llb). The interior of the tower has an apsidal dome. The exterior door is covered by a perfectly preserved slot machicolation. The depressed arch over this outer portal has pivot housings on the interior to accommodate double doors. The doors were secured by a crossbar bolt. In contrast, the inner door of the gatehouse is jambless. Directly above this door is the frame for a now missing dedicatory inscription.

Separating the south bailey from the central bailey is the magnificent three-story donjon H.[25] To the east and west, H is protected by steep cliffs (pl. 287c). At the south the donjon is separated from the bailey by a broad scarped trench which is over 15 m in length (pl. 288c). In medieval times a removable drawbridge provided access into the keep. On the north side of H there is a more narrow scarped and masoned trench (pl. 16b) whose floor is not flat like that at the south, but reaches a tapered point in the center. Its steep sides resemble the gable of a roof[26] and require the party approaching the central bailey to walk atop the central spine. The apex of the gable is only 80 mm in width with a sheer 180-m drop on either side. In plan, the interior of the keep can be *roughly* divided into three parallel units that run on an east-west axis. Only the south unit rises to three stories; the central and north divisions are one story. When viewed from the west, east, or north, the donjon resembles the throne of a mythical giant (pl. 288c). This exterior plan is almost identical with the smaller Crusader-built donjon in the nearby fortress of Amuda.[27] On the exterior of H there are two distinct building periods with different types of masonry (pl. 12). Throughout the first story and most of the second the outer facing consists of an unusually large (almost cyclopean) ashlar with each polished stone precisely fitted into neatly aligned courses. On the south and west walls many of the faces have broad, neatly drafted margins with the slightly raised center brought to a flat, relatively smooth surface. The ashlar on the east flank is identical in size and consistency, but its outer face is uniformly smooth. The north wall of the lower level has collapsed, but a few fragments of the facing are also uniformly smooth (pl. 16a). Sometime after the initial construction a disaster (perhaps an earthquake)[28] caused considerable damage to the donjon and necessitated its partial reconstruction. The masonry of the repair period is rusticated, which is typical of Armenian construction. But here it is used in a somewhat inconsistent fashion that is not seen anywhere else in this fortress (pl. 288b). In the Armenian-built circuits of the south and central baileys (which I credit to Tʿoros I) the types V and VII are laid in a very consistent and regular fashion. In the donjon the repairs are made with great haste, so that spoils are intermixed with the roughly laid types V and VII stones; occasionally the stuccoed margins of the type VI are visible.[29] This is especially evident on the west side and is also seen at the north and east. The reconstruction of H did not change the original plan. Because donjons and the masonry used in the first building period of H are *completely* unknown in Armenian military architecture, I believe that this complex is from the Crusader period of occupation (ca. 1098–1110) and perhaps is the first known example of Crusader military construction in the Levant.[30]

All previous commentators on the fortress of Anavarza have concluded that the donjon was first constructed by Baron Levon II. A now mutilated inscription, which is presently in situ on the south face, records his dedication in 1187.[31] However, a close examination of the inscription block (pl. 12) reveals that it is fitted into a preexisting rectangular space from the first period of construction; the points of junction are rough and crude.[32] Levon II, who was about to move his baronial headquarters to Sis in 1187, merely carried out hasty repairs to the donjon. At this time Levon was very concerned with his *persona* because he was trying to persuade the European powers to sanction his coronation. In order to enhance his image, he erected this rather boastful inscription. This may also explain why Levon posted a dedicatory inscription on the Byzantine island castle of Korykos.

Today the only passable entrance into H from the south is through a breach in the southwest corner (pl. 12). The original entrance was probably the door that is further to the north in the west wall (pl. 13b). A wooden drawbridge must have attached to the sill of the door, since the area above is the only point of the donjon equipped with machicolations. The south unit of the donjon, which constitutes the tower, jogs inward at the southwest corner. At the widest point the donjon measures over 12 m in width (from east to west). From its foundation to its top the tower is over 15 m high. The east wall of H bends inward slightly. The damage that necessitated the repair of the donjon must have left the south unit in a weakened condition, for the Armenians blocked the lower-level embrasures and large sections of the lower-level vaults here (pl. 13a)[33] with a mortar cement. It is difficult to determine the exact arrangement of rooms in the lower level of the south unit; two vaulted chambers of smooth ashlar masonry are set on a north-south axis. The embrasured loopholes of the Crusaders are designed quite differently from those of the Armenians. The Frankish embrasures are squareheaded and have unusually large arrow slits (pl. 14a). The first level of the south unit had two embrasures; the second level had at least one. The two upper-level floors in the south unit (pl. 12) are each covered by a single vault (on an east-west axis) and opened by roundheaded doors/windows in their north walls.

The narrow central unit of H consists of two adjacent chambers. The westernmost room is covered by a groined vault and now is fully opened at the north and east (pl. 14b). In the west wall is the beautifully executed door that was once connected to the wooden drawbridge. The south wall of the west central chamber is opened by a small breach (pls. 12, 13b). Separating this room from the chamber to the north (that is, the west chamber of the north unit) is a diaphragm wall. The lower-level support for this wall (perhaps stone and wood) has fallen away; a piece of that support is still visible at the east. The diaphragm wall has settled slightly and separated noticeably from the contours of the pointed vault over the west chamber in the north unit. The interesting feature about this diaphragm wall is that on the south side two corbels are placed below a large square opening in the apex of the wall (pl. 14b). At one time the west room of the north unit could be sealed off (probably by a bolted door below the diaphragm wall), and a defender on the interior of that room was capable of protecting the west door through the square opening. The west portal has a slightly pointed top, and its double wooden doors were secured by hinge housings and a crossbar bolt. The groined vault over the west chamber of the central unit appears to have been repaired (pl. 14b). The large east chamber of the central unit is a rectangular hall covered by two adjacent, half-groined vaults. These vaults and the south wall of this room belong to the period of Armenian reconstruction. The north wall seems to have been enlarged and buttressed with type VI masonry. This addition partially covered an embrasured arrow slit in the east wall. The present north and east walls of the east chamber in the center unit are Crusader. A broad, slightly pointed door in the north wall of this room gives access to the long east room of the north unit (pl. 16a).

This east room is a magnificent example of Crusader architecture. A perfectly executed groined vault covers the center of the chamber (pl. 15a). A transverse arch articulated by three tori joins the groined vault to a now collapsed, slightly pointed vault at the east (pl. 15b). In Cilicia the Armenians *never* sculpture the surface of their transverse arches, but such decoration is common in Crusader construction.[34] Unfortunately the east wall of this chamber and the east half of its north wall have completely collapsed. In the surviving section of the north wall there is still evidence of a squareheaded door/window (pl. 16a). At the west a staircase rises over the vault of the west room of the north unit (discussed in the preceding paragraph) and leads to the open grassy top of the central and north units (pl. 288c).

North of H across the ditch is the complex of rooms (I and V on the plan) that forms the southern tip of the central bailey (pl. 287c). During my visits some of the rooms were blocked. We must wait for excavations to improve on Gough's description of this area.[35] What is readily apparent from the study of the masonry is that this area of the central bailey has undergone numerous phases of reconstruction. At the west end of this complex there is a long vaulted room that is made entirely of brick. The arches in its wall once formed a graceful arcade that was blocked in a later period of construction. Unlike the brick used in the *opus listatum* near tower B, the construction here consists only of brick tiles laid in regular courses with thin intervening beds of mortar. The masonry in the bathhouse of Anazarbus is identical; Gough characterizes this construction as pre-Byzantine.[36] It is likely that this brick vault in the central bailey is from the Roman period and perhaps the first construction on the outcrop. The south edge of the central bailey, which rises from the north face of the fosse, and parts of the flanking circuit have been built with a recycled ashlar in the Arab fashion (pl. 16b). Later this ashlar was repaired with an exterior facing of type IV masonry, which characterizes the post-Arab Byzantine construction in Cilicia. This type IV is also prominent in corridor J. Armenian construction with types V and VII masonries dominates along the west flank from I to P (pl. 289b). Embrasured loopholes periodically open this west circuit. The interior side of the west circuit is flanked by at least eight separate rooms (pl. 287c). Southeast of tower T the exterior of the Armenian circuit may have been repaired during the Mamluk occupation.

There are no inscriptions in the central bailey. The only evidence of relief sculpture is in the Armenian chapel K[37] and in the surviving north end of corridor J. The two reliefs in the corridor are on the springings of a transverse arch. On the east flank there is a highly stylized depiction of the upper half of a man's body (pl. 17a); a bead and reel frames the top of this relief. On the west side the springing is decorated with two flanking palms and two centrally placed rosettes. Here the upper band consists of dentils (pl. 17b).

Most of the rooms in the central bailey are attached to the circuit wall (pl. 289b). The only exception at the south is a tholos-like chamber. The inner and outer facing stones of the lower half of this structure are a type III masonry (pl. 18a), while the few traces of the upper half have an interior of brick and an exterior of type III. The beds of mortar between the bricks are irregular and filled with numerous rock chips. Occasionally blocks of ashlar are used with the brick but not in an *opus listatum*. It cannot be determined if the upper part formed a rotunda. The north and south sides of the lower half are opened by symmetrical holes that may once have been doors. Although the inner facing of the lower half has been stuccoed, it does not appear that the chamber functioned as a cistern. In fact, joist holes on the interior may indicate that this circular room was divided into two floors.

Most of the west wall in the north half of the central bailey is a Byzantine construction (pl. 18b). At the far north, Armenian masonry is prominent in the horseshoe tower M (pl. 289c).

[1]During my four visits to this site (1973, 1974, 1979, 1981) I did not attempt to resurvey the circuits but relied on the plan of M. Gough. I have reproduced his plan here with a few minor modifications. Only the south bailey (A–H) and the central bailey (H–M) are shown on that plan. The north bailey with its single lengthy

circuit wall remains unsurveyed. The contour lines on the plan are at approximate intervals of 2 m.

[2]As with Gough (85 note 1), I use the name Anazarbus to indicate only the ancient city; the medieval castle and modern village are called Anavarza. In Byzantine texts this site is frequently referred to as Anabarza or Anazarbos (cf. Stephen of Byzantium, *Ethnica*, ed. A Meineke [Berlin, 1849], 91 f; Zonaras, ed. M. Pinder et al., CSHB, III [Bonn, 1897], 149). The Armenians prefer Anavarz, Anawarza, Anarzaba, or Anarzap, while the Arabs use Nāwarzā and ʿAyn Zarbah. Wilbrand calls this site Naversa.

Anavarza appears on most modern maps of this region, including *Adana (2), Cilicie*, and *Marash*.

[3]My intent here is merely to offer a brief summary of the major historical events. The only thorough study of this site was undertaken by Michael Gough in 1949. For other significant discussions or comments on Anazarbus/Anavarza see: J. Hirschfeld, "Anazarba," *RE*, 2101; M. Canard, "ʿAyn Zarba," *EI*², 789f; M. Gough, "Anazarbos," *PECS*, 53 f; M. Streck, "Aynzarba," *IA*, 74 f; Jones, 204–7; Magie, I, 275, 473, II, 1151; Lilie, 109 f, 160, 162, 170, 397 note 47, 404 note 101, 496 note 117; Ramsay, 291, 311, 341, 348, 350, 365, 374, 381–86, 415, 451; Barker, 54–56, 275, 283; Schultze, 315–20; Ritter, 56–67; King, 234–36; Cuinet, 92; Heberdey and Wilhelm, 34–38; J. Keil and A. Wilhelm, "Vorläufiger Bericht über eine Reise in Kilikien," *Jahreshefte des Österreichischen Archäologischen Instituts in Wien, Beiblatt* 18 (1915), 52–58; Davis, 152–57; Texier, 580–83; Ēpʿrikean, I, 170–74; Sevgen, 62–64; Schaffer, 41 ff; Alishan, *Sissouan*, 272–83; Langlois, *Voyage*, 434 ff; "Anarzaba," *Haykakan*, 375 f; Mikaeljan, 96 ff; Yovhannēsean, 140–57; Hellenkemper, 191–201; Boase, 153; E. Hicks, "Inscriptions from Eastern Cilicia," *JHS* 11 (1891), 238–42; R. Mouterde, "Inscriptions grecques et latines du musée d'Adana," *Syria* 2 (1921), 287; P. Verzone, "Città ellenistiche e romane dell'Asia Minore: ʾAnazarbus," *Palladio*, n.s. 7 (1957), 9–25; E. Michon, "Sarcophage d'Anavarza," *Syria* 2 (1921), 295–304; Budde, II, 73–86; O. Taşyürek, "Anavarza, 1972," *AS* 23 (1973), 15–17; idem, "Cilician Excavations and Survey, 1973," *AS* 24 (1974), 26 f; idem, "1973 Yılı Kilikya Araştırma," *Türk Arkeoloji Dergisi* 22.1 (1975), 117 f; M. Altay, "Anavarza Mozayikları Hakkında Ön Rapor (I–II)," ibid., 15.2 (1966), 49–54; Imhoof-Blumer, 431–33.

For Arab and Greek contacts at this site see: G. Schlumberger, *Un empereur byzantin au Xe siècle, Nicéphore Phocas* (Paris, 1890), 191 ff; Le Strange, *Palestine*, 387 f; idem, *Caliphate*, 129; Al-Balāḏurī, 264; Ibn Ḥawqal, 163, 165, 180, 185, 188.

[4]Theophanes, *Chronographia*, ed. K. de Boor, I (Leipzig, 1883; rpr. Hildesheim, 1963), 171, 235; Malalas, *Chronographia*, ed. L. Dindorf, CSHB (Bonn, 1831), 418.

[5]At this same time Anazarbus still had a metropolitan, who was under the jurisdiction of the patriarch of Antioch. See: G. Zacos and A. Veglery, *Byzantine Lead Seals*, I, pts. 2 and 3 (Basil, 1972), 720, 1667 f; S. Vailhé, "Anazarbe," *Dictionnaire d'histoire et de géographie ecclésiastiques* (Paris, 1914), 1504–6. Cf. E. Honigmann, *Evêques et évêchés monophysites d'Asie antérieure au VIe siècle*, CSCO 127, Subsidia 2 (Louvain, 1951), 78.

[6]Gough, 98.

[7]Canard, 806–8; Matthew of Edessa, 5; Honigmann, 68, 93, 96; Canard, *Sayf al Daula*, 138–40, 349, 383, 392.

[8]Oikonomidès, 259, 265–67, 356, 359; idem, *Byzantine Lead Seals* (Washington, D.C., 1985), 10, fig. 18, 24; G. Zacos, *Byzantine Lead Seals*, II (Bern, 1984), 347 f. In the mid-11th century Anazarbus again was the seat of a Greek metropolitan who held authority over all of Cilicia east of Korykos; see F. Conybeare, "On Some Armenian Notitiae," *BZ* 5 (1896), 124.

It seems unlikely that Anavarza was actually incorporated into the kingdom of Philaretus, which was based near Maraş. See: M. Canard, "ʿAyn Zarba," *EI*², 789; Hellenkemper, 193. About this same time (the late 1080s) Seljuk raiders may have briefly occupied the town, but there is no indication that the fortress was captured. Consult *La geste de Melik Dāniṣmend*, I, intro. and trans. I. Mélikoff (Paris, 1960), 82.

[9]Edwards, "Donjon," 54. Matthew of Edessa (30, 31 note 5) calls it the Crusader's "new Troy."

[10]Vahram of Edessa, 499; Samuel of Ani, 449; see also M. Čevahirčyan, "Mijnadaryan kositaṙan Ṙubinyannerə," *Patmabanasirakan Handes* 95.4 (1981), 182–94.

[11]Tritton and Gibb, 73, 276; Michael Italikos, 253; Choniatēs, 25–27; Cinnamus, 16–18, 180; Ibn al-Aṯīr, *RHC, HistOrien*, I, 424; Nersēs of Lampron, 577; Gregory, 152.

[12]Bar Hebraeus, 462.

[13]Dardel (60 f) gives a brief view of the Armenian occupants in a happier moment.

[14]Gough, 85–150, esp. 89–91. For a summary of the recent scholarship on the pre-Arab church architecture at Anavarza see Hild and Hellenkemper, 198–201. Cf. G. Bell, "Notes on a Journey through Cilicia and Lycaonia," *RA*, Ser. 4, 7.1 (1906), 12–29.

[15]On the plan, Gough has actually shown them as semicircular towers.

[16]Edwards, "Second Report," fig. 27. The Byzantine aqueduct in Anazarbus, which dates roughly to the early 6th century, has an *opus listatum* with three layers of brick tiles. This construction resembles that in the original entrance (north of tower B).

[17]This identification was first made by Gough (121).

[18]This masonry also occurs at the *extreme* west end of the south wall.

[19]See above, Part I.7, note 23.

[20]The rounded tops of the merlons on towers B and C have been broken off. The Armenian merlons are generally solid except atop the towers east of B, where they are occasionally pierced by embrasured loopholes.

[21]It is possible that the Mamluks constructed these merlons in the east walls.

[22]Gough, 122.

[23]For a complete description of church E and the parapet chapel, see: Edwards, "First Report," 156–61 and "Second Report," 128–30, 131 f; V. Langlois, "Voyage dans la Cilicie, Anazarbe et ses environs," *RA* 13.1 (1856), 366–68; J. Strzygowski, *Die Baukunst der Armenier und Europa*, II (Vienna, 1918), 740 f.

[24]The roundels do *not* appear to be column drums used as headers but merely round, smooth blocks of stone. These are probably recycled pieces, and thus their use by Armenians is unprecedented.

[25]The depiction of H on the plan does not accurately show the internal divisions of this complex.

[26]Gough, 123.

[27]Cf. Hellenkemper, 200.

[28]Ibid., 199.

[29]Even the rusticated masonry south of gate G is laid with greater care than in the donjon.

[30]Later the Crusaders used an *identical* masonry on their fortifications in Syria and Palestine; see: Müller-Wiener, pls. 14–15 (Sahyun), 71 (Crac des Chevaliers), 87 (Jebail); Edwards, "Anavarza," 53–55.

[31]Langlois's translation (*Inscriptions*, 16–17; idem, *Rapport*, 43–47):

> L'an 636 de l'ère arménienne . . . .
> . . . .
> Roupène fils de Léon
> son frère le pieux, avait . . . .
> près le mont Taurus, son noble séjour . . . .
> terrestre pouvoir . . . commencé par Ochin . . . .
> des places d'Anazarbe, Djenkia, Hada. . . . .
> . . . de ce pays . . . .
> il a bâti ce mur. . . . . il a établi .
> . . . .

Hellenkemper (291) published the most recent translation (by Manian), which I cite below for the convenience of the reader:

> Im Jahr 636 der armenischen Zeitrechnung (i.e., 1187) verdunkelte sich die Sonne so, dass die Sterne sichtbar wurden und der Türke eroberte die Heilige Stadt Jerusalem. In jenem Jahr starb Rupen, Sohn des Stephané und auf seinem Thron folgte der fromme Leon. Unter seiner Herrschaft standen Kilikien mit dem Berge Taurus und dem 'Schwarzen Berg' und die Ufer des Meeres bis Adalia und im zweiten Jahr seiner Regierung begann er diese Gla (Burg) in Anavarza, der Mutter der Städte, zu bauen. Er spaltete stärker mit Eisen diesen Felsen und baute auf festen Fundamenten die Mauer mit schweren Steinen und festigte sie mit Eisen und Blei und es wurde binnen eines Jahres vollbracht.

Some of the information in this inscription is cryptic and misleading. For example, it is stated that the masonry is anchored with iron clamps and lead. However, I found no evidence of such construction anywhere on the donjon. To my knowledge, the only Armenian construction with clamps is in the church of St. Sophia at Sis; see Edwards, "First Report," 168–70 and "Second Report," 134–40. Because of extensive mutilation, I was unable to make my own transliteration of the inscription.

[32] Unfortunately the masonry above the epigraph has collapsed, so it is impossible to tell how far down the south face repairs were needed. I did examine the mortar around the inscription, and it is identical to the binding and fill materials used in the period of repair (pl. 14a).

[33] The lower half of the second level seems to be intentionally blocked with debris. There may be an intervening floor between what now appears to be the first and second levels.

[34] Müller-Wiener, pls. 82 f (Crac des Chevaliers).

[35] Gough, 123 f.

[36] Ibid., 105.

[37] For a description of the chapel see Edwards, "First Report," 168 and "Second Report," 132–34.

Andıl

Today Andıl Kalesi[1] can be reached with a jeep only by driving north from Kozan for about 45 km. At first the trail follows the smooth, though undulating, contours of a very dry valley. After veering to the east the track ascends the precipitous side of Andıl Dağı. This mountain, which rises to almost 1,510 m, guards the west flank of the important north-south road from Kozan to Feke and Saimbeyli. The road up the Dağı does not end at the summit but at a ledge about 200 m below. Here, Andıl Köy, a small village of about forty homes, is perched atop a shelf of limestone on the southwest flank of the mountain. To the southeast of the village the upper slopes of the Dağı have been terraced for farming (pl. 19a). Interspersed among the wheat fields are fig and olive trees. The terraces are supplied with water by wells, spring runoff, and frequent storms in the summer.

We have no certain historical information about this fort.[2] Occasionally Andıl Dağı and Kalesi are listed on the maps of the region.[3] C. Favre and B. Mandrot, two nineteenth-century explorers of Cilicia Pedias, mention only that the site of "Andıl" is a fort visible from Sis.[4] In the area of the village, the farmers have uncovered numerous fragments and complete

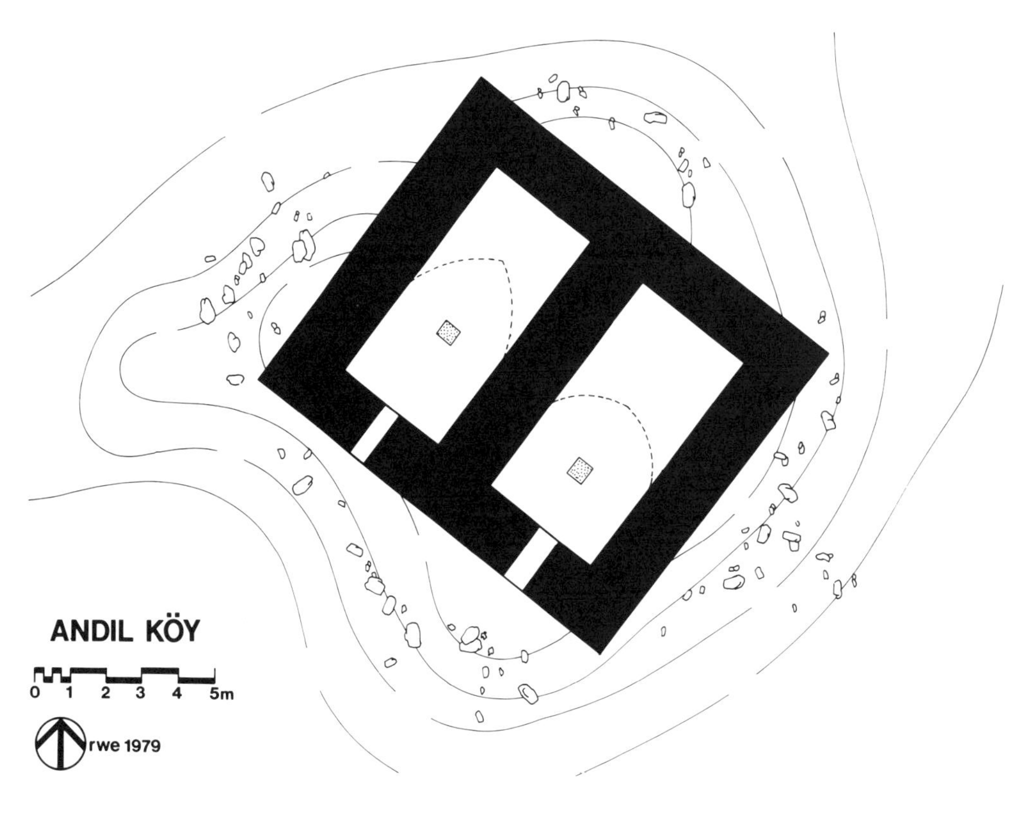

Fig. 9

sections of Byzantine epigraphs.[5] The nature or extent of the Greek occupation at this site cannot be determined. The few medieval coins discovered by the villagers in the fort and its environs date to the reigns of Kings Het'um II and Levon II.

The medieval site of Andıl has a fortified estate atop the summit (fig. 10) and a simple vaulted structure on the terrace, just south of the village (fig. 9). Because of the masonry and the peculiar aspects of design, it is certain that the two structures are Armenian.

Andıl Köy

The small building at Andıl Köy stands atop a low, gently sloping hill (fig. 9). It consists of two vaulted chambers encased in a single rectangular unit. All of the exterior masonry is type VII. The interior masonry is a high-quality type V. The core is poured in the typical Armenian method (that is, pouring the core at each course level) and consists of limestone mortar and rock fragments. All available evidence indicates that there was one period of construction. Each chamber has a straight-sided entrance at the southwest (pl. 19b). A single lintel-block covers each entrance. The most northerly of the rooms is covered by a pointed vault that is pierced by a single hatch (now blocked). The other chamber is quite similar, except that its vault is almost barrel in shape. There are no distinguishing features on the interior of these rooms. Because the walls extend for four courses above the vaults of the rooms, the roof probably served some utilitarian function.[6] There are no indications that this second level was divided by a wall or vaulted. The upper-level entrance may have been over the door to the southernmost room.

Andıl Kalesi

The easiest way to reach the top of Andıl Dağı is to hike northeast around the southeast flank of the summit until the trail circles 180 degrees and terminates under tower D (fig. 10; pl. 20a). The fort, which rests on the highest point of the oblong summit, is essentially a fortified estate house. Aside from the now unattached tower K at the south (pl. 20b), Andıl Kalesi consists of eleven rooms that are built at two levels and encased in a rectangular frame (pl. 20a). Only a single centrally placed tower at the west breaks the flat facade. The Armenian engineers also placed a thin talus along the western wall. The emphasis on defense was at the west because it was here that the ascent was relatively easy compared to the steep cliffs at the north, east, and south.

The exterior masonry of this fort generally consists of a combination of types V and VII (pl. 21a). Type VII is less frequent. Type IX, sometimes with a bossed face, is used around the frames of windows (pl. 21b). A large section of the north wall of room C and almost all of the east walls (except for the southern end of A) of A/F, B/G, and C/I are repaired. In some areas the repairs involved the use of a poorer quality masonry (for example, type IV). On the interior of the fort the masonry consists of types III and IV with the latter predominating (pls. 22a, 22b).

The lower level has the three rectangular chambers of A, B, and C and a small room in tower D. Today the latter is shattered at the west end; the apsidal end of its slightly pointed vault has also disappeared. Room D had no doors and was opened by a single square hatch in the ceiling. Since the walls were plastered, this room probably functioned as a cistern. A similar function can be assigned to room A (pl. 22a). In addition to the hatch in the ceiling of A, there is a small high-placed window in the east wall for ventilation. The north wall of A is built atop a long oval rock. This monolith does not protrude into the space of B. To the north, room B is opened by a single high-placed door in the east wall. This is the only door in the first level. On the exterior side, the voussoirs of the door's rounded top have the joggled joints that are so common in Armenian construction. Directly below the door are two corbels; the one at the south has almost completely fallen away. The corbels probably anchored a retractable platform and stairway. This portal has accommodations for a double door. Room B probably served as a storage chamber. In the southeast corner of B is a curious niche with two small asymmetrical openings through the east wall. Just west of this chamber and above is a rectangular passage that leads to the second level. Unfortunately it is now blocked with debris. To the north of B is the vaulted chamber C which is identical to A except that its alignment is slightly to the west of A and B, and its high window at the east is splayed on the interior. It seems likely that C also functioned as a cistern. What all three chambers have in common is that the centers of their west walls are pierced by passages that were later blocked with masonry and debris. In B the passage turns to the north, and in A it turns in two directions. The floors of these conduits, which are parallel to ground level, are barely wide enough to permit the passage of a man. Today their function remains a mystery.

The upper level of the complex has seven chambers. Here the upper level of D consists of a hexagonal room pierced by a window in the north side and one

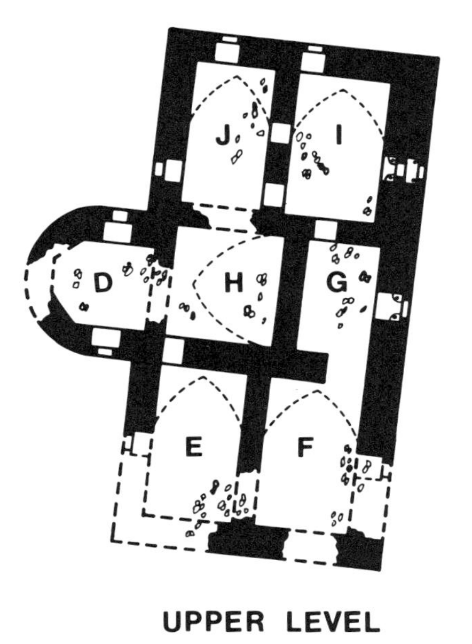

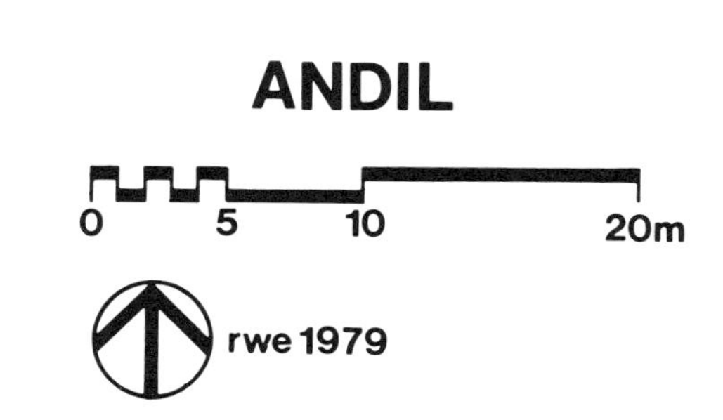

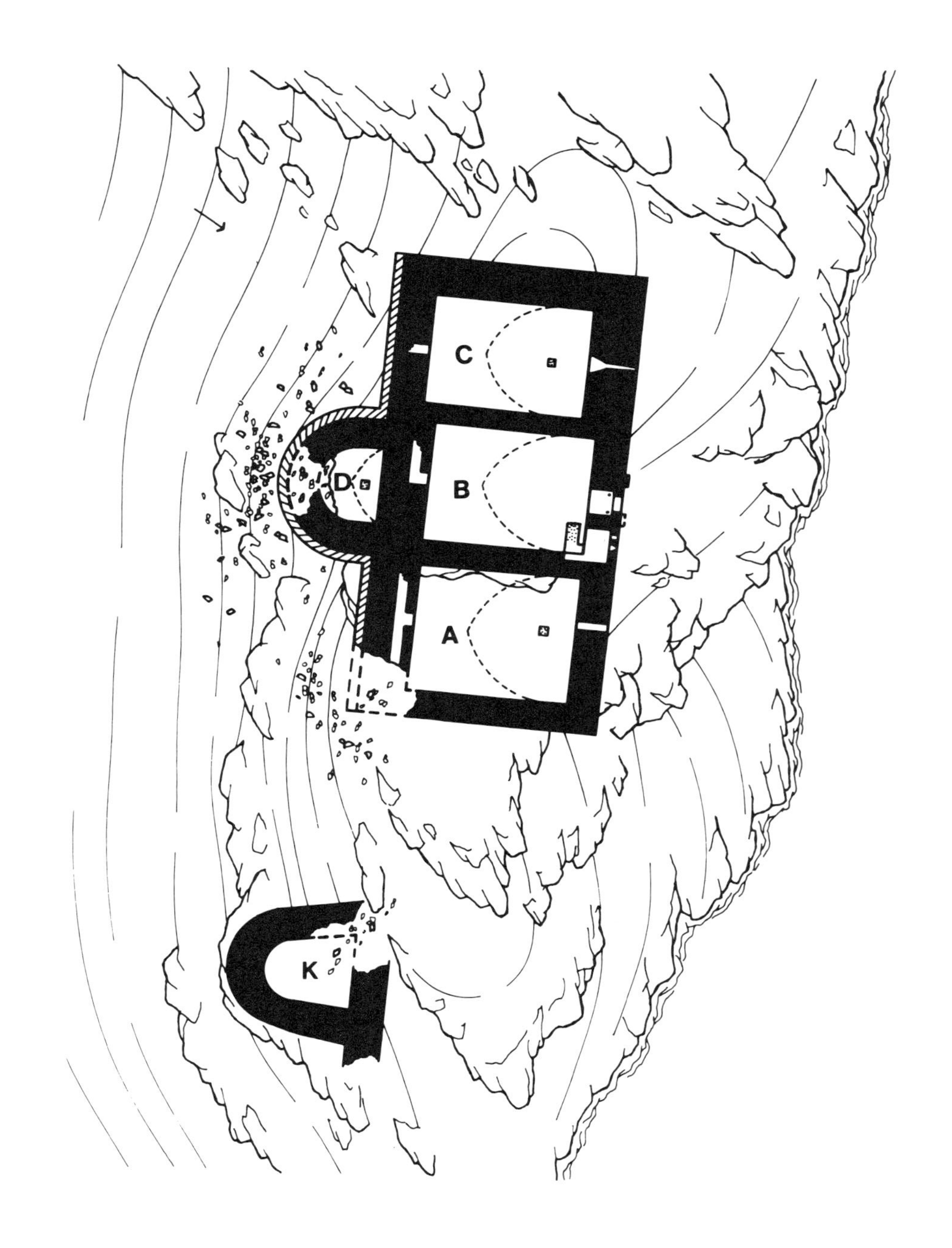

Fig. 10

in the south side; both windows are limited by jambs. A now missing door probably opened the east wall of the tower. Only a few sections of the upper level of D stand to more than three courses in height (pl. 23a). This degree of destruction is repeated throughout the upper level where most walls barely stand to the level of the springing course. The floors of all chambers are strewn with the debris of the collapsed vaults. Of the four internal doors only the two roundheaded portals between I and J remain undamaged though barely visible amid the rubble. The jambs of all the windows that open onto the exterior are drilled in a fashion that is common in other Armenian forts (pl. 22b).[7] These windows appear to be squareheaded, at least on the exterior side. Of particular interest are the east windows in G and I; both openings were reconstructed after the collapse of part of the east wall (pl. 23b). In I the addition of a new wall did not destroy the jambs (at least on the south side) or lintel of the original opening. A second lintel and jambs were merely added, though this time without pivot housings. The window in G has only a single lintel with jambs (pl. 22b), which indicates that the original frame was probably lost in the catastrophe. Below the north window in I there are the plugged sockets for at least three corbels; these supported some sort of removable breastwork.

From the present unexcavated remains it seems that this estate house had at least three periods of construction. By comparing the masonry and wall junctions, it appears that room C was built first as a single unit. Later, in the second and largest period of construction, chambers A and B were appended slightly to the east of C (no doubt due to the demands of the topography). At this same time the upper level (D–I), tower K, the talus, and a completely uniform exterior facing were added.[8] A large hatch was built in the southeast corner of B to communicate with G. After the completion of this second period of construction, either an earthquake or displacement caused the weakening and partial collapse of the east wall (especially around window G). Before the time of the collapse, the east wall was *not* built in a straight line;[9] when moving from south to north, the wall probably jogged inward after passing B/G to conform to the more western alignment C/I. After the east wall was damaged, it was completely refaced (the third period of construction) except for the southeast corner of A/F, which possesses the uniform masonry of the rest of the exterior (pl. 21a). In order to add stability to the new wall, the architects expanded the east wall of C/I to the east so as to be flush with the east walls of A/F and B/G. The seam in the northeast corner of C/I is quite visible (pl. 21b). The enlarging of this wall explains why the stones on the inward side of the seam have a neatly squared boss (as if they were once an exterior facing) and the presence of two sets of jambs in the east wall of I. The area of repair in the north wall of C is certainly after the second period and may be from the third period of repair. In some period subsequent to the second stage of construction, the conduits in the west walls of A, B, and C were blocked.

The now destroyed tower K was probably attached to the southwest corner of the estate house (pl. 20b). The interior of K had a simple apsidal room.

What is certain is that Andıl Kalesi was *not* a garrison fort. The remoteness of the locale is impractical for the stationing of a large body of troops. Its position and size indicate that it was a residence, retreat, and watch post. It certainly has intervisibility with Sis to the south and Alafakılar to the east. Its role in communication was obviously an important one. However, this point should not be overstressed, since weather conditions frequently reduce visibility to a few hundred meters, even in summer.

[1]The plans of this hitherto unsurveyed site were executed in August 1979. On the plan of Andıl the contour lines are separated at intervals of one meter. On the survey of Andıl Köy the interval of distance between the contours is 50 cm.

[2]Father Alishan (*Sissouan*, 66, 264) mentions Andıl twice, but he does not locate the site on his map. He reports that Andıl is near Sis and that some (unspecified) scholars believe it to be the ancient site of Davara. Alishan himself associates Andıl with the "hermitage" of Andul. The latter may have functioned as a summer retreat for King Hetʿum I and his wife in 1238. It certainly served as a monastic scriptorium in the 13th century, where both Grigoris of Sis and Vardan Arevelcʿi worked. See: Ēpʿrikean, I, 180; Pʿ. Antʿabyan, "Vardan Arevelcʿu 'Žłankʿə'," *Banber Matenadarani* 8 (1967), 157–80; Part I.7, note 34, above.

Because modern Turkish legends associates Andıl with hawks, it is quite possible that this site is the famous "Castle of the Sparrowhawk" in the medieval romance of Melusine (J. K. Anderson drew my attention to this possible association). See: M. Letts, *Mandeville's Travels* (London, 1953), II, 311; A. Bryer and D. Winfield, "Nineteenth-Century Monuments in the City and Vilayet of Trebizond: Architectural and Historical Notes, Part 3," *Archeion Pontou* 30 (1970), 249 f; idem, *The Byzantine Monuments and Topography of the Pontos*, DOS 20 (Washington, D.C., 1985), 105 f.

Kʿēlēšean mentions that the 19th-century Armenians of Sis call Andıl the fortress of Tʿoros. This unspecified Tʿoros supposedly settled Arabs near his fortress.

[3]This site appears on the following modern maps: *Kozan, Malatya, Marash*. It seems that Hellenkemper never visited Andıl but merely transcribed its approximate location on his map from earlier sightings (e.g., Kiepert). Cf. *Handbook*, 323.

[4]Favre and Mandrot, 148, 150; cf. Langlois, *Voyage*, 408.

[5]Pieces of sculpture also found in the area are stored in the one-room schoolhouse of Andıl Köy. No Byzantine coins were reported found by the denizens. I was not permitted to photograph or transcribe the inscriptions.

[6]It is not inconceivable that the twin chambers functioned as tombs; see A. Machatschek, *Die Necropolen und Grabmäler im*

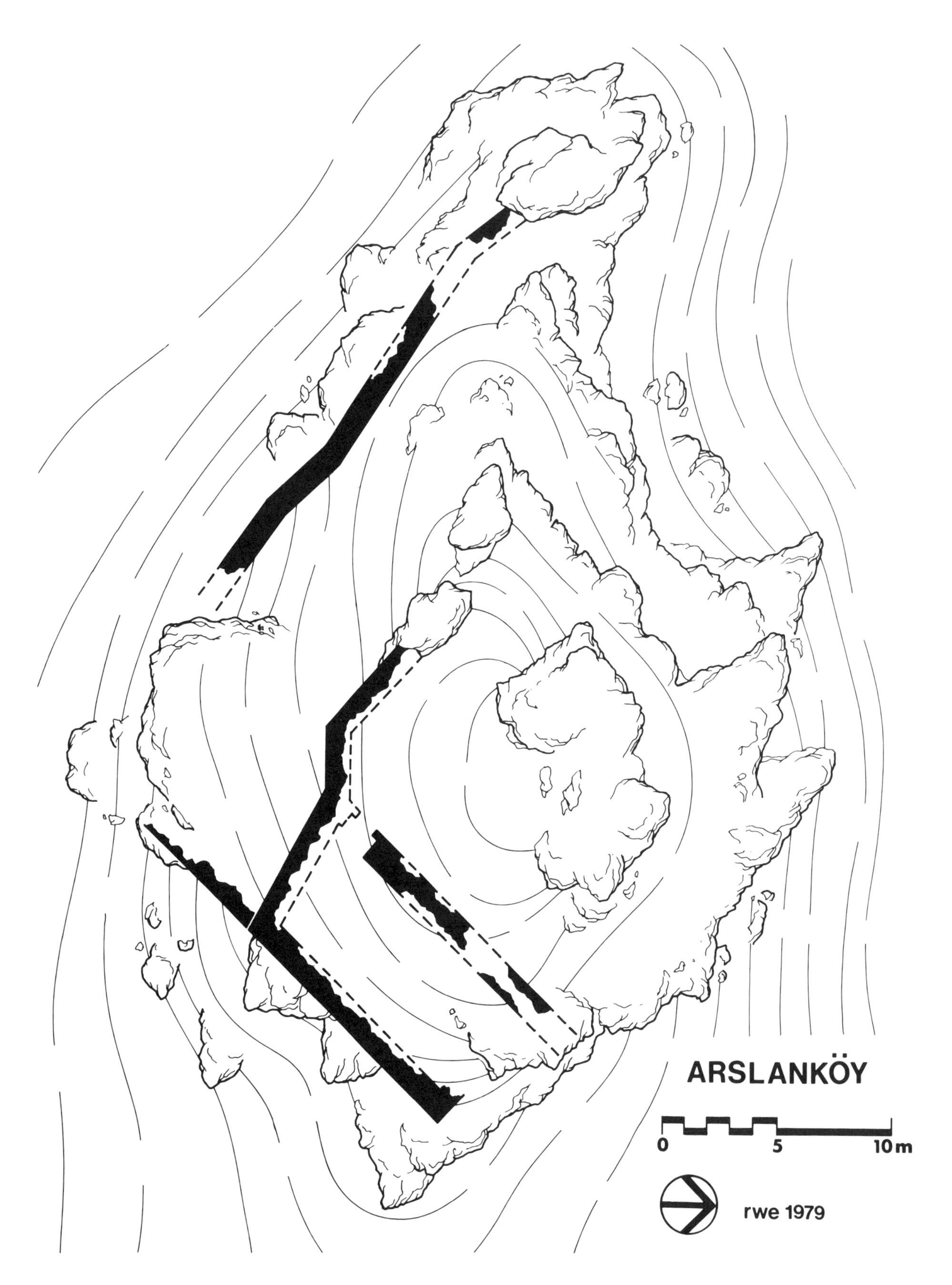

Fig. 11

Gebiet von Elaiussa Sebaste und Korykos im rauhen Kilikien (Vienna, 1967), pl. 21.

[7]Refer to my discussion below of the windows in building O at Meydan Kalesi (Catalogue).

[8]The west and east walls of C were replaced; the north was covered over with the new facing.

[9]The west wall in the second period of construction was built in a straight line because the west walls of E, H, and J were placed in alignment. Only tower D (also from the second period of construction) breaks the line of the facade.

Arslanköy

The small garrison fort and village of Arslanköy[1] are isolated in a moist Highland valley on the eastern fringe of Cilicia Tracheia, more than 70 km northwest of Mersin.[2] The fortified outcrop is located about 3 km south of the modern settlement (pl. 24a). North of Arslanköy is the mighty barrier of the Taurus Mountains. Except for the strategic trail that snakes northwest from the village into the region of Ereğli, this wall of rock and ice is almost impenetrable. To the east and the southeast of Arslanköy three roads fan out in the direction of the coast and the Cilician plain. Since the region has not been fully explored, it is impossible to determine if Arslanköy Kalesi has intervisibility with other forts. Although unrecognizable by its modern name in any of the medieval chronicles, this site, with its commanding view of the entire valley, must have been of some importance.

Considering the simplicity and limited extent of surviving construction, the masonry and architecture here can be discussed simultaneously. The natural terrain plays an important role in the defense of the fort, and the walls are envisioned as a supplement to protect the easiest points of access at the south (fig. 11). Today the circuit at the southeast is badly decayed and stands to less than 2 m in height. The northeastern end of the east wall is constructed exclusively with an exterior facing of type IV masonry. To the southwest the rest of the east circuit is built with type V masonry. At the corner where the wall turns sharply to the northwest a few type VII stones are visible; this northwest wall continues with type V masonry. Below the corner at the southwest there is a revetment wall which is set at a much lower level and constructed of type IV masonry. Six meters to the northwest of this corner are the remnants of a single door; a jamb is still preserved on the northeast side (pl. 24b). One of the upper-level stones over the jamb of the door is curved like a voussoir; this may indicate that the door was covered by a collapsed arch. The freestanding circuit at the west has an exterior facing of type V masonry. Since most of the site is badly damaged and buried, further observations about its plan are difficult to make. However, it is possible that the exterior facing of type IV represents an initial phase of Byzantine construction, while types V and VII are from the Armenian period.

Arslanköy Kalesi is like Rifatiye I in two respects. First, its size and simplicity of design belie the value of the site; second, the walls show what may be a Byzantine as well as an Armenian period of construction. Concerning the second point, it is not uncommon for Armenians to construct or reoccupy small garrison forts and watch posts at the opening of a strategic trail into Cilicia in order to protect a large baronial castle (in this case, Çandır) to the south (cf. Saimbeyli and Vahga; Fındıklı and Geben).

[1]The plan of this previously unattested site was completed in June 1979. The contour lines are separated by intervals of 50 cm.

[2]The village of Arslanköy is listed on the following maps: *Central Cilicia, Mersin (1), Mersin (2)*. An alternate spelling for the name of this site is Aslanköy.

Ayas (Yumurtalık)

The fortified port of Ayas[1] was the conduit for most of the trade funneled through Armenian Cilicia. Ayas, which is frequently called by its official name of Yumurtalık, is situated on the west flank of the Bay of Iskenderun.[2] Unlike the abandoned ports at Silifke, Tarsus, and Pompeiopolis, the deep harbor of Ayas did not serve as an estuary for one of the silt-bearing Cilician rivers. Today it is a prosperous resort community and the port for a number of fishing vessels.

The history of Ayas in the twelfth century is obscure.[3] Although this site was in the confines of the Armenian kingdom, it did not have a representative at the coronation in 1198/99 (see below, Appendix 3). This may imply that Ayas was administered directly by the king, just as Anavarza and Sis were his personal possessions. Ayas did not take on preeminent importance for the Venetian and Genoese traders until the Crusaders lost control over a number of Levantine ports in the last quarter of the twelfth century. Numerous treaties have survived in which the Armenian kings grant various privileges to the Italian merchants.[4] There are few references to permanent Armenian gar-

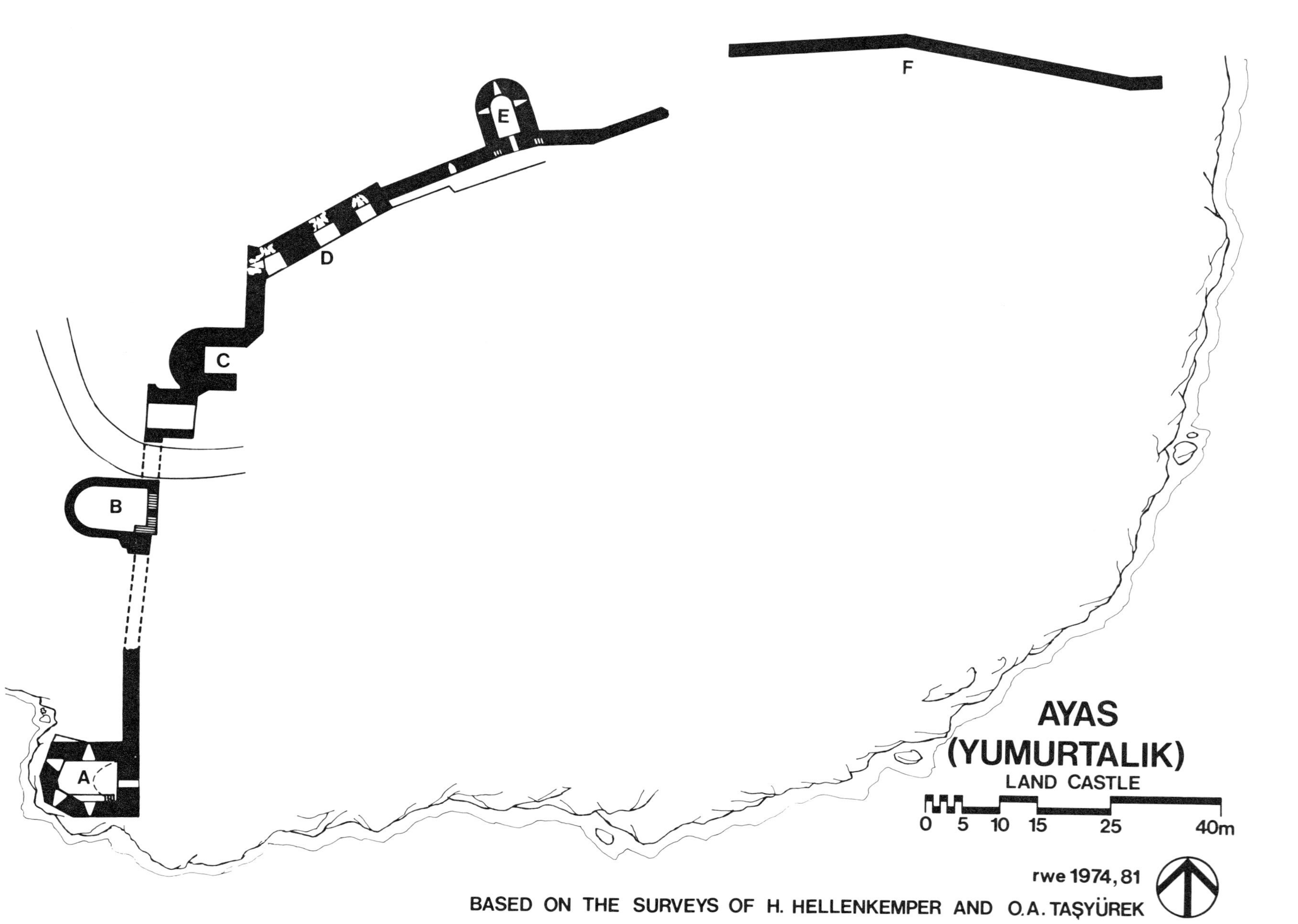
F
E
D
C
B
A
AYAS
(YUMURTALIK)
LAND CASTLE
0 5 10 15 25 40m
rwe 1974, 81
BASED ON THE SURVEYS OF H. HELLENKEMPER AND O.A. TAŞYÜREK

Fig. 12

risons at Ayas; there is no evidence for a standing Armenian navy.[5] In 1266 and 1275 the Mamluks briefly plundered Ayas. During the 1280s two attacks by Turkmen bandits were repulsed. On or about 1282 a small castle was built on the island in the harbor.[6] In 1305 and again in 1320 Armenian forces dispatched from the north prevented a Mamluk attack on the port. However, in the spring of 1322 a huge force of Egyptians burnt the town and severely damaged the land fortress. With the financial assistance of Pope John XXII the land castle at Ayas was rebuilt in the following year. By 1337 the Mamluks took permanent possession of Ayas.[7] The town was garrisoned and maintained by the Egyptians who were ill-prepared in the fall of 1367 to repulse an invasion by Peter I of Cyprus.[8] The port faded into obscurity until it was rebuilt by Süleyman the Magnificent and served as a minor base for his fleet. Today Ayas is still remembered as the point of embarcation for Marco Polo and others who left Europe en route to the Mongol Empire.

As with Korykos to the west, Ayas has land and sea castles, but here the two are not joined by a dike. A breakwater, which extends from the shore (west of the land castle), curves to the east creating a small bay (pl. 25a). The island with the sea castle is over 400 m east of the shore (pl. 26b). A large part of the modern town of Yumurtalık is built in and around the land castle. The topography around the land fortress is relatively flat and barren, punctuated only by the rolling hills. To the north and west of the land castle and the present town are the remains of the classical-late antique city of Aegae.[9] There are the remains of a stone bridge and a gigantic structure that is constructed out of brick tiles.

The only substantial post-Arab military construction outside the land and sea castles is a large watchtower that was built in the first half of the sixteenth century by Süleyman I (pl. 290a). This polygonal structure is located 1.3 km west of the land castle and has a commanding view of the sea. The masonry of the tower is made entirely from recycled stones. Column drums, which are used as headers, are found throughout the construction. The tower has a single squareheaded entrance at the first level. Both the first and second levels of the tower are covered by stone vaults; the third level is an open terrace with battlements. The crenellations at the top are wide enough to accommodate small cannons. All three floors are equipped with identical embrasured loopholes. The design of these loopholes is quite unlike any shooting ports that were built in the period of the Armenian kingdom. Typically, the Armenian loophole is quite long and often has a narrow splayed base that looks like a flattened stirrup. In this Ottoman tower the slit is rather short, and the stirrup is so greatly enlarged to receive a gun that it constitutes about 40 percent of the total length of the loophole. This design is exactly like that of the loopholes in the Ottoman fortress at Payas.[10]

Land Castle

Today the single wall that constitutes the land castle is curved in plan to encompass the tip of a small peninsula (fig. 12). When viewing the fortress from the breakwater at the south, it appears that the seaward side is without defenses (pl. 25a). However, the rough plan of the site by Alishan shows that sea walls were present in the late nineteenth century.[11] Sometime during the last eighty years they were dismantled; only a few traces of the foundation for a south circuit are visible east of tower A. The modern town occupies most of the west half of the enceinte that was created by the land walls. What survives of the land walls today is buttressed by four towers.

After a careful inspection of the masonry, I could find *no* evidence of Armenian construction in the land castle. In fact the limestone and basalt masonry is the same as that in the polygonal watchtower. Recycled ashlar blocks of various colors are neatly fitted into regular courses with thin, but visible, stripes of mortar in the interstices (pl. 26a). In some areas over 50 percent of the exterior facing consists of type VII stones. This masonry was probably plundered from medieval Armenian structures. Periodically mixed with the ashlar and almost always of a dissimilar height are column drums that are used as headers. In a few cases they have been arranged systematically for aesthetic effect (pl. 27b). It is also apparent that the present circuit has no prior foundation and represents (except for a few minor repairs to the facade) one period of construction. Although most of the embrasured loopholes have been demolished, enough fragments of their exterior frames survive to show that they were constructed exactly like their counterparts in the nearby watchtower and at Payas (pls. 26a, 290a). It now appears likely that what survives of the land castle of Ayas is from the Turkish period of occupation. Without excavation it is impossible to determine the extent or nature of the pre-Ottoman fortress on the mainland of Ayas.

The largest bastion in the land fortress is tower A (pl. 25b). All three stories of this polygonal tower are opened by embrasured loopholes with casemates (only the second level is shown on the plan) (pl. 26a). The single embrasure in the first level is located in the north

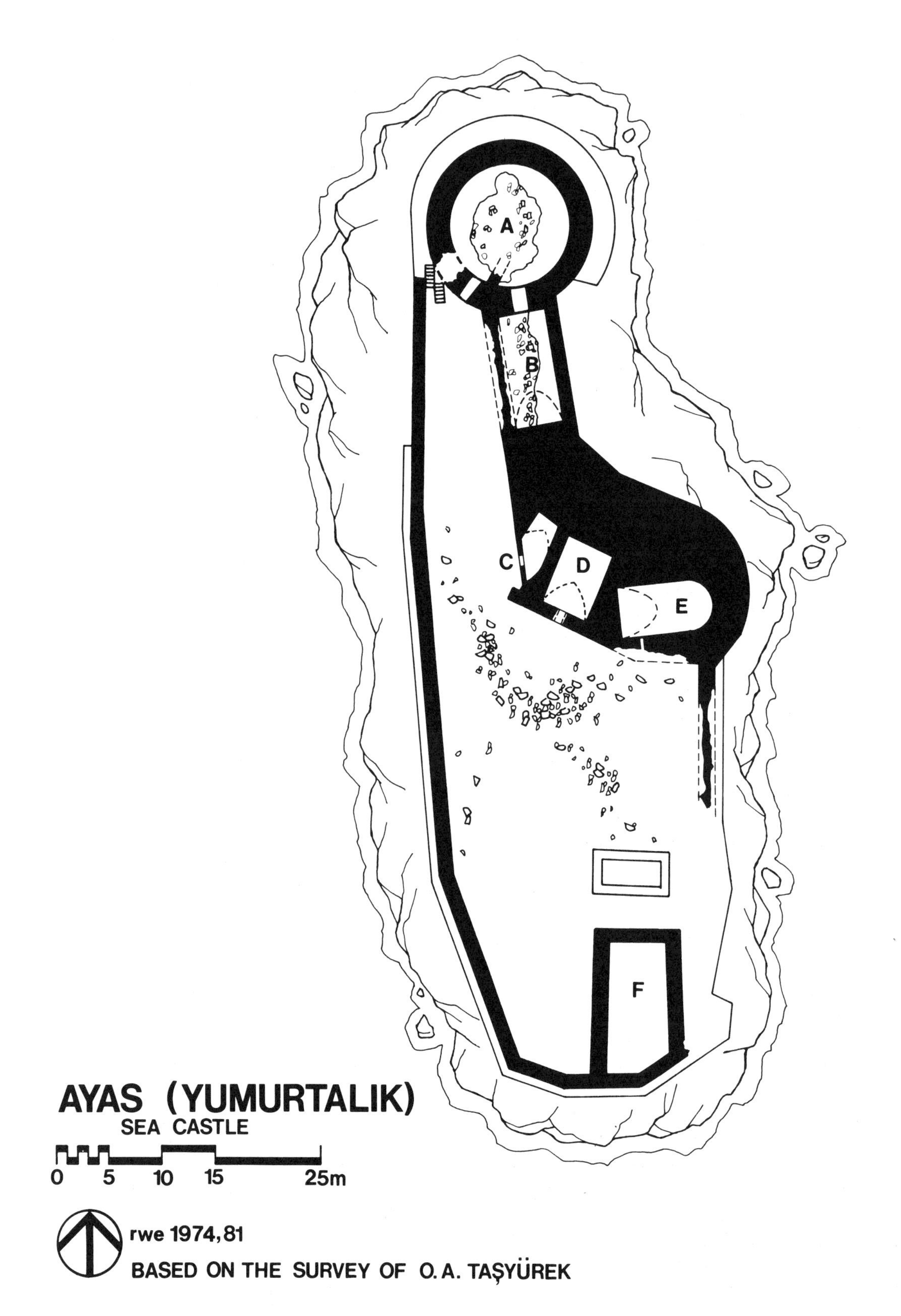
A
B
C
D
E
F
AYAS (YUMURTALIK)
SEA CASTLE
0 5 10 15 25m
rwe 1974,81
BASED ON THE SURVEY OF O.A. TAŞYÜREK

Fig. 13

wall and is preserved only in the base of the loophole. The only entrance into the first-level barrel-vaulted room is at the east. The second level has a similar entrance and vault, but it has four embrasures to command the harbor. The badly damaged third level seems to have had an equal number of embrasures. The wall connecting A to tower B has suffered considerable damage. Today tower B has been repaired and adapted into the living space of an adjoining house (pl. 25b). On the south flank of B a buttress-room was recently added. A modern paved road cuts through the circuit just north of tower B (pl. 27a). On the other side of the road, tower C and what may be an attached gate to its south have been covered by modern constructions. The main mosque for the village is just east of tower C and battery D. On the top of tower C are the remains of a shattered embrasured loophole with casemate and merlon (not shown on the plan). The adjoining battery D has three casemates with embrasures (pl. 27b). Today the embrasures are shattered on the exterior. The easternmost of the three casemates is the smallest (pl. 28a). The westernmost casemate is the largest; this opening has two separate embrasured loopholes (pl. 28b). The ceiling of the casemate is in part constructed with recycled corbels from the earlier land fortress. The wall connecting D to E (not accurately surveyed on the plan) has a single embrasure with casemate (pl. 29a). Tower E appears to have a cistern in its first level; the second level is opened by three embrasures with casemates. The third (or terrace level) of E has a few shattered fragments of embrasures. Except for a 14-m gap, the circuit of the land castle continues east until it reaches the beach (pl. 26b). A few abandoned buildings within the circuit have been constructed with medieval spoils (pl. 29b); just when these buildings were erected cannot be determined.

Sea Castle

The sea castle of Ayas, which consists of chambers A through E, occupies about 25 percent of the limestone mass that makes up the island (fig. 13). A badly damaged circuit wall extends south from A and E to encompass most of the open space on the island.[12] Surprisingly, the limestone masonry of the standing structures is very consistent (pl. 291a). The exterior facing is a uniform type V. The interior facing of the walls consists of types VIII and IX, while types III and IV are more common in the vaults. This factor, the slightly pointed vaults over B, C, D, and E, and the curving outer face of the complex seem to indicate that A through E are Armenian constructions. I found no evidence of repairs or reconstruction in the medieval-period rooms. It is likely that Italian merchants commissioned and financed the building of this site by Armenian masons. However, this is not the first period of construction on the island. Northwest of room F there are numerous and carefully carved dovetail sockets. These sockets are visible in the scarped rock and limestone blocks that once constituted the floor of a late-classical-period building.[13] Amid the rubble there are also substantial traces of cornices with classical motifs, inscriptional fragments, and pieces of triglyphs. Careful cataloguing and sorting of the debris as well as excavation of the southern half of the island may reveal substantial late-classical-period foundations.

The largest and most unique of the rooms on the island is the circular hall A (pl. 291a). This room was entered from the southwest and had a connecting door with chamber B. Most of the circular vault over A has collapsed. It appears that a wall divided this room into two halves and supported the ceiling. This covering is not a mere cupola that rises on pedentives, but more of a rotunda construction. Most of the slightly pointed vault over the adjoining room B has collapsed. The vaulted rooms C and D have both lost their entrances at the south. The two units are separated by an arcade of two arches (not shown on the plan). When I visited the island in 1974 the entrance to room E was blocked. Room F is barely visible at its foundations. There is no evidence of embrasured loopholes in any of the island constructions.

[1] I visited this site in 1974 and 1981. During my second visit I was able to reassess both forts in light of the plans published by Hellenkemper. Because of the relatively flat topography, contour lines do not appear on the plans.

[2] Ayas is the most frequent Armenian spelling for this port. The Byzantines refer to this site by its classical name, Aegae. The most common Italian designations are Laiazzo (Lajazzo) and la Giazza. Frequently the Franks label the site Laias. See also Abū'l-Fidā' (27) for the Arab spelling Āyās. Ayas appears on the following maps: *Adana (1), Adana (2), Cilicie, Malatya.*

[3] The best historical survey of this site was published by Hellenkemper (155–60) who relied in part on the earlier accounts of: Alishan, *Sissouan*, 426–51; Langlois, *Voyage*, 425–31; F. Taeschner, "Ayas," *EI*², 778 f; Schaffer, 97; Heyd, II, 73–92. See also: J. Schmidt, "Aigai," in *RE*, 945; Barker, 265; Heberdey and Wilhelm, 14; M. Čevahirčyan, "Ayas navahangistkʿałakʿə kilikiayun," *Patma-banasirakan Handes* 60.1 (1973), 97–110; M. Balard, *La Romanie génoise*, II (Paris, 1978), 727 ff, 859; Sanjian, 75, 79; *Handbook*, 139–44, 218 f, 693 f; Flemming, 26, 34 f, 84; Al-Jazarī, *La chronique de Damas*, trans. of extracts J. Sauvaget (Paris, 1949), 36; Canard, "Le royaume," 240 ff; Ramsay, 385 f; "Ayas," *Haykakan*, 339; R. Irwin, *The Middle East in the Middle Ages. The Early Mamluk Sultanate 1250–1382* (Carbondale, 1986), 68 f, 145 f; A. Atamian, "Āyās," *Dictionary of the Middle Ages* 2 (New York, 1983), 19; Gaudefroy-Demombynes, 88, 96, 98, 217, 248; Le Strange, *Palestine*, 405; Cuinet, 107 f; B. Darkot, "Ayas," *IA*, 42 f; Beaufort, 300 f; Schultze, 326 f; Ritter, 115 f; Boase, 155; Rey, 348 f; Langlois, *Rapport*, 50–52; Mas Latrie, I,

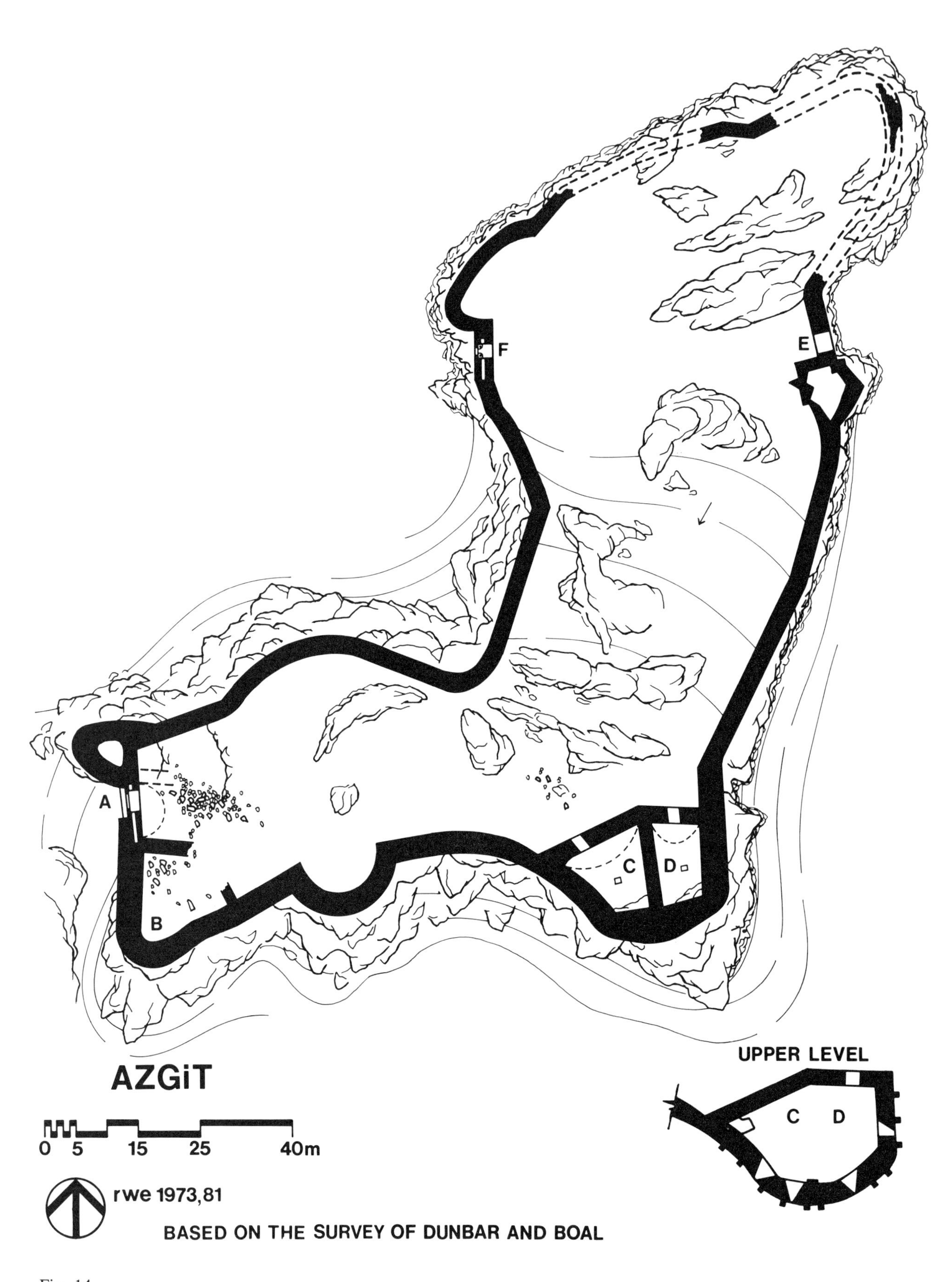
F
E
A
B
C
D
AZGiT
0 5 15 25 40m
rwe 1973,81
BASED ON THE SURVEY OF DUNBAR AND BOAL
UPPER LEVEL
C D

Fig. 14

377 f, 324, 371, 394, 400, II, 84; G. Thomas, *Der Periplus des Pontus Euxinus. Nach münchener Handschriften. Ingleichen der Paraplus von Syrien und Palästina und der Paraplus von Armenien (des Mittelalters)* (Berlin, 1863), 283 f; J. Richard, *Chypre sous les Lusignans* (Paris, 1962), 36-49; Hierocles the Grammarian, *Le synekdèmos d'Hierokles*, text, intro, and comm. E. Honigmann (Brussels, 1939), 38; Imhoof-Blumer, 423–28; R. Edwards, "Ayas," *The Seventeenth International Byzantine Congress, Abstracts* (Washington, D.C., 1986), forthcoming.

[4]Langlois, *Cartulaire*, 4 ff; Bedoukian, 25 ff; Alishan, *L'Armeno-Veneto*, I, 13 f, 46, II, 9 f, 22, 30, 40, 117, 124, 134 ff; *Diplomatarium Veneto-Levantinum*, ed. G. Thomas, I (Venice, 1880; rpr. New York, 1966), 21 f, 55–58, 72 f, 75, 176–81, 206, 234 f, 237 f, 255–57, 310 f; *Felice de Merlis, prete e notaio in Venezia ed Ayas (1315–1348)*, ed. A. Bondi Sebellico, Fonti per la storia di Venezia, sez. III—Archivi notarili, I (Venice, 1973), II (Venice, 1978); *Urkunden zur älteren Handels- und Staatsgeschichte der Republik Venedig*, ed. G. Tafel and G. Thomas, III (Vienna, 1857), 374–76; E. Ashtor, *Levant Trade in the Later Middle Ages* (Princeton, 1983), 43 f, 54 ff, 104; D. Jacoby, "L'expansion occidentale dans le Levant: Les Vénitiens à Acre dans la second moitie du treizième siècle," *Journal of Medieval History* 3 (1977), 234 ff; Edwards, "Şebinkarahisar," 35–38, esp. notes 46, 50, 53, 59, 60; C. Desimoni, "Actes passés en 1271, 1274 et 1279 à l'Aias (Petite Arménie) et à Beyrouth par devant des notaires génois," *AOL* 1 (1881), 434–534; C. Trasselli, "Sugli europei in Armenia: A proposito di un privilegio trecentesco e di una novella del Boccaccio," *Archivio storico italiano* 122 (1964), 471–91.

[5]See: Sümer, 54 ff; the exaggerated view of Armenian seapower in Ju. Barsegov, "Borʾba Kilikiĭskoĭ Armenii protiv piratstva v sredizemnom more," *Patma-banasirakan Handes* 62.3 (1973), 72 ff; *RHC, DocArm*, II, 205, 212, 325, 331–33.

[6]Bar Hebraeus, 465. There is a reference on the destruction of Ayas in a letter to King Edward I (1281); see *A Crusader's Letter from 'the Holy Land'*, trans. W. Sanders, PPTS (London, 1888), 12.

[7]See A. Luttrell in Boase, 137–44, esp. notes 89–119.

[8]Dardel, 36 note 4. Peter captured and briefly held the island castle; see J. Delaville Le Roulx, *Les Hospitaliers à Rhodes jusqu'a la mort de Philibert de Naillac (1310–1421)* (Paris, 1913), 160. His troops laid waste the undefended town, but were unable to take the land castle; cf. Makhairas, I, 192–94, II, 124, 141. In the 15th century Ayas became a district in the Mamluk province of Aleppo; see Popper, 17.

[9]Hellenkemper (163 f, pl. 80) believes that a *substantial medieval* settlement covered the same area. However, I could find little physical evidence for this. Budde's plan (I, 62) shows only the general dimensions of the late antique city. See also Jones, 200 ff. Cf. Alishan, *Sissouan*, 438 f.

[10]Cf. Edwards, "Şebinkarahisar," 27, 57 f.

[11]Alishan, *Sissouan*, 433.

[12]The depiction of the circuit wall on the plan is *very* hypothetical as are the dimensions given to rooms A through F. Most of the circuit wall is made from spolia and probably pre-dates the thirteenth century construction.

[13]Neither the Armenians nor any other medieval builders in or around Cilicia erect fortifications with clamps. The only known Armenian use of clamps is in the church of St. Sophia at Sis; see Edwards, "Second Report," 134 ff.

Azgit

On the west flank of the strategic road that links Göksun and the Cappadocian plain to Kadirli and Cilicia Pedias is the small garrison fort of Azgit.[1] Azgit Kalesi is also at the junction with the Maraş trail which passes Kalası and Dibi Kalesi.[2] Minor tributaries of the Ceyhan, which flow east and south of Azgit, keep the nearby valleys green the entire summer. This fort crowns a kidney-shaped outcrop of limestone and has a commanding view through the entire vale (fig. 14). This site is only a few kilometers north of Andırın and is readily accessible from the paved road. Azgit Kalesi has been the object of two surveys in the last thirty years. The first plan was made by J. Thomson.[3] In his brief description he concludes that the fort is a Byzantine construction. Some years later, J. B. Dunbar and W. W. M. Boal completed a more exhaustive survey of Azgit Kalesi.[4] In general the survey conducted by Dunbar and Boal is good, and my intention here is merely to offer supplementary information.[5] I accept their conclusions that this complex is of Armenian origin and that its medieval name is unknown.

Because of the sheer cliffs around the north half of the outcrop the easiest line of access is at the southwest. The architects placed there the major entrance, gate A. The single bailey of the fort has an irregular circuit that closely follows the sinuosities of the rock. There are two small postern gates at the north (E and F).

In general the masonry is used in a manner consistent with Armenian traditions and appears to be the result of one period of construction. The exterior facing consists of a roughly cut type V.[6] The facing around the west postern (F) has bossed centers with drafted margins. The interior facing consists of a combination of types III and IV with the latter being confined to the lower courses. A darker-colored limestone masonry in the upper part of the circuit does not represent a different period of construction (pl. 290c), but was probably taken from a different quarry site than the gray limestone in the lower half. Where the cores of the walls are exposed, they seem to be a fairly consistent mass of fieldstones bound in a heavy matrix of mortar. Only in a few cases are the stones obliquely set in regular courses, resembling a herringbone pattern.

A very narrow pathway at the southwest leads to gate A and its flanking tower. This tower may have functioned as a small cistern. The gate is a typical Armenian construction in that it has an outer pointed arch, which is separated from the jambs by a narrow slot machicolation (pl. 290b). By 1973 the arch on the interior side of the jambs had collapsed. This door was secured by a crossbar bolt. A short vaulted(?) passage on the interior side of the door has also collapsed. The south wall of this passage probably joined a second wall at the southeast to form the separate room B in the extreme southwest corner of the fort.

At the southeast is the most complex construction in the fortress. It consists of a large salient whose lower level is divided into the two vaulted chambers C and D. These rooms are vented through high-placed holes in the north walls and probably functioned as cisterns. Water was drawn from above through a hatch in each vault. The upper level, which was entered through a squareheaded door at the north, was probably covered by some sort of wooden roof. Four embrasured loopholes open the walls of this level. On the exterior of the upper level at the south and east there is a series of corbels that once supported some sort of wooden brattice. This platform could be reached by two openings in the wall (not shown on the plan) above the level of the embrasures. A heavily fortified salient was placed here because of the line of access to gate A. I would not call the multistoried complex C–D a keep.[7] By definition a keep is an independently defensible unit within the fort. If this salient were a keep, it would not have three jambless doors on the north side nor would the brattice level be connected with the wall walk at the west. It is likely that the corbels would have extended around to the north side if this were the place of final retreat.

Farther to the north as the topography descends inside the fort there is the east postern E. Postern E is flanked at the south by a now collapsed enclosure. The badly damaged postern shows no evidence of jambs. The remains of the circuit in the north half of the fort show considerable evidence of joist holes. It appears that many wooden buildings were accommodated on the interior.

Like E the west postern at point F is protected by a salient. F is the best-preserved door in the circuit (pl. 290c). The jambs of this door are covered by a lintel on the exterior; on the interior side there is a vault and pivot housings behind the jambs to accommodate double wooden doors. These doors were secured by a crossbar bolt. There are the remains of three incised crosses on the frame of this door. Each cross is surrounded by a circular border, and its four arms have tapering ends.[8]

A variety of features marks this complex as an Armenian construction. Among these are the design of gate A, the irregular plan, the avoidance of corners in the circuit wall, and the masonry. The southern neighbor of Azgit, Ak Kalesi, is an Armenian fortification that has an identical kind of masonry.[9]

[1] On the plan the contours are separated at intervals of 50 cm.

[2] Azgit may appear as "Dunkale" on the *Marash* map. Azgit is marked merely as "Kale" on the maps of *Gaziantep* and *Maras*. See also the map in B. Atlay, *Maraş Tarihi ve Congrafyasi* (Istanbul, 1973).

[3] Fedden and Thomson, 46 f.

[4] Dunbar and Boal in Boase, 84–91.

[5] The geographical description offered by Dunbar and Boal (Boase, 85–86) is accurate, but I should emphasize that the area around Azgit has not been fully explored. It is impossible at present to assign historical names to any of the sites in this region. Cf. Alkım, map 3.

[6] Boase, pls. 35–38.

[7] Ibid., 89.

[8] Ibid., pl. 39. Dunbar and Boal call these reliefs "Greek crosses," but it should be noted that such a design is common on Armenian buildings after the 6th century (cf. Işa in Edwards, "Second Report").

[9] We also know that Ak is an Armenian construction because of the unique chapel in its compound; see Edwards, "First Report." The chapel immediately south of Azgit Kalesi is an Armenian construction; see Edwards, "Second Report," 123 note 2.

Babaoğlan

This site lies on the fringe of Cilicia Pedias, almost midway between Amuda and Kum Kalesi.[1] This garrison fort has clear intervisibility with Çardak, Bodrum, and probably Anavarza. Babaoğlan[2] is the guardian of a strategic road that gives access to the Cilician plain from the north. Today the fort and the village of Babaoğlan (the latter lies immediately south of the fortified outcrop) are easily reached by driving north on the road between Bodrum and Amuda and taking the designated turnoff to the right. This site has never been the subject of a scholarly investigation. In 1947 H. Bossert and U. Alkım published a small selection of photos of this site but no commentary on the fort, except for noting that "the oldest parts of the citadel are clearly of a very antique period."[3] This site has no recorded history and consequently the name of Simanagla assigned to it by Hellenkemper and its supposed twelfth-century date are purely speculative.[4] The location of Simanagla in Cilicia cannot be determined.

The easiest approach to the castle is from the southeast, where a trail eventually winds around the south spur of the elongated outcrop and surmounts the summit at the northwest. This limestone outcrop emerges from the rolling, shrub-covered hills like the back of a serpent (fig. 15). The walls fortify the entire summit.

To the northwest at the base of the outcrop are the scarped remains of what appears to be a chapel-apse (the detail "A" on my plan; pl. 30a). The masonry that once surmounted the scarped walls and constituted the entire nave has been removed. Only a few ashlar blocks remain in situ.[5] Coursed rubble is visible around this site. The scarped apsidal wall is relatively well preserved. On the south side are two scarped niches. The easternmost has a pointed cover and is located in the apse proper, while the other is actually positioned in the nave at the point of junction of the apsidal and nave walls. The niche in the nave is set at a lower level and rounded on the interior. A corresponding niche may once have been located in the shattered west end of the north apsidal wall. This curious arrangement of niches indicates that this chapel is of Armenian construction.[6] In the center of the apsidal wall at the base of the scarped section there is a large, seminatural depression (not shown on the plan) that extends around the interior of the apse for more than 2 m. Only excavations can determine whether this is a passage to a subterranean chamber or a tomb. Just north of the apse there is a narrow scarped cavity with long straight sides; its function is unknown. Two meters north of this opening is a scarped oblong hole, probably a grave. This area, as well as the east side of the outcrop, has numerous examples of such tombs, most of which have been plundered. About 9 m west of the apse is a fragment of a revetment wall (not shown on the plan) that probably supported the flattened terrace of the chapel.

After ascending to the summit of the outcrop, a large relief on the face of an almost vertical cliff becomes visible (the relief is north of the castle, and its location is not shown on the plan). The relief depicts a rearing horse with rider and a spear-bearing attendant(?) restraining the horse with his right hand. The attendant-soldier has a sword(?) strapped to his waist. The frame around the relief has a gabled top. The clothing of the figures and the style of the relief are quite similar to Roman/late-antique depictions in Anatolia.[7]

About 60 m to the south of the relief is the fortified complex. The fort consists of a large summit-bailey (D–F), which is preceded at the north by the bailey-outworks B and C. The latter are somewhat reminiscent (although on a much smaller scale) of the outworks at the Euryalos fort in Syracuse. At the extreme north is a fighting platform that guarded the approach from the northeast.

The masonry of the castle is difficult to evaluate. The inner facing for all of the walls is type IV masonry (occasionally type III). The south (or outer) face of the south wall of ward C consists of a combination of type IV and poorly hewn fieldstones. With the exception of the outer facing of the northeast wall of ward B and the south wall of ward C, the exterior facing stones in this castle consist of a crude type IV masonry (pl. 30b). In a few cases along the west circuit mortar has been used to stucco the exterior face of the type IV. The upper sections of the north walls of B and C have been repaired with type III masonry. The inner face of the south wall of ward C and tower D has a well-cut type V masonry (excluding some repairs at the far west; pl. 31a). The core in the type V has a large number of brick fragments that do not appear in the cores of the type IV walls. Because of the proximity of the tripartite gate to tower D (pl. 32a), the Armenians are probably responsible for the wall and the adjoining tower D. The outer face of the northeast wall of B is rather complicated (pl. 32b). It appears that the poured core was faced with a relatively smooth ashlar masonry in the lowest levels. The two lowest courses of the facing consist of rather large stones that form a broad socle. The next four courses are a large, well-cut type IX. Above the type IX there are six courses of a small crude ashlar. Above this the rest of the wall consists of a well-coursed type IV masonry. It is possible that the crude type IV used throughout the castle as an exterior facing belongs to the Byzantine period of occupation. Just who is responsible for the unusual combination of masonry types on the outer face of the northeast wall of B is a mystery. It seems unlikely that it dates to the period of the kings of Castabala (first century A.D.),[8] since no similar masonry styles are seen in Cilicia before the third century A.D. It is quite possible that the relief was carved before the construction of the fort.

The most formidable structure guarding the line of access into the main bailey is tower D and the adjoining gate (pls. 31a, 32a). This bastion is placed midway between the summit of the outcrop at the southwest and the first entrance to the castle (that is, the now missing gate at the east end of ward B). On the interior tower D has a single, hexagonal chamber that is entered from the south through a door covered by a pointed vault. The room is opened by three windows (pl. 31b). On the interior the central window is covered by a round vault, while the lateral windows are pointed. On the exterior the three windows are topped by a lintel. The frames below these lintels are bored to accommodate crossbars. The sills of all three windows have a splay for archers. This room in tower D is covered by an apsidal vault of crude stones. The battlements that once stood atop this tower have collapsed.

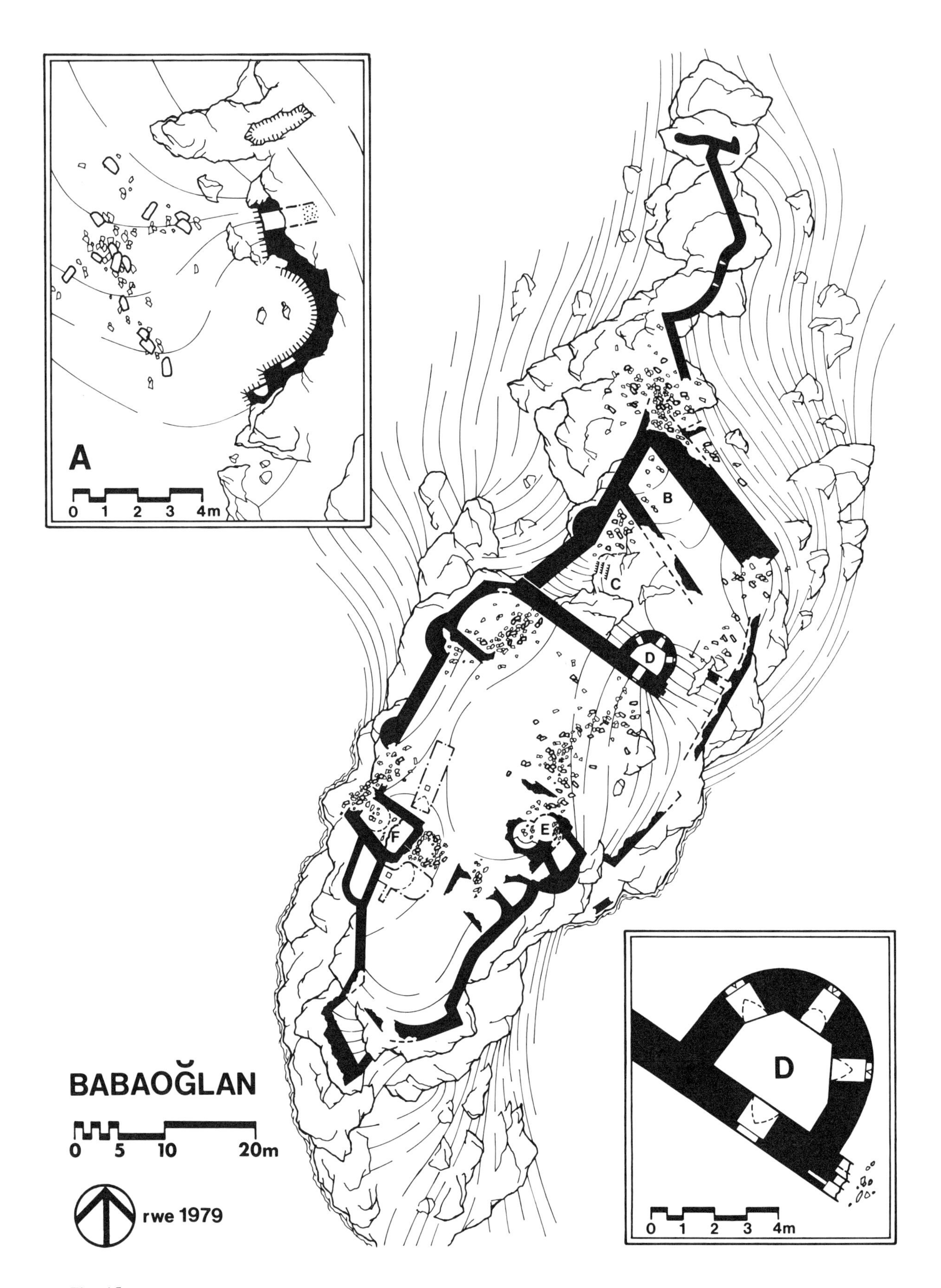

Fig. 15

The gate on the east flank of D has also collapsed, leaving only a section attached to the tower. This is the typical Armenian gate with the crossbar bolt and the slot machicolation (cf. Anahşa Kalesi).

Once past tower D, the approaching party enters the main ward of the fort. Because of extensive damage and heavy undergrowth many of the buildings are difficult to locate. The Armenian-type of glazed pottery abounds in the castle. When excavations are permitted, the richest area archeologically may be complex E. It consists of two circular rooms that appear to have had tile floors at one time.[9] The westernmost room has fragments of a stone bench attached to its south wall. Near the bench I located large sections of two different types of mosaics. One type has small square stones that measure 7 mm on a side. These tesserae are colored pink, white, and navy blue. In an area east of the bench I located larger, perfectly square stones, measuring on their side an average of 32 mm. These pieces of mosaic consist of a very fine red and white marble. Amid this rubble I found large fragments of mother of pearl. The latter and perhaps the marble are not indigenous to Cilicia and would have to be imported. Also visible directly outside of complex E are large fragments of plaster covered with a dark blue pigment. These finds indicate that Babaoğlan Kalesi was not a mere garrison fort but doubled as a residence for a wealthy lord. Mortar analysis has shown that chemically the binding material of the mosaics is much more like the mortar in the type IV walls than the mortar in the type V facing of tower D.[10] If the type IV is a Byzantine masonry, then complex E may date to that period.

The only other structure of significance is cistern F (pl. 33a). It appears that this natural depression was partially scarped and masoned walls were added to complete the cistern. Its walls were stuccoed, and a drain pipe was added at the south. Most of F's barrel vault has collapsed. To the northeast and to the southeast of F are two subterranean cisterns that are in a better state of repair. Both of the flanking cisterns are opened by a small hatch and covered by a vault.

[1]The plan of this hitherto unsurveyed site was executed in July 1979. The contour lines are separated by intervals of 50 cm. Because of the limitations of scale, chapel A could not be included in the main survey but appears as an enlargement.

[2]This site appears on the *Kozan* map, where a "Kale" is located about 5 km east of Karakaya. See also the map in Bossert and Alkım and Alkım, map 3.

[3]Ibid. Hellenkemper's account (135–36) is derived merely from the photos of Bossert and Alkım.

[4]Hellenkemper, 136.

[5]The area in solid black on my survey of "A" indicates where the masonry stood atop the scarped rock.

[6]This may be a Byzantine chapel later converted by the addition of niches to Armenian use. This chapel is not discussed in my "First Report" or "Second Report."

[7]Cf. the analysis by Bossert and Alkım, 24. I do not see any similarity between the Neo-Hittite reliefs at Karatepe and the relief at Babaoğlan. Similar reliefs in Cilicia are found at Korykos and Kanlıdivane. Cf. Keil and Wilhelm, pls. 41, 45.

[8]Bossert and Alkım, 24 f, pls. 22 f.

[9]At this time there is insufficient evidence to conclude that E was part of a bathhouse.

[10]See below, Appendix 2.

Başnalar

Başnalar Kalesi[1] is a small garrison fort that crowns the top of a gently sloping hill not far from Kuzucubelen (fig. 16).[2] Başnalar can be reached by turning off the paved highway near Kuzucubelen and driving east-northeast along a winding road for about 15 km. From the end of this road to the fort is an uphill hike of about twenty minutes (pl. 33b). Today there are few signs of habitation in the area immediately around the fortified outcrop. One farmer is using a well at the base of the outcrop for irrigation. I could not determine if Başnalar Kalesi had intervisibility with the watch post at Kuzucubelen. There is a clear view from the former site into the upland valleys at the north and west (pls. 34b, 35b). The view to the east is blocked by a mountainous outcrop of limestone. At an altitude of 550 m there is also clear intervisibility with the Mediterranean at the south and the southeast. It appears that Başnalar Kalesi guarded an abandoned trail that once led from the Mediterranean and joined the road to Arslanköy north of the village and fort of Kuzucubelen. No historical names can be associated with this site.

This fort is a rather symmetrical structure. It consists simply of a single circuit wall, two open salients, and two towers that enclosed rooms. One of the latter (at the south) is squared on the exterior. This factor, as well as the exposed angles of the circuit and the relatively crude masonry, distinguishes this fort as a Byzantine, rather than an Armenian, construction.

The difference between the exterior and interior facing is that type IV is used on the facade, while type III seems to be common on the inside (pls. 34b, 35a). Both types of masonry are especially crude, being little more than roughly cut fieldstones. Their broad interstices are filled with small rocks and much mortar. As in the other Byzantine forts of Cilicia, there is no con-

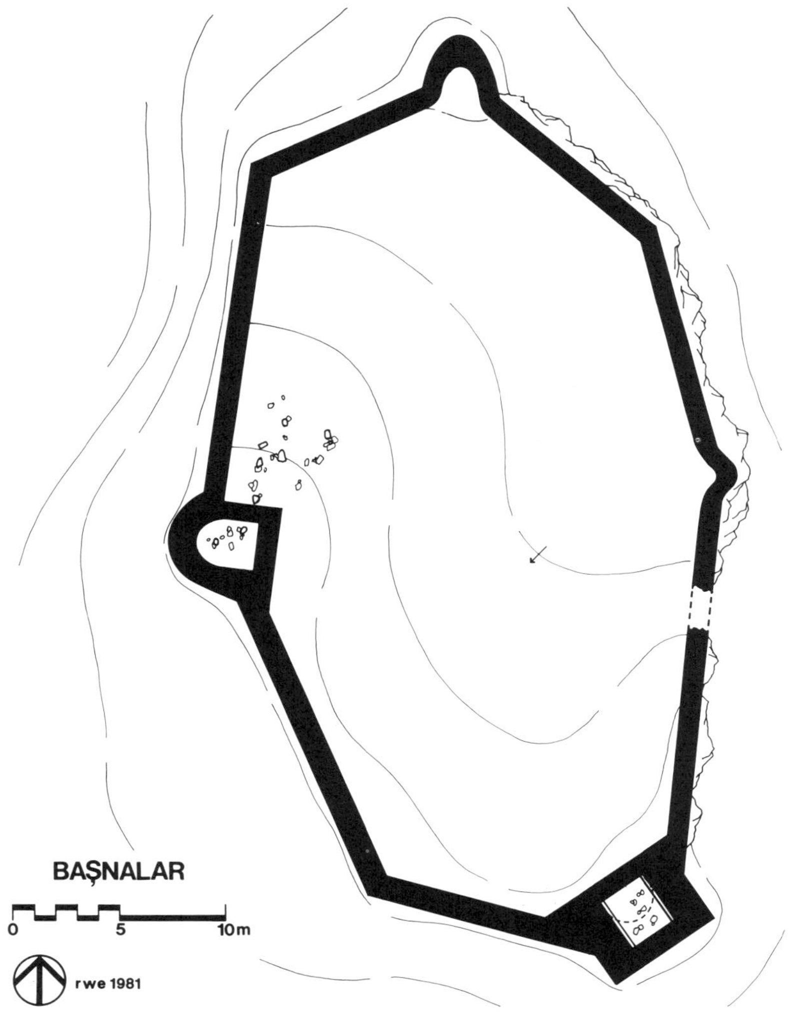

Fig. 16

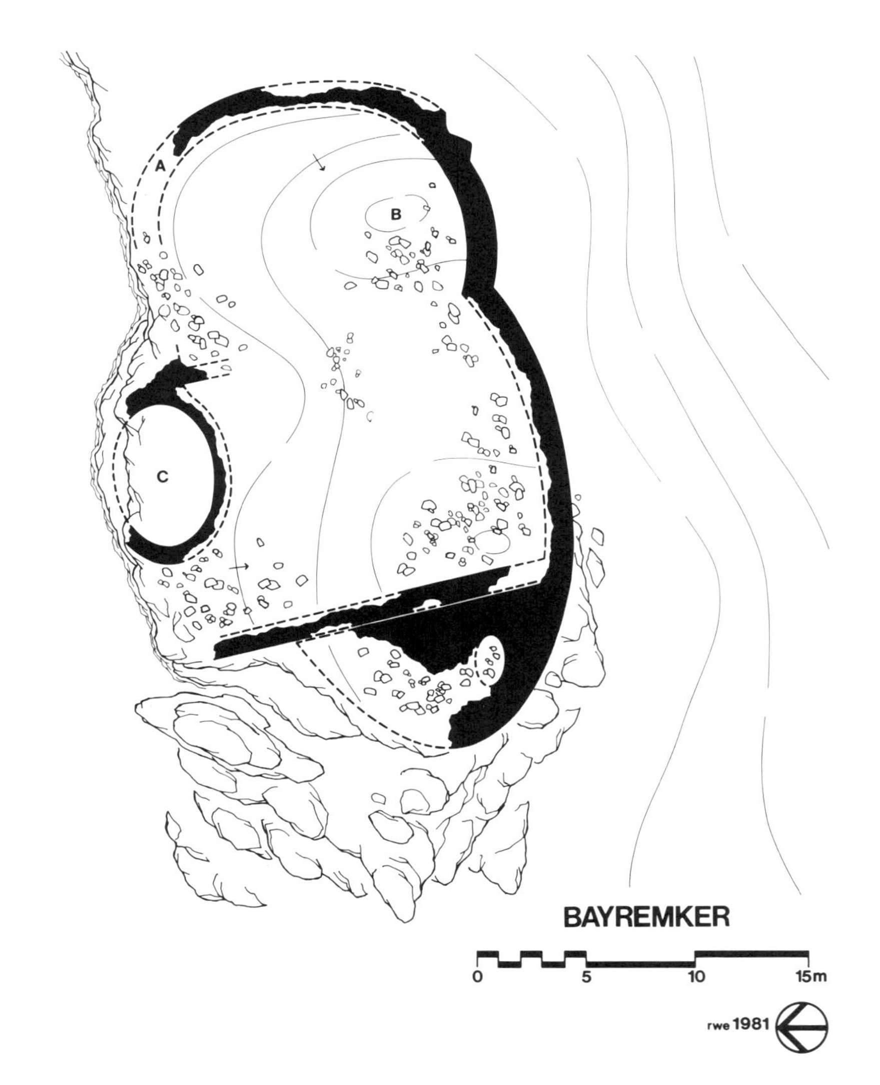

Fig. 17

scious attempt to taper the inner side of each block. Occasionally, very large, neatly squared stones are placed at regular intervals, or there is a single course of such stones found in the lower half of the wall (pl. 35b).[3] These oversized blocks act as headers. The exposed angles of the circuit wall are constructed with quoins (pl. 34a).

Today most of the circuit wall is in a relatively good state of preservation with the average height at over 2 m. Because of extensive undergrowth it is difficult to determine if there are any stone buildings on the interior. At the south the square tower enclosed a split-level room. It appears that the lower-level room was topped by a ceiling of wood. This roof was anchored on shelves in the east and west walls. The upper-level room seems to have been covered by a vault. Today only a few of the springing stones of that vault are still in situ. The south wall of the upper-level room may have been opened by a broad window (not shown on the plan) since parts of a flat, vertical frame are still evident.

The tower in the west circuit enclosed a single room that was two stories in height (pl. 35b). There seem to be no horizontal or vertical divisions in this apsidal room. The shallow salients at the north and northeast do not enclose rooms (pl. 34b). However, we know that there were wooden rooms at the north end of the bailey because there are numerous joist holes to support horizontal beams. Since the only opening in the circuit is a small breach on the east side, this was probably the site of a gate.

[1] The plan of this previously unsurveyed site was executed in June 1981. On the plan the contours are separated by intervals of 75 cm. The "Manascha Kale" of Schaffer (60) may lie between Başnalar and Belen Keşlik; cf. Heberdey and Wilhelm, 40 f.

[2] The site of Başnalar appears on the *Mersin (1)* map. Cf. *Handbook,* 195.

[3] Judging from Lamas and Korykos, Byzantine builders have a tendency to use headers to stabilize the walls.

Bayremker

Bayremker Kalesi[1] is a small garrison fort that guards the strategic road between Bostan and Mansurlu. This is one of the two roads that link Sis to Kayseri (Caesarea). South of the fort the outcrop dips, creating a natural pass to the neighboring valley at the east and consequently a junction with the road that leads to Maran. I could find no solid evidence that this site has intervisibility with any other medieval station. However, large parts of the immediate area still remain unexplored. The road to the west of the fort flanks the Inderese Çay, one of the tributaries of the Seyhan River. Exactly why the locals call the fort Bayremker is a mystery. Only two villages are relatively close to the site: Çardak Köy at the southwest and Bulhaniye to the northwest (pl. 36a). No historical names can be associated with this fort.

The easiest way to reach this site is to hike northeast from Çardak Köy. When approaching the fort from the east some fragments of ashlar masonry are visible just below the top of the outcrop. On one of the fragments I observed the faint but certain traces of a cross: ☩ . One of the locals said that the cross marked a grave, but there is no evidence to substantiate this claim.

Essentially two types of limestone masonry are used in construction of this fort. On the exterior of the walls a very crude type V masonry is employed (pl. 37a). In a few cases the rusticated face has drafted margins. These blocks are laid in regular courses and measure on the average 47 cm in length and 30 cm in height. The interstices are filled with more rock chips than usual, and erosion has removed much of the mortar. In the center of the south wall an area seems to have been repaired with poor-quality stones. The core anchoring the facing consists of a white sandy mortar and large fieldstones (pl. 37b). On the interior the facing stones are crudely cut and seldom rectangular (pl. 36b). These stones, which constitute a large type III masonry, are anchored in regular courses by smaller boulders and an abundance of mortar.

The rounded asymmetrical circuit of the fort conforms to the contours of the outcrop (fig. 17). There is no evidence of doors or windows in the small shattered circuit. The easiest point of access is through a hole in the east wall (pl. 37b). Most of the facing stones on the curving east end have fallen away. The south wall is in a better state of preservation (pl. 37a). Much of the bulging projection at the southwest has collapsed. This salient was actually the lower half of the west defense; a straight wall placed atop the east end of the salient constitutes the upper half of the barrier. It seems that the west salient is solid. Because of landslides the entire north circuit is missing. The southern two-thirds of the circular room C will soon be another victim. Attached to the east flank of C there are the traces of a thin wall that may have cut across the fort, no doubt creating a separate room. There is no other evidence of divisions within the fortification. Because the south side of the fort is constructed on a steep slope it appears

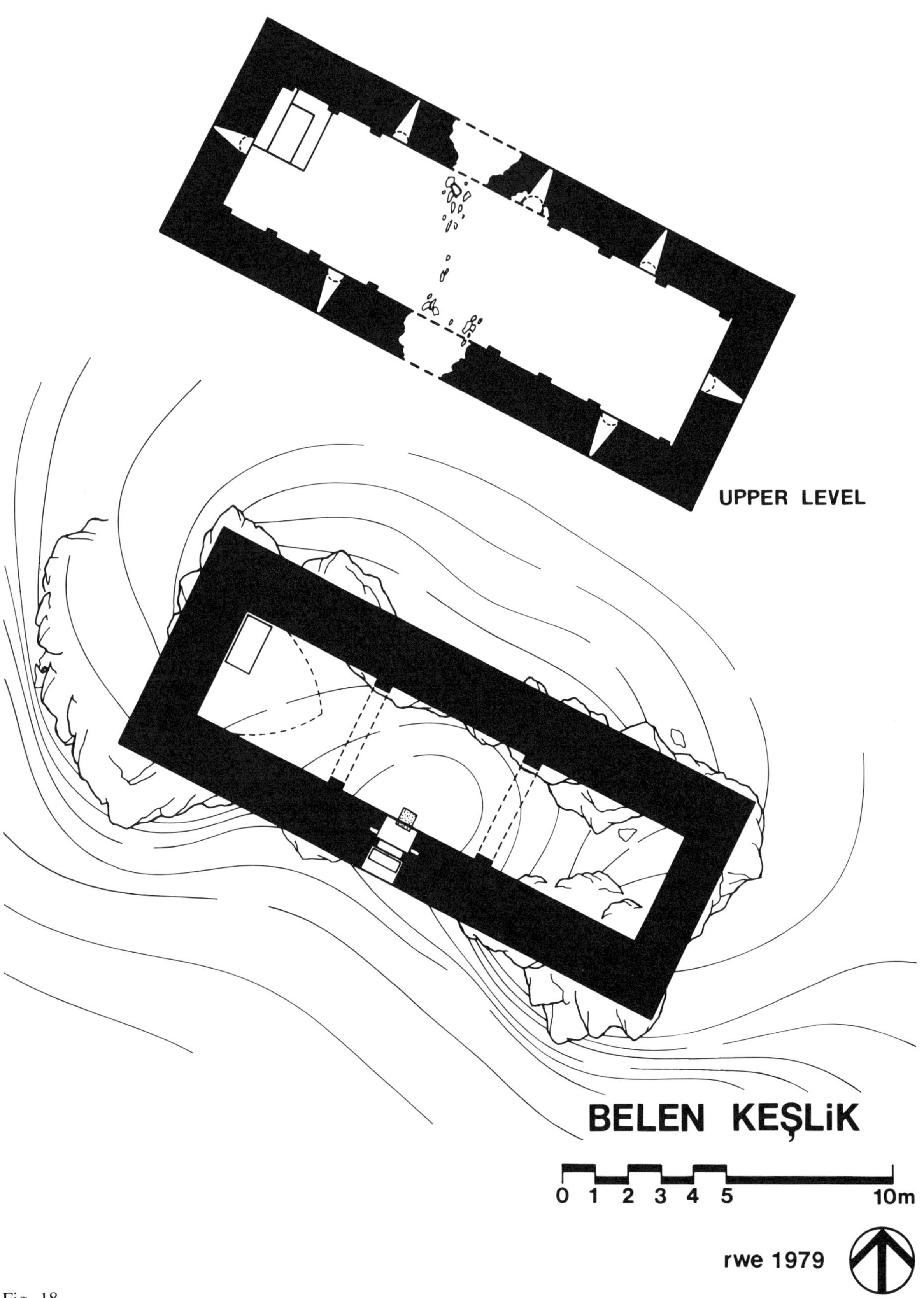

Fig. 18

likely that some lower-level rooms were present. There is some evidence for this at point B where what appears to be the floor level is actually caving in (pl. 36b). Perhaps the hatch-opening to a cistern below has collapsed. On my plan the thickness of the wall was measured at the upper level.

The design and masonry of this fortification indicate that it is of Armenian construction. The tendency in Armenian architecture to round the corners on the outside of the circuit and employ a rusticated exterior facing is evident at Bayremker Kalesi.

[1]The plan of this hitherto unattested site was executed in July 1981. The contour lines are spaced at intervals of 75 cm.

Belen Keşlik

Belen Keşlik Kalesi[1] is a small fortified estate house that is located 5 km southwest of Gösne, near the divide of two major north-south roads. The adjacent village of Belen Keşlik is a tiny agricultural community.[2] Small streams to the north and south supply the village with water. There is no evidence of a cistern within the estate house; the closest source of water is a well about 110 m to the south. The area around the settlement and the fort is relatively flat and is dotted with small limestone outcrops. The fort sits on one of these protrusions (fig. 18). Since the medieval name of this site is unknown, so is its history.

This estate house is a rectangular two-story keep (pl. 38a) that is neither surrounded by a bailey wall nor protected from attack by towers or overhanging machicolations.[3] Today the lower floor is in a perfect state of preservation (pl. 38b), while the roofless second level has large breaches in the north and south walls. There is a single lower-level door at the south.

On the exterior the masonry is a fairly uniform type VII with occasional large blocks of type VIII at the lower levels. The largest grouping of the type VIII is just southwest of the door and may represent an earlier foundation. Natural rock has been skillfully integrated into the foundations of the north, east, and south walls. The stones in the walls gradually and uniformly become smaller toward the top. The masonry on the interior of the lower level is a type VI (pl. 39a), whereas type IV predominates as the inner facing of the upper level (pls. 40a, 41a). Type IX is used to construct the ribs and major openings.

The door at the south is raised over one meter above the current ground level (pl. 38b). Today a pile of crude stones beneath the sill gives access to the door. In medieval times a removable ramp was used. The tripartite door is equipped with jambs and provisions for a large crossbar bolt. Directly southwest of the jambs and at a much higher level is an internal machicolation. This is slightly different from the typical slot machicolation, which is simply a narrow gap between the plane of the jambs and an exterior arch (cf. Anahşa). Here the slot between the exterior facing of the south wall and the wall above the plane of the jambs is carefully framed like a hatch with well-coursed stones. The resulting opening is slightly smaller than the normal slot machicolation. The section of the door directly northeast of the jambs is covered by a single monolithic lintel. On the inside of the door there is a second machicolation (pl. 39a). This square hatch, which is now blocked, allowed the defenders above to attack a party who had breached the door.

The interior of the lower level consists of a single rectangular room that has a gently sloping floor (pl. 39b). This floor is occasionally pierced by chunks of living rock. This chamber is covered by a pointed vault that is supported with two transverse arches. These arches rise on rounded corbeled blocks that project from the center of the wall. In the northwest corner of the room there is a large hatch that opens into the second level. At this upper level the hatch is framed by two shelves. No doubt a retractable wooden ladder was anchored on these shelves.

On the upper level seven narrow slits with stirrup bases are visible. These are the openings for the round-headed embrasured loopholes (pls. 40a, 40b, 41a). Considering the somewhat symmetrical placement of the extant embrasures, it is possible that an eighth was present in the center of the damaged south wall. Surrounding the perimeter of the upper-level interior wall are fifteen small corbels that extend from the center of the wall at the same height as the top of the embrasures (pl. 41a). The corbels are too numerous to have supported transverse arches. They probably carried a flat wooden ceiling. Joist holes above each corbel accommodated wooden crossbeams. No doubt this ceiling was the floor of a third-level fighting platform.

[1]The plan of this hitherto unattested site was executed in June 1979. The interval of distance for the contour lines is 20 cm.

[2]The village appears on the map of *Mersin (2)*. Cf. *Handbook*, 237.

[3]Corbels may once have been placed over the door in the center of the now-shattered south wall (pl. 38b; cf. Sinap [near Lampron]).

Bodrum

Bodrum Kalesi[1] is a garrison fort that crowns the end of a spiny, snake-like outcrop of limestone on the east flank of the ancient city of Hieropolis/Castabala.[2] This site has a commanding view of eastern Cilicia Pedias and has intervisibility with the forts at Toprak, Tumlu, Babaoğlan, and Çardak (and probably with Anavarza and Yılan). The fort at Bodrum is at the junction of two very strategic and ancient roads that link Osmaniye to the Kadirli-Sis route and to the Geben-Göksun road. Today the site is easily reached from Osmaniye by a new paved road. A major tributary of the Ceyhan River flows south past Kum Kalesi and turns abruptly to the west at a point about 2 km south of Bodrum. Smaller tributaries, which flow west from the Nur Dağları, join the Ceyhan Nehri near Bodrum. Today substantial remains of the city of Hieropolis/Castabala still survive.[3] We are not certain of the medieval name of this site.[4] J. Laurent, who translated and edited the *Peregrinatio* of Wilbrand von Oldenburg, associates Wilbrand's city of Canamella with Castabala.[5] The design and masonry of the fort testify to its Armenian origins. The only serious study of Bodrum Kalesi was published in 1976 by H. Hellenkemper.[6]

Today the two baileys of Bodrum Kalesi show considerable damage (fig. 19). The south wall of the south bailey has collapsed, taking with it all traces of the gate-entrance. Part of the south wall of the higher north bailey and almost all of its east wall are missing.

At Bodrum Kalesi normally type VII is used as the exterior facing stone; the interior consists of a combination of types III and IV. The unique feature about the masonry at Bodrum is that some ashlar blocks from buildings in the ancient city below have been recycled into the circuit walls (pl. 291b). The proportion of recycled stones to the freshly cut limestone masonry is quite small. These older blocks are much lighter in color (probably a cheap grade of marble) and have a uniformly smooth surface. Their appearance in the walls contrasts sharply with the rust-colored rusticated blocks of Armenian origin. The reluctance of Armenian masons to recycle the ancient stones may be an indication that they believed type VII masonry was better suited for military structures. There are only a few traces of brick tiles amid the interior masonry of Bodrum Kalesi. This is surprising when one considers the large amount of brick construction at Hieropolis/Castabala. The core of the walls consists of fieldstones, sherds, and fragments of brick bound in a gray limestone mortar.

Because the slope on the west side of the outcrop is fairly gentle (pl. 291b), the architects positioned there the four major towers of the fort as well as the highest walls. The east flank of the fort is positioned on the top of the cliff, and there is little need for extensive construction (pl. 291c). In order to blunt an enemy assault from the north a fosse was cut in medieval times across the entire width of the outcrop. The scarped walls of the fosse are vertical.

In the lower (or south) bailey the principal construction is the south tower. This structure is not opened by any windows but was once crowned by battlements. On the interior of this tower there are two levels. The lower level (shown on the plan) was a rectangular cistern that has stuccoed sides and was covered by a wooden ceiling. The upper room is a vaulted chamber. This latter unit had an open north side and probably some sort of wooden extension, since the walls are studded with numerous joist holes. An almost identical design is used for the southwest tower in the upper bailey, which has a lower level cistern and a partially opened upper-level chamber. Like the south bailey, all traces of the entrance into the north bailey have vanished.

Although the west wall of the south bailey does overlap the southwest tower of the upper bailey, I found no evidence that the north and south baileys represent distinct periods of construction.

[1]On the plan the interval of distance between the contour lines is one meter.

[2]The site at Bodrum appears on the following maps: *Adana (2), Kozan, Marash*. In addition, see: *Handbook*, 310 f; Schaffer, 91, 110; Schultze, 323–25; Boase, 157.

[3]For a specific discussion of the archeology and history see A. Dupont-Sommer and L. Robert, *La déesse de Hiérapolis Castabala (Cilicie)* (Paris, 1964). The claims of F. Stark (*Alexander's Path* [New York, 1958], 8–11) that the fortress is constructed by the Knights of St. John is without any support.

Further information on the history, topography, and architecture of the late antique city of Hieropolis/Castabala can be found in: M. Gough, *PECS*, 392; Hellenkemper, 137 note 3; W. Ruge, "Kastabala," *RE*, 2335 f; Ramsay, 66, 342, 460; Magie, I, 275, 377, II, 1073, 1151 f; Heberdey and Wilhelm, 25–31; J. Keil and A. Wilhelm, "Vorläufiger Bericht über eine Reise in Kilikien," *Jahreshefte des Österreichischen Archäologischen Instituts in Wien*, Beiblatt 18 (1915), 50–52; J. Bent, "Kilikia," *American Journal of Archaeology* 6 (1890), 547–50; idem, "Recent Discoveries in Eastern Cilicia," *JHS* 11 (1890), 243 f; E. Hicks, "Inscriptions from Eastern Cilicia," ibid., 243–54; G. Bell, "Notes on a Journey through Cilicia and Lycaonia," *RA*, ser. 4, 7.1 (1906), 5–9; P. Verzone, "Città ellenistiche e romane dell'Asia Minore: Hieropolis-Castabala," *Palladio*, n.s. 7 (1957), 54–57; Jones, 202–4; Davis, 117, 121, 128 ff; Tomaschek, 39; Imhoof-Blumer, 446–48; idem, "Zur Münzkunde Kilikiens," *Zeitschrift für Numismatik* 10 (1883), 267–90; Frech, 578, fig. 18.

As early as the 4th century Castabala had a bishop who was dependent on the metropolitan of Anazarbus. See: Le Quien, *Oriens Christianus*, II (Paris, 1740; repr. Graz, 1958), 901 f; R. Janin, "Castabala," *Dictionnaire d'histoire et de géographie ecclésiastiques* (Paris, 1953), 1410.

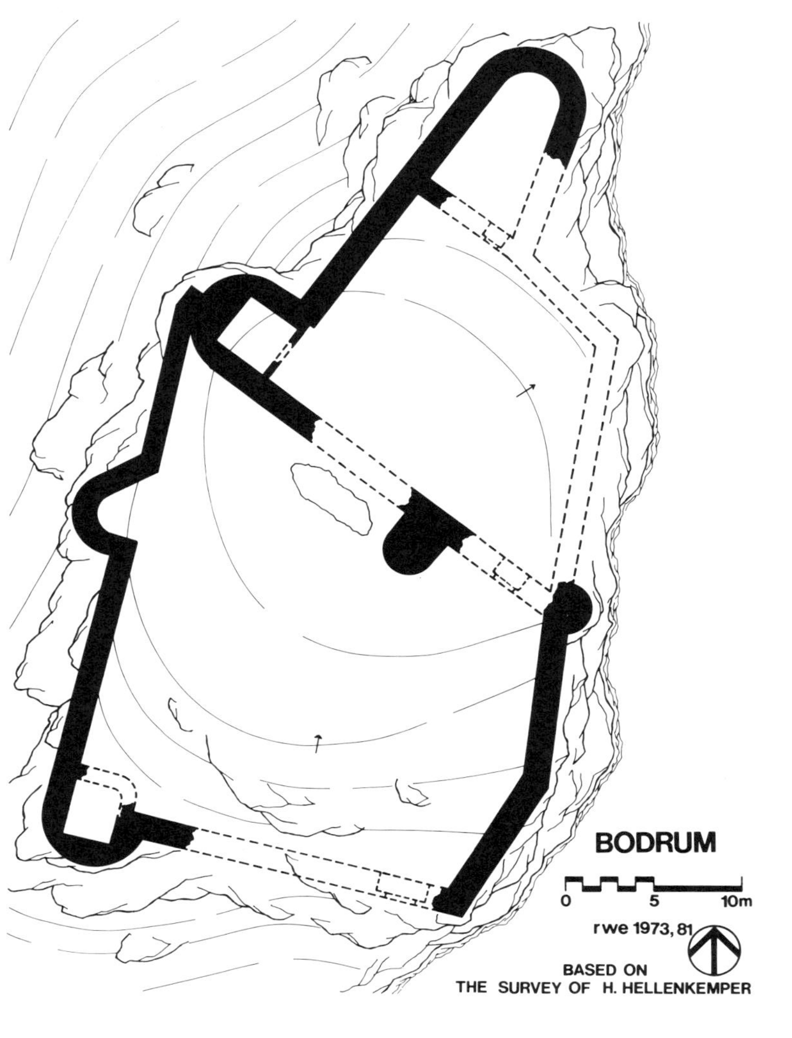
BODRUM
0
5
10m
rwe 1973, 81
BASED ON
THE SURVEY OF H. HELLENKEMPER

Fig. 19

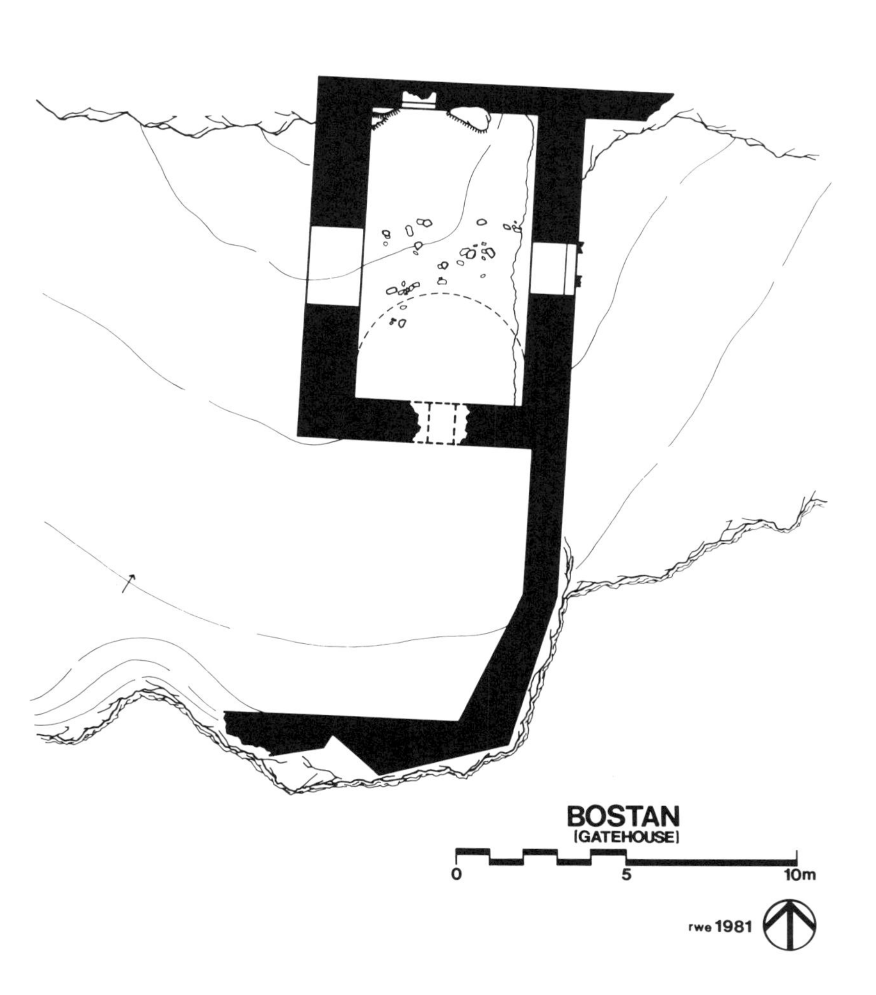
BOSTAN
(GATEHOUSE)
0
5
10m
rwe 1981

Fig. 20

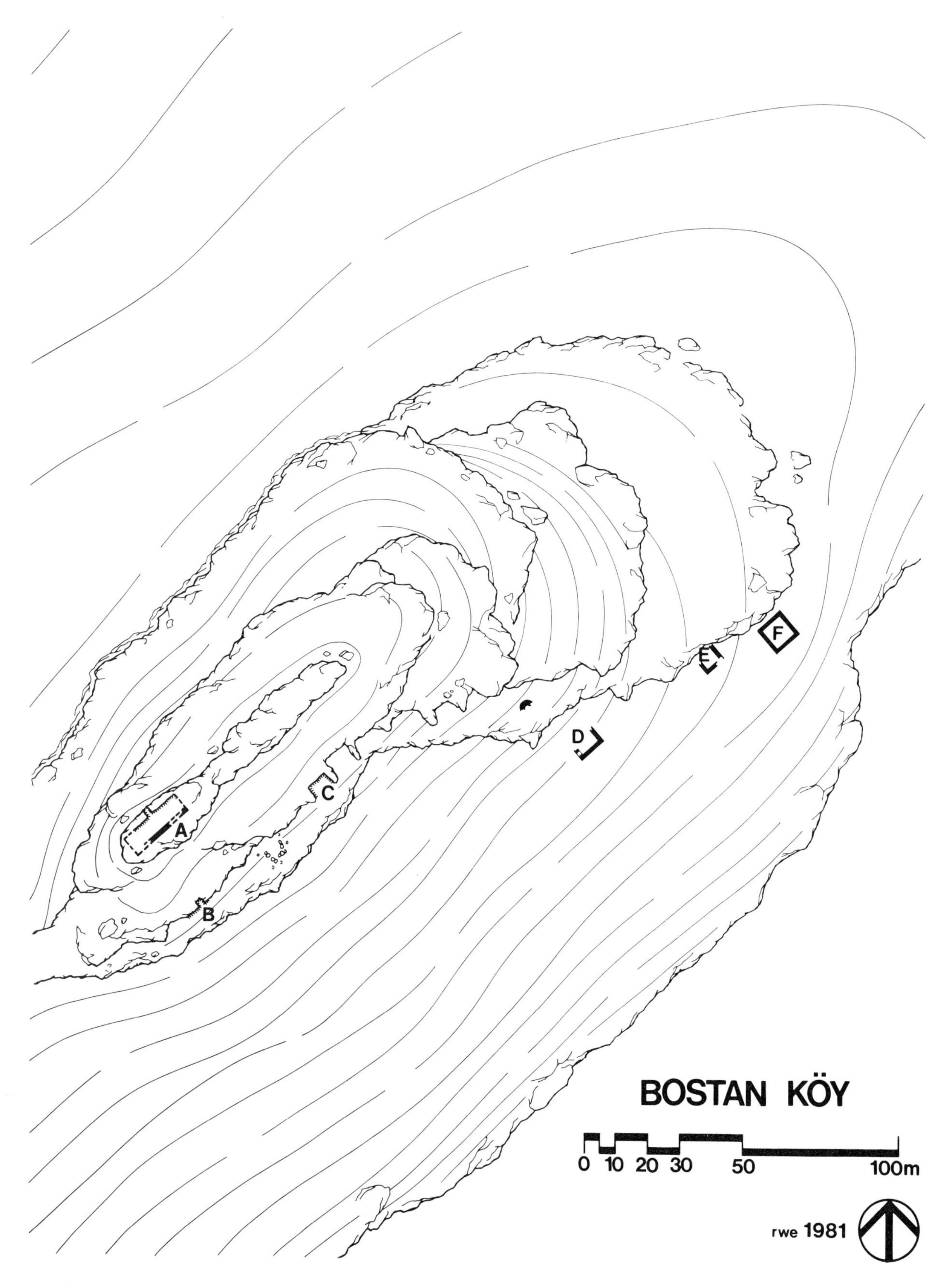

Fig. 21

[4]The origin of the modern name of Bodrum is quite uncertain; see Alishan, *Sissouan*, 228–30.

[5]Wilbrand von Oldenburg, note 98; *RHC, DocArm*, I, xxx. Cahen (149 f, 208 note 8) identifies Canamella with a site on the coast, Ḥiṣn at-Tīnāt. Cf. Deschamps, III, 70 note 2.

[6]Hellenkemper, 137–39. He associates Canamella with Payas (ibid., 107). Cf. Cahen, 148; Rey, 332 f; Langlois, *Voyage*, 472. For a summary of the recent scholarship on the pre-Arab ecclesiastical architecture at Bodrum see Hild and Hellenkemper, 194–98.

Bostan

Bostan Kalesi and the adjacent Bostan Köy (Marangeçele)[1] are two of the most dramatic sights in Cilicia.[2] The fort is situated on the flat top of a limestone pinnacle that rises on the west flank of a deep river canyon (pl. 41b). The road from Kozan (Sis) follows the lower level of the canyon and leads to Mansurlu and Kayseri (Caesarea). The stretch of road from Bostan to Mansurlu is guarded in addition by the small fort of Bayremker. It is very near the site of Bostan that three mountain streams[3] merge to form the Seyhan (pl. 46b), one of the two great rivers of Cilicia Pedias. Since medieval trails are usually associated with canyons carved by rivers, it seems likely that Bostan was a major junction. I did locate a second road that leads northwest from Bostan. The southern stretch of the Kozan–Bostan road is guarded by a watch post at Alafakılar.

The Fort

The fort of Bostan is south of the small modern village of Yeni Köy and southwest of the abandoned site of Bostan Köy (pl. 41b). The hike from Yeni Köy to the top of the citadel is difficult. The easiest approach is to leave this village and hike southwest, slowly ascending the saddle that connects the fortified outcrop to the mountainous spur at the west. Near the summit the unique geology of the site becomes quite apparent. The bulk of the outcrop consists of a poor-quality limestone that is layered in horizontal patterns. The very crown of the summit is a hard gray limestone that has weathered more slowly and has sheer vertical walls. Since the fort is essentially a natural creation, there are no circuit walls but only a gatehouse. On the south side of this hard limestone crown there is a series of natural shelves that lead to the gatehouse at the southeast.

The limestone masonry of the gatehouse is typically Armenian. The exterior facing stones consist of types V and VII (pl. 42a). The stones have been executed with great care and have margins that show few traces of mortar or rock chips. There is a subtle alternation in the height of courses. At the point where the outwork angles to the southwest there is a sharp vertical line to mark the shift. For reasons of structural security it is typical in Armenian architecture that the junctions between stones never occur at the angle where the wall changes course (cf. Kız [near Dorak]). On the interior of the gatehouse the masonry has the same dimensions and quality as the exterior facing, but in some areas the ashlar blocks are smooth (type VIII). The rusticated blocks tend to be on the interior of the north and south walls (pl. 44a). The extant fragments of the barrel vault show that its interior face once consisted of a fine ashlar of smaller dimensions than the stones in the wall below. The avoidance of sharp protruding right angles on the exterior face of the outwork is typical of Armenian construction (pl. 42a).

The gatehouse complex is a rectangular structure that has a curved outwork attached to its southeast corner (fig. 20; pl. 42b). The outwork functions to close off any path of ascent to the west, leaving only a single controlled entrance. The gatehouse was covered by a now collapsed barrel vault. There is a single high-placed window in the south wall which overlooks the courtyard of the outwork. Because most of the voussoirs of the roundheaded jambless west door have fallen away (pl. 42b), little can be said about its construction. Fortunately the east door is in an excellent state of preservation. On the interior the door is covered by a round arch that consists of two adjacent sets of five perfectly fitted voussoirs (pl. 43b). The springing course for the barrel vault of the gatehouse consists of long, monolithic blocks; these are set just above the level of the door's arch. The two extant blocks of the springing course are each almost 2 m in length. Separating these two blocks from the apex of the east door's arch is an oversized, joggled jointed keystone. The exterior of the east door is covered by a single, monolithic lintel (pl. 43a). In the center of the lintel there is a cross incised in a circle. The sides of the door are not distinguished by jambs, but the springing and the lower voussoirs of the door's arch extrude to such a degree that they provide a firm abutment for a wooden door. Sockets carved in the base of each of the springing stones were probably used to anchor the wooden frame for the door. There is no evidence of a crossbar bolt. Just how the door was locked is unknown. Above the east door are two corbels that probably supported a removable wooden brattice. The only unusual feature on the interior of the gatehouse is a shallow, rectangular niche in the north wall (pl. 44a). The niche is covered by a monolithic lintel. Curiously, the natural rock is allowed to protrude through the lower half of the north wall. The north wall of the gatehouse acts as a revetment. The wall that rises above the vault level of the north wall provides our only evidence that the gatehouse had a second floor.

The gatehouse is the most impressive building in the complex. Because of limited time and the fact that the remaining evidence of medieval habitation is less substantial and widely dispersed over the summit, it was impossible to draw a plan for the entire site. A broad flat trail leads northeast from the gatehouse and gently ascends to the very top of the outcrop. Flanking this circular trail at the northeast and north are flat areas that are covered with hundreds of tons of fallen masonry (pl. 44b). In most cases the remnants of walls stand to no more than one meter in height. The types of masonry range from III through V. Because no inscriptions or other readily identifiable evidence has been found, it is impossible to determine whether the collapsed buildings date from a late antique/Byzantine period of occupation or, like the gatehouse, belong to the period of Armenian occupation. What is certain is that the summit of the outcrop protected a sizable community. The possibility that it also functioned as a monastic retreat should not be ruled out. Unfortunately its history is unknown.

Bostan Köy

Bostan Köy (also called Marangeçele by the locals) is an abandoned late antique/Byzantine village that is located approximately 4 km northeast of Bostan Kalesi (fig.21). This village covers the east flank of an outcrop and can easily be reached by hiking west from the modern Yeni Köy. From the summit of Bostan Köy there is a commanding view into the passes and areas at the northeast and west, which are not visible from Bostan Kalesi. Because of this position it is likely that the inhabitants of Bostan Köy also controlled the road. Like the word "Bostan," the designation "Marangeçele" is modern and tells us nothing about the history of the site.

The geology of the outcrop and consequently the masonry are quite unique. Instead of limestone, a lower-grade metamorphic rock, known as phyllite, was quarried from the ascending tiers of the outcrop to construct the buildings (pl. 46a). Since phyllite has a cleavage similar to slate and schist, no extensive cutting was necessary to construct the rooms. In a number of cases the cliffs were simply scarped, and masoned walls were added. The type of masonry used here is atypical of Armenian construction. This masonry has neither a poured core nor any mortar whatsoever to bind the stones in the walls. Because the walls are relatively weak, quoins are used for support.[4] The stones in this masonry are often randomly picked from the surface debris, and their average dimensions are impossible to determine.

In many cases the buildings of this abandoned village have deteriorated to such a degree that they are impossible to survey accurately. However, six structures are distinct enough to warrant comment. At the summit is the long, partially scarped, rectangular building A (pl. 45a). The scarped walls of A, which are prominent at the north, stand to over one meter in height. The exact function of this room is unknown. To the south and below at point B is a rock mass that has been scarped and adapted to now missing walls. In what would have been the northwest corner of this room is a Greek inscription (pl. 45b) cut on the scarped rock:

ΧΑΙΡΕΦΙΛΙΠΠΕ
ΝΕΩΝ. . . . .

Greetings Philip
Of the new (?). . . .

To the northeast room C has only its scarped foundation. Farther to the east, room D is the best preserved on the outcrop (pl. 46a). It has a jambless door at the southwest, and its masoned south wall is over 2 m in height. The east and west walls become progressively shorter as they ascend the slope. At the extreme east end rooms E and F have deteriorated greatly. Near room F is a well-preserved corn grinder(?) and some sort of milling wheel (pl. 46b). Today there are no apparent sources of water on the outcrop, not even a cistern. This may explain why the Turkish residents do not inhabit the site. The fields to the northwest and northeast of this outcrop are still farmed today.

[1] The site of Bostan has two quite separate units: the medieval village of Bostan Köy and Bostan Kalesi. The two are separated by a distance of about 4 km. The two plans of this hitherto unsurveyed area were completed in July 1981. On the plan of the fort, which depicts only the gatehouse, the contours are spaced at intervals of 75 cm. On the plan of Bostan Köy the separation between contour lines is 5 m.

[2] Bostan Köy is listed on the *Kozan* map.

[3] The easternmost of the three streams and the one that flows directly below (east of) Bostan Kalesi is the Gök Su. West of the fortress the Inderese Çay merges with the larger tributary, the Zemanı Su. The latter, which flows from the northwest, is also called the Yenice Irmağı (cf. the maps of *Central Cilicia* and *Kozan*).

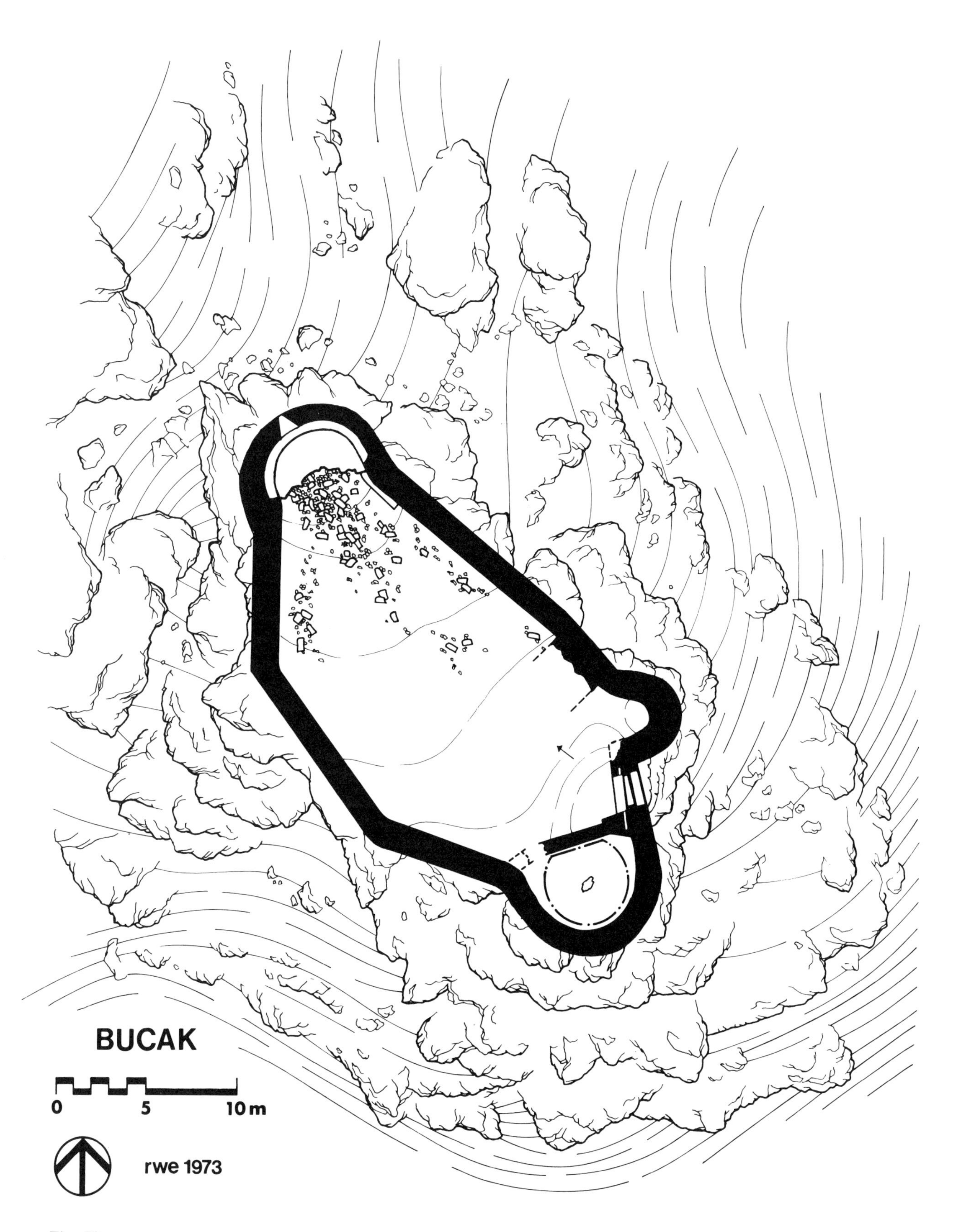

Fig. 22

[4]This masonry technique is quite unlike that used by the modern Turkish residents of Bostan, who employ small fieldstones anchored by vertical wooden headers.

Bucak

The small garrison fort of Bucak[1] lies 26 km east of Kozan and approximately 34 km north of Anavarza.[2] This site can be reached by driving north from Ceyhan through the hamlets of Mercimek and Çukurören. A modern village, which is situated southeast of the fortified outcrop, bears the same name as the fort. Just south of the outcrop a number of streams bearing the spring runoff merge into the principal irrigation canals of the northern plain. Bucak Kalesi stands at the southern edge of the Taurus Mountains and guards the exit of the alternate route that runs south from Vahga (the primary road terminates at Kozan). Even though Bucak is physically close to Sis, the nature of the topography allows intervisibility with only the more distant forts of Anavarza and Amuda.[3] Its position, size, and orientation at the extreme northern tip of the plain indicate that the fort's main function was to house a garrison. Secondarily, it could communicate information on troop movements to the larger forts at the south. From the physical remains there is no evidence that Bucak served as the medieval residence for even a petty baron. We have no historical or literary references for this site.

Bucak Kalesi rests atop a limestone outcrop at the southern end of a large descending spur. The irregular trace of its circuit wall follows the outlines of the gently sloping summit (fig. 22). The ascent to the fort is made from the southeast, and eventually rock terraces (some having scarped faces) lead to a gate at the south.

The circuit is constructed with type VII facing on the exterior and types III and IV on the interior (pl. 292a). Repairs with type V masonry were made at a later date to the upper section of the south tower's outer face. It is also evident that the north wall of this south tower was mended with a crude form of masonry (type IV). The tripartite gate at the south is constructed with type IX masonry.

The machicolation of this gate is slightly wider than normal, and the jambs are somewhat asymmetrical. Today the arches over the gate have collapsed, leaving only the springing stones in situ (pl. 292a). Evidence indicates that the easternmost (or exterior) of the two arches was pointed. A sliding bar locked the double doors beneath the posterior arch. Directly south of the gate is a split-level tower. Because of extensive damage it is unknown whether the upper level was covered by a vault of stone or a wooden roof. The first story of the south tower consists of a subterranean cistern that is covered by a dome. The dome and the cistern walls are plastered to prevent the seepage of water. A hatch, which opened the center of the dome, has collapsed. In placing a cistern and gate at the south end of the fortress, the architects took advantage of the downward slope so that the rain would collect in a central area. An enemy breaching the door would have to fight his way uphill.

On the interior, the areas north of the gate and its flanking tower have suffered considerable damage. Just north of the flanking tower the inner face of the circuit wall shows a few springing stones of a collapsed vault. Because of its relatively flat arc it could not have spanned the width of the fort. The vault probably created a small undercroft against the eastern wall. At the northern end of the circuit is a rounded corner. The upper level of this salient, which has a wall only two-thirds the thickness of the lower level, consists of a fighting platform and a single squareheaded embrasured loophole. Its lower level contains a chamber that is covered by a partially collapsed apsidal dome. An ascending ledge just southeast of the dome probably served as a stairway to the top of the wall and the fighting platform. The interior of the circuit wall southwest and southeast of the northern salient is dotted with a number of joist holes. These openings are about 2 m from the current ground level; they undoubtedly supported horizontal beams for wooden structures.

On the basis of design and masonry, this garrison fort is definitely Armenian; its date of construction is impossible to determine.

[1]My plan of this site was executed in summer 1973. The contour lines are spaced at intervals of 50 cm. Hellenkemper published his survey of this site in 1976.

[2]Hellenkemper, 214 note 1. An alternate name for this site is Sombas. Bucak appears on the *Kozan* and *Marash* maps. On the latter it is referred to as "Bujak Kale." Cf. *Handbook*, 300.

[3]Fedden and Thomson, 12.

Çalan

Çalan Kalesi[1] is situated on an oblong outcrop of limestone astride a major east-west pass in the Nur Dağları. This route, like the Belen pass 15 km to the south, affords a convenient point of access into the gulf of Alexandretta from Syria and the plain of Antioch. Çalan can be reached by driving either northwest from Kırıkhan or east from the road junction north of Iskenderun, known as the Pillar of Jonah ("Portella") in medieval times. Just as Trapesak in the principality of Antioch guards the eastern end of this road, so Sarı Seki and the Pillar of Jonah at the edge of the Armenian kingdom guard the western end. When Paul Jacquot visited Çalan in April 1922, he described a very arduous journey over difficult roads.[2] Today the trip can be made wth relative ease over roads that consist of alternating sections of asphalt and packed gravel. Most of the place names that Jacquot lists on his itinerary are still in use today. The group of valleys to the southwest of the fort, which are separated from the fortress outcrop by a rocky spur, still bear the name Dermen Deresi. The scattered village immediately below the outcrop at the south and southeast is called Çınar (pl. 47a). At an altitude of over 1,200 m, this castle has a commanding view into the interlocking valleys at the southwest and southeast as well as intervisibility with the trails from the north which empty into this valley (pl. 52b).

It is highly unlikely that the Çalan pass played an important role in Alexander's encounter with Darius at Issus.[3] Sometime in the twelfth century the Templars constructed the fortress of Çalan. At present there are no indications that they built over an earlier structure. However, there is a problem in determining the medieval name of this site. C. Cahen, E. G. Rey, R. Dussaud, R. Grousset, and the *Guide Bleu* have offered confusing advice on the names of the various Crusader forts at the south end of the Nur Dağları.[4] The most recent comment on the subject came from Paul Deschamps, who hypothesized that Çalan is La Roche Guillaume, a site that he also identifies with the Arab Ḥağar Šuğlān.[5] Deschamps places La Roche de Roissol south of Hınzır Burnu ("the promontory of the pig"), which is near the ancient site of Rhosicum at the southeastern tip of the gulf of Alexandretta.[6] Cahen believes (as does this writer, though somewhat reluctantly) that Çalan is La Roche de Roissol, and that La Roche Guillaume is a separate site in its vicinity. Unfortunately we know little about the history of either site and have no accurate information on their location. In June 1981 I attempted to visit the fort south of Hınzır Burnu, but I was turned back by a military patrol that had secured the region. I did explore the area north of the military barrier. Atop the promontory there is only an abandoned lighthouse (built in the 1930s), but at the base (just above sea level at high tide) there was a shelf that had been scarped flat. There were a few fragments of coursed masonry.[7] I located definite traces of cut trenches to accommodate walls. Undoubtedly there was some sort of fortified port. Considering that a number of medieval fortifications still remain undiscovered in the southern Anti-Taurus, it is impossible to determine with certainty the Frankish name of Çalan.[8]

The easiest way to reach the fortress from the base of the outcrop is to begin the approach from the southwest, ascending the edges of the rock mass in a clockwise direction. At the north about two-thirds of the way up the outcrop, there is a series of caves (common in limestone formations) which seem to have been adapted for habitation. Near these caves are mounds of glazed sherds; the medieval inhabitants in the castle above must have dumped their debris in this area. Continuing on the trail one eventually encircles the entire outcrop and reaches the lower bailey at a point west of E, joining with what Jacquot calls the "chemin d'accès."[9] However, I saw no evidence of Jacquot's "ancienne passerelle." At the southeast end of the lower bailey and just west of E a steep but accessible slope leads to the first level of the summit (pl. 47b). There is no sharp vertical division between the lower and upper baileys (fig. 23). The summit constantly ascends until it reaches its highest point at the east (pl. 48a). This eastern third of the summit, which constitutes the upper bailey, is relatively flat and is separated from the lower level by the west walls of F and G.

One of the striking features about this fortress is the masonry. Unlike the Armenian forts, which usually employ a quality ashlar for the exterior facing of the walls and a poorer-quality masonry on the interior, here small, poorly cut stones are used as both the interior and exterior facing. The stones of this masonry are anchored by an abundance of mortar and rock chips. In structures that are two stories high, usually the stones at the lower level consist of a larger, poorly hewn ashlar. This masonry is similar to that seen in the nearby Crusader forts at Bağras and Trapesak. The frames for the windows and doors usually consist of finely cut ashlar.

Today the lower bailey is strewn with the rubble of collapsed buildings and the circuit wall. These pieces of coursed masonry came not only from structures in the lower bailey but from the deteriorating

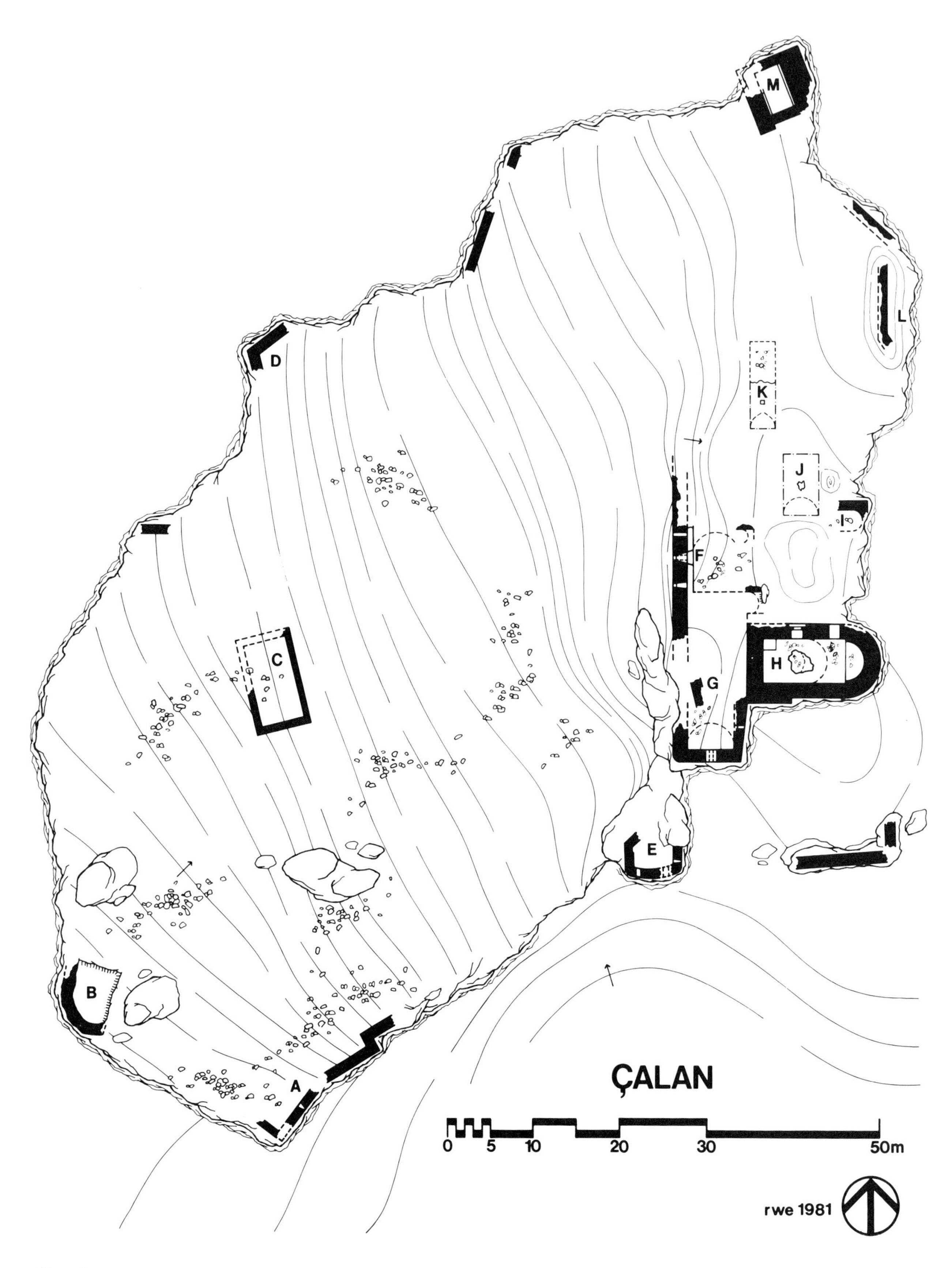

Fig. 23

walls of the upper bailey. Undoubtedly the edges of the lower bailey were once crowned entirely with a fortified circuit. Only fragments of that wall stand today; the largest sections are at points A and D. At the southwest there are still the traces of one embrasured window at A (pl. 47b). Northwest of A is the foundation for chamber B; the north and east walls of this room are scarped. Because B is located at the lowest level in the bailey, it is quite possible that it functioned as a cistern. To the north is the large rectangular chamber C. Its walls stand to about 1.5 m in height. The area of the lower bailey east and northeast of wall D has less evidence of coursed masonry and no visible remains of buildings.

The small outwork E is at a point below the level of the upper bailey, yet isolated from the lower bailey on the crown of a projecting mass of rock (pl. 48a). The south wall of E has two straight-sided windows; the east wall has a single embrasured window. Directly east of E and below there is a fragment of a circuit wall that is perched precariously on a small mound of rock. This too was an outwork, but, considering the nature of the sheer cliffs below E, G, and H, it is difficult to see how it was manned.

There is no obvious entrance to the upper bailey from the lower bailey (pl. 48a). Today the easiest path of ascent is around the north end of room F. Unfortunately only the west wall of that chamber stands today. Springing stones at the top of the second level of that wall indicate that F was covered by a vault. The first floor of F was covered by a wooden ceiling. In a place where it seems likely that the east wall of chamber F would stand, there is the foundation of a small vaulted hollow. This may have been a niche in the east wall or a door. The only opening in the first level of the west wall of room F is a large embrasured window (pl. 49b). On the interior it consists of a broad, tall splay covered with a slightly pointed top. On the exterior a monolithic block of unusually large dimensions forms a flat top for the window. This large embrasured window is reminiscent of the openings at Bağras and is typical of Templar architecture in this region.[10] The west wall of F at the second level is opened by two windows (see the plan; pl. 49a). Like the lower-level opening, the second level windows are embrasured and round-headed on the interior. On the exterior they are topped by a flat lintel. Each is about 60 percent smaller than the lower-level window. The northernmost of the two upper-level windows is centered almost directly over the lower-level opening. A third opening in the second level at the far north appears to be a straight-sided ventilation shaft (for the sake of convenience, I have noted on the plan the position of all four openings on the lower-level west wall of F). A shelf is present below the second-level windows because the lower-level wall of F is substantially thicker than the upper level.

Immediately adjacent and to the south of room F is G (pls. 48b, 50a). Jacquot believes that this area was occupied by a cistern "dans laquelle on peut descendre au moyen d'une corde."[11] Today all but the south wall and portions of the east and west flanks of G are missing. It seems unlikely that G would have functioned as a cistern because its relatively high position would not permit the collection of runoff and, like F, its west wall was probably opened by windows to enhance the defense of the upper bailey. There are indications that at least the south wall of G has been completely rebuilt. On the exterior the base of the southwest corner and what remains of the west wall consist of squared stones that show moderate traces of mortar in the interstices. The upper half of that corner and the rest of the south wall of G are made with smaller stones that are bound in thick beds of mortar. The division between the two types of stones is very irregular. It is likely that the latter represents a later period of Templar repair. In the upper half of the south wall there are two superimposed windows. The lower appears to be an arrow slit, but since this opening is buried in debris, it cannot be determined if the window was embrasured. Directly above there is the sill of a second, wider and straight-sided window. There is also a straight-sided window in the east wall. It is clear from the springing atop the surviving fragments of the southeast corner that room G was covered by a vault and probably consisted of one floor. Because of the extensive destruction, little else can be said about this room. A large section of its collapsed west wall is lying inside of G on the surface.

To the east is chapel H, the best-preserved structure in this fort (pl. 50b). It seems that a door connected G to the single entrance of H at the north. Since the door from G is lower than the portal to H, stairs must have been present. The south and west walls of H, which have both deteriorated greatly, show no signs of having been opened by windows. The facing stones on the rounded east end of H are falling away rapidly. The north wall of the chapel is opened by a wide round-headed window at the east and a door at the west (pl. 52a). The door has jambs and a porch. Both the door and the porch are covered by slightly pointed arches. The outer face of the jambs has that depressed beveled edge that characterizes much of the Templar construction in the Levant.[12] On the interior small portions of the chapel vault are still in situ (pl. 51a). Like the door, it is constructed with perfectly fitted smooth ashlar. In the center of the nave recent (unauthorized) excavation has revealed the presence of a subterranean vaulted crypt (pl. 51b). This may have been the resting place for one of the great knights of the Templar order. There also seems to be a small rectangular opening in the floor near the northwest corner of the nave. Its function is unknown. The apse of the chapel, which is clearly defined by semi-pilasters of ashlar, has almost completely collapsed. There is no evidence of an apsidal window; niches are absent in the walls of the apse and nave.[13]

The area of the upper bailey north of F and H is partially covered by a forest of oak trees. By its location on the edge of the cliff it appears that I is part of a collapsed tower. This salient may once have joined with the fragment of the circuit wall at L. J and K both appear to be vaulted cisterns whose apices are opened by a single hatch. M is a square tower that has lost its vaulted ceiling. The northwest corner of M has completely collapsed; the position of doors and windows cannot be determined.

Although we cannot unquestionably associate Çalan with any historical site, the masonry and architectural features are Templar. Historically, we know that the Templars did try to secure the southern Anti-Taurus in order to create a permanent buffer for the principality of Antioch.[14]

[1] The plan of this site was completed in June 1981. The contour lines are placed at intervals of one meter.

[2] Jacquot, 114–27.

[3] See my discussion of the Pillar of Jonah in the Catalogue; see also Engels, 131 ff; Janke, 34.

[4] Cahen, 141–44, 512, 539; E. G. Rey, *Les colonies franques de Syrie aux XIIme et XIIIme siècles* (Paris, 1883), 350; Dussaud, 433; R. Grousset, *Histoire des croisades*, II (Paris, 1935), 828 note 3; *Guide Bleu de Syrie* (Paris, 1932), 188.

[5] Deschamps, III, 70–71, 363–65; Boase, 92, 159. Considering the location of the fortress and the information supplied by our extant sources, it is difficult to associate Çalan with Ḥaǧar Šuǧlān. The apparent phonetic affinity between Çalan and Šuǧlān is misleading. The modern designation *Çalan* is actually derived from the earlier Turkish *Čivlan*. According to al-Yāqūt (*Muʿǧam al-buldān*, s.v. "al-Ḥaǧar al-Aswad," ed. F. Wüstenfeld, *Geographisches Wörterbuch*, II [Leipzig, 1867], 214, lines 3 ff), Ḥaǧar Šuǧlān is "a fortress in the Ǧabal Lukkām, near Antioch, a high place overlooking lake Yaǧrā (*mušrif ʿalà ʾl-buḥayrah Yaǧrā*), and it is owned by the Templars (*Faranǧ*), who devote themselves to killing Muslims, being celibate, and keeping (their Order) divided into monks and knights." Çalan is in the Ǧabal Lukkām (the modern Anti-Taurus or Nur Dağları), but neither this fortress nor the summits of the adjoining mountains have a view to any lake. Al-Yāqūt (s.v. "Buḥayrat al-Yaǧrā," ed. Wüstenfeld, I [1866], 516, lines 20 ff) says that the lake, which is also called as-S(a)llūr from its indigenous fish, is situated between Antioch and Cilicia Pedias and is filled by the rivers that flow south from Maraş. This body of water is the present Amık Gölü, and the only fortified site in close proximity is the Templar castle of Bağras, which is not likely to be the rock of Šuǧlān. Cf. Le Strange, *Palestine*, 447; *RHC, DocArm*, I, 303 note 1; R. Hartmann, "Politische geographie des Mamlūkenreichs," *Zeitschrift der Deutschen Morgenländischen Gesellschaft* 70 (1916), 33 notes 8–10, 34 note 4; Gaudefroy-Demombynes, 89; N. Elisséeff, *Nūr ad-Dīn, un grand prince musulman de Syrie au temps des Croisades*, I (Damascus, 1967), 194 f; Popper, 17.

[6] Hınzır (or Domuz) Burnu is the Arabic Raʾs al-Ḫinzīr. For information on the latter and La Roche de Roissol (Roissel) see: *Handbook*, 153, 514; Rey, 332 f; Heberedey and Wilhelm, 20; Deschamps, III, 70 notes 7 and 8, 71; Boase, 177; Dussaud, 416, 440–44, 447. La Roche de Roissol has been incorrectly identified with Savranda in *RHC, DocArm*, II, 772. Both La Roche Guillaume and La Roche de Roissol seem to have remained in the hands of the Templars after the campaign by Saladin in 1188, but they were *briefly* captured in November 1203 by Levon I, who later was asked to pay reparations for damage inflicted on both sites. See: Innocent III, Patrologia latina 215, ed. J. Migne (Paris, 1891), 504, 689–94; William of Tyre, *La continuation de Guillaume de Tyr (1184–1197)*, ed. M. Morgan (Paris, 1982), 58 f, 87.

[7] Little can be said about the origin of the masonry. The geology at Hınzır Burnu is very interesting, for the promontory consists of Dunite, Olivine, and Serpentine. Also, minor amounts of Chromite are present. All of these are an outflow of deep-crusted rocks.

[8] I did not locate the "pierre avec inscription" sighted in 1922 by Jacquot (120). It is possible that the Templars abandoned Çalan when they retreated in 1266–68 from the other forts in this region; see: Ibn al-Furāt, 230 note 6; Mas Latrie, I, 210; Cahen, 717; Eracles, *RHC, HistOcc*, II, 457.

[9] Jacquot, 119.

[10] Compare the construction at Trapesak, the donjon at Anavarza (in the Catalogue), and room AA at Bağras (in Edwards, "Bağras").

[11] Jacquot, 120.

[12] Edwards, "Bağras," 429.

[13] Edwards, "First Report," 164 f.

[14] Edwards, "Bağras," 431 f; C. Alagöz, "Coğrafya Gözüyle Hatay," *Ankara Üniversitesi, Dil ve Tarih-Coğrafya Fakültesi Dergisi* 2.2 (1944), 205 ff; J. Riley-Smith in Boase, 92–117.

Çandır

The fortification of Çandır,[1] which is the medieval Armenian site of Papeṙōn (Papeṙon, Barbaron), stands about 40 km north of Mersin in the Taurus Mountains.[2] The castle covers the entire summit of a lofty plateau that rises to an altitude of 1,450 m, just southwest of the village of Çandır. Geographically, it is a most strategic point, since two major routes from central Anatolia meet just north of this castle. A northwestern trail leaves modern Ereğli and winds southeast through the mountainous defiles past Evciler until it merges with a second road from Lampron to form a single southern route. South of Çandır this road to the sea is protected by three other forts: Sinap, Gösne, and Belen Keşlik.

There is also a road east of the village to the cloister at Kız. The valleys around the plateau of Çandır are very fertile and grow a variety of crops including wheat and peaches. The amount of arable land is being increased since the once great forest in the Highlands of this area is being reduced by clear-cutting. Because of the proximity of the mountains, there is a constant supply of water in the summer to feed the rivers north and east of the plateau. At the castle there is no evidence of springs; runoff was stored in natural and man-made cisterns.

The history of this site is one of the richest and most complex in the Armenian kingdom.[3] Its story begins long before the arrival of the Armenian immigrants. If Alishan is correct, Çandır is actually the Byzantine fort of Παπίριον or Παπούριον.[4] This Byzantine site comes into prominence during the reign of Emperor Zeno, who appeared there after his departure from the throne. Joshua the Stylite reports that the emperor entrusted the castle to his companion, Illus, who was to maintain there a treasury and supplies for emergencies.[5] In 479 Prince Marcinus, the younger son-in-law of Verina, was banished to the castle. Five years later the unfaithful Illus retreated there and was executed. However, the identification of the site at Çandır with the Greek Παπίριον is far from certain, as Gottwald recognized. Late antique historians could distinguish between Κιλικία, Καππάδοκες, and areas in Isauria. Παπίριον does not always seem to be associated with Cilicia.[6] If excavations at Çandır someday reveal pottery and numismatic finds from the fifth century (or even inscriptions), then Alishan's identification will be credible.

Concerning the initial Armenian occupation of Papeṙōn, we learn the folowing from Samuel of Skevra's quasi-reliable biography (ca. 1190) of Nersēs of Lampron:[7]

> Emperor Alexius I dispatched the Armenian prince Aplłarip from his seat in Vaspurakan to the province of Cilicia with the command to take charge of Tarsus and Misis. Aplłarip found at the foot of the Taurus mountains and in the Highlands above Tarsus two unassailable castles, one being Lampron and the other being Papeṙōn. Both were surrounded by villages and plots of farm land. Aplłarip maintained Papeṙōn as a treasury and constructed there a large church as a retreat for himself and his family. But he gave Lampron to his trusted and beloved prince Ōšin, who accompanied him from Armenia. Aplłarip died and was buried in Papeṙōn.

Since Aplłarip had no male heirs he bequeathed the castle to Isaac, brother-in-law of his friend Ōšin. By the late eleventh century it fell into the possession of the Het῾umids. One of its prominent owners, Baron Smbat, died before the walls of Misis in 1151/52 while fighting alongside the Byzantine army against his rival, the Rubenid Baron T῾oros.[8] Smbat's son, Bakuran, later (ca. 1169) gave refuge to his sister Rit῾a and her two sons, Ruben and Levon, during the usurpation of Mleh.[9] Bakuran must have lived to a ripe old age since he is still master of Papeṙōn when Levon I is crowned in 1198/99.[10] Later this site falls into the possession of his brother Vasak, then to the latter's son Constantine, and eventually passes to Constantine's son, Smbat the Constable. By 1296 T῾oros, the brother of King Het῾um II, had become lord of the castle.[11] Papeṙōn came into royal hands only when its lord, Ōšin of Korykos, was assassinated in 1329.[12] In the mid-fourteenth century the site was briefly occupied by the Karamanids. The only previous Moslem incursion at Papeṙōn was in 1245 when the Sultan of Konya, Kayẖusraw II, and his disgruntled ally, Constantine of Lampron, torched the district on their march to Tarsus against Het῾um I.[13] For a brief period in 1309/10 this castle (or perhaps Lampron) may have served as a prison for Henry II of Cyprus. In the late fourteenth century under the Mamluks it probably became the headquarters for the district of Balsalūç.[14]

From any approach, the sight of the castle and its plateau inspires awe (pl. 292b). The sides of the plateau consist of almost vertical cliffs of limestone. The architects took special care to seal even remotely possible points of access to the top with short walls (fig. 24). The cliffs, not the walls, were the front line of defense. Since this natural defense was impregnable, towers were not constructed at this site. This exploitation of the terrain is a common feature in Armenian military architecture. Gottwald is right when he says that this fort "keinen festungsmässigen Charakter hat."[15] The walls at the edge of the summit functioned more as revetments. Because these walls were not subject to direct attack, their exterior facing (as well as most of the interior facing) consists of type VIII and occasionally of type IX masonries. As is expected, the only place where type VII appears on the exterior face is the low-level terrace wall at the northwest (just north of F).[16] Below the summit at the north, a single short wall (not shown on the plan) is constructed with an exterior facing of type IV masonry. This type IV may represent one of the remnants from an earlier Byzantine(?) period.

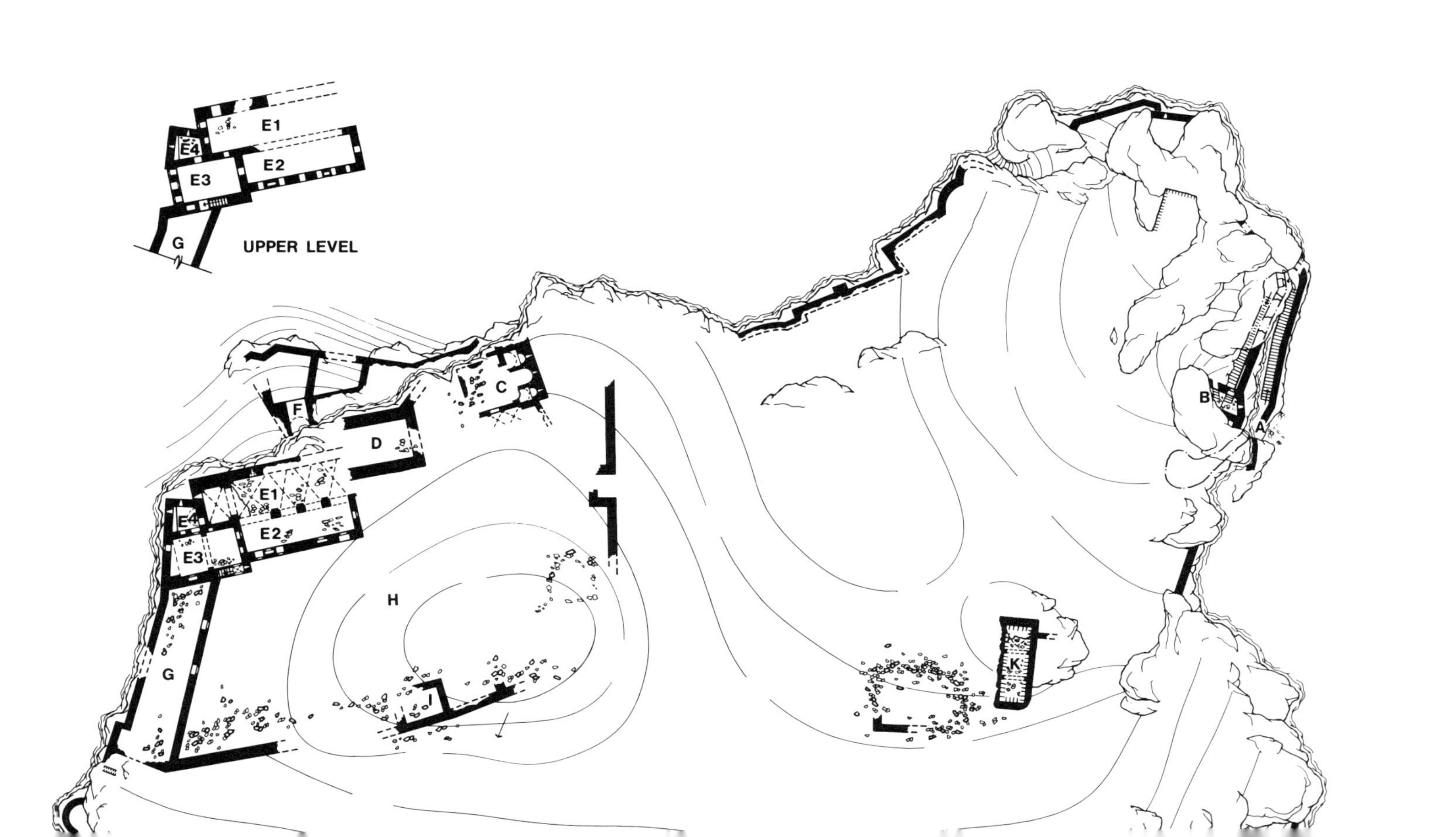
E1
E4
E2
E3
G
UPPER LEVEL
F
C
D
B
A
E1
E4
E2
E3
H
G
I
K

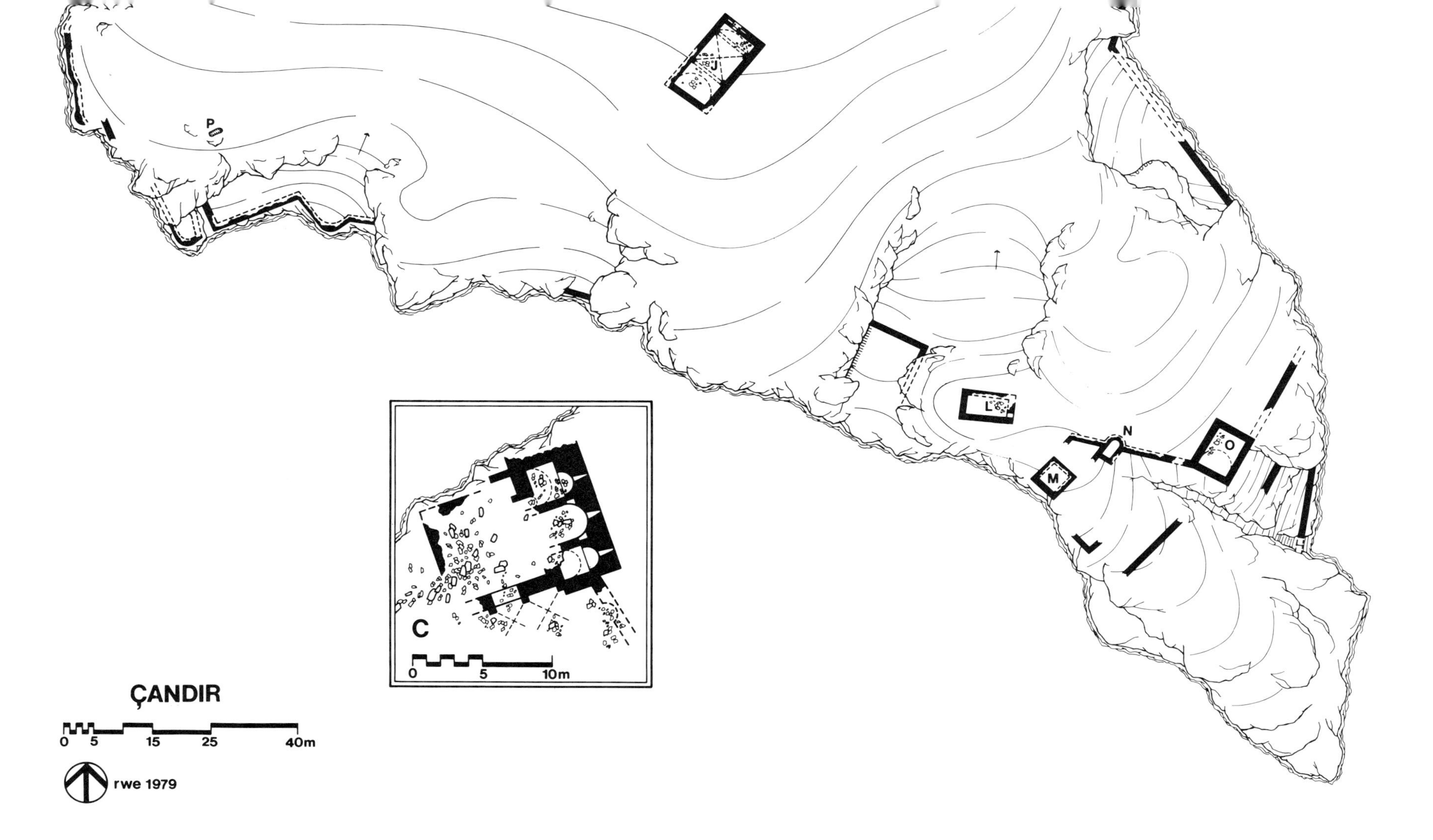

Fig. 24

At various points along the base of the plateau, natural cavities have been enlarged by excavation. It appears that wooden buildings were once adapted into these openings. Along the northeast face of the plateau there is a labyrinth of natural and scarped caves (their postions are not indicated on the plan). During my brief exploration I found no point of access to these openings. In medieval times a portable ladder may have connected the caves to the ground level or summit. At present the ground level is 8 m from the nearest cave opening.

Except for the gate, only one area of the summit appears to have any defenses. At the extreme northeast tip the Armenian engineers placed a platform for archers across a cleft. To support the platform they constructed a transverse arch and set a wall with an embrasure directly on top.

The most important feature on the exterior of the castle is the gate complex A and B at the northeast (pl. 292b). Here two long flights of stairs zigzag up the face of the outcrop. At the base of the first flight (point A) there is a small shelf that is 1.3 m above ground level. The rubble in the area may indicate that a wall and door once protected this shelf. There are no masoned or scarped steps to give access to point A, and a retractable ramp must have been used. At the southwest end of the shelf there is a partially scarped and masoned door that leads along a series of scarped steps to a long natural sink. This cavity, which extends 14 m into the body of the plateau, was converted into a cistern and fed by scarped drains along the sides of the cliff and along the stairway. Directly north of shelf A, the sixty-three steps of the first flight of stairs are terminated by a sharp hairpin turn to the left. The fifty-nine steps of the second flight lead to platform B.[17] When ascending the first flight two important features become apparent: the east flank of the stairs is protected by a short retaining wall (pl. 53a), and a gutter was constructed on the west flank of the stairs to carry water into the cistern. At the lower and central levels of the first flight, the width of the stairs is fairly consistent, while the width of the gutter fluctuates with the jagged face of the rock. The average length, width, and height of each step is 122, 34, and 23 cm respectively.

At the top of the first flight of stairs there is a small platform that leads to a roundheaded door with jambs and to the second flight of stairs (pl. 53b). Just to the north of the door on the outside there is a circular niche. Its function is unknown. Above the door in the thickness of the wall is a small, flatheaded embrasured window (not shown on the plan). There appears to be no accomodations on the interior above the door to man this opening. Its sole purpose must have been ventilation and light. Once past the door, it becomes apparent that the lower two-thirds of the second flight is a covered corridor (pl. 54a). When seen from the base of the plateau it appears that the east retaining wall of the second flight is freestanding and rises to a considerable height (pl. 292b). However, this wall actually abuts against an overhanging cliff. Except for two breaches, the retaining wall is in a good state of preservation. The southernmost of the two breaches is probably a collapsed window. A window here would be a convenient place to fire on a party that was approaching the first door. The wall for the southern or upper third of the stairs is shorter and partially damaged. It is in this area that the overhanging cliff turns sharply to the west, leaving the rest of the stairs exposed to the sky (pl. 54a). Since rain would now enter this otherwise dry corridor, the steps in the southern third of the second flight incline slightly to the east so that water can flow into a drain and eventually into the gutter of the first flight of stairs. This drain is carefully constructed between the steps and the retaining wall and is only 26 cm wide. At the twenty-fourth step from the top, the drain abruptly stops.

At the top of the second flight is the small gatehouse B. The entrance from the stairs to the gatehouse is roundheaded and limited by jambs (pl. 54a). Once inside of B, a sharp right turn leads through a second door and into the plateau itself. The only other opening in this gatehouse is a small squareheaded embrasured window in the east wall. The roof over this gatehouse has collapsed entirely. Most of the entrance complex is constructed out of type VIII masonry, with the doors and windows executed in type IX. This elaborate combination of corridors and double bent entrance occurs nowhere else in Armenian Cilicia.[18]

Within the castle the surface of the ground is highly irregular; small limestone outcrops and a thick undergrowth of foliage prevent a clear view from one side to the other. A few of the rock protrusions have been scarped to accommodate wooden buildings. West of the gate complex is the largest group of buildings (C through I). Just southeast of the church is what may be the east wall of courtyard H. This wall and its central gate isolated the northwestern complex from the rest of the castle. This area appears to be the private quarters of the resident baron.

West of the Armenian church C[19] are the remains of the large rectangular building D. There is evidence that D was once vaulted, but the shape of that vault cannot be determined. D was probably connected to complex E because of the latter's proximity and the similarity in the styles of the masonry.

One of the most stunning accomplishments of Armenian architecture is the residential building E (pl. 292c). This structure consists of eight rooms at two levels. Unlike the house/keep at Kız (near Dorak), where each internal unit is almost self-sustaining and defensible, at Çandır the defenses of the plateau and supplementary walls were considered adequate, and thus the rooms in E freely open onto one another (note that jambs are absent in the upper-level doors in E4, E3, and E2; complex E has no crossbar bolts). There is also an absence of cisterns and of corbels to support a brattice. Because of their close interrelation, the units in building E are not given separate letter designations on the plan. One of the most impressive features of E is the quality and exactitude of its construction and masonry. Throughout the building, types VIII and IX are the dominant forms of masonry. Occasionally, a few stones have a bossed exterior face—no doubt for aesthetic effect (pl. 293c). In the two extant vaults, type IV is used in the typical Armenian fashion (pls. 292c, 293b); these vaults did not detract from the general appearance of the rooms since they were once stuccoed and painted. In a few areas *small* stones in the style of type VIII appear (pl. 56a); these stones are prominent in the top half of the upper-level north wall of E3 (pl. 56b). The placement of the small type VIII may represent a repair carried out after an earthquake or a hasty attempt to complete these apartments when funds ran short. Except for the latter, the junctions of the adjoining units and the consistency of the core indicate that E is the result of one period of construction. Unlike church C, the rooms in complex E are covered with mason's marks. The catalogue of these marks appears below (Appendix 1).

When approaching building E from the east, the most direct path leads to the shattered east end of E1. The lower level of E1 is a magnificent hall that is divided from E2 by an arcade of four pointed arches (pl. 55b). The arch at the far west is twice the volume of the other arches. Today the floor level of dirt and debris has risen to the springing of the arches. Five groined vaults, which once rose from the north faces of the piers formed by these arches, covered the width of E1. Of the five groined vaults, only one is extant at the far west, and it is separated from the others by a transverse arch, which to a minor extent divides the west end from the rest of E1 (pl. 293b). The west end has a single roundheaded embrasured window in the northwest corner. Two other roundheaded windows are in the north wall of E1 (pl. 293c). Both are executed with unblemished precision; the larger is broadly splayed on the inside with a fan-like hood and the thinnest of openings on the outside. The second window is smaller on the inside and set directly above the other window. This superimposed window has a sill slanted toward the interior and an opening on the exterior that is three times the width of the one below. Like most of the ceiling, the north wall east of the windows has collapsed. Because of the extensive damage at the lower level of E1, the nature of much of its upper level is a mystery. The north wall west of the superimposed window has most of its upper level intact. From the base of the plateau, the division between floors is made clear by a thin horizontal chase on the outside wall (pl. 293a). This chase, which continues around the exterior, appears to have been inlaid with wood.[20]

Today the only extant area of the upper level of E1 is at the west end. There were no doors connecting it to the upper levels of E4 or E3. There is only one high-placed window between E1 and E3. No doubt access was from the upper level of E2 and the lower level of E1. In the north wall of this upper-level west end there is a large squareheaded window (pl. 294a). Much of this window's frame and lintel is damaged, but enough survives to show the skill of Armenian artisans. The sides of the frame have been carefully drafted and fluted with long vertical bands. Eight meters east of this window the north wall stops abruptly and appears to be squared off (pl. 292c). This was not the terminus for the upper level but merely the side of a now vanished window; there are indications of springing stones for the top of the window and a transom slot. In the west wall of the upper level of E1 is a niche, which has stylized stalactites (Muqarnas) carved on the interior of its pointed hood.[21] This again reflects the artistic investment of the builder as well as the appearance of motifs that are so common in Greater Armenia. One meter north of this niche is a tall window, near the junction with the north walls. This opening is roundheaded on the interior and limited on the exterior by jambs and a transom. The sides of the jambs have parallel holes that may have held crossbars or a wooden frame for shutters (cf. Meydan Kalesi, building O).

South of and parallel to E1 is a second and slightly smaller hall E2. Unlike E1 the first floor of E2 is not covered by groined vaults. A horizontal chase near the top of the west wall of E2 indicates that the floor for the upper level was supported by simple, wooden crossbeams (pl. 55a). Today all trace of the floor has vanished. The arcade, which serves as the north wall of E2 (pl. 55b), is articulated by a horizontal groove at the springing level of each pointed arch (directly below the spandrels). The east and west walls of E2 are opened by single roundheaded doors that are limited by jambs. The east face of the door in the west wall is now shattered. Under the springing stones for the west door's arch there are small notches that probably held a removable transom. Above the east and west doors in the upper level, there are tall, roundheaded openings that have transoms below the springing stones of their arches. The high-placed opening in the east wall served as a window, while the one in the west wall was a door of access between the upper levels of E3 and E2. The south wall of the ground floor of E2 has a myriad of openings. Beginning at the east end of this wall there is a roundheaded door that is set relatively high and would require wooden stairs on the interior for comfortable access. The curious feature about the door is that the jambs are on the inside. The significance of this arrangement is unknown. Directly to the west and set closer to ground level is an arched fireplace (pl. 55b). Further to the west is a large squareheaded door (or window) with jambs on the exterior. Adjacent to this door is the second fireplace in E2 which today is barely visible above current ground level. To the west of the fireplace is a squareheaded window (pl. 55a). Directly above these openings and just under the level of the first floor ceiling are four small squareheaded windows (not shown on the plan) placed at regular intervals. The south wall of E2 in the upper level is opened by six squareheaded windows. The flues for the two fireplaces continue into the upper level.

West of E2 is the rectangular room E3. The now absent floor of the upper level of E3 was not supported by a vault below but with wooden beams anchored in a horizontal chase (pl. 56b). Below the level of the chase there is a second horizontal groove which is very shallow and no doubt was inlaid with decorations. The wooden floor of the upper level of E3 was supported below by a centrally located transverse arch (now collapsed). The transverse arch at the west end of the lower room is set askew and supports the three upper-level west windows as well as the west end of the floor (pl. 56a). On the ground floor of E3 the east wall is articulated by a single door, which opens into E2, and a niche near the south corner. In the thickness of the south wall near the east is an impressive stairwell. The antechamber of the stairway can be entered through slightly pointed doors (pl. 294b). Both doors are limited by jambs. Not only do both of these doors have slots at their springing levels for transom bars, but so does every door and window in both levels of E3. A groined vault with the typical Armenian cross-shaped keystone covers the antechamber (cf. gatehouse at Yılan). Today pink and white lichens grow in profusion throughout the chamber. The stairway, which ascends to the second level, is not vaulted but is covered with a series of lintels. Toward the western end in the south wall of E3 is a large fireplace with a shattered frame (pl. 295a). Just above the fireplace are ten carefully sculptured shallow niches with stylized foliage. Directly above these is the same horizontal chase that is seen in the north wall (pls. 56a, 56b). The tapered flue for the fireplace is carried through the second level. The lower level west wall of E3 is pierced by a door/window that is covered by a depressed arch and limited by jambs. Like many of the architectural elements in this fort, this door has been disfigured by graffiti. Above the west door there are three graceful roundheaded windows on the second level.[22] The central window is slightly taller, and, unlike the two flanking windows, it has its jambs on the inside of the room (pl. 56a); this same arrangement is followed for the placement of a monolithic transom block (pl. 294c). Behind all three transom blocks, neatly incised horizontal slots run to the opposite ends of the doors. Undoubtedly they supported some sort of decorative wooden shelf. Of the three tympana created by the transom blocks, only one in the north door is filled with neatly cut masonry; at one time the other two were probably filled. Directly below the windows on the interior there are alternating corbels that helped to anchor the floor. The lower level of E3 at the north has two openings flanking the central transverse arch (pl. 56b). The one at the west is the roundheaded entrance to the lower level of E4. The opening to the east of the central transverse arch is a roundheaded niche. On the exterior near the top of the niche, round decorative indentations have been carefully incised. Like the niche in the east wall of this room, the sides have shallow slots to hold shelves. Directly above the lower-level door from E3 to E4 there is a roundheaded entrance to the upper level of E4. This door is constructed like the central upper-level window in the west wall of E3 and the upper-level door between E3 and E2. The only other opening in the upper level of the north wall of E3 is the window to E1 mentioned earlier; there may be a sill of a second window west of the upper-level door to E4, but extensive damage makes this difficult to confirm. Like the upper-level rooms of E1 and E2, there is no extant roof over the second level of E3. Since there is no indication of stone vaults, the covering may have been wood.

Although E4 is the smallest unit in the complex,

it is one of the most carefully constructed. The interior of the lower level is covered by a pointed vault. Except for the courses at the springing level, the vault is constructed with type IV masonry and was probably stuccoed. The end of the vault is set back from the west wall about 68 cm (near the south end). The resulting space is asymmetrical and continues through the upper level. No doubt it was used for ventilation. The single door for this room is flanked by two roundheaded niches. The niche in the south wall is badly shattered, and it is the only one here that has a slot accommodation for a shelf. In the northwest corner of the room are two embrasured windows that have sills whose interior halves slant downward. The west window is squareheaded, with its north side slightly askew. The north window is roundheaded, with two very small, shallow, square holes carved on the slanting half of the sill. The upper level of E4 is partially shattered. Aside from the door in the south wall, there is a splayed window in the north and west walls. The window in the north wall is squareheaded and in excellent condition.

North of complex E and outside of the summit at a much lower level are the vaulted chamber F and attached walls. Most of chamber F is stuccoed, and it seems to have functioned as a cistern. Attached to the chamber is a series of walls that display a variety of masonry types. In one section type VII is surmounted by type IV, which may be a repair carried out after the Armenian period.

South of complex E is the long, rectangular structure G. Because of extensive damage, nothing can be said about the interior or the nature of the vaulting. It seems that the upper-level platform of the stairway in the south wall of E3 had access into this building. The east wall of G is opened by two roundheaded doors that are constructed in the same style as the windows and doors in the upper level of E3.[23] The position of windows in the west wall of G cannot be determined.[24] Both of the east doors open onto courtyard H. This area is littered with fallen masonry, and someday excavations will reveal other buildings. The south end of the courtyard is defined by a circuit wall and what may be a cistern in building I.[25] Between building I and the east entrance of the courtyard there are the remains of a curious rectangular building. This vaulted structure is not shown on the survey because the thick jungle and treacherous potholes prevented free access. The building is partially apsidal and consists of scarped stone and masonry. Since much of the masonry is stuccoed over, it may be a cistern(?).

South of courtyard H is the large vaulted structure J. Type IX is the predominant form of masonry in this building; only the vaults, which were stuccoed and painted, are built of type IV masonry (pl. 54b). Two transverse arches divide the building into three unequal units. Only the south rib is standing today. The largest unit in the center was once covered by a groined vault that sprang from corbels that are still attached to the sides of the transverse arches. Each triangular springing tapers to a fine point and has a total length of 190 cm. The areas north and south of the transverse arches are covered by a single barrel vault. At the north and south ends of the lower-level west wall of J is scarped rock (not shown on the plan). The scarped sections vary in height from 1.3 m to 3.5 m and are polished as smoothly as the faces on the type IX masonry. J is opened by a single door at the northwest. The door is square-headed and topped by two side-by-side monolithic lintels. The door is 136 cm high, 106 deep, and 74 wide (from the jambs the width is 67 cm). The jambs are placed toward the interior, and pivot holes in the lintel indicate that double doors opened outward. A crossbar bolt could lock the doors from the outside only. The exact reason for the construction of this door and the function of the well-executed windowless hall are a mystery. J could not be a cistern since the walls were not stuccoed.

Northeast of hall J is cistern K. The interior of this rectangular building is carved entirely out of living rock. Traces of masonry cover its exterior perimeter. The roof consists of a pointed vault supported by two ribs. All but the north end of the vault has collapsed.

South of K a number of rocks have been scarped with post holes for wooden buildings. This may have been a residential area for the nonbaronial residents of the castle. There is also a large amount of coursed masonry strewn throughout the thick underbrush. In the southeast corner of the plateau there is a small group of utilitarian buildings, L through O. The apsidal building N is stuccoed on the interior and was probably a cistern. Unfortunately the wall levels of most of the buildings in this area are so low that the positions of doors and windows cannot be determined. Near the southwest end of the plateau is the small scarped pit P. P may have been a sarcophagus.

[1] No systematic survey of this site had been undertaken prior to my own work in summer 1979. The contour lines on my plan are separated by intervals of one meter.

[2] This site appears by its modern name on the *Mersin (2)* map. The easiest way to reach Çandır is to drive on the asphalt and gravel road from Mersin via Belen Keşlik and Gösne.

Gottwald ("Kirche," 86–100) appears to be one of the few 20th-century travelers to have explored this site. Hellenkemper (237–40) did not visit Çandır but merely relied on the findings of

Gottwald. Father Alishan (*Sissouan,* 72 ff) and Davis (43 ff) may have relied in part on the unpublished accounts of Sibilian and Ancketill. See also: Schaffer, 59; *Handbook,* 246; Tēr Łazarean, 15 f.

[3] In this incomplete historical summary I draw heavily on the sources cited by Gottwald ("Kirche," 87–89) and the information supplied by Alishan (*Sissouan,* 72–76, 42, 65) and Yovhannēsean (81–91). A genealogy of the lords of Papeṙōn can be found in Toumanoff, 279–81 and Christomanos, 146. Cf. "Papeṙon," *Haykakan,* 131.

[4] Alishan, *Sissouan,* 72; discussed by Gottwald, "Kirche," 88–90.

[5] *The Chronicle of Joshua the Stylite,* trans. and ed. W. Wright (Cambridge, 1882), 9, 12.

[6] Cf. Gottwald, "Kirche," 89; Ramsay, 382.

[7] Gottwald ("Kirche," 87 note 1) gives a German translation of this passage; Alishan (*Sissouan,* 73) provides a French translation.

[8] Smbat, 619; cf. Gregory the Priest, 168.

[9] An interesting description of Bakuran is in *RHC, DocArm,* II, 623; see also Vahram of Edessa, 509; Smbat, 622–23; Smbat, G. Dédéyan, 50 f.

[10] See Appendix 3.

[11] Smbat, 655 f. Tʿoros succeeds Levon, the son of Smbat the Constable, and is followed in 1298 by the latter's grandson, another Smbat.

[12] Ibid., 670 f. Prior to Ōšin's death it is uncertain if this fief briefly became the possession of Joan of Anjou; see: Simon of St. Quentin, *Histoire des Tartares,* ed. and notes J. Richard (Paris, 1965), 88 note 1; C. Kohler, "Lettres pontificales concernant l'histoire de la Petite Arménie au XIVe siecle," *Florilegium ou recueil de travaux d' érudition dédiés à Monsieur le Marquis Melchior de Vogüé* (Paris, 1909), 318 f.

[13] Smbat, 649. The extent of any damage to Çandır Kalesi is impossible to determine.

[14] Hellenkemper, 240; Gaudefroy-Demombynes, 102.

[15] Gottwald, "Kirche," 86.

[16] Gottwald's claim (ibid., 98 f) that "Bossenwerk" is absent from the walls and buildings at Çandır is generally accurate.

[17] The number of steps listed here for the first flight of stairs differs significantly from that counted by Gottwald (ibid., 92) and may be an indication of the continuous destruction of this site. Due to the requirements of space, my plan does not show the exact number of steps.

[18] Vahga Kalesi is the closest parallel; it has a corridor with turns.

[19] The history and architecture of the church are detailed in Edwards, "First Report," 161–64. This church was dedicated in 1251.

[20] This wood may have added flexibility to the walls during an earthquake. It should be noted that the chase never extends into the core of the wall to act as a header; see Edwards, "First Report," 172 and "Second Report," 135.

[21] The Muqarnas design here is similar to the one in the south door of the church; on the origin of this design see A. Ödekan, *Osmanlı Öncesi Anadolu Türk Mimarisinde Mukarnaslı Portal Örtüleri* (Istanbul, 1977), 19 ff.

[22] These very distinctive openings also appear at the Armenian chapel of Kız (near Gösne). This design is peculiar to the Armenian architecture in Cilicia.

[23] Cf. Gottwald's picture of the courtyard ("Kirche," pl. 4).

[24] Ibid., 93.

[25] Ibid. Unfortunately I am not able to establish any sort of chronology for the construction of the buildings on the outcrop (except the church). Although the site was occupied by Armenians for at least 300 years, we have no references in our extant chronicles regarding construction. Reliable answers to these questions will have to await excavations.

Çardak

Çardak Kalesi[1] surrounds the summit of a pine-covered mountain 12 km southeast of the city of Osmaniye. At an altitude of 740 m, the fort not only has a commanding view north into the highway leading to the Amanus pass (Savranda being the principal guardian of the latter), but it guards the junction of four different trails across the Anti-Taurus Mountains (pl. 59a).[2] This fort has clear intervisibility with Haruniye to the northeast and Toprak to the west.[3] Below Çardak Kalesi to the northwest is a small village that bears the same name. Some of the inhabitants of this village and the surrounding area refer to this site as Gavur Kalesi. Çardak Kalesi has never been consistently located on any of the published charts.[4] There appear to be no wells or springs within the fortified complex; the closest sources of year-round water are the streams in the fertile valleys below. We do not know the date of construction nor the medieval name of this site.

There is a possibility that the fort is the medieval castle of Hamus (Arabic: Ḥamūs, Ḥāmīṣ). This identification is a weak one, depending solely on the name of a stream in this area at the beginning of the twentieth century.[5] In January 942 the army of Sayf ad-Dawlah ravaged the towns near Hamus. The first *known* occupants of Hamus were the Byzantines; later it was seized by the Armenians, then the Mamluks. Aside from the assassination of Stephen, the brother of Tʿoros II, the most notable event at Hamus was an earthquake in 1268/69.[6]

Of all the large garrison forts in Cilicia, Çardak is the most symmetrical in plan (fig. 25). Its almost rectangular circuit abruptly crosses the sinuosities and flow of the topography in a style reminiscent of the Roman castrum. Its circular towers (that is, E and H) are rare in Cilicia, being repeated only at a few Byzantine sites (for example, Evciler). While the circuit wall from A to H is in a relatively good state of preservation, the section from I to A is badly shattered and is difficult to measure. It is likely that the main entrance into the fort was once situated between I and A. The progress of my own survey was hindered by a lack of water at the site and the dense pine forest, which engulfs the entire fort.[7] Excluding A and B, the highest point in the interior is northwest of chapel J.

The interior and exterior facing used throughout this fort is a large type IV masonry, which is layered in uniform courses (pls. 57a, 57b, 58a). On the exterior the stones are slightly larger in size than on the interior, and the margins of the exterior masonry are plugged to a greater degree with rock chips. A small

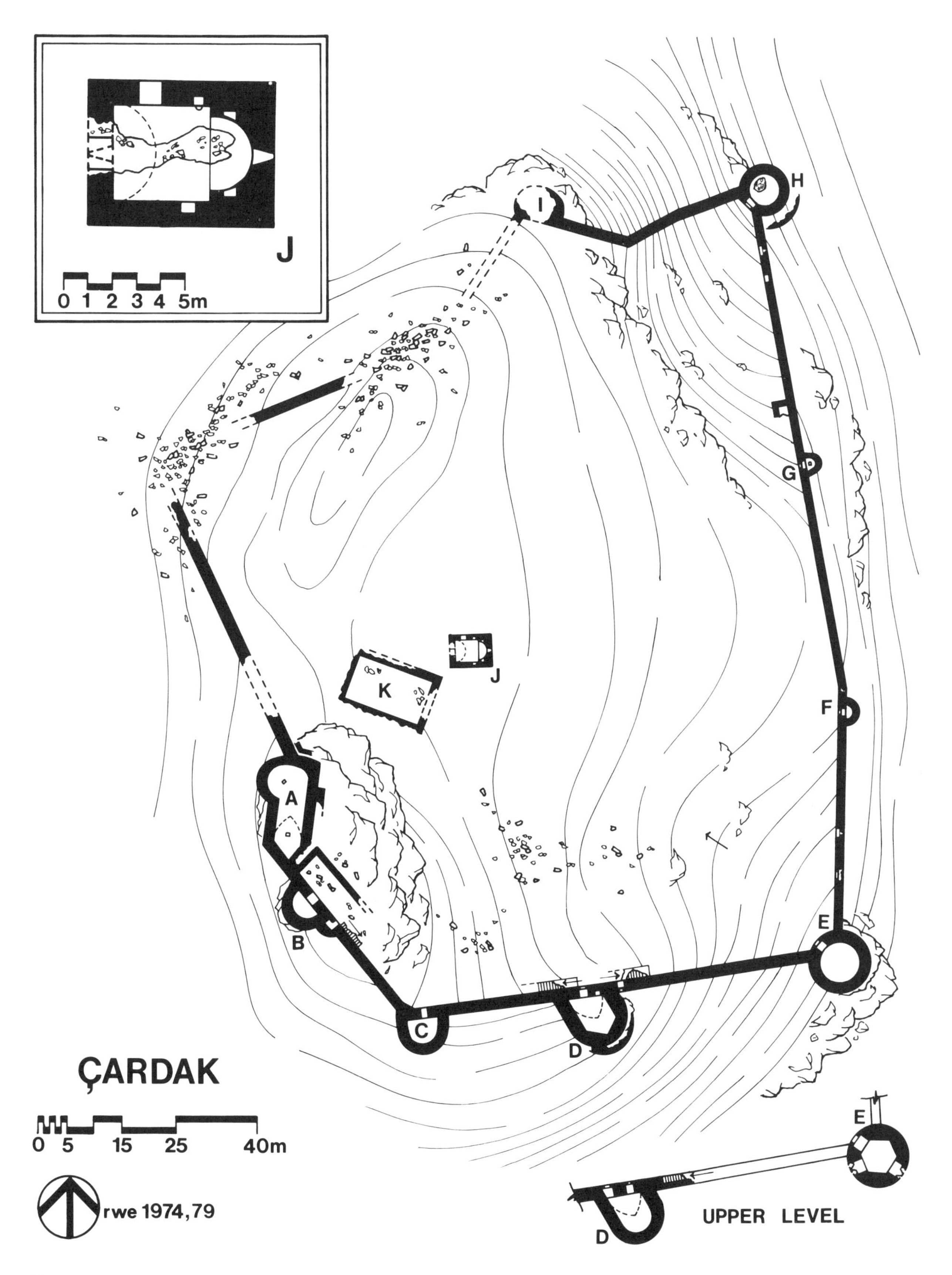

J
0 1 2 3 4 5m
H
I
G
F
K
J
A
B
E
C
D
ÇARDAK
0 5 15 25 40m
rwe 1974,79
E
D
UPPER LEVEL

Fig. 25

form of type IX masonry is used for the frames of doors, windows, and embrasures (pl. 58b). Chapel J in the center of the fort employs a slightly larger size of the ashlar around its openings, and its walls are made with a small, poorly coursed type V masonry.[8] Except for chapel J, the masonry indicates that most of the fort is the result of one period of construction.

Building A, which in plan is circular at the north and pointed at the south, is somewhat of an enigma. It is designed like a cistern with stuccoed interior walls and numerous clay pipes. Its pointed vault, which splays into a semidome at the north, is pierced by two square hatches. These may have served as draw holes. A problem arises in that A is constructed on a high rock mass, which means that the area of possible runoff into the cistern is quite small. Normally cisterns are placed at the lowest point in the enclosure to catch the maximum amount of rain. There is the possibility that water was channeled from B and the adjoining wall walk into A. Directly northeast of A is the unroofed building K, which is a more likely candidate for a cistern. Just south of K and east of A at the base of the rock mass, there is a pit (not shown on the plan).

Northeast of tower B on the interior side of the circuit wall there are the crumbled remains of a small rectangular room. Just south of B at the base of the stairway, a small elbow-shaped chamber is located in the thickness of the wall. This chamber is covered by a pointed vault and may once have been a garde-robe for the rectangular room. Towers B and C are each opened by a single jambless roundheaded entrance (pl. 57a). Their chambers are covered by tall apsidal vaults. Unlike the single level of B, C has an identical upper story (not shown on plan), which also seems to be windowless (pl. 58a). Tower C marks a pivot point where the circuit wall turns abruptly to the east and the rock mass supporting A and B ends. The height of the wall drops sharply between C and D (pl. 57a). Undoubtedly a now missing stairway was incorporated into the descending wall walk to handle traffic. Today no section of the circuit preserves traces of battlements or merlons. Two splayed windows in the wall between C and D (not shown on the plan) were blocked at a later date.

Tower D is unlike the circular or apsidal salients in the rest of the fort because it juts out at an angle diagonal to the plane of the circuit wall in order to anchor its foundation on a rock ledge. This somewhat curious shape is accented at the base by a revetment wall (on the southeast flank of D) (pl. 57b). On the interior of the circuit wall flanking tower D there are the broken remains of two corbeled stairways, which were designed to give access to the wall walk (pl. 57a). Under the stairway at the east there is a curious niche that is 2 m wide and 1.6 m deep. Because the wall between D and E was set at a much lower level than the section between C and D, a stairway was built on the wall walk of D to give access to D's upper level (pl. 59a). None of the southern towers at any level have stairways to facilitate entrance through their high-placed doors. In medieval times access was probably made with a retractable ladder, lest an enemy who had breached the wall quickly capture the entire fort. A small postern gate, the only extant door through the circuit wall, is situated on the east flank of tower D. Today most of the voussoirs that framed this opening are missing. The postern measures only 98 cm in width.

The lower level of tower D, which is entered through a tall shattered door, consists of a pentagonal room. There are no other openings inside this room, except for a small niche in the north wall. The upper level of D is apsidal in shape, and, like the lower level, it is covered by a slightly pointed vault. The small roundheaded door of the upper level opens into the northwest corner of the room. The only other opening is a high-set squareheaded window near the northwest corner. The ceiling in the lower level of D is black with the soot of recent campfires. The locals report that smugglers from Syria frequently use Çardak Kalesi as a storage depot.

Tower E is probably the most complex structure in the fort. Both its lower and upper levels are opened by roundheaded doors. The circular wall of the first level is pierced by no other openings and is covered by something resembling a beehive dome. In the upper level an identical dome rises from the hexagonal walls. This level is pierced by two embrasured loopholes with casemates; the exterior frames of both windows have collapsed (pl. 59a). On the interior the casemate measures 1.75 m in width. Above the upper level is a fortified terrace (not shown on the plan) with the remains of four shattered embrasured loopholes and their casemates; two of these point to the southwest, one to the south, and the other to the northeast. The openings in the fortified terrace of H are identical in shape and size; the width and depth of their casemates are 1.15 m and 0.70 m respectively, and the inner side of the loophole (that is, the splay) measures 0.60 m in width. The fortified terraces of E and H may once have been covered by wooden roofs.

From tower E the circuit pivots directly north until it reaches the small salient F. On the interior of the circuit between E and F there are two openings: an embrasured loophole with casemate and, closer to E, a strange portal (niche?) that is blocked on the outside and has jambs on the inside. Tower F has a simple windowless chamber which is covered by an apsidal vault and opened by a single roundheaded door. The circuit wall from F to G gently descends and turns to the northwest. Tower G is opened at the west by a high-placed squareheaded door. The stuccoed walls and the absence of windows indicate that it functioned as a cistern. Its apsidal vault is pierced by a single draw-hatch. Six meters northeast of G and attached to the interior side of the circuit wall is the extant north half of a small rectangular room. This chamber is covered by a half vault that springs from a low level in its west wall. In the circuit wall south of tower H there are two openings: the most southerly appears to be a straight-sided niche with a pointed hood, and the one at the north is an embrasured loophole with casemate. Accumulated debris makes an accurate assessment of the former impossible.

Tower H, which is anchored on a rocky ledge at the northeast and supplemented by a reveted wall (perhaps a talus) at its south flank, is the principal salient in the north half of the fort. Its interior is opened by a large roundheaded door (pl. 58b); today no other openings are present, except for a gaping hole in the beehive dome of this room. Above this level there is a terrace similar in design to the one atop tower E. Two of the embrasured loopholes with casemates are still extant in the terrace of H at the north and northwest. Their broadly splayed loopholes are not Armenian in character. From tower H the circuit wall ascends to the west toward the shattered salient I and eventually is lost in rubble and trees. Building J in the center of the fort meets every criterion for an Armenian chapel in Cilicia.[9]

From the nature of its masonry and design, I conclude that Çardak Kalesi is one of the largest and most impressive Byzantine forts in Cilicia.[10] This fort is also one of the few Byzantine structures whose plan was not erased by Armenian reconstruction (cf. Gökvelioğlu and Anavarza). In fact, the only archeological evidence of Armenian occupation at Çardak Kalesi is chapel J.

[1] In March 1974 I began my study of this hitherto unsurveyed site. The plan was completed in June 1979. The approximate distance between the contour lines is 75 cm. Due to the limitations of time, the exact alignment of the north and east circuit has been approximated on the plan.

[2] Today two of the trails have been enlarged and slightly altered by the Turkish Forestry Service. One still remains the chief route to Yarpuz. Presently, my own charting of the roads into this area remains incomplete. See also Cahen, 146 f; Deschamps, "Servantikar," 381 ff.

[3] Hellenkemper, 109 note 4.

[4] Ibid., note 1. Hellenkemper discusses this problem and cites the relevant maps. On Kiepert's chart (published by Humann and Puchstein) there appear in the vicinity of "Tschardak Kalesi" two other forts: Frenk Kale and Karafrenk Kale. See my discussion of Karafrenk in the Catalogue and of Frenk in Edwards, "Second Report," 123–25 note 3. Çardak also appears on the *Kozan* and *Malatya* maps and may be labeled as "Frenk" on the *Marash* chart. Cf. Schaffer, 95; Frech, 578; *Handbook*, 454 f.

[5] Hellenkemper, 109 f, esp. note 2 on 109; cf. Cahen, 145 ff; Tēr Łazarean, 176. From Hellenkemper's brief narrative, it appears that he did not explore this site or its environs. See also: Abū'l-Fidā' 29; Le Strange, *Palestine*, 543; Vasiliev, 295; Boase, 166. Honigmann (121 note 2, map II) places the "Ḥamuṣ-ṣuyu" *north* of Savranda. Cf. Canard, map opp. 240, 279 f; idem, *Sayf al Daula*, 78.

[6] Smbat, G. Dédéyan, 48, 75; Smbat, 621 f; Appendix 3. There is a *very* remote possibility that Çardak was first established as the Arab castle of al-Muṯaqqab; cf.: Le Strange, *Palestine*, 510 f; Alishan, *Sissouan*, 236; Hellenkemper, 110 note 11. Unfortunately there is not sufficient information in the medieval chronicles to identify this site. Refer to note 3 in my discussion of Gökvelioğlu (the Catalogue). See also E. Honigmann, "Neronias-Irenopolis in Eastern Cilicia," *Byz* 20 (1950), 58–61.

[7] A more thorough survey may reveal the existence of other internal buildings.

[8] Edwards, "First Report," 166 f.

[9] Ibid.

[10] A handful of similar Byzantine forts can be found just outside the perimeters of Cilicia. See W. Ramsay and G. Bell, *The Thousand and One Churches* (London, 1909), 274–94, 489–502.

Çem

Çem Kalesi[1] stands on the east flank of a very strategic road that winds from Kadirli to Göksun (pl. 65a). From Memetli the road eventually turns to the northeast and ascends a narrow river valley that leads to the Bağdaş and Mazdaç passes.[2] Çem, the guardian of the first pass, is situated just southeast of the tiny village of Katarlı Köyü. The locals reported that a second fort (Esende Kalesi?), about 35 km to the northeast, guards the Mazdaç pass. This site is presently unsurveyed. Çem Kalesi girdles the top of a large limestone outcrop (fig. 26). The fort is easily approached by a trail (passable by jeep) that leads south from the village. Below this track at the west, the steep sides of the gorge fall to the river below. Just north and northeast of the fortress there is a small, relatively flat area of land. This fortress-outcrop rises abruptly from this level, and its almost vertical sides at the northeast show some signs of being scarped (pl. 60a). The east side of the outcrop is articulated by a small, low-level protruding terrace of rock. The northeast end of the terrace has been blocked by wall G to prevent enemy incursions. The

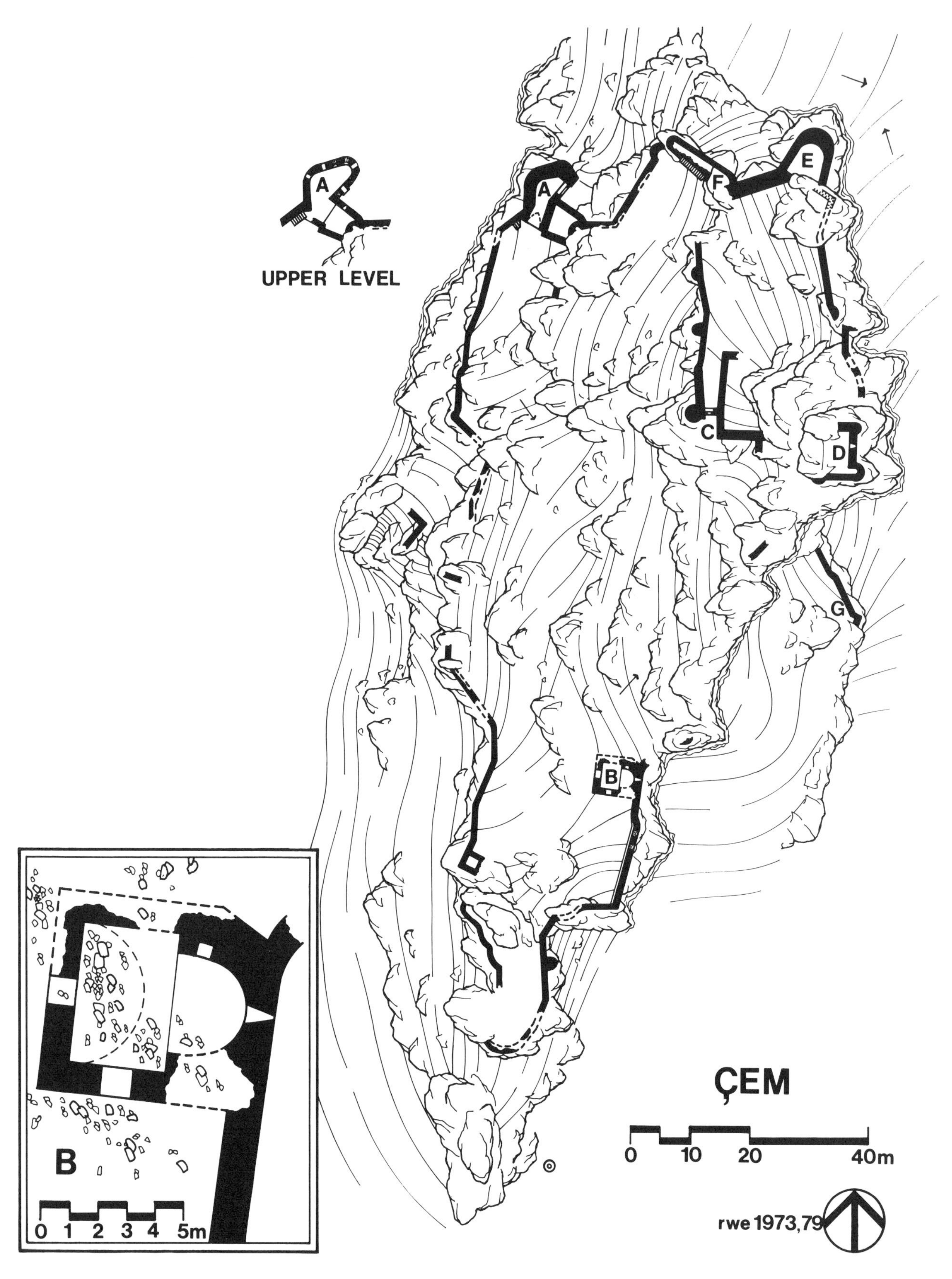

Fig. 26

areas on the south and southwest sides of the outcrop have more gentle slopes. At the northwest the sides of the outcrop become more steep. There is no evidence of cisterns inside of the fort, but there is a large well northeast of tower E and a natural cistern (unsurveyed) near the southern tip of the site. We have no certain historical references to Çem Kalesi; the medieval name and date of construction are unknown.[3] This castle appears on one of the maps of Cilicia.[4] We have two very brief published accounts of the site.[5] Today the only comfortable approach into the fort is at the north. A path leads to the bent entrance, gatehouse A.

The masonry on the exterior of this entrance complex is a curious mixture of the Armenian type VII and some smooth ashlar that was recycled from the adjacent Byzantine city (pl. 60b). Often reliefs and large sections of Greek inscriptions were used as facing stones for decorative purposes. Among the reliefs are a standing lion, a prancing stag, numerous Greek crosses with and without circular frames, and a triglyph(pl. 59b). Especially interesting are the interproximal spaces between the upper-level windows, where the three courses that constitute their height are made with smooth ashlar(pl. 60b). Because the rusticated masonry predominates elsewhere, the visual effect is one of a broad stripe crossing the windows (cf. Fındıklı). The smooth recycled ashlar is used exclusively on the exterior wall that separates the outer door of A from the rock mass at the east (pl. 59b). In general the interior facing stones of the gatehouse consist of a crude heavily mortared type VIII in the lower-level wall with types III and IV used in the lower-level vault and throughout the upper level (pl. 61a). The recycling of the Greek inscriptions is haphazard; in many cases the epigraphs have been inverted.

Gatehouse A is designed to take full advantage of the twisting rock mass (pl. 65a). The curving wall of this gatehouse thrusts abruptly to the north and envelops the projecting edge of the cliff, like a fist in a stone gauntlet. The lower level of A consists of a corridor covered by a pointed vault. (pl. 61a). Even on the interior, fragments of inscriptions abound on the faces of the walls. The south end of the corridor is simply a jambless opening, (pl. 62b), but at the northeast there are provisions for an arched door with jambs and a crossbar bolt (the latter is not shown on the plan; pl. 59b). Instead of the traditional exterior lintel over this east portal, a bulbous monolithic block carved with a human figure between two attacking lions occupies its position and most of the tympanum (pl. 295b). Bossert and Alkım noted that there is no way of determining if this relief was a recycled piece or executed in medieval times for the door.[6] However, in the opinion of this writer, the gap created between the high pointed arch over the jambs and the tympanum block is certainly an indication that the latter was recycled. The medieval masons filled the gap with coursed stones.

Gatehouse A has an upper level which functioned as a gallery (pl. 62a). This level was opened by seven squareheaded windows (four of which were intentionally blocked at a later period) and two doors. The door at the east is roundheaded (pl. 61b), flanked by a small window at the south (now plugged), and once opened into a fighting platform (and probably machicolation) directly above the east entrance of the first level (pl. 59b). This platform was in part supported and protected by now collapsed corbeled walls. On the exterior side this upper level east door is flanked by recycled jambs, inscriptional reliefs, and a handsome lintel articulated with moldings. The now collapsed second door of the upper level was near the southwest corner at the top of the stairway that led from the lower to the upper level (pl. 62b). All of the lintels that cover the seven straight-sided windows consist of a single squared block (pl. 62a), except for the second window from the west, which is surmounted by two blocks. The lintel over the fourth window from the west, obviously a recycled piece, bears the following Greek inscription:[7]

By God's help	ΑΓΑΘΗΤΥΧΗ
?.	ΚΟΥΝΔΕΙ
?.	ΚΑΝΤΗΑ

The window south of the upper-level east door does not have a lintel that extends beyond the impost, as do the other six windows. (pl. 61b). Because of the consistent placement of joist holes, it is certain that this upper level was once covered by a wooden roof. This roof also created a platform from which the troops could man the merlons of the terrace level (pl. 65a). These merlons are well preserved at the north and east.

Adjacent to the southeast end of A there is an open (unroofed) boxlike enclosure that joins A to the rock mass at the east. Except for its north flank, the lower level of this chamber consists largely of the scarped walls of natural rock; the masoned upper level probably had a wooden floor to allow the defenders to use the embrasured window in its north wall.

South of gatehouse A the circuit snakes along the sinuosities of the rock. In general these walls are built like the exterior of A, except that type V masonry is used frequently. The interior of the castle is difficult to assess because it is covered by a thick, almost impenetrable undergrowth of shrubs and trees. The most important structure at the south end of the fort is chapel B. This is the typical Armenian chapel with a barrel-vaulted nave and an apse pierced by niches and a narrowly splayed central window.[8] The presence of a door at the west end of the nave is certain. A door at the south, placed at a higher level, is possible (and appears on my survey). Excavations could reveal that the gap in the north wall of the nave was a door and that the high opening in the south wall was merely a window. Attached to the chapel at the southeast are the remains of the circuit wall. A shelf on the west side of the wall, two plugged windows, and an assortment of smaller holes (not shown on the plan) are visible in this area. Sections of this wall are topped by merlons and crenellations. The heavy use of type III masonry as an exterior facing gives the southern circuit walls a Byzantine character.

The highest point on the fortified outcrop, the platform D, is flanked at the north by the walls of the upper bailey (pl. 64a). Today the only apparent entrance into this upper ward is gate C (pl. 63a). This gate, which stands at the south end of the entrance corridor, has accommodations for a double door and a crossbar bolt (pl. 63b). The interior side of the door is covered by an arch that consists of four almost Cyclopean voussoirs. Except for this door, which is constructed out of smooth ashlar, the rest of the circuit is built with a crude type V masonry (pl. 63a). There is no evidence today of a slot machicolation over the gate. In the southeast corner of the upper ward, platform D is supplemented with two salients and an embrasured window on the east flank. The interior of the window is broadly splayed in the Byzantine fashion; crude plates of stone and mortar form the window's rounded hood (pl. 64b). Much of the east wall of the upper ward has collapsed, and the scarped trenches, which once functioned as the socle for the walls, are still visible. The north flank of the upper bailey, which descends to gate A, is built on an entirely different scale. In salients E and F, types V and VII masonries are used to build large sections of the exterior walls in the same manner as gatehouse A. Today the interior of E is filled with dirt and shrubs. The stairway that once flanked F is quickly falling into ruin. The first few courses of the wall separating E and F are constructed with a poor-quality type IV, while the upper courses are built with the finer masonry seen in salients E and F.

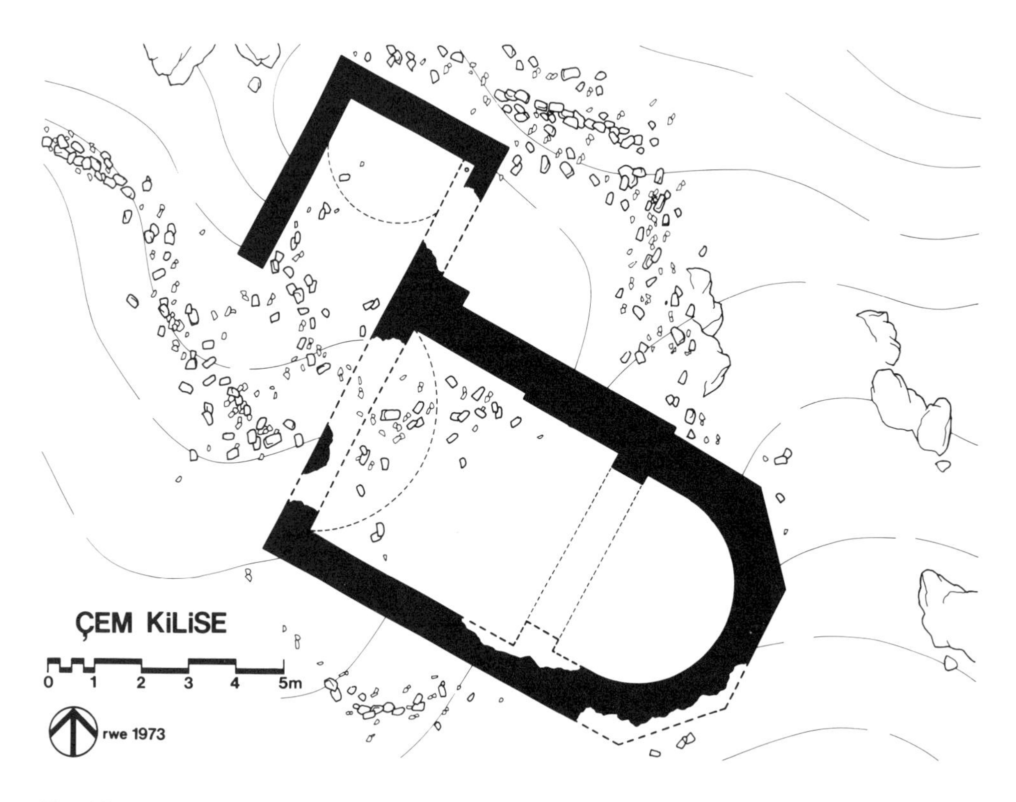

Fig. 27

D, G, and areas south of B appear to be Byzantine by reason of their crude exterior facing. The Armenians certainly refortified this Byzantine site on the critical north face with their typical round bulging towers and their well-coursed rusticated masonry. The fact that the Armenians added the chapel and refused to refurbish the splendid Byzantine chapel in the adjacent city (preferring to use some of its masonry in their rebuilding of the fort) may indicate that the site of the Byzantine city to the east was abandoned in the Armenian period of occupation.

Çem Kilise

Because of circumstances, I was unable to survey the Byzantine city. The city is separated from the east flank of the fort by the intervening gorge. The city is surrounded by a now collapsed wall. On the interior there is evidence of cisterns and the remains of numerous buildings. The jambs of doors still stand to two or three courses in height. I was able to survey one of the best-preserved structures in the city, a Byzantine chapel (Çem Kilise), located at the north end of the settlement (fig. 27). Today its vaults have collapsed, but parts of its walls are standing to at least seven courses in height, and they reveal a structure that is distinctively Byzantine (pl. 60a).[9] The masonry consists of superbly coursed blocks of Cilician marble(?). Although the length and depth of an individual block may vary from 19 to 80 cm, the blocks in each course have an identical height. An extremely thin and irregular poured core separates and binds an inner and outer facing of stone. In a few areas the core consists simply of rubble without a mortar binding agent. This type of smooth ashlar is atypical for the Byzantine chapels of Cilicia.[10] Another unique feature of this chapel is that it has a vaulted narthex at the west. While the narthex extends over 3 m north of the body of the nave, it seems that it did not extend to the south, since the exterior of the southwest corner of the nave is neatly squared off. Amid the present rubble the southern terminus of the narthex could not be determined. On the interior of the narthex its west wall is separated from the springing course by a molding with a flat cavetto. A small jog articulates the north interior wall of the nave. It seems likely that a transverse arch separated the apse from the nave. What remains of the interior of the apse is not pierced by windows or niches; its exterior is polygonal. The exterior of the north wall of the nave has two jogs not present in the south wall. The Byzantine city of Çem deserves a full archeological survey.

[1] I first visited the fortress and the late antique city of Çem in fall 1973. At that time the plan of the chapel was completed and the survey of the fort was undertaken. The plan for the latter was not completed until a subsequent visit in July 1979. On the plan of the fortress the contours are roughly spaced at intervals of one meter. On the survey of Çem Kilise the interval of distance between contours is 25 cm.

[2] See Geben (note 4) and Haçtırın (note 2), both in the Catalogue.

[3] Edwards, "First Report," note 71.

[4] This site does appear on the *Malatya* map. Cf. Alkım, 214, maps 2 and 3.

[5] Bossert and Alkım, 24, pls. 19 f; Hellenkemper (216 f) based his account *entirely* on the photographs and the half-page description of Bossert and Alkım.

[6] Bossert and Alkım, 24, pls. 19 f.

[7] Although Greek inscriptions abound in the upper level of the gatehouse, this is the only epigraph that can be easily approached and transcribed.

[8] For a detailed description of this chapel, see Edwards, "First Report," 167 f.

[9] Refer to my discussion on Korykos for the typology of Byzantine chapels in Cilicia; Edwards, "First Report," 173.

[10] Ibid. However, the Byzantine churches from the 5th through 7th centuries are executed with well-coursed ashlar.

Dibi

Dibi Kalesi[1] rests on a small outcrop north of Torlar Köyü at the side of the auxiliary trail from Maraş to Azgit/Andırın. Refer to my discussion of Kalası, Dibi's closest neighbor, for details on this road. We have no historical references to this small garrison fort.

The defenses at Dibi (fig. 28) are more complicated than the structure at Kalası. It consists of two enclosed units or baileys. The west bailey, which has the remains of two large towers on its south wall, is built on the descending slope of a hill. The north half of the east ward embraces the slope at a point where its rounded northeastern tip is anchored firmly on a rock mass (pl. 65b). The south half of the east ward ascends and surrounds the top of the rocky summit (pl. 66a).

The interior and exterior facing throughout the fort is a combination of types III and IV masonries. The latter is especially crude; typically, the stones are anchored to the poured core by mortar and an abundance of rock chips (pl. 67a). At the corners quoins are used in the usual Byzantine fashion to anchor the walls (pl. 66b).

The entrances into the fort are impossible to determine from the present remains. In the west bailey there may have been a door in the west wall or perhaps one between the towers at the south. There is some

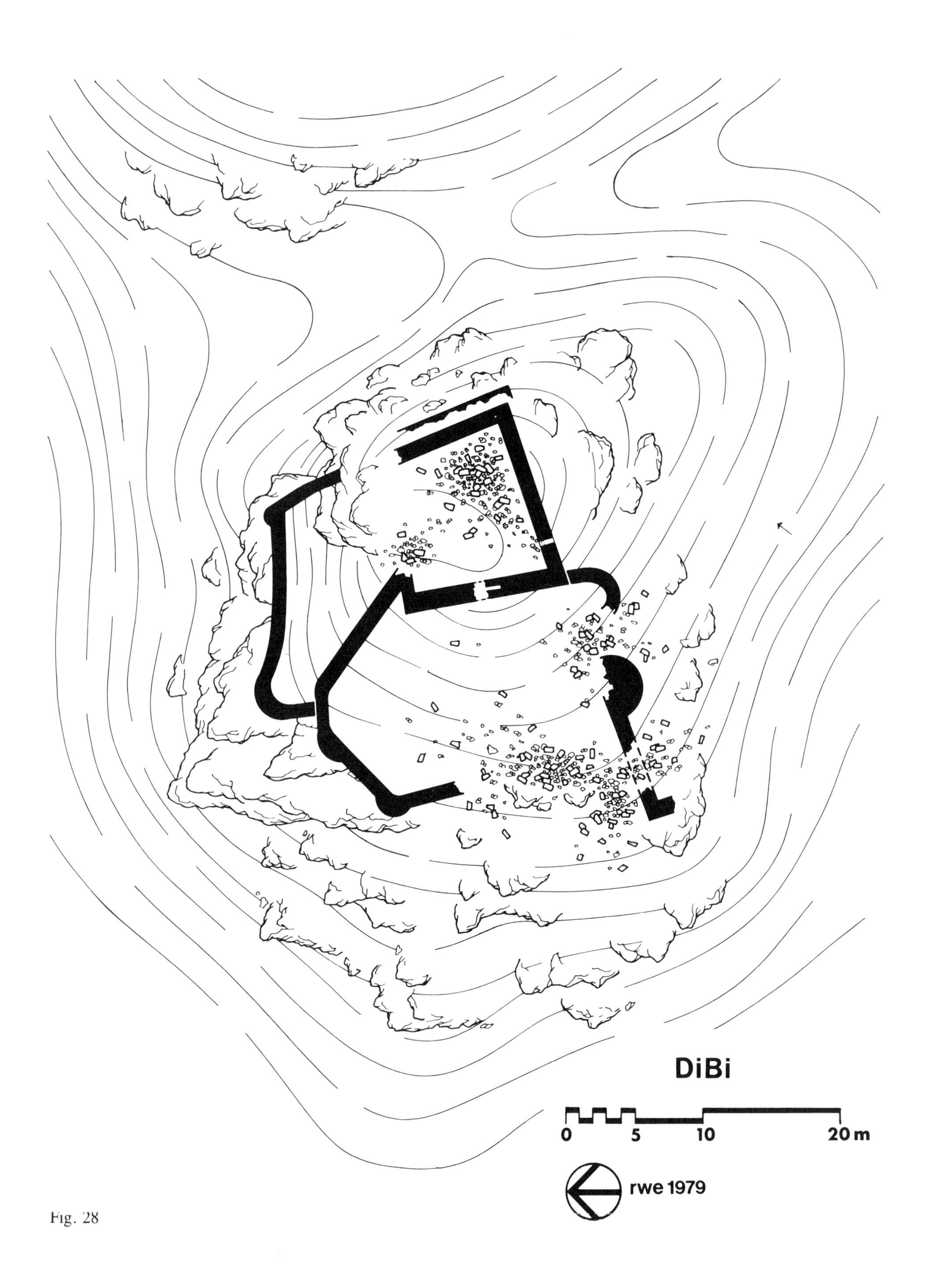

Fig. 28

evidence that the gap in the east wall of the west bailey was once an opening into the higher east ward (pl. 67b). The south side of the gap has a square hole in the core of the wall. This may be an indication of a crossbar bolt.[2] The east wall of the east bailey may also have had a door at one time. The only extant window in the fort is at the west end of the south wall in the east ward (pl. 66b). Today the interior half of the small squareheaded window is buried in debris. Just below the southeast corner of the east ward there are the remains of a thin revetment wall. It is curious that the designers of this complex placed rounded corners at the north and sharp angular ones at the south end of the east ward (the southwest corner of this ward is considerably higher than the adjoining south tower of the west ward; pl. 66a).

From the present remains, it appears that this fort represents one period of Byzantine construction. The local villagers admitted to finding no coins at this site. As with Kalası, it probably had a period of Armenian occupation.

[1]The plan of this hitherto unsurveyed site was completed in July 1979. The contour lines are spaced at intervals of one meter.

[2]There is the possibility that the builders laid horizontal pieces of timber in the core of the wall during construction to add stability (cf. Evciler).

Evciler

Northeast of Hisar Kalesi is the small garrison fort of Evciler.[1] The fort stands on a hill just off the road that runs from Çandır to the villages of Evciler and Arslanköy. Evciler Kalesi guarded this alternate route from Arslanköy to the plain and was just a few kilometers east of the junction with the road to Ereğli. Today the fort can be reached by driving 5 km north from Kızıl Bağ.[2] The closest source of water to Evciler is a stream that flows alongside the east-west road. Like its neighbor, Hisar, the medieval name and history of this site are unknown.

The fort consists of a simple, relatively symmetrical bailey and a keep (fig. 29). The keep is at the summit of an outcrop, and the bailey walls descend down the south flank (pl. 70a). Today in the bailey there is evidence of a single door in the east wall. The entire circuit of the bailey has only one round tower in the southwest corner. The tower is hollow and shows no evidence of windows or doors. The present circuit varies in height from 1 to 4 m, depending on the topography and extent of destruction. At the northeast a second wall, which is parallel to the wall of the bailey, emerges from the corner of the keep and appears to have a rounded east end. The remains are insufficient to determine if this was a chapel. On the exterior the bailey wall is faced with type IV masonry, while the inner face of the wall has a mixture of types III and IV.

The three-story keep is constructed in a similar way, except that the exterior faces of the east and west walls have huge semicircular headers and occasional rectangular blocks with neatly bossed faces (pls. 68a, 68b). The semicircular blocks appear to be recycled pilasters. Also, the corners are fitted with massive bossed quoins. This bossed masonry is *not* similar in style to the Armenian type VII because it has six neatly squared sides and is twice the average size of the Armenian stones.[3] This style of bossed masonry was known to the late antique world. It is likely that the builders borrowed these quoins and headers from older neighboring structures to reinforce the keep. It seems probable that the only breach in the lower level of the keep was the original location for the entrance (pl. 69b). On the interior the headers are not visible (pl. 69a). Occasionally, the stones in the regular courses of the crude interior masonry are placed in a herringbone pattern.

Today the ceiling and the interiors of all three floors of the keep have collapsed along with most of the third-story walls. The lower floor is divided into two units (pl. 69b). The smaller room at the west does not appear to have been vaulted, nor does it have doors, windows, or the stuccoed walls of a cistern. Podia and joist holes supported a wooden ceiling over this room and the floor of the second-level room. The only interesting features in the second level are the two thin embrasured windows in the west wall (shown on the plan). The large first-level room at the east was covered by a now shattered barrel vault. The unique feature of this windowless chamber is that its walls and covering were constructed before the rest of the keep.[4] Today the displacement of the land is causing the two units to separate neatly at their points of junction. The wall that divides the two first-floor rooms is badly damaged and may not have extended through the central and upper levels. There is evidence all around the interior that the wooden floor of the third level was braced in the thickness of the wall. An interesting find at this site are the well-preserved pieces of wooden planks in the cores of the walls (pl. 69a). The long planks acted as horizontal headers. No doubt they added flexibility in earthquakes. The Armenian tech-

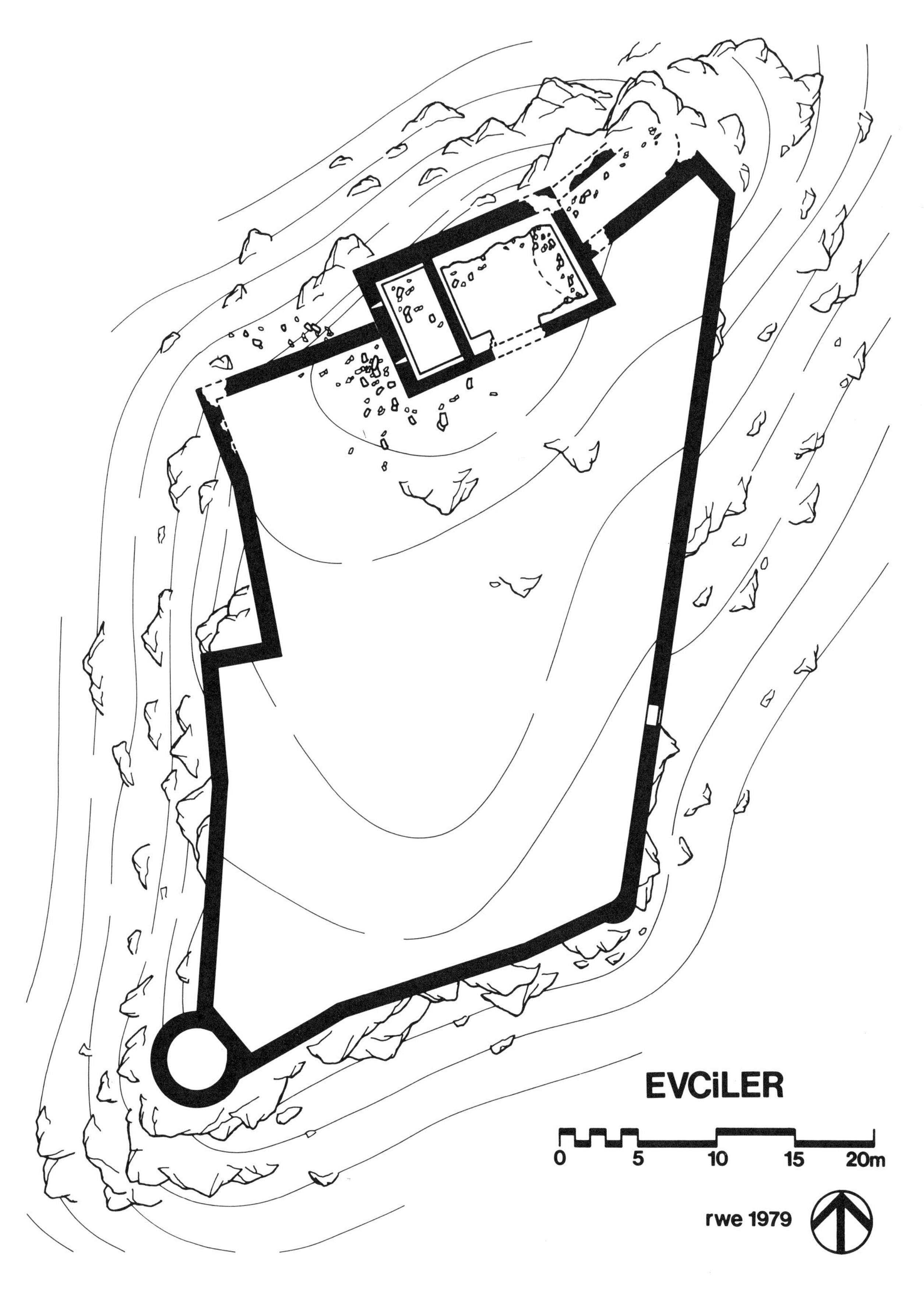

Fig. 29

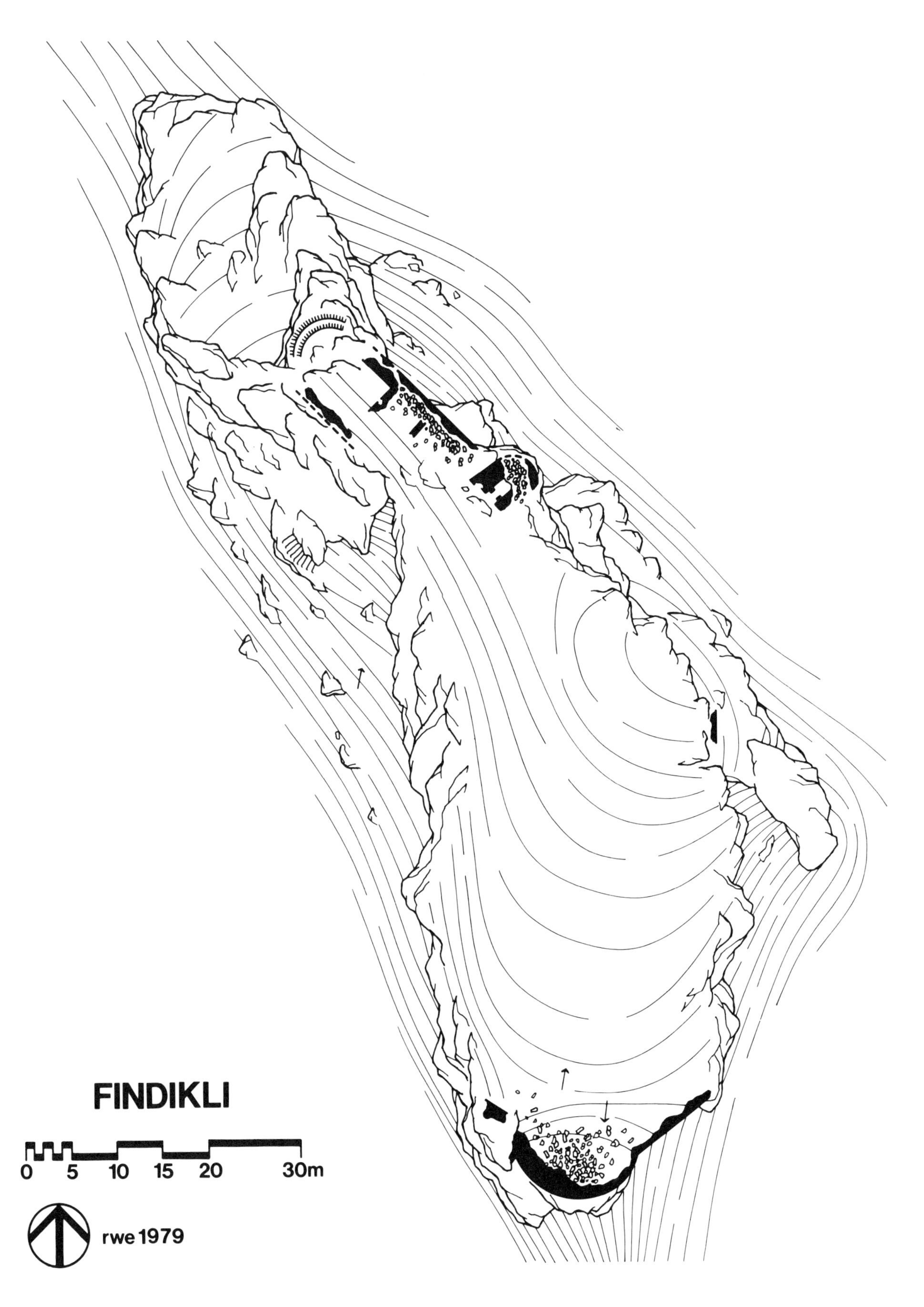

Fig. 30

nique of employing a wooden horizontal chase in the exterior facing (not in the core) may be related to this Byzantine technique. What is certain is that the symmetrical plan, keep, and masonry of Evciler are anathemas in Armenian construction but common in Byzantine. It is highly probable that the Byzantines are the builders of Evciler Kalesi.

[1]The plan of this hitherto unsurveyed site was executed in June 1979. The interval of distance between the contour lines is one meter. My depiction of the keep at Evciler shows the ground level; the two embrasured windows in its west wall are openings in the second floor.

[2]Kılıl Bag and/or Evciler appear on the following maps: *Cilician Gates, Cilicie, Mersin (2)*. The road from Mersin to Ereğli is depicted on the map published by Davis (opp. ix).

[3]Since the inner side was not pointed, this masonry was not expected to bind with a poured core in the Armenian fashion.

[4]There is the possibility that the vault was built *after* the completion of the keep, but in any event it is not contemporary.

Fındıklı

Fındıklı Kalesi[1] is perched on a high outcrop about 2 km west of the strategic road from Göksun to Kadirli (pl. 295c).[2] This fort is one of the northern guardians of the Meryemçil pass, while Geben protects its southern end. Approximately 18 km east of Fındıklı in the village of Değirmendere is a fortress known as Akpınar.[3] In recent times this site was partially leveled by the locals who recycled the stones in their homes. Conditions did not permit my examination of Akpınar. The village of Fındıklı, officially called Fındıklı Koyak Köyü, is about 3 km north of its fort. There are numerous springs around the base of the fortified outcrop. Obviously, the Turkish name "Fındıklı" is unattested in our chronicles. By reason of its proximity to Geben and Göksun, it may be the Armenian fortress of Kanč᷾ (Gantchi).[4] This conclusion is purely speculative.

The approach to the summit of Fındıklı Kalesi is a dangerous track that is situated on the west side of the massive outcrop. In the northern half of this west face there are the remains of scarped steps (fig. 30). Upon reaching the north end of the summit it becomes evident that the walls and internal buildings have suffered the most severe damage. Only the outline of a few buildings and towers can be seen amid the accumulation of dirt and fallen masonry. In a shattered tower at the northeast is the foundation for a door. Farther north there are two scarped circular recesses. Undoubtedly these served as socles for a tower.

The only sizable structure in the fort is the massive tower at the south. It too has suffered severe damage, and only its southern face is standing today (pl. 71a). All of its exterior facing, like the collapsed structures at the north end, consist of type VII (pl. 70b). However, we see an interesting variation in that the courses of stones around the coat of arms and the three courses below it have had their bosses shaved off. This motif is also used at the Armenian site of Çem. The coat of arms is high up in the center of the tower and completely inaccessible for close view. If the shield in the center of the frame (that is, the escutcheon) had figures in relief, they appear to have been erased by weathering.[5] The plan and masonry of this site appear to be Armenian.

[1]The plan of this hitherto unsurveyed site was completed in July 1979. The interval of distance between the contour lines is 1.5 m.

[2]Cf. my discussions of Geben and Haçtırın in the Catalogue.

[3]The village of Fındıklı and its neighboring fort, which is also referred to as Kızıl Kale, appear on the maps *Elbistan (1)* and *Elbistan (2)*. On the *Marash* map, Fındıklı Kalesi may appear as Marianchil Kale. Sterrett (253 and map 2) finds in "Deïrmen Deresi" a Greek inscription of A.D. 107 and lists a "Kale" at Fındıklı. Cf. *Handbook*, 383; Alkım, map 3.

[4]Alishan, *Sissouan*, 213–15, map; Smbat, G. Dédéyan, 76 notes 40 and 43, map; Aghassi, 21, map. Kanč᷾ should not be confused with the site of Čanči, which appears to be farther to the northeast. To reach Fındıklı from the greater Fırnıs (Fawr̃naws) river valley is a half-day's journey by foot to the west. If the site of Kanč᷾ is clearly associated with that valley, then its identification with Fındıklı is weakened. See Appendix 3, note 5.

[5]Cf. Robinson and Hughes, 198 f.

Fındıkpınar

Deep in the mountainous Highlands of Cilicia Tracheia is the small village of Fındıkpınar and the neighboring fort[1] that bears the same name.[2] This village is nestled in the basin of a large valley (pl. 71b), which is defined by the surrounding outcrops and two merging tributaries. The resulting river flows south past Tece Kalesi and empties into the Mediterranean near Seymenli. Much of the road, which once followed the course of this river, is now abandoned and passable only by foot or horse. A number of upland streams flowing into the vale of Fındıkpınar have carved and consequently oriented adjacent valleys at the north and east to this road. Today Fındıkpınar can be reached by the paved road

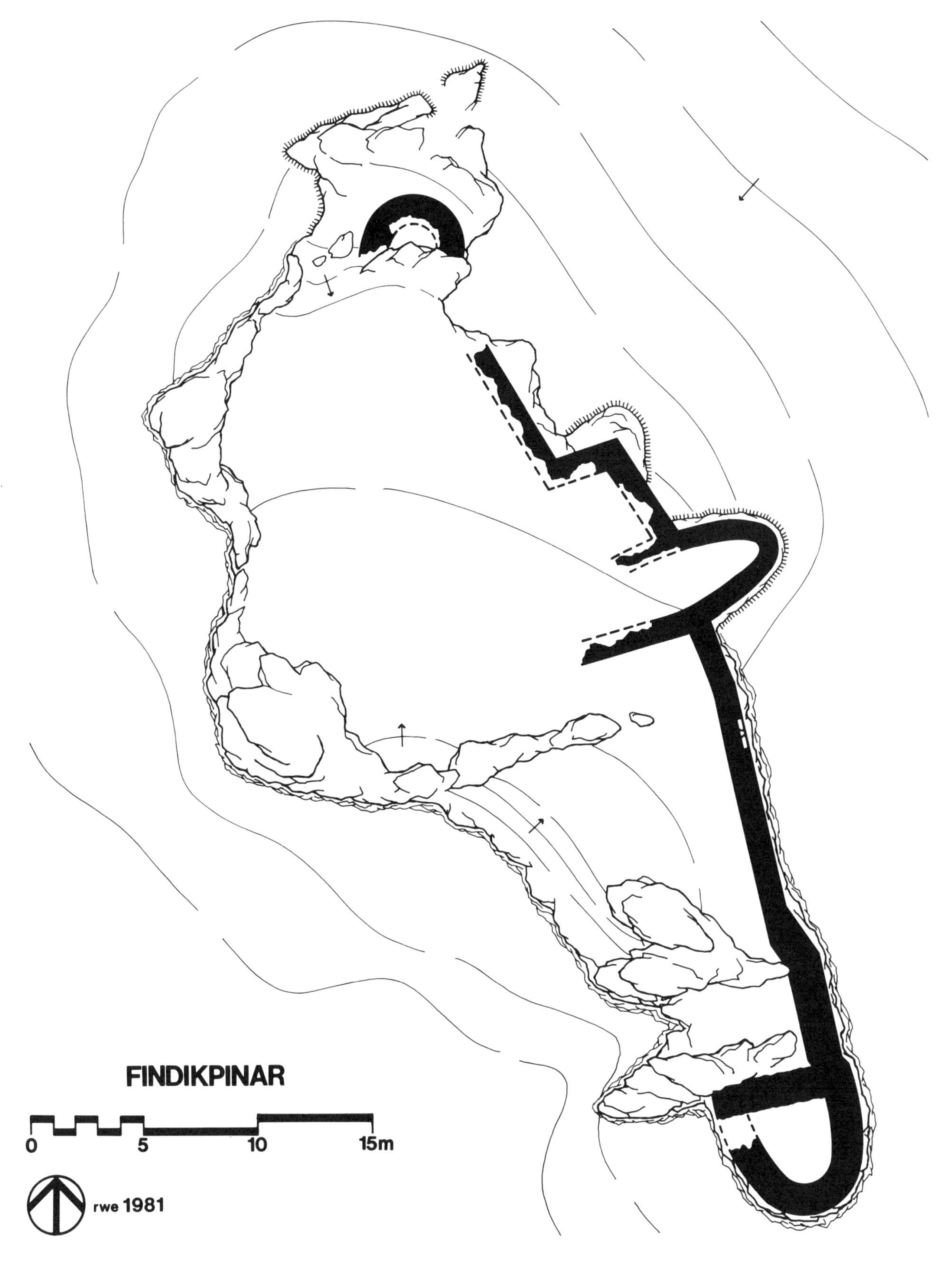

Fig. 31

that links Kuzucubelen to the Mediterranean. The fort, which is situated on an elongated outcrop of limestone east of the village, has a clear view into all the adjacent valleys (pl. 72a). Fındıkpınar Kalesi is carefully adapted to the irregular folds of the rock (fig. 31). Only at the east, where the cliff is not too severe, do circuit walls and towers appear along the edges of the outcrop.

Since most of the cut stone above ground level has been recycled recently into modern constructions, the masonry can be evaluated only from the foundations and the revetted faces along the east circuit. The majority of the exterior facing stones consist of a neatly cut type V (pl. 73b). The masoned walls rise on the carefully carved rock. In the center of the east circuit the masoned tower simply continues the shape of the scarped salient below (pl. 72b). Between this tower and another scarped salient (3 m to the north) a short masoned wall rises from ground level to join the two elements and seal off any line of access. The smaller of these two salients (at the north) is not topped by a rounded bastion but simply by an angled jog in the course of the circuit wall. The protruding angle does not violate Armenian architectural traditions since it only crowns a round salient of rock. In a few areas along the circuit, type VII masonry is substituted for type V. The nature of the interior masonry is more difficult to determine, but it appears that it was a combination of types III and IV (pl. 73a).

Because of extensive destruction it is difficult to evaluate the architecture of the fort. The fort is confined to the northern tip of the outcrop and probably held a rather small garrison. Other than the east and south towers there is no evidence of rooms constructed with stones nor any indication of a cistern. Today the only evidence that the circuit at the east could accommodate archers is the stirrup base of a loophole south of the central-east tower. The stirrup design is very common in Armenian forts. Only the base of the tower at the far south is still in situ (pl. 73a). Unlike the east and south towers, which enclosed rooms, the core of the north tower seems to be solid rock. Just north of the north tower there is a vertical wall of scarped rock. It seems that the gap between this tower and the scarped wall was the entrance into the bailey. There was probably some sort of gate to control the entrance.

Immediately to the east and northeast of the fortified outcrop (pl. 72a) many of the protruding rocks have been scarped to accommodate wooden buildings (not shown on the plan). Square joist holes in the faces of this scarpment are common. Also, the foundations of many buildings, which are constructed with type IV masonry, are visible. In medieval times there appears to have been a sizable community below the east flank of the fort.

Because of the asymmetrical plan of the complex, the rounded towers, and the masonry, it appears that this garrison fort is an Armenian construction.

[1]The plan of this hitherto unsurveyed site was completed in June 1981. The contour lines are at intervals of one meter.

[2]This village appears on the *Central Cilicia* and *Mersin (2)* maps. See Heberdey and Wilhelm, 41, map; Schaffer, 61; *Handbook*, 194 f.

Geben

In northeast Cilicia almost midway on the strategic road from Andırın to Göksun stands the fortress of Geben.[1] A modern village bearing the same name is located 3 km to the southeast of the castle.[2] The medieval site surrounds a tiered outcrop on the east flank of the road just before the northbound traffic ascends the Meryemçil pass (pl. 80a). This thoroughfare is broad and, judging from the weathering of the scarped rock, is a path of considerable antiquity. Recently the Turkish Forestry Service has widened sections of the road to accommodate the large logging trucks around the hairpin turns. Once over the pass, the course to Göksun is relatively easy. North of the Meryemçil pass, another trail from Rifatiye and the Mazdaç pass joins the Göksun road about 9 km south of Fındıklı. A second defile, which is northeast of the castle, has a trail that is completely inhospitable to vehicular traffic and should be attempted by only the most experienced climbers (the path does not appear on my map, fig. 2). This trail eventually joins the road from Göksun to Maraş (via Şadalak). The vale of Geben covers about 130 sq km and is traversed by two other trails, one at the east and the other at the west. The western track begins in the village of Geben and moves to the southwest until it crosses a narrow pass and terminates in the valley of Çokak, northwest of Andırın and 15 km northeast of Rifatiye. About 7 km south of the castle of Geben the eastern track breaks off from the main north-south road and exits the vale through a wide pass. In this pass the track divides, with one branch snaking south to Maraş and the other moving directly northeast to join the Şadalak trail from Maraş to Göksun. Neither the east nor the west track through the vale of Geben appears to be guarded by forts. At the extreme south end of the valley is the small garrison fort of Azgit, which stands on the west side of the north-south road. From December to late March the snows halt all vehicular traffic on the three roads. The melting snows and the summer thunderstorms provide a constant supply of water for the fertile valley. The streams near Geben form part of the Andırın Suyu and flow south into the tributaries of the Ceyhan River. These water courses have cut the strategic route from the plain to Göksun.

The immediate topography around the fortress of Geben is spectacular. To the south and near the perimeter of the outcrop are the rolling fields of wheat and sorghum. To the north, east, and west (pl. 74a), almost vertical walls of pinkish-gray limestone rise from a sloping base to form an alpine canyon. The visual impact of these cliffs, which tower over 400 m above the castle-outcrop, is broken only by occasional pine trees anchored to the ascending ledges.

So strategic a site certainly played an important role in medieval history, but unfortunately it is not well attested in the non-Armenian chronicles. In fact, this site is also ignored by many modern cartographers and mislocated in the mountains. Fedden and Thomson place Geben southeast of Coxon (Turkish: Göksun), when in reality it is actually southwest of that city.[3] Hellenkemper calls "Keben" (that is, Geben) a strategic station in the area of modern Faziye "am Meryemçil-Pass über Göksun."[4] Dunbar and Boal confuse the site of Geben and other forts in this district and go so far as to state that there is no fortress near the village of Geben.[5] What is apparent is that this castle was an important Byzantine stronghold that eventually fell into Armenian hands.[6]

One of Geben's most important roles in history is the job that it did *not* play during the First Crusade. Prior to entering the Levant in 1097 the Crusaders divided their forces at Heraclea, with Tancred and Baldwin passing south through the Cilician Gates, while Godfrey of Bouillon and the main army, guided by the Byzantines, moved through Göksun to Maraş by a more tedious and dangerous route.[7] S. Runciman poses the question of why they did not take the safer and "usual" road to Maraş.[8] There are two trails leading south from Göksun.[9] Of the two trails, the easiest and most direct course to Maraş is through the Meryemçil pass via Geben and Azgit. If the castle at Geben were in the hands of the Danişmendid Turks (and consequently hostile to the Byzantines and Crusaders), then the only alternative would be the hazardous southern route via Şadalak, which lies 28 km east of Geben.

The medieval chronicles record some of the significant events in the castle's history. In 1139/40 the Danişmendid Emir Muḥammad b. Ġāzī took Vahga, Geben, and certain small forts that had been occupied (ca. 1138) by John Comnenus in the region of the Red Mountains.[10] Seven years later it was recaptured by the Armenians. Like Vahga, Geben was an important baronial seat for the Rubenid family because it guarded a major trade route into Cilicia.[11] In fact, it was the young Baron Levon II, the first king of Armenian Cilicia, who received the castle as his fief in 1182. On the Coronation List (see below, Appendix 3) a baron by the name of Tancred is cited as lord of the castle. If Tancred was a Frank,[12] then he probably commanded an Armenian garrison, since no Crusader orders were ever granted fiefs in the mountains. On the same Coronation List, Grigoris is cited as the bishop of Geben.[13] By 1215 Baron Levon, a relative of King Levon I, is mentioned as master of the castle in the trade agreement with the Genoese.[14] One year later the Sultan of Konya, Kaykāʾūs I, laid siege to Geben. The Grand Baron Constantine, accompanied by both Hetʿumid and Rubenid lords as well as a large Armenian army, quickly marched his forces from the south and met the Turkish army in a pitched battle at the base of the castle-outcrop. The Armenian troops were crushed, and Constantine, along with other barons, was taken prisoner.[15] Surprisingly, Baron Levon and his castle were not captured. In the following year the sultan lifted his siege and withdrew. Later, in 1296/97, when King Hetʿum II journeyed to Constantinople to attend a wedding, his brother Smbat, whom he had left in charge of the kingdom, usurped his throne. A certain Constantine, the lord of Geben, temporarily subdued Smbat on behalf of Hetʿum.[16] Geben's final appearance in the history of the Armenian kingdom is a sad one. In 1375 the Armenian kingdom had collapsed; the fortress of Geben was the last to surrender to the Mamluks after a nine-month siege. One tradition has it that King Levon V took refuge here at that time and was captured along with his family, the principal nobles of Cilicia, and Geben's garrison.[17] Thereafter, the wife of a slain Hetʿumid general dispatched troops in order to seize the fortress for her son, Gēorg, who then resided in Geben as its lord well into the fifteenth century.[18] In the nineteenth century Geben had a sizable Armenian community and a church.[19]

Today the castle covers almost the entire outcrop and shows two periods of construction: Byzantine and Armenian (fig. 32). On the southwest side where the slope is less severe, multiple circuits have been employed. When making the ascent on this flank, the approaching party enters through the only opening in the lower circuit wall at point A. No doubt a gate once stood at this spot. Just northeast of this opening are the

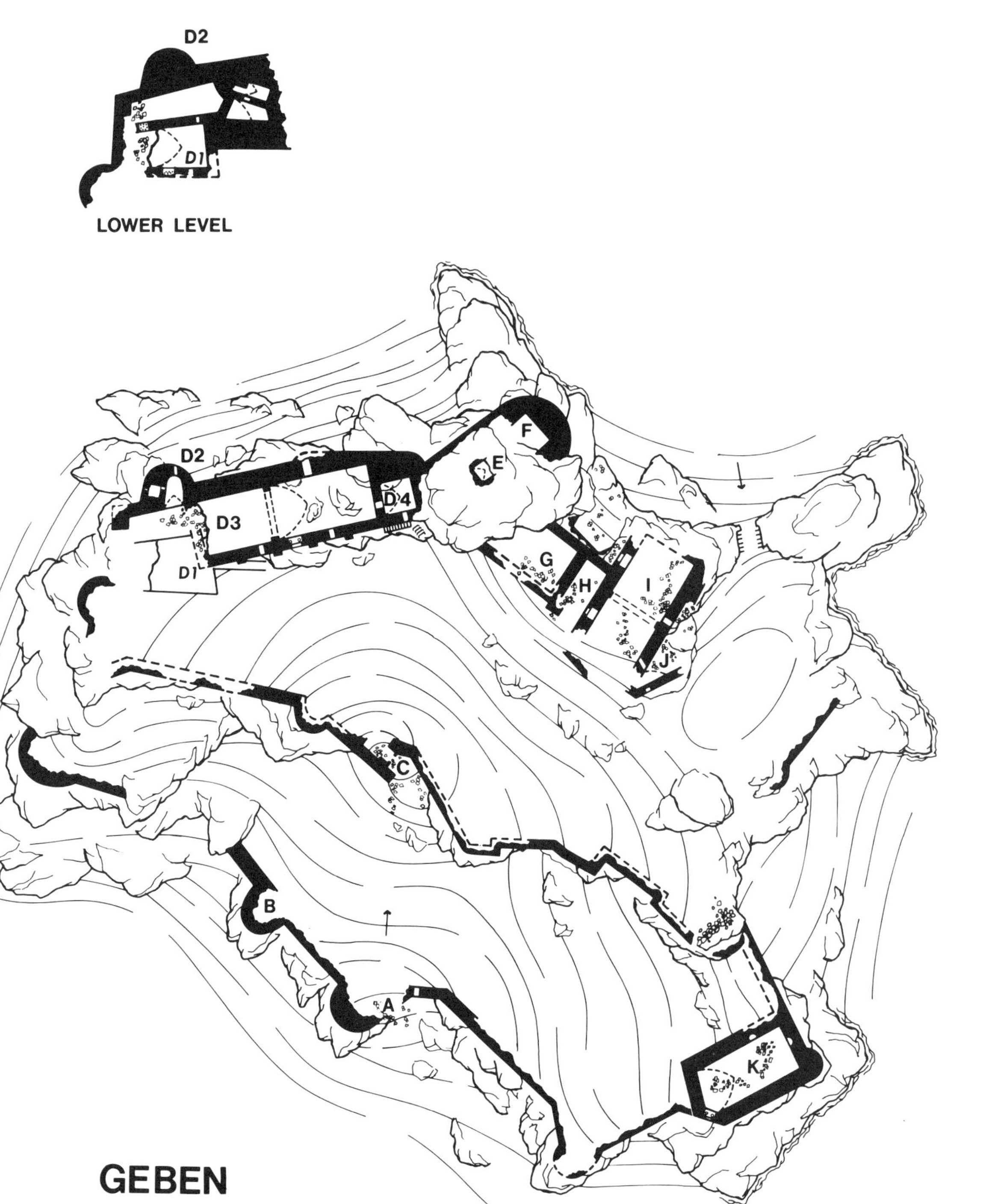

0 5 15 25 40m

Fig. 32

remains of a window. This lower circuit is guarded by four rounded towers. The southernmost and northernmost towers are now in ruins. The entire lower circuit wall is badly damaged. This destruction may be due to the siege of the Mamluks or perhaps to earthquakes. The northernmost tower in this lower circuit is carefully integrated into the projecting rocks and is more of an outwork than an integral part of the wall. The two central towers, A and B, stand to over 2 m in height. This lower circuit wall appears to represent one period of Armenian construction; a single, continuous core binds the type VII masonry of the exterior facing to an interior facing of type IV. The bailey between the lower and central circuits is devoid of any structures. The peculiarities of building K will be discussed later.

The central circuit wall is pierced by a single gate at point C (pl. 74a). As with its lower-level counterpart, this gate is severely damaged, but here there is still some evidence of jambs. Judging from the thickness of the wall, compound arches probably covered gate C. The wall of this central circuit carefully follows the course of the rock ledge and shows two periods of construction. In the lower sections the Byzantine architects consistently employed a large type IV masonry in the exterior facing. Later the Armenians built over this with a combination of types V and VII masonries. The areas immediately around gate C and the southeast end of the central circuit are devoid of towers. Instead, there are sharp jogs in the wall with exposed corners. This technique of construction is uncustomary for Armenian forts and indicates that the Armenians retained the original Byzantine plan.

Once inside gate C, the impressive complex of building D dominates the upper bailey (pl. 74b). This collection of rooms constitutes the main apartments in the castle. Building D is a rectangular structure that consists of three units on the main level. A path ascends the steep slope from gate C to the southeast end of D4, where a few scarped steps are still visible. These steps lead immediately to a cantilever stairway that is attached to the south face of D4. In turn, the stairway gives access to a roof that once was a third level covered with rooms.

Because of extensive damage, the only evidence of construction today at the third level is a part of the north wall. There are also the remnants of a window here above the door in the north wall of D3 (the third, or upper, level is not shown on the plan). The masonry of this upper level is a harmonious blend of types V and VII in the exterior facing and types VIII and IX on the interior; it is identical with the types used in the main level (D2–4). The only difference is that the walls of the main level have a thin socle of type III masonry. Both levels are contemporary and probably Armenian.

The main (or central) level of building D is in an excellent state of preservation. The south wall of D3 is pierced by one roundheaded entrance that is limited by jambs and equipped with two post holes to accommodate double doors. This south facade is opened also by two small squareheaded windows set high in the wall. Due to the nature of the descending topography, the architects were unable to construct the south wall with the same thickness as the north wall. In order to support an upper level, four buttresses, which extend about 75 cm from the thickness of the wall (pl. 74b), were added to the south facade. The interior of chamber D3 shows a similar adaptation to the topography in that massive monoliths protrude into the body of the chamber at the east end and are incorporated into the thickness of the walls (pl. 76a). Only small amounts of the surface areas on the exposed rocks are scarped. This room is covered by a single pointed vault with only one transverse arch (this rib is west of the door) of neatly cut voussoirs to add support. The entire vault is stuccoed. At the springing level of the vault, square sockets are still visible. These probably supported the wooden crossbeams of the centering; after construction the beams were left in position to add stability to the walls. The only opening in the north wall of D3 is a large roundheaded door. The base of the door is about one meter from the ground level of the room. This door opens onto a small circular porch that has now collapsed. At the east end of D3 is a roundheaded door that leads into chamber D4. The only other opening in D4 is a small squareheaded window in the south wall. The entire northwest corner of this small vaulted chamber is filled with the rock mass that occupies the northeast corner of D3.

At the western end of D3 the vault and the west wall have fallen away (pl. 75a). Undoubtedly there was once a passageway leading from D3 to D2. D2 is a rounded tower constructed with the same types of masonries as rooms D3 and D4. The apsidal room of the tower is covered by a slightly pointed vault and is pierced at the north by one highly placed squareheaded window (pl. 77a). To the west of the apsidal room is a small vaulted chamber that is entered through a single roundheaded door. The only other opening in this chamber is a small roundheaded embrasured window in the west wall (pl. 77b). On the interior side the width of the window is 30 cm.

The lower level of complex D consists of four chambers. Two of the rooms are directly under the western half of D3 (pl. 76a). Unlike room D3 on the main (or central) level, both of these lower-level chambers are on a northwest-southeast axis. There is only a single door connecting the two chambers. The southernmost of the two rooms is trapezoidal in shape and covered by a partially collapsed half vault. The northernmost is rectangular with a niche(?) at the southeast near the connecting door. This room is covered by a single pointed vault. Both chambers are constructed with type IV masonry, like the lower level of the central circuit wall, and may be Byzantine. Since neither of the rooms is stuccoed, they were not used as cisterns; they may have served as storage areas with access from above through hatches. The area east of these chambers is solid rock. Both rooms functioned as a support to maintain the central-level floor of D3.

Southeast of these rooms is the lower-level vaulted chamber D1 (pls. 75b, 76b). Like the two lower-level rooms under D3, the north wall of D1 is constructed with type IV masonry and has a roundheaded window at the east and an arched passage at the west (now blocked). The window at the east is distinctively different in design and execution from the other arched openings in chambers D2–4 in that a number of small voussoirs are used to form its vaulted covering. This window may be Byzantine. The south wall of D1 has undergone major reconstruction in the Armenian period (pl. 75b) when a slightly depressed roundheaded door (now partially blocked) and a squareheaded window were added. The door and window are almost identical in size and masonry to the openings in the south wall of D3. This door in D1 was once limited by jambs; today the salient corners of the jambs are shattered. The interior of the slightly pointed vault over D1 was built and stuccoed like its counterpart in the central level of D3. All of the west wall and most of the east wall in this chamber have vanished. From the center of the east side of D1 a low-level wall of type IV masonry projects eastward for about 9 m (pl. 75a). Part of the south wall and floor in the western half of D3 (central level) is built over the low-level wall but on a different axis. The result is that the low-level wall projects beyond the foundation of the central level of D3, creating a bulge. The Armenian engineers refaced this revetment wall with type V masonry, giving it the appearance of a talus. Directly north of D1 is a sloping pentagonal corridor that may have provided a measure of communication between D1 and D2. Only excavation can reveal the true complexity of the lower level. What is apparent is that the Armenians integrated their own plan into preexisting Byzantine structures. Whether an earlier Byzantine building occupied the entire space of complex D is unknown. Except for the lower-level chambers, complex D represents one period of Armenian construction.

Moving east from complex D, we find the small vaulted cistern E, which rests atop a truncated pinnacle in a natural sink. The north, south, and west walls of the cistern are flanked with masoned stones, while the east wall is naturally flat. The Armenian masons girdled the center of this pinnacle with a round tower of type VIII masonry. The interior of this tower (F) has a small unroofed rectangular chamber to accommodate guards. The flank of tower F continues southwest until it abuts against building D. When seen from the east, this tower appears to emerge from the living rock and today stands as testimony not only to the dexterity of Armenian masons but to the military principal that every potential point of access must be guarded.

To the south of tower F stand the shattered remains of buildings G and H (pl. 79a). Types VII and VIII masonries are used exclusively in the construction of these buildings. Both structures are rectangular, although the interior of building G is somewhat irregular due to the abutment with the southeast face of the truncated pinnacle. Just below the northeast facades of buildings G and H are the remains of revetment walls constructed out of types III and IV masonries. These walls may have been part of an earlier Byzantine building that tumbled down the cliff. Perhaps buildings G and H were set back from this earlier foundation to avoid the problem of landslides.

Immediately adjacent to building H is the vaulted hall I. Like complex D, I may also have been an apartment. Today only its east wall, which is built of types VIII and IX masonries, is standing; a lower section of its shattered vault is precariously balanced on the east wall. From ground level the east wall reaches a total height of more than 7 m. Judging from the thickness of the walls, a second floor was probably accommodated above the first-level vault. Like undercroft D3, only a single transverse arch (rib) added support to the vault. In the east wall just north of the rib is a small, slightly roundheaded window with a slanted sill (pl. 78a). A second window appears in this wall at the south end. This window, which is squareheaded and set diagonally, is located in the springing level of the vault. Two square joist holes are still visible in the east wall. The foundation of the west wall of hall I has two doors that are limited by jambs (pl. 79a). One door gives access to building H, while the other opens onto a low-level porch. Eight meters northeast of hall I and at a considerably lower level is a carefully scarped passage between sections of jutting rock. If wood or masoned ramparts were lodged in this area they are absent today. Undoubtedly some sort of blockage prevented access from below to the upper bailey.

Directly southeast of hall I is the curious triangular structure J. Because of the slippage of the cliff, only the south wall of J is presently standing (pl. 74b). The exterior of this wall is built exclusively with type VII masonry, while the interior is a mixture of types VII and VIII (pl. 78b). The only oddity on the exterior of the south wall is a shallow squareheaded niche with a thinly incised rounded hood on the exterior. The niche is over 2.5 m from ground level and access is impractical even from the flanking rock. The function of this orifice is a mystery. This south wall carefully ascends the face of a rocky projection and rises almost to the height of the east wall of hall I (which is also the west wall of building J). It is clear that the south wall of building J and the east wall of hall I were connected by a quadrant vault since springing stones are still visible on the exterior of the east wall (pl. 78b). The roundheaded window in the center of the east wall of hall I is about 1.5 m below the springing for the quadrant vault of building J. At the north end of the east wall of hall I a salient protrudes to the southeast. This salient was once connected (by means of a now collapsed third wall) to the south wall of building J. This interesting arrangement is again due to the sloping topography and indicates that level space on this very rocky site was at a premium.

Southeast of building J is a small portion of the upper-level circuit wall that protected the east flank. This wall may have joined with building J and the central circuit. From building J there is a gentle sloping path to building K (pl. 80a). The only entrance into K is a door at the south. Although building K joins the lower and central circuit walls it is elevated above the level of the lower bailey and has no direct access to it. Building K does not function as an outwork. Internally the structure is almost rectangular. Today its slightly pointed vault has collapsed as well as parts of the northwest and northeast corners. There are indications that a second story was positioned over the vault. The single entrance at the south is limited by jambs; two post holes accommodated a double door. All of building K, with the exception of the south door and the wall immediately flanking it, is constructed with type IV masonry and is probably Byzantine (pl. 79b). The arches over the two roundheaded niches in the east and north walls of K are built with small voussoirs and are identical to the arch over the window in the north wall of D1 (pl. 76b), which is also attributed to the Byzantines. The south door and its flanking wall are constructed with type IX masonry on the interior and type VII on the exterior. This door has a depressed arch and is identical to the Armenian doors in the south walls of D3 and D1. Unlike complex D, the Armenians salvaged most of the Byzantine building K. The function of the two large niches, which are about 1.6 m from the floor level of this room, is unknown. It seems highly unlikely that this building was a chapel for either the Byzantines or the Armenians.

[1] The plan of this previously unsurveyed site was begun in early fall 1974 and was completed on my second visit in July 1979. The contour lines are spaced at intervals of one meter.

[2] In Armenian this site is referred to as Kapan, Gaban, Gabon, Gabnubert, and Geben. One Greek designation (undoubtedly derived from the Armenian toponym), Καπνισκέρτι Φρούρια, is found in Cinnamus (20); see: Honigmann, 130; *RHC, DocArm,* I, xxv; "Gaban," *EI*², 970; "Geben," *IA,* 761 f. In Syriac the fort is known as Gabnûpîrath. The Armenian word *gaban/kaban* means "narrow pass."

[3] Fedden and Thomson, 12; cf. H. Ter Łazarean, 34; Texier, 586; Ritter, 36; Alkım, 211 ff, map 3.

[4] Hellenkemper, 263; *RHC, DocArm,* I, xxvi note 4. See "Archaeological Research in Turkey, 1949–50," *AS* 1 (1950), 19; cf. U. B. Alkım, "The Road from Sam'al to Asitawandawa: Contributions to the Historical Geography of the Amanus Region," *Anadolu Araştırmaları 2. 1–2 (1965), 1–41; Handbook,* 383.

[5] Boase, 85. Geben Kalesi appears as Saransak Kale on the *Elbistan (2)* map; the village of Geben is accurately located on the chart of *Central Cilicia.* A less precise position for Geben is given on the *Marash* map.

[6] Alishan, *Léon,* 14.

[7] Tomaschek, 86; *Gesta Francorum,* trans. and notes R. Hill, ed. R. Mynors (London, 1962), 26 f; Albert Aquensis, *RHC, HistOcc,* IV, 357 f; H. Hagenmeyer, *Chronologie de la première croisade (1094–1100)* (Paris, 1902), 99 f; Petrus Tudebodus, *Historia de Hierosolymitano Itinere,* ed. J. and L. Hill (Paris, 1977), 61 f.

[8] S. Runciman, "The First Crusade: Constantinople to Antioch," in K. M. Setton, ed., *A History of the Crusades,* I (Madison, 1969), 298.

[9] A third and very rough trail via Elbistan is highly impractical and may not have existed in medieval times.

[10] Cinnamus, 20; Bar Hebraeus, 266; Michael the Syrian, J. B. Chabot, III, 248; Smbat, 616 f; Gregory, 152–54; Der Nersessian, 637; Cahen, 360. The fact that Cinnamus refers to Geben by its Armenian name indicates that the site was a Rubenid possession before 1137. Neither Cinnamus nor any other source provides the original Greek toponym for the fortress.

[11] Vahram of Edessa, 513 note 2; Alishan, *Léon,* 73; Smbat, G. Dédéyan, 57.

[12] Considering the tendency during Levon's reign to adopt Frankish customs and names, it is remotely possible that "Tancred" is an Armenian.

[13] Geben must have maintained an Armenian population of at least moderate size since bishops were assigned there through the 14th century; see *RHC, DocArm,* I, lxviii.

[14] See Langlois, *Cartulaire,* 126 f.

[15] Hetʿum, 483; cf. Vahram of Edessa, 513; Alishan, *Sissouan,* 209–11; idem, *Léon,* 297–300; Cahen, 623; Smbat, G. Dédéyan, 92 note 13.

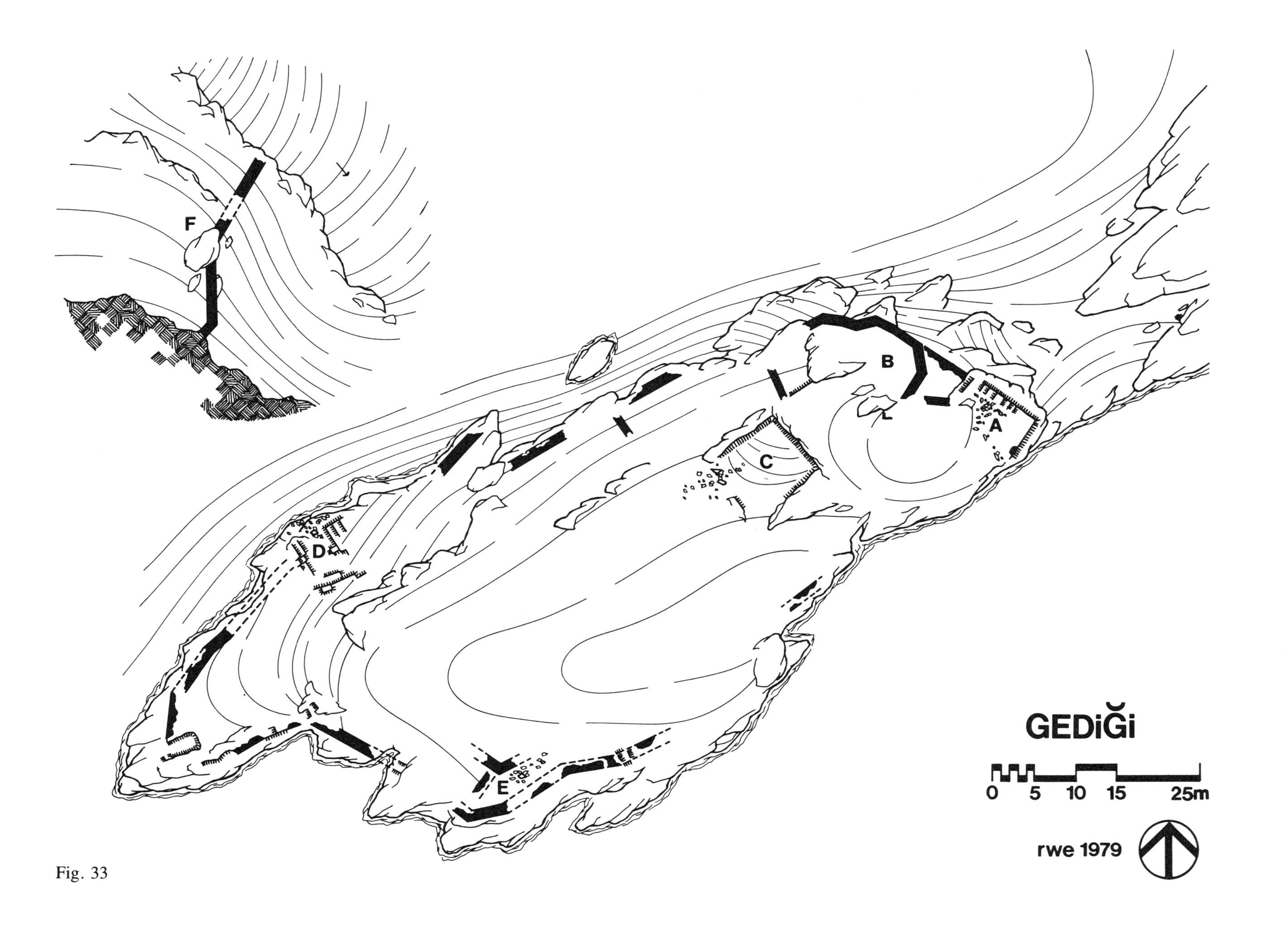

Fig. 33

[16]Samuel of Ani, 465; cf. Smbat, 656. In 1307 the Baron Ōšin is listed as the lord of Geben, but we know nothing of the events in his life; see Dulaurier, 313f; *RHC, DocArm,* I, lxxxv, cxiv. In 1309 Geben served as a valuable troop station on the frontier; see Mas Latrie, I, 299. For a genealogy of the lords of Geben see Toumanoff, 282 f and Christomanos, 148.

[17]*RHC, DocArm,* I, 686 note 3; 719 f.

[18]Alishan, *Sissouan,* 198.

[19]Alishan, *Sissouan,* 209–11; "Central Turkey Mission," *The Missionary Herald* 99 (April, 1903), 161 f. Aghassi, 22, 255–60. A comprehensive summary of Geben's history can be found in Yovhannēsean, 159–66; cf. Cuinet, 243 f.

Gediği

The medieval cloister of Gediği[1] is located 10 km south of the village of Yavca on the road from Hisar to Arslanköy. A second route on the south flank of Gediği Dağı leads to the Mediterranean. Aside from watching these two routes, the site has an overwhelming view to Mersin and to the upland valleys at the northeast and northwest. Gediği has intervisibility with Hisar, Arslanköy(?), Fındıkpınar and Evciler. To the southwest of Gediği is a second hill which is surmounted by a series of terraced caves (pl. 81a). A number of these natural limestone cavities have been sealed by manmade walls. Our schedule did not permit a formal survey of this site, but the general impression is that Gediği's neighbor was also a monastic complex.

Of all the sites that I have visited in Turkey, Gediği is the most inaccessible to climbers.[2] This cloister is perched atop the summit of a mountain that rises to an altitude of 2,210 m (fig. 33). Once at the base of the mountain, it takes an experienced climber the better part of the afternoon to reach the top.

The only possible approach to the summit of Gediği is from the north. About two-thirds of the way up the mountain is wall F and a possible gate. This wall is actually an outwork that runs downhill from under an overhanging cliff; it is designed to control the approach. The masonry of the wall is a mixture of types VII and VIII. The north section is almost exclusively type VII, while the upper section at the south is mainly type VIII. This mixture does not occur frequently on the summit; I am hesitant to label it a separate period of construction. A hole in the wall was once a door. Because of the rough topography of this site it was impossible to determine precisely the location of point F in relation to the summit. The trail from this outwork to the top passes a number of empty caves. The winding path eventually turns from its southwesterly course to the northeast end of the summit, where the ascending ridges and revetment walls lead to what may have been a gate between points A and B (pl. 80b).

Once inside the enclosure it becomes apparent that none of the masoned walls stand to more than six courses in height and that the entire summit is covered in a thick blanket of vegetation. This plant life conceals the foundations of numerous buildings. The extensive destruction here may be the result of earthquakes. Tons of coursed stones have fallen down the sides of the cliffs. Wooden buildings may have been adapted to the numerous scarped areas on this site. The few traces of exterior facing stones appear to be type VII. The function of units A and B cannot be determined from surface remains. The interior of B may have had a number of rooms. West of B there are recent signs of treasure hunters probing through the foundations of rooms. The units A, C, and D have only the skeletons of their scarped faces and scattered rubble. The curious alignment of walls at E was probably not for defense, since the cliffs on the south and west sides of the summit provide adequate protection. There is no evidence of a cistern at this site.

Because of the isolation of this complex, it could not have served as a refuge for the local population, nor could it be an efficient station for troops, since access is difficult and horses would have to be maintained at the base of the mountain. It is also apparent that Gediği is not constructed like a fortification, for there is no complex entrance, towers, or any divisions in the bailey. Gediği was probably a monastic retreat. It has military value as an observation post and signal station.

[1]The plan of this previously unsurveyed site was executed in June 1979. The contour lines are placed at intervals of one meter. The lower-level wall F and the summit complex were too far apart to survey on a continuous drawing. F appears as a separate drawing on the same plan as the summit.

[2]This site is listed on the *Mersin (2)* map.

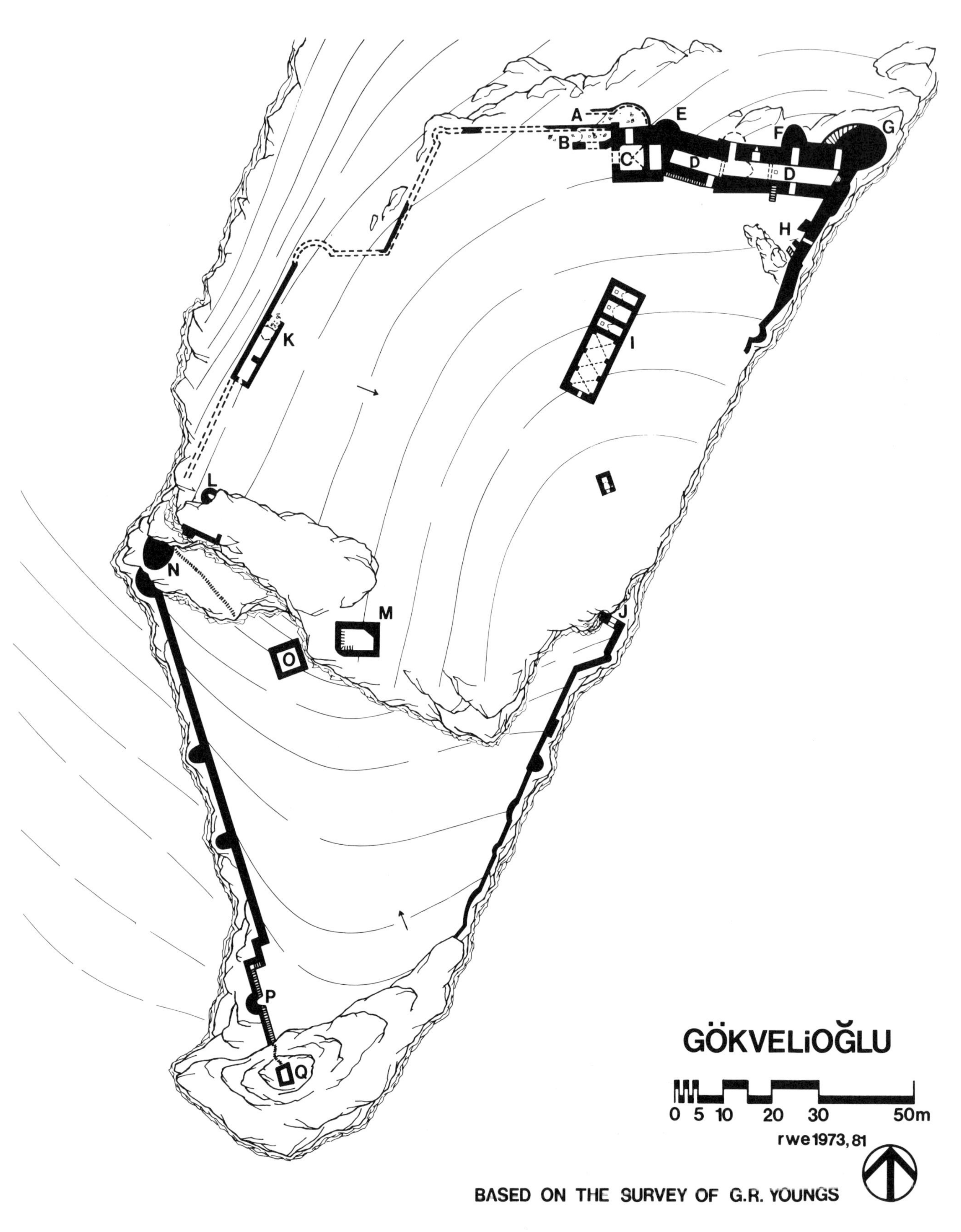
A
B
C
D
D
E
F
G
H
I
J
K
L
M
N
O
P
Q
GÖKVELİOĞLU
0 5 10 20 30 50m
rwe 1973, 81
BASED ON THE SURVEY OF G.R. YOUNGS

Fig. 34

Gökvelioğlu

Gökvelioğlu Kalesi[1] is a large garrison fort that crowns the southernmost spur of the Cebelinur Mountains. From an altitude of 735 m, this site has intervisibility with Misis (Mopsuestia), the Mediterranean, Adana, and a 30-km stretch of the southward-flowing Ceyhan River. The principal function of this fortress was to guard the strategic road between Ayas and Misis. Both of these cities are about 15 km from Gökvelioğlu. On the west flank of this site, where the Misis-Ayas road skirts the base of the fortified outcrop, is the small village of Güveloğlu (pl. 87b).[2] We do not know the medieval name of this fort, and consequently its history is unknown.[3] In 1965 G. R. Youngs published a very competent survey and general description of this site.[4] In my comments I will repeat in a summary form only enough information to maintain a continuity with my descriptions of other sites. I will add some new observations based on the results of my own field survey.

Except for the north end and the southwest flank, the two baileys that constitute this fortress are defined by almost impregnable cliffs (fig. 34). The southern bailey is 60 m below the level of the upper bailey and is separated from it by a vertical wall of rock. At the southern end of the lower bailey there is a pinnacle of rock (pl. 87b); the latter and the south cliff of the north bailey resemble a gigantic pommel and cantle when viewed from a distance. There is almost no evidence of construction along the edges of the cliff at the southeast. However, in response to the more gentle slope below the north and northwest edges of the upper bailey (pl. 81b), towers and a curtain abruptly rise from the limestone mass.

When approaching the north bailey from the northeast, the natural contours of the rock lead directly to ramp A and gatehouse C. Only fragments of the north retaining wall of ramp A survive today, but it shows that the exterior facing was a combination of types V and VII masonries. The south flank of the ramp consists of another retaining wall at a much higher level; the facing stones of this wall consist of a rough type III. The south retaining wall of A once supported the foundation of the now collapsed north wall of the vaulted chamber B. At the east end of ramp A there is part of the foundation of the semicircular platform that preceded the lower-level north door of gatehouse C (pl. 83a). The north side of this door has collapsed, but the facing stones that remain on the lower-level north wall of C consist of types VII and VIII. Above, there was a second-level north door in gatehouse C (pl. 83b). This door was preceded at the north by a now collapsed vaulted room (second level is not shown on the plan). The exterior facing stones of the upper half consist of types V and VII. The lower level may represent a first building period, and the upper a second period of construction.

One interesting piece of evidence to support this conclusion can be found in the northwest corner of C, which also functions as the east wall of the vaulted chamber B (pl. 83a). The level of this wall, which is equal in height to the type VIII on the exterior of the north wall of C, is made of type IV masonry. Quite abruptly and in an *irregular* pattern types V and VII facing surmount the type IV, completing the upper section of B's east wall. The types V and VII correspond to the course levels where they appear on the exterior of the north wall of C. The south wall of B, which is opened by a tall pointed arch, is constructed also with types V and VII masonries (pl. 296a).[5] Thus chamber B, like the upper level of gatehouse C, is the result of a later period of construction.

But is the south wall of B contemporary with the upper level of C? The junction of the south wall of chamber B with the northwest corner of C is highly irregular. Here the type V masonry of B and its upper extension of type III (the latter being part of the springing of B's collapsed vault) are crudely adapted to the type IX masonry (some with bossed centers) of the west opening of the gatehouse (pl. 85a). Obviously the south wall of B was built *after* the second period of construction in the gatehouse. The interesting point about the first building period in C is that some type VIII does appear. The type IV of chamber B was an interior facing and was not subject to direct attack.

The exact nature of the vaulted overhanging chamber which preceded the present upper-level north wall of the gatehouse (not shown on the plan) is difficult to determine. The vault that once covered the upper level extended out beyond the limits of the lower-level north wall. This probably indicates that some sort of overhanging machicolation or loophole was constructed above the lower-level door (pls. 82b, 83b). The west wall of the second-level room appears to extend farther to the north. The door opening into the second-level overhang from the interior of the gatehouse is roundheaded except for the monolithic lintel over the jambs at the south (pl. 84a). This door was closed and secured from the interior of the overhanging room, which meant that the defenders could hold out there if the enemy breached the gatehouse. I accept the earlier conclusions of Youngs that the rough, crude masonry (type III) in the lower levels of the north circuit and towers E through G is a result of the first Byzantine period of construction and that the types V and VII in the upper levels are the Armenian additions (pls. 82a, 82b).

I cannot determine with certainty the exact sequence of construction of the two floors on the interior of gatehouse C.[6] Excavations will someday reveal the lower half of the first floor and new information could certainly alter the following conclusions. On the interior of C, different types of masonry are used and blended harmoniously. In the opinion of this writer, the interior of the gatehouse should be the result of two periods of Armenian construction. When the upper level was added (or rebuilt) later, a new exterior facing was added to the southwest corner and the south wall of C from the level of bedrock to the roof (pl. 296a). This new facing was a combination of types V and VII. The exposed sections of the core in the south and east walls show an identical combination of gray mortar, small pebbles, fragments of brick, and fieldstones.

A common, though not universal, feature of Armenian bent entrances is that the interior opening into the fort is a tall pointed arch with no accommodations for a wooden door (cf. Yılan and Tumlu, with Savranda and Vahga). At Gökvelioğlu this seems to be the case at the west end of both floors. The function of the top floor of C is uncertain (pl. 84a). Perhaps the now collapsed ceiling of the first floor was opened by machicolations that were manned with men at the second level (cf. Vahga).

There are two features on the interior of the gatehouse that warrant attention. The masonry of the lower level, including the springings of intersecting vaults, consists of a very fine quality type IV (pl. 84a). Most of the upper level has a superbly executed combination of types V and VIII (some of the latter have bossed centers; pl. 84b). Normally, in Armenian architecture the larger stones are placed at the lower level to give greater stability to the walls. But in this case the addition of a new exterior facing to the walls of the lower level (from the second period of construction) certainly created a firm foundation for the larger stones. With regard to the vault over the lower floor (over the large west room), only the two triangular springings of the east side are still visible (pl. 84a). Unfortunately what *may* have been the west half of a groined vault has collapsed today. It is significant that what survives of the springings of this vault at the east is constructed in the typical Armenian manner with stones in parallel courses forming a pointed apex.[7] The upper level (not shown on the plan) is essentially covered by a pointed vault, running on a north-south axis. At the east (pl. 84b) this vault meets another vault over the lower-level cistern, triangular junctions are created. Unlike the lower level, the stones here do not meet at the apex in parallel courses but overlap in a herringbone pattern. This motif is common in Crusader architecture and may reflect its influence on Armenian masons. The fact that we have springings for vaults constructed in two different patterns and associated with two different masonries may indicate again that there are two different periods of Armenian construction in gatehouse C. As mentioned earlier, the homogeneous exterior facing on the south wall and southwest corner of the gatehouse are from the second period. Until excavations are completed and a more thorough analysis of the core can be made, my conclusions about the two Armenian building periods in the gatehouse are speculative.[8]

From the available evidence it appears that all of gallery D is an Armenian construction. The exterior facing of its north and south walls consists of a combination of types V and VII (pls. 82b, 85b). These masonries are used for the interior facing in the first five courses from the present ground level and are surmounted by two or three courses of type VIII (pl. 86a). The slightly pointed vault over D consists of type IV masonry. The south wall of this angled gallery is opened by two jambless doors (pl. 85b) and has two exterior stairways that lead to the top of the gallery. Both doors are covered by slightly pointed vaults; the door at the west is constructed like the upper-level window/door in the south wall of gatehouse C with a series of smooth-faced voussoirs. Above the west door is the broken frame for a now missing inscription. The north wall of D is opened by three doors that give access to the ramparts atop towers G, F, and a now collapsed baston between E and F. The north wall of gallery D is opened by only one embrasure with a casemate. There is a single embrasured window in the east wall of D (pl. 86b), which was plugged after the initial period of construction. The window, which simultaneously opens the *base* of the west wall of D and the *top* of the upper-level east wall of gatehouse C (pl. 84b), was not suitable for communication and light in D, but may have aided in ventilation (this west window is not shown on the plan). Since the bottom of this window was partially blocked on the east side by the apex of the subterranean vault below D, we can safely assume that gallery D was constructed after the upper level of the gatehouse. The subterranean vault is the floor of the west end of D. This vault covers a small cistern (not shown on the plan).

Today three towers flank the north wall of D (pl. 81b). Tower G appears to have once had a cistern on the interior, which was plugged. Aside from refacing parts of the original Byzantine towers, the Armenian masons added a bulging talus on the west flank of tower G (pl. 82a). On the east flank of tower E and in the adjoining wall to the east are two typically Armenian arrow slits with stirrup bases (pl. 82b). Since both of these appear to have been blocked by the construction of gallery D, it appears that the Armenians repaired the Byzantine towers and circuit with types V and VII before the construction of D. In comparison with the Armenian masonry, much more of the Byzantine type III facing has fallen away because its non-tapered inner sides did not provide a firm bond with the core. Considering the sequences of construction, it appears that the Armenians had no master plan but added to the north complex as the need arose.

The east flank of the upper bailey at the north end shows both Byzantine and Armenian construction on the exterior of the circuit. At point H there are solid projections of masonry that resemble internal buttresses. One simple squareheaded window, which is covered by an oversized monolithic lintel on the interior, is the only sign of defenses along the short east wall.

To the southwest is cistern I. The north half is a Byzantine construction that consists of three rectangular chambers. Each chamber is covered by a vault and opened by a single hatch. The south half of the cistern complex I is a slightly larger Armenian addition that is covered by three abutting groined vaults. The walls in the south half have been carefully plastered; there is only a single high-placed door in the south wall. On the exterior the distinction between Armenian and Byzantine construction is quite prominent, with the former having a perfectly cut type V. The Byzantine masonry is decidedly inferior. Fifteen meters south of I there is a small barrel-vaulted cistern. From the cliffs at the southeast there is an excellent view of gate J below. This portal appears to be the only entrance into the lower bailey from the base of the outcrop. Because of the vertical nature of the cliffs, access to the upper bailey from this gate is impossible.

On the far west side of the north bailey is undercroft K. This rectangular structure is made out of a very poor quality type IV masonry and is covered by a slightly pointed vault (pl. 87a). Part of its south wall and all of its north wall have collapsed. The west wall of K rests on the retaining wall which is one of the surviving fragments of the thin west circuit. The only other constructions of importance in the upper bailey are the isolated tower L, which is built with a perfect example of type VI masonry, and the foundation of building M with its partially scarped and partially masoned walls. The latter was not a cistern and was constructed out of type III masonry. Its function remains a mystery.

About 25 m below the extreme southwest corner of the upper bailey is tower N. This solid bastion, which guards the treacherous west descent to the lower bailey, appears to have once been topped by battlements. A series of small rock-cut steps leads southeast from tower N to cistern O. This appears to be the only direct line of access between the lower and upper baileys. During my visits to Gökvelioğlu Kalesi in 1973 and 1981, I did not examine cistern O. The lower bailey, which is about half the size of the upper bailey, is devoid of other buildings. The exterior facing of the west circuit and its solid towers consist of types V and VII masonries. Flanking tower P and continuing to the south is a stairway that leads to the summit of the rock mass and tower Q (pl. 87b). Tower Q is a solid square bastion that functioned as a lookout post.

[1] On the plan of Gökvelioğlu the contour lines are placed at intervals of approximately 2 m.

[2] This site appears on the maps of *Adana (1)* and *Malatya*. See also: Fedden and Thomson, 12; the map in Favre and Mandrot; Cahen, 151; E. Forrer, "Kilikien zur Zeit des Hatti-Reiches," *Klio* 30 (1964), 155.

[3] Hellenkemper (168 f) believes this site to be the medieval Vaner. King Levon I granted this fortress to the Knights of St. John in 1214. Unfortunately there are no specific references on the location of this site. Hellenkemper's conclusions are highly speculative at best. See: Alishan, *Sissouan*, 66, 424; idem, *Léon*, 358 f; Delaville Le Roulx, 164 f; Smbat, G. Dédéyan, 77; Yovhannēsean, 200–205.

However, an important (although still very tentative) association can be made between this fortress and the Arab site of al-Muṯaqqab. If we assume that the Byzantine construction here dates before the mid-9th century, then the rough coordinates given by Arab geographers for al-Muṯaqqab could actually fit the site of Gökvelioğlu. The earliest mention is in al-Balāḏurī (*Kitāb futūḥ al-buldān*, ed. M. J. de Goeje, *Liber expugnationis regionum* [Leiden, 1866], 166, lines 20 f) who says that al-Muṯaqqab is located in the Ṯaġr (the zone of frontier fortresses) and it was built (more likely repaired) by the Caliph Ḥišām b. ʿAbd al-Malik. A more specific idea as to the location of the fortress is given by al-Maṣʿūdī (*Kitāb Murūǧ ad-Dahab*, ed. and trans. C. Barbier de Meynard and (A.) Pavet de Courteille, *Les Prairies d'or*, I [Paris, 1861], 264, line 2). He notes that "Ḥiṣn al-Muṯaqqab is situated at the foot (or "on the slope") of the Ǧabal Lukkām" (. . . *Ḥiṣn al-Muṯaqqab, wa ḏalika fī ṣafḥ Ǧabal al-Lukkām*). In his attempt to chart the limits of the Mediterranean al-Maṣʿūdī cites (ibid., 275, lines 1 ff) Aḥmad b. aṭ-Ṭayyib as-Saraḫsī (who in turn drew his information from the 9th-century writer al-Kindī) and lists in succession Antioch, al-Muṯaqqab, the coast near Misis, Tarsus, etc. (*min . . . Anṭākiyyah waʾl-Muṯaqqab wa sāḥil al-Maṣṣīṣah wa Ṭarsūs* . . .). It is clear that al-Muṯaqqab is near Misis, and while it does not have to stand on the shore, it is still close enough and of such importance to give its name to the neighboring coast. In his narrative al-Iṣṭaḫrī does not list it among the fortresses of the Ṯuġūr, nor does he qualify it as a fortress on the coast, such as al-Iskandarūnah. Rather al-Iṣṭaḫrī calls it (*Kitāb al-masālik waʾl-mamālik*, ed. M. J. de Goeje, *Viae regnorum*, BGA I[Leiden,

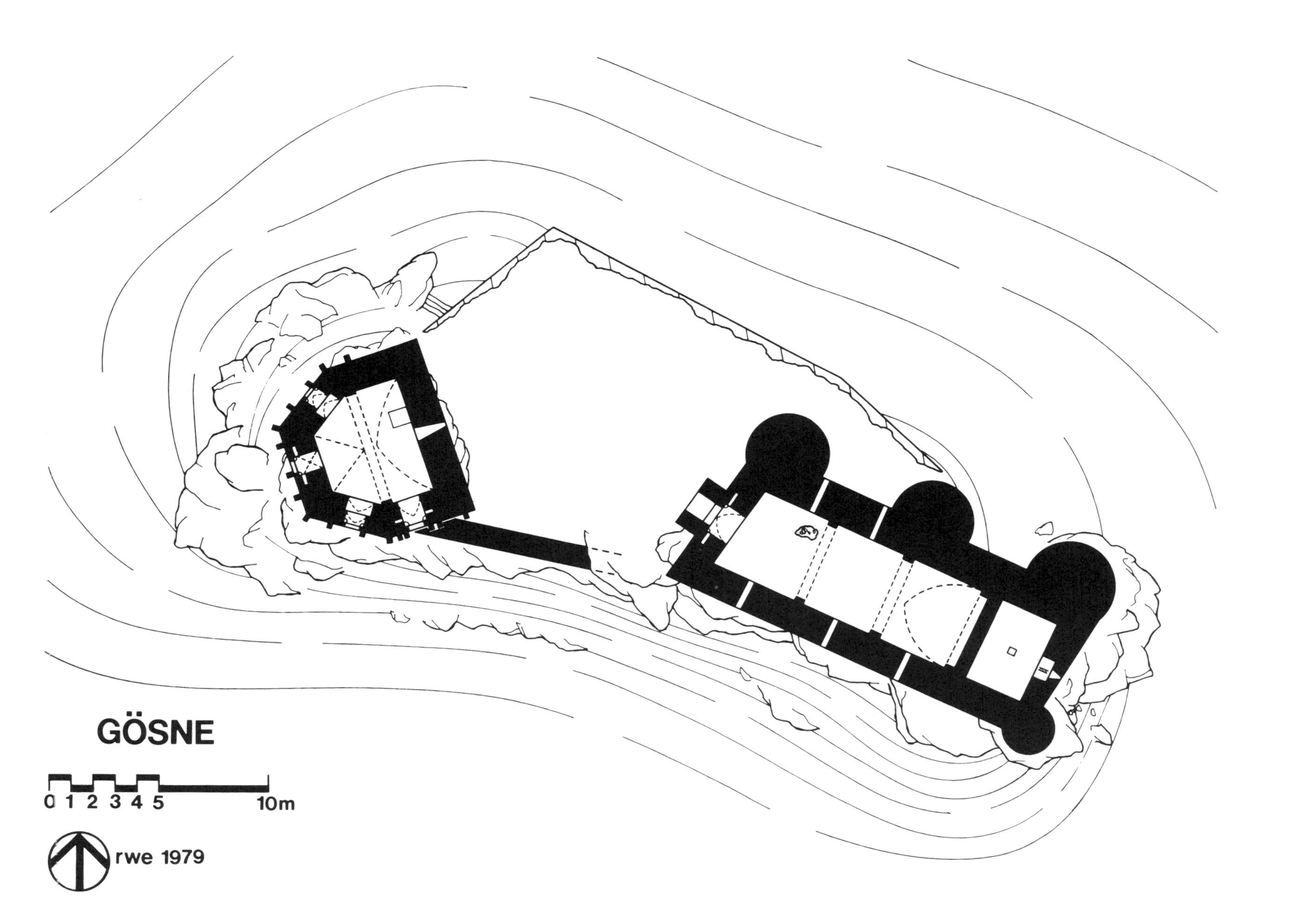

Fig. 35

1870], 55, lines 13 ff) "a small fortress built by ʿUmar b. ʿAbd al-ʿAziz, and it has a mosque (*minbar*) with a copy of the Qurʾān. . . ." It lies between at-Tīnāt on the shore and Misis (ibid., 65, lines 14 ff). At-Tīnāt is located on the coast north of the modern Payas (near Çalbı Köyü). Ibn Ḥawqal repeats the same arrangement (*Kitāb ṣūrat al-arḍ,* ed. J. Kramers, *Liber imaginis terrae,* BGA II.1 [Leiden, 1938²], 186, lines 1 ff) and says that al-Muṯaqqab is a fortress close to the coast of the sea (ibid., 182, lines 16 ff). He adds that it was once inhabited by the noble tribe who descended from ʿAbd aš-Šams. Al-Idrīsī (*Kitāb nuzhat al-muštāq,* IV.5, 21, ed. L. Veccia Vaglieri, *Opus Geographicum,* VI [Naples-Leiden, IUON-IsMEO, 1976], 652, line 11–653, line 3) notes that one travels in succession from al-Bayās, at-Tīnāt, al-Muṯaqqab, the river of Misis (the Ceyhan), the river of Adana (the Seyhan), Tarsus, etc. "All these sites are in succession along the coast of the sea." The implication is that al-Muṯaqqab is east of the Ceyhan River and certainly has a view to the sea. He also mentions (ibid., IV.5,8; 646, lines 3 ff) that at-Tīnāt is separated from al-Muṯaqqab by at least 8 miles. The 13th century account of al-Yāqūt (Muʿǧam al-buldān, s.v. "al-Muṯaqqab," ed. F. Wüstenfeld, *Geographisches Wörterbuch,* IV [Leipzig, 1869], 414, lines 20 f) states that al-Muṯaqqab is "a fortress on the coast of the sea (*ʿalā sāḥil al-baḥr*) close to (*qurba*) al-Maṣṣīṣah (Misis). It bears this name because it stands in the mountains (*fī ǧibāl*), all of which are pierced (*muṯaqqabah*), and in it (*fī hi*: "in the fortress") there are large windows." Gökvelioğlu Kalesi is the only fortified site known today that fits all these criteria. Excavation alone must determine the extent of Arab construction at this site. Çardak Kalesi is a less likely candidate to be al-Muṯaqqab. Cf. the translations in: al-Balāḏurī, 257 f; Ibn Ḥawqal, 180, 184.

[4] Youngs, 118–25. In Hellenkemper's later monograph (165–68) there is a digest of Young's description of this site. When Hellenkemper reproduced Young's survey of the fort (pl. 90) the directional (or north arrow) appears to have been reversed.

[5] The surviving sections of the vault over B are made with type III masonry. It is common in Armenian architecture to use a type III vault on finely masoned walls.

[6] The lower level of C (shown on the plan) has two components: a large room at the west, which is covered by intersecting vaults, and a narrow cistern that was closed off by a solid wall. This wall (and thus the cistern) did not extend to the second level.

[7] Edwards, "Bağras."

[8] The only problem with my theory is the type III facing on the revetment of the ramp. Since the ramp was exposed to frontline attack, the presence of type III would be inconsistent with Armenian theory. This ramp may be part of the original Byzantine construction.

Gösne

The town of Gösne, which is about 29 km northeast of Mersin, is built along the sides of an upland valley. The altitude of this site is 1,090 m. The valley is continuously watered by mountain streams and is now becoming an important summer resort. The fortified estate house at Gösne[1] stands on a small limestone spur southeast of the town and guards an important north-south trail.[2] Like its neighbor to the south, Belen Keşlik, the medieval name and history of this fort are unknown. It received only the briefest description in an article by Gottwald.[3]

This site consists of two fortified chambers with connecting walls (fig. 35). The wall to the north is a revetment of modern construction that was added some years ago when the locals converted this site into a park. The south wall is a medieval construction, but it was recently surmounted by a thin parapet of cinder blocks (pl. 88a). Both units of the complex are in an excellent state of preservation.

The buildings are faced with a superbly cut type VII masonry on the exterior and type VIII on the interior (pl. 89a). In some cases the interstices of the type VII are almost flush. Type IX is used only for the ribs and for the frames around the doors and windows. The unique feature about this masonry is that the limestone in this region is extremely brittle, and frequently just the face of the type VII blocks have been sheared away to about a depth of 60 mm (pl. 88a). Such long regular cleavage planes are abnormal in other Cilician limestones, and this may indicate the presence of additional minerals in the indigenous limestone.

The easternmost structure is a rectangular hall with four protruding towers.[4] The only entrance into the building is on the west flank; walls extrude from the face of the west facade to form a porch (pl. 88b). The double doors of this opening are limited by jambs with now shattered hinge housings (pl. 90b). A cross-bar bolt secured the doors. The sill of the entrance is less than a meter from the present ground level. Because of extensive damage, it cannot be determined if a slot machicolation was incorporated into the upper half of the porch. The jambs and the interior side of the door are covered by pointed arches. The interior of the east hall is enclosed by a pointed vault and three transverse arches (pls. 90a, 90b). These ribs rise from ground level[5] and almost divide the building into four equal bays. Five thin squareheaded shafts (three in the south wall and two in the north wall) bring light and fresh air into the hall. These openings are slanted downward. Gottwald's assertion that this is a windowless mausoleum or the only fortified chapel in Armenian Cilicia is unwarranted.[6] In this quadripartite hallway there is a solid wall flush with the east transverse arch. The wall seals off the fourth bay at the east end. This east room has no doors and only a thin highly placed embrasured opening with casemate, which is not practical for an archer. In the sill of this opening is a narrow scarped trough, whose function is unknown. The *only* entrance into this room is through a hatch in the vault. On descending through this opening I found no evidence of stucco; this room was not a cistern. Just why it has a small window and no door is a mystery.[7]

Fig. 36

Judging from the masonry, towers, and the design of the west door, all four bays appear to represent one period of Armenian construction. The wall that seals off the east room is a slightly inferior form of type VIII masonry, and may have been added later (pl. 90a).

West of this hall is the intricate hexagonal tower with three windows and two doors (pls. 88a, 89a). The masonry of the tower is identical to the stones in the hall at the east. At the top of the tower on the south, west, and north sides are fifteen projecting corbels that supported machicolations and/or fighting platforms. Two projecting corbels are set below the southeast door and probably supported a retractable entrance platform (the westernmost corbel is now broken). The southeast portal, like the other three openings to the northwest, has jambs, a slightly pointed top, and hinge housings to accommodate double doors (pl. 89b). On the interior side of the jambs are holes for a crossbar bolt. Directly to the west is a window with an exterior aperture of roughly half the size of the door; the sill of the window and the adjoining bench are about 60–80 cm from the ground level of the tower. To the northwest of this window is the second door which differs from the first in that its interior is covered by a miniature groined vault. To the north is the second window which is almost identical to the one just described. The third window is a simple roundheaded embrasured slit that is set high up in the east wall. The ceiling of the chamber is divided by a single transverse arch that rises from the floor level. On the west side of the arch are three flat facets; the east side is covered by a pointed vault. In the latter is a single hatch that opened onto the upper-level platform.

The exact purpose of this rather elegant complex is unknown. No doubt in time of peace it functioned as the residence for either a petty baron or a small retinue of soldiers.

About one kilometer southeast of the fort at Gösne is an Aramaic boundary inscription.[8] Montgomery suggests that the mysterious district of "RNL," for which this stone is a marker, can be identified with the site of Lampron. Modern topographical studies show that Lampron is much farther away than the district stronghold of Çandır. Unlike Lampron, Çandır's recorded history may go back to classical times. In any event the ambiguous "RNL" is difficult to identify with any of the strongholds that were prominent in medieval times.

[1] The plan of this hitherto unsurveyed site was executed in June 1979. The interval of distance for the contour lines is 50 cm.

[2] The site of Gösne appears on the maps of the *Cilician Gates* and *Mersin (2)*.

[3] Gottwald, "Burgen," 96 f. This site was also mentioned by Schaffer (59). Cf. *Handbook,* 197.

[4] Gottwald's assertion ("Burgen," 96 f) that this building has six towers is inaccurate.

[5] This is quite different from the corbels that spring from the center of the walls at Belen Keşlik. However, similarities between these two sites may suggest that in one or both cases Armenian masons are building these estate houses with plans submitted by the Crusader owners. The two sites may have been built at different times and consequently reflect the changing tastes in architectural motifs.

[6] Gottwald, "Burgen," 96 f. Unfortunately this conclusion was recently repeated by commentators who did not visit this site but merely relied on Gottwald's account; see Hild and Hellenkemper, 283.

[7] This east bay would have made an ideal prison.

[8] J. Montgomery, "Report on an Aramaic Boundary Inscription in Cilicia," *Journal of the American Oriental Society* 28 (1907), 163–67.

Gülek

About 2 km southwest of the Cilician Gates, Gülek Kalesi[1] crowns the rather flat oblong summit of an imposing spur of the Taurus Mountains (fig. 36)[2]. Like the nest of some prehistoric bird, the masoned walls and rough cliffs create an awesome enclosure almost 1,600 m above sea level. At the north and northeast the natural scarpment provides a secure defense. Below these cliffs the side of the mountain briefly forms a long thin slope before an almost vertical descent to the road below. At the south and southwest (pl. 92b), where the ascent is less severe, the fortified walls extend a barrier around the entire summit. Today, as in medieval times, the approach is made by a winding, but comfortable, road from the village of Gülek Köy, a few kilometers to the southwest.

Despite its strategic importance, the history of this site is not too detailed. The coins discovered by local treasure hunters show that the fort had prolonged periods of Byzantine, Arab, Armenian, and Mamluk occupation. The bulk of the surface pottery revealed in the areas of erosion appears to be the Armenian glazed ware. During the period of the Armenian kingdom, this fort was probably kept in royal hands and administered by a trusted lord.[3] On the coronation list of 1198/99 a certain Baron Smbat is the lord of Gülek.[4] A few passing references in the non-Armenian chronicles stress the strategic importance of this site as a toll station and guardian of the road to Tarsus.[5] In the nineteenth century there is mention that squatters had taken up residence in the fort.[6] For a brief period in 1838–39 Ibrahim Paşa occupied this site during his revolt from Ottoman rule.[7] In 1973 I found two permanent(?) abodes; the larger, which is built of stone, is north of gate B (pl. 92b), and the smaller is a wooden shack east of tower G. The present inhabitants are employed by the Turkish Forestry Service.

Prior to my work, the only modern scholar to survey this site was Theodor Kotschy in 1855.[8] While his plan neglects many architectural features and is not drawn to scale, the general dimensions of the fortified outcrop are portrayed.

All of the unexcavated walls in this complex, with the exception of sections near A and B, appear to represent one period of Armenian construction. Typically, types V and VII masonries are used as the exterior facing stones on the walls, while the doors and vaults are constructed with type IX (pl. 91a). Type IV occasionally appears as the interior facing on walls, but type V is more common. Just why there is no tangible evidence of Byzantine and Arab construction cannot be explained. Perhaps they thought the outcrop sufficiently strong without fortified walls, or their walls were removed by later Armenian builders. The corridor on the interior of gate A and the north and west walls of B were recently constructed with recycled masonry (pl. 91b). These new walls do not have poured cores. This style of construction resembles that seen in the two forts built near Pozantı by Ibrahim Paşa in the late 1830s. Similar walls here may represent an attempt by the Paşa to refortify Gülek.

At present a pathway leads to gate A from the southwest (pl. 91a). Gate A is built in the typical Armenian pattern, except that the slot machicolation is not framed by an arrière-voussure but by a single arch and lintel. The inner element is actually a segmented lintel resting on the jambs (pl. 92a). The center of this tripartite lintel has a convex keystone. The interior half of the lintel has a stepped soffit with pivot holes to accommodate double doors. The doors were secured by a crossbar bolt (not shown on the plan). The upper course of each jamb has a conical projection (cf. gate A at Meydan Kalesi) that protects the corners of the wooden doors from attack. The rather broad outer segment of the door is covered by a high, pointed arch. Undoubtedly a now missing wall surmounted the lintel—a wall that equaled the height of the outer arch and also permitted the use of the intervening machicolation. An outer wall surmounted the arch and may even have held corbels or embrasures. Quite recently a fine cement was placed over the apex of the arch to keep the voussoirs firmly in place. It is probable that this door is part of a now vanished gatehouse. As mentioned earlier, the inner passage of the gate, which is made in part with the recycled type VII, is a recent (nineteenth-century?) construction.

This same recent construction is used on the interior of tower B (pl. 92b), where it was decided not to rebuild the shattered south face but simply to seal the gap with a cross-wall on the interior and to add a new north front. This new construction also follows the course of the original circuit wall for about 18 m to the west.

Tower C is the only square bastion in the fort and one of the few in Armenian military architecture. To the west the facades of towers D and E and the adjoining wall are relatively well preserved (pl. 93a). The rock ledges supporting these towers and others along the south circuit have been scarped to extend the vertical planes of the projecting bastions. On the exterior of tower D the masons appear to have used the type VII in a pseudo-isodomic fashion with two oversized courses in the center that form a broad stripe. On the interior of tower D, what was once a large chamber is partly visible in outline. The remains of a door, covered by a monolithic lintel, are present in the north wall of tower E (pl. 93b). Like tower E, the lateral walls of tower F are rapidly deteriorating (pl. 94a). From the southwest corner of the outcrop the circuit pivots at tower G and moves directly north. Only a few fragments of the western circuit stand in situ.

On the interior of Gülek Kalesi coursed rubble is visible in all areas. There is substantial evidence of collapsed buildings with scarped foundations. The two most prominent are H and I. Room H appears to have been cut into a natural cleft; unfortunately the upper-level masoned walls have collapsed. Room I is actually wedged between two protrusions of limestone. The south wall of this chamber is built with type V masonry. A centrally located arcade, which consists of three slightly pointed interconnecting arches, divides the room into two unequal sections. The arcade probably supported a roof of stone (cf. E1–2 at Çandır).

The east unit of the south circuit, which extends from gate A, has suffered heavy damage. The best-preserved section is between A and J, where the quality of the types V and VII is quite good, but the courses vary greatly in size. Recently the southwest face of tower J collapsed, exposing the interior side of its north door. The south circuit east of J is visible only in fragments. About 40 m northeast of tower K is the subterranean cistern L (unsurveyed).

[1] The plan of Gülek was begun in August 1973 and completed on my second visit in August 1979. The interval of distance between the contour lines is roughly one meter.

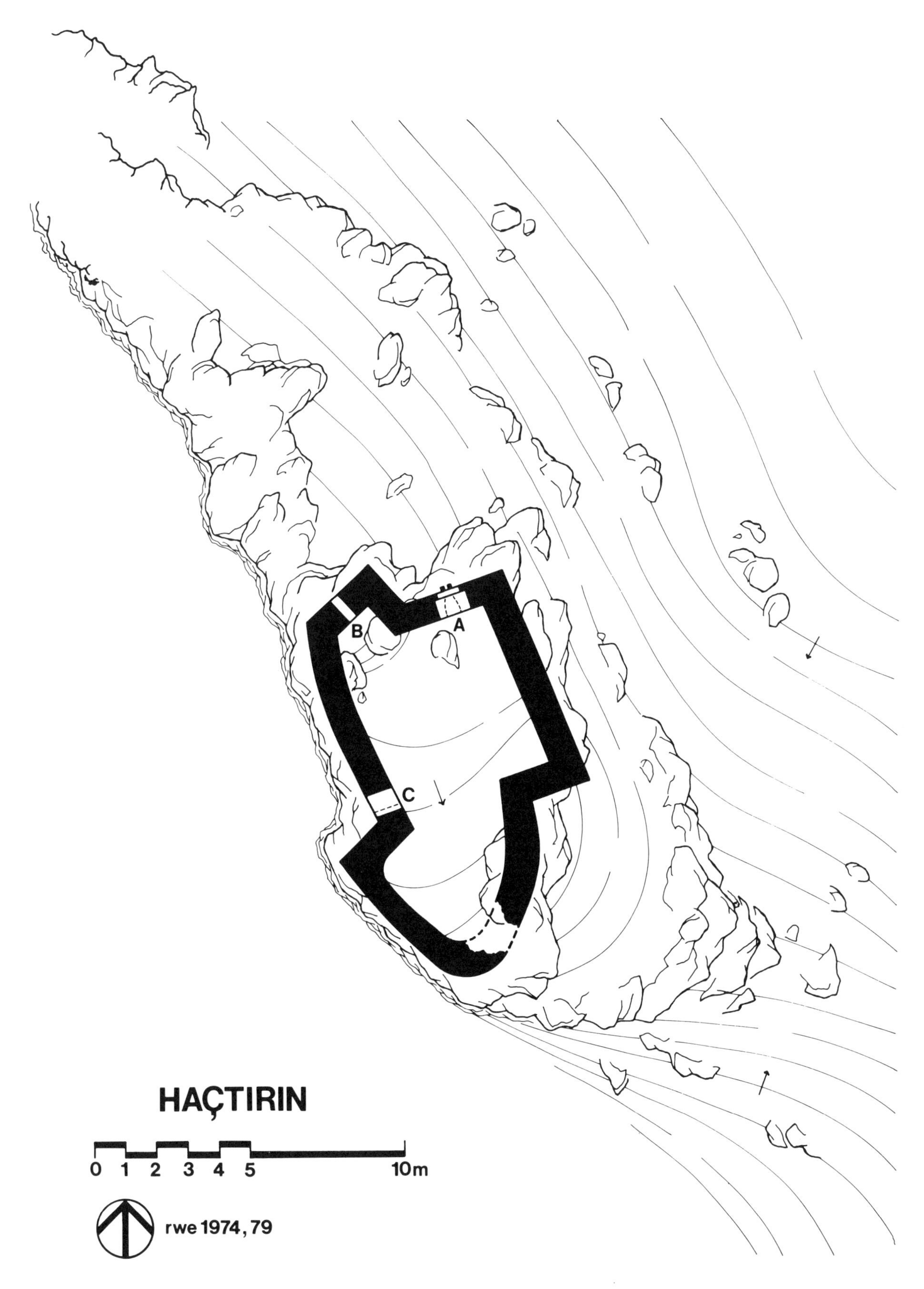

Fig. 37

[2]This garrison fort is known as Guglag, Kuklak, and Gogulat in Armenian and Kawlāk in Arabic. The Latin and Frankish designations are Cogelaquus, Cogueloch, and Cublech. In the 19th century it was referred to as Kulek-Boghaz, Kulek Kalesi, and Güllek Gala. Regarding the strategic importance of the pass see: Kotschy, 204 ff; Taeschner, I, 139 f; *Handbook*, 55–58, 240–42; Canard, 283; Vasiliev, 79, 87, 138 f, 318; Canard and Grégoire, 14; Texier, 726 f; H. Treidler, Πύλαι Κιλίκιαι, *RE, Supp*. 9, 1352–66; Le Strange, *Caliphate*, 133 f; W. Ruge, Κιλίκιαι πύλαι, *RE*, 389 f; Cuinet, 49; Tomaschek, 80, 84; V. Langlois, "Voyage en Cilicie, La route de Tarse en Cappadoce," *RA* 13.2 (1857), 482–89; idem, *Rapport*, 38 f; Fevre and Mandrot, 135; Boase, 165; Hild, 57–59; Ritter, 273 ff; Janke, 90, 97–111.

This site appears on the maps of *Cilicie* and *Ulukişla*. On the chart of the *Cilician Gates* the fort is marked as Nimrud Kale.

[3]The earliest Armenian occupation of this site may date to the period of Baron T'oros I in the second decade of the 12th century; see Mathew of Edessa, 98. Cf. Yovhannēsean, 97–102; B. Darkot, "Külek," *IA*, 1075–78.

[4]Smbat, 637; below, Appendix 3; Christomanos, 146.

[5]Hellenkemper, 227 notes 2 and 5; Brocquière, 102, esp. note 2; Gaudefroy-Demombynes, 101 f, 218; Honigmann, 42, 82 f.

[6]Kotschy, 204–7; Sümer, 22 ff.

[7]M. Canard, "Cilicia," *EI*², 38.

[8]Kotschy, 202 ff and his attached map; cf. Langlois, *Voyage*, 48, 150 ff; Hellenkemper, 226 f; Alishan, *Sissouan*, 132–37; Ainsworth, 77; Schaffer, 55. Most recently Hild and Restle (263 f) have made the very plausible association of this site with Ḥiṣn Bwls, which was a Greek fort captured by the Arabs in the first decade of the 8th century. Compare R. Edwards, "The Garrison Forts of Byzantine Cilicia," *Abstracts of the Eighth Annual Byzantine Studies Conference* (Chicago, 1982), 45 f. For the association of Gülek with Qaḏaydiyya/Karydion/al-Ğauzāt see Vasiliev, 82 f, 82 note 1; Honigmann, 82 f.

Haçtırın

South of Ak Kalesi along the road that leads from the Meryemçil pass to Kadirli is the watch post of Haçtırın,[1] also called Kale Bakımanı by the locals (pl. 94b). The area of Kum Kalesi (adjacent to the neo-Hittite site of Karatepe) has often been associated with the portion of the road near Hactırın.[2] My own topographical research indicates that Kum and Karatepe guarded one of the two auxiliary trails to the south (the trail to the east of Kum links Haruniye to the Kadirli–Meryemçil road). Kum is at least 6 km south of the major north-south road and was not intended as a guardian for that route. The importance of Kum and Karatepe lay in the fact that they are situated at a point where the Ceyhan River first became navigable.[3]

Haçtırın Kalesi sits atop the south end of a narrow, steep outcrop. The west side of the outcrop is almost vertical, while the opposite flank slopes gently to the northeast. A mountain stream, teeming with European brown trout, flows past Haçtırın at the southeast. Today a fenced enclosure at the northwest end of the outcrop serves as a station for the Public Road Service (Karayolları 56. Şube Şefliği). At least five other fortifications (Fındıklı, Geben, Azgit, Ak, and Anacık) guard this road from the Meryemçil pass to the Cilician plain.

The fort of Haçtırın is a very uncomplicated structure (fig. 37). Its masonry, which is consistent throughout and probably represents one building period, is type III (pl. 95a). Sections of the walls, especially on the interior, are stuccoed (pl. 95b). In all of the corners quoins are used. The only entrance (A) is at the north. The jambs and the rounded head of the door are deteriorating rapidly. Above the door is the sill of a splayed window. The window narrows from its widest point of 80 cm at the south to 48 cm at the north end. Projecting 24 cm from the sill of the upper window are two corbels. The corbels did not support a breastwork but probably some sort of shield or machicolation that was manned from the top of the wall. To the west of A is the narrow opening B. B probably functioned as some sort of straight-sided shooting port. Near the center of the west wall is the large roundheaded window C. The defenders were probably able to pass through this opening and shoot at attackers below. Above this window and slightly to the south there is a small straight-headed window. Just how access was made to this window is difficult to determine. At the south end of the fort there is a large breach in the wall. The possibility that there was another opening in this area should not be ruled out. There is no consistent placement of joist holes on the interior of the walls to indicate that the fort had a second story. Judging from the shape of the existing walls, this watch post was probably covered by a wooden roof. The sharp exposed corners of the circuit are atypical for Armenian architecture. Considering its design and masonry, it is likely that the builders were Byzantine. We have no historical references to this site.

[1]The plan of this hitherto unsurveyed site was executed in May 1974 and was revised in July 1979. The contour lines on the plan are placed at intervals of 30 cm.

[2]U. B. Alkım, "Karatepe," *Belleten* 14 (1950), 658 f; I. Winter, "Problems of Karatepe: The Reliefs and Their Context," *AS* 29 (1979), 134 f; Alkım, map 3.

[3]This explains the presence of reliefs that depict sailors and ships in the palace of Karatepe.

Haruniye

By A.D. 785/86 Hārūn ar-Rašīd had built this fort to secure part of a strategic road that links Maraş to the Cilician plain. Haruniye Kalesi[1] is about 3 km northeast of the village that still bears the name of the Abbasid caliph.[2] This Arab fort stands atop an almost pointed outcrop that towers above the Maraş trail. Today this route, which is positioned in a cleft of the Nur Dağları, is passable only by jeep. The mountains, which form the east flank of Haruniye, supply water to the stream at the base of the fortified outcrop. This fort does not guard the immediate approaches to the Amanus pass but does have intervisibility with Çardak Kalesi.

The history of this site spans a considerable period of time.[3] It seems that Haruniye Kalesi was conceived as a strategic link in a small chain of Arab forts that ran along the Ǧabal Lukkām (Nur Dağları). This fort and its adjoining walled town were a day's march from Maraş. According to al-Yāqūt, the Byzantines under Nicephorus Phocas captured the fifteen hundred Moslem residents of Haruniye during their campaign against the Arabs of Syria in 959. The Greeks probably inflicted considerable damage on the fort since Sayf ad-Dawlah financed its reconstruction along with the neighboring town in 967. Ibn Ḥawqal, who visited the "heavily fortified" site shortly after the Byzantine attack, reports that the population in the town of Haruniye was restored. By 1198/99 (and probably much earlier) we know that the fort is an Armenian possession, since its baron, Levon, assisted in the coronation of King Levon I (see below, Appendix 3). Godfrey followed Levon as the Lord of Haruniye. On 22 January 1236 King Het'um I and his wife, Zapēl, bestowed on the Teutonic Knights the site of Haruniye as recompense for services rendered to the Armenian kingdom. The reference to the "civitatem Haronie" in the deed of transfer probably alludes to the village and fort of Haruniye.[4] None of the architectural features in the garrison fort indicate that any significant German construction is present.[5] By the end of the thirteenth century Haruniye had fallen into the hands of the Mamluks. In the mid-fourteenth century the Egyptians made Haruniye the administrative headquarters of east Cilicia.

The plan for the fortification is unlike any other structure in Cilicia (fig. 38). The design is much too compact to be Armenian.[6] What we have at Haruniye is essentially an elongated keep with a tower at the east.[7] The engineers have placed continuous galleries on the north and west flanks where the general grade of the hill is less steep. There is no substantial rise in altitude from gate A to tower E.

What is also unique about this fort is the use of three types of masonry. The original fort, which is identical in plan to the present structure, is built entirely with type IX masonry of limestone and basalt. On the average, these blocks are slightly longer and not as high as the type IX seen on Armenian structures; the interstices here are somewhat wider, and many are stuccoed. Occasionally the inner faces of these stones are slightly tapered but not enough to categorize the masonry as type VIII. Extensive repairs were probably carried out in 967 with a type of masonry that is identical in size and color to the original type IX, except that its outer face is bossed(pls. 96a, 96b).[8] This surface is unlike type VII masonry because the drafted margins at Haruniye are much wider and the boss is quite thin. The interstices of these second-period stones have a wide stripe of mortar on the exterior (much of the mortar has fallen away today). In a few cases the original ashlar that had fallen away from the wall was recycled and mixed with the masonry of this first repair period. At some later period a third type of masonry, a well-cut type III, is used to repair only the exterior of tower E and areas above gate A (pls. 96a, 97b). This use of rather crude masonry is reminiscent of some Teutonic construction at Kum and Amuda. The type III masonry is probably the German contribution to the fortress.[9] It could also belong to an unrecorded period of Byzantine occupation or perhaps to the repairs of the Mamluks. There is no tangible evidence of *any* Armenian construction at this site.

Today, when one approaches the castle, the trail begins at the south flank of the fortified outcrop and eventually reaches the summit at the southeast before it turns abruptly toward gate A (pl. 97a). Entrance A, which appears to be the main gate for the fort, is positioned between the square salient ends of the northwest and south walls. At the ends of the salients and in the square forecourt, much of the original smooth type IX is present. These stones are framed by broad margins of mortar (margins characteristic of the later rusticated type IX), which probably indicates that they were recycled during the reconstruction of 967 or more likely that they were stuccoed on the exterior at that time (this process was repeated in other parts of the fort). On the exterior of gate A we have smooth ashlar of very fine quality whose interstices are filled with thin margins of mortar. It is typical to use the finer quality stones on the exterior of the gatehouses. In the third period of construction type III masonry was used to repair the top of the gate. Today gate A is badly

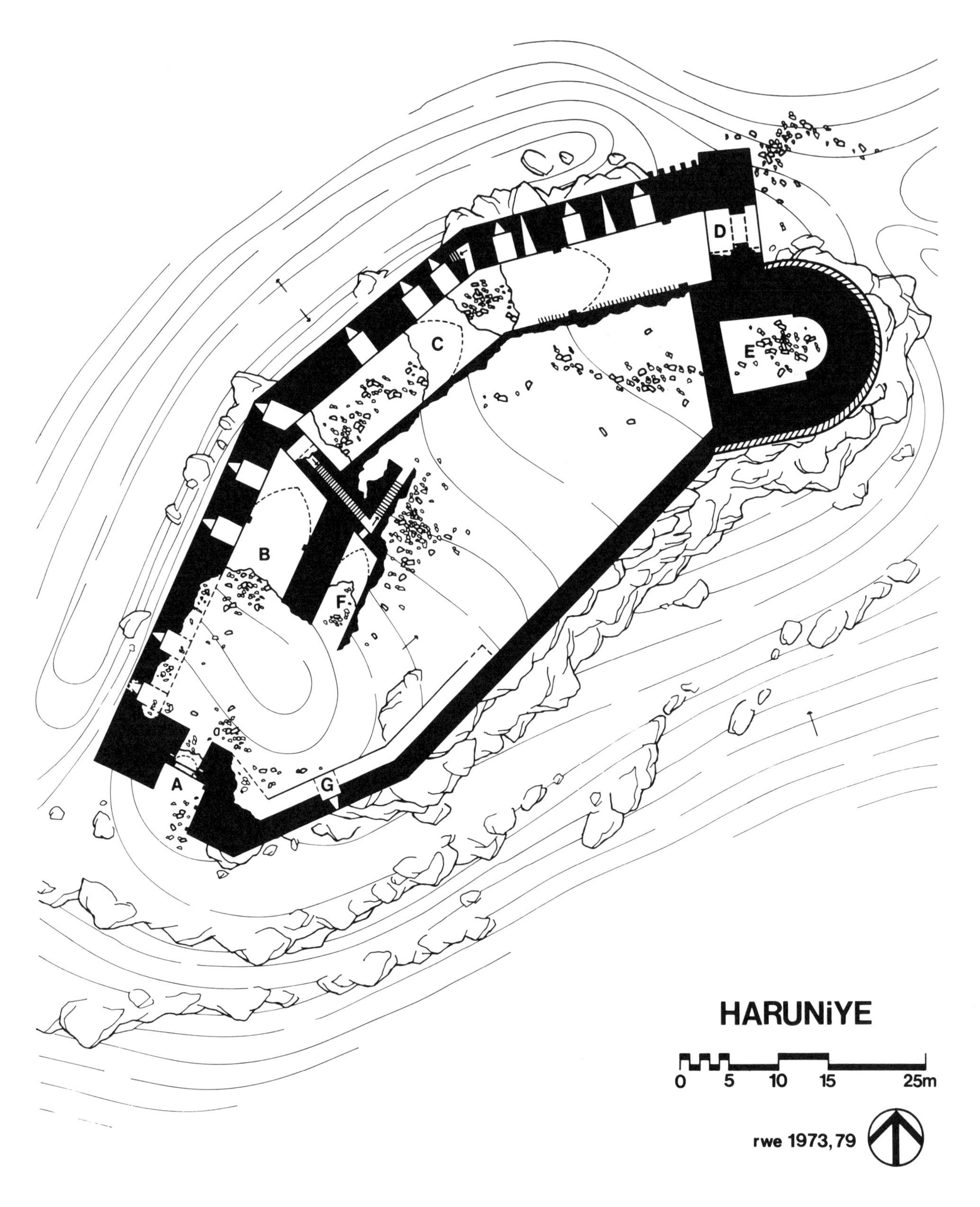

Fig. 38

damaged (pl. 97b), but the rounded shape of its vaulted cover on the interior is still visible. Behind its shattered jambs are accommodations for a crossbar bolt. On the interior side the gate has collapsed, except for a lower-level wall of smooth ashlar.

The two embrasures with casemates immediately to the northwest of A are shattered (pl. 98a); it appears that all of the interior facing and part of the core of the circuit between the extant section of vault B and gate A have collapsed. There is no evidence today that vault B once covered these twin casemates. The remains of a thin parapet rise above the level of the two embrasures. The southernmost of the two embrasured openings is barely visible, while the one to the immediate north is better preserved. The interior of this embrasure has a broadly splayed arrow slit covered by a depressed arch. The adjacent casemate was covered by a pointed vault set at a higher level. This design is identical to the casemates and adjoining embrasures in B and C and is our only certain example of Arab embrasures in Cilicia. All of these openings are constructed out of well-cut ashlar. This design differs from the Armenian model, where the sides of the embrasured loophole are not as deep nor as widely splayed and where the arched cover is more round. In Haruniye's arrow slits the Arab builders employ a stirrup that has rounded sides (pl. 96b). This is quite unlike the Armenian tradition, where the stirrup base has short, straight, triangular sides. Along the exterior of the northwest wall, all of the embrasures seem to belong to the first period of construction, while an identical embrasure (G) in the southwest wall seems to be from the period of reconstruction in 967. G differs from the other embrasured openings in that its exterior facing consists of rusticated stones and its flanking casemate is not set entirely in the thickness of the wall but was positioned on a narrow shelf.

The north end of undercroft B is well preserved. The walls are made with the same well-cut ashlar of the embrasures. Its pointed vault is constructed with smaller blocks whose interstices are filled with an abundance of rock chips and mortar. Two low-level corbels project from the lateral walls. These did *not* support a transverse arch; their function remains a mystery. Dividing B from the angled undercroft C is a wall and a jagged opening; the latter was probably a door. In the center of the wall is a narrow stairway that leads up to chamber F and eventually to the court southwest of tower E. Chamber F is a narrow room covered by a pointed vault.

In C there are seven embrasured openings with casemates in the north wall. A narrow passage and broken stairway are lodged in the thickness of that wall. The stairway probably led to a now missing upper-level wallwalk. The vault over C (pl. 99a)is identical in construction and shape to the one in B. At the east end the lower sections of the south wall of C consist of scarped rock (pl. 99b). Also at this end are two pairs of low-level corbels (shown on the plan). About 85 cm above these corbels are five pairs of smaller parallel corbels (not shown on the plan). Their specific function is unclear; they may have supported the wooden floor of a cramped second level or some sort of centering during the initial construction of the vault.[10] The only lateral openings for a potential upper-level floor are two very narrow embrasured windows in the northeast wall of C. Both are covered by pointed arches, and their sills angle downward. On the exterior of this wall at the east are five projecting corbels that no doubt supported a removable brattice.

At the east end of C are the remains of a second gate (D), probably a postern (pl. 100a). The fragments of the jambs are barely visible today. There is no evidence of a crossbar bolt. On the exterior of the gate smooth ashlar predominates as the facing stone; in this same flank the dark basalt alternates with a lighter limestone for aesthetic effect. The upper and lower courses of the wall were repaired with a bossed ashlar. Gate D is the only place in the fort where a large number of mason's marks are visible (see below, Appendix 1). The exterior side of gate D has two unique features. First, at the lower end of the northeast wall, three one-meter-long sections of alternating courses project from the wall. Their function is unknown. A similar feature occurs only on the south flank of tower E (neither of these corbeled sections is shown on the plan; pl. 96a). And second, some sort of arrière-voussure (without a machicolation) seems to have been used in the covering over the door. The two sets of facing stones for the double arch are visible only in the top of the door; the two facings are separated by only a thin line (7 cm) of mortar.

Adjacent to gate D is the large tower E (pl. 100b). On the exterior it is constructed with smooth ashlar of basalt. A thin decorative band of white limestone delineates the upper third of the tower (pl. 96a). Repairs are made with type III masonry on the southeast face of the tower. At the base of the tower is a thin talus. On the interior of E is a trapezoidal room with an apsidal east end. Present evidence indicates that this is not a chapel. On the interior of that room a small ledge (not shown on the plan) projects from the walls as if to support a wooden floor. There is a possibility that the interior had two stories and that the lower floor was a cistern. The interior walls of the tower are constructed with a crude form of type IV masonry. The tower was

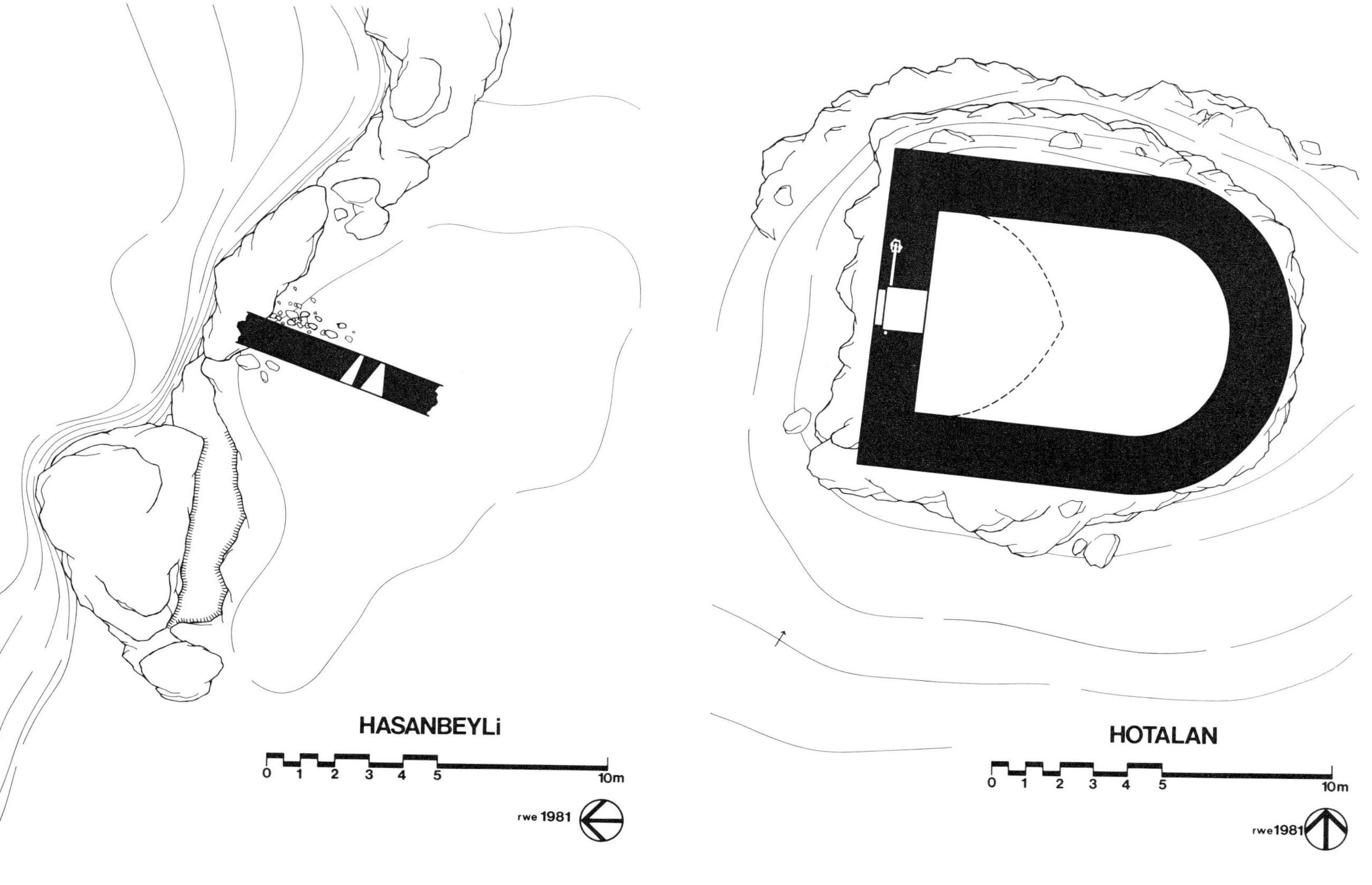

Fig. 39

Fig. 41

probably covered by a wooden roof. A solid west wall (with no door visible above ground level) separated the tower from the interior of the court (pl. 98b). The court is covered with shrubbery and undoubtedly is hiding a few buildings. There is no evidence of other cisterns.

[1] The survey of Haruniye Kalesi was undertaken in late spring 1973 and was completed on my second visit in August 1979. To my knowledge, no previous plan of the fort had been attempted. On my plan the contour lines are at intervals of 50 cm.

[2] In Arab chronicles this site is referred to as al-Hārūniyya(h). The Armenian name for the fort is Harun (or Harunia); the Franks call it Haronia (or Aronie). Haruniye appears on most modern maps, including: *Central Cilicia, Gaziantep, Maras, Marash, Malatya.* The map that Alishan published with his *Sissouan* places Til Hamdoun (Toprak Kalesi) at the site of Haruniye. The name of this site appears on Arab coinage as early as 785; see: H. Lavoix, *Catalogue des monnaies musulmanes de la Bibliothèque Nationale,* I (Paris, 1887), 169 ff; M. Bonner, "Hārūnābād, al-Hārūniyya and al-Khalīfa al-Marḏī," *American Numismatic Society, Museum Notes* (forthcoming).

[3] Textual references and historical summaries can be found in: T. Weir, "al-Hārūniya," *EI*[2], 234 f; idem, "al-Hārūnīya," *EI*, 272; idem, "Hârûniye," *IA*, 303 f; Le Strange, *Caliphate,* 128–31; Le Strange, *Palestine,* 449 f; al-Balāḏurī, 264; Alishan, *Léon,* 174; Alishan, *Sissouan,* 236 f; Cahen, 145–49; Hellenkemper, 116–19; Ibn Ḫurdāḏbih from Honigmann, 43; Fedden and Thomson, 36; al-Yāqūt, *Muǧam al-buldān,* ed. H. Wüstenfeld (Leipzig, 1886–70), V.945; Canard, 279 note 552, 799; Ibn Ḥawqal, 163–65, 179, 185; Gaudefroy-Demombynes, 59, 101, 218; Canard and Grégoire, 196; Bar Hebraeus, 166; Canard, *Sayf al Daula,* 37 f, 44, 51, 61, 128, 381, 392.

[4] The mid-9th-century Arab geographer al-Balāḏurī (*Kitāb futūḥ al-buldān,* ed. M. J. de Geoje, *Liber expugnationis regionum* [Leiden, 1866], 170 ff) *specifically* refers to Haruniye as a fort. Subsequent accounts (see note 3 above) describe it as a fort and/or a city. The "civitatem Haronie" in the Latin charter of 1236 includes the Arab-period fortification. See: Alishan, *Sissouan,* 236 f; *RHC, DocArm,* I, xlix note 1; Forstreuter, 61–63, 235 f; Röhricht, 277; Langlois, *Cartulaire,* 141–43; E. Strehlke, *Tabulae ordinis Theutonici ex tabularii regii Berolinensis codice potissimum* (Berlin, 1869; rpr. Toronto, 1975), 65 f, 127; J. Riley-Smith in Boase, 114 f.

[5] Hellenkemper, 118 f, 123–31, 134. Kum and Amuda, the German forts in Cilicia, are completely different from Haruniye with respect to their design. The masonry of a third period of construction at Haruniye does resemble some of the masonry at the two Teutonic sites.

[6] The Armenians would have employed multiple circuits (creating at least two baileys) and numerous bulging salients.

[7] Today there is no evidence of the outer circuit and iron gate that al-Yāqūt mentions in the 13th century.

[8] There is an important link with the military construction in Aleppo. In 1979 I was fortunate enough to visit the excavations in the citadel of Aleppo when the first period of construction under Sayf ad-Dawlah (A.D. 962) was uncovered. The earliest facing stones consisted of type IX masonry with a rusticated face and drafted margins.

[9] Some years ago a now missing inscription was found in the village of Haruniye (see Hellenkemper, 119 and note 1). The epigraph, which dates to 1244, probably records the construction of a German tower inside the village. Today all traces of this tower have vanished.

[10] Cf. Hellenkemper, 117.

Hasanbeyli

Today the watch post of Hasanbeyli[1] consists of a single wall that is perched on a small rocky outcrop southeast of the village that bears the same name (fig. 39; pl. 101a).[2] The old road that links Osmaniye and the Cilician plain to Gaziantep passes through this village. From the east flank of the watch post the road leads directly to Fevzipaşa (pl. 101b); an alternate route from Hasanbeyli carves a northeasterly course to Karafrenk before descending to Fevzipaşa. Hasanbeyli is almost at the midpoint in the Amanus pass. Savranda is the guardian at the west end of the pass, a distance of about 8 km from Hasanbeyli.[3] The Bahçe pass farther to the north is a distinct and separate route through the Nur Dağları. The modern four-lane highway to Gaziantep follows the routes of neither the Bahçe nor Amanus passes, but is situated between the two arteries. A track that is suitable for pack animals and four-wheel-drive vehicles leads south from the fort and joins the hiking trails to Yarpuz and Islâhiye. Hasanbeyli Kalesi has intervisibility with Savranda but not Karafrenk. Unfortunately we have no historical references for this site to determine the medieval name or date of construction. Hellenkemper believes that the watch post here is the "Black Tower" (a Crusader toll station) that is mentioned in one of the Armenian chronicles.[4] However, this association is purely speculative. The fort at Karafrenk may be more appropriate as the "Black Tower" since it is farther to the east and thus closer to Crusader territory.[5]

Architecturally, all that survives at Hasanbeyli Kalesi is a part of one straight wall (pl. 102a). Judging from the size of the summit of the outcrop, the watch post here was quite small. Also, Hasanbeyli Kalesi could not have had a square plan. An irregular circuit, which followed the broad scarped trench in the surface of the rock at the north, seems likely. There is no evidence of any salients at this site.

The masonry of the inner and outer facing consists simply of type IV. Only a few pieces of neatly squared ashlar are used in the construction. The sole peculiarity occurs on the interior facing where the margins of the stone seem to have been stuccoed (pl. 102b). This stucco occurs in the lower 60 percent of the wall.

The surviving wall of the fortification is two stories in height. The upper level is opened by a round-headed embrasured window. The first floor has a small squareheaded window near the top; this window has one beveled side at the north (both windows are shown on the plan). The division between the two stories is quite distinct on the interior, where large square joist

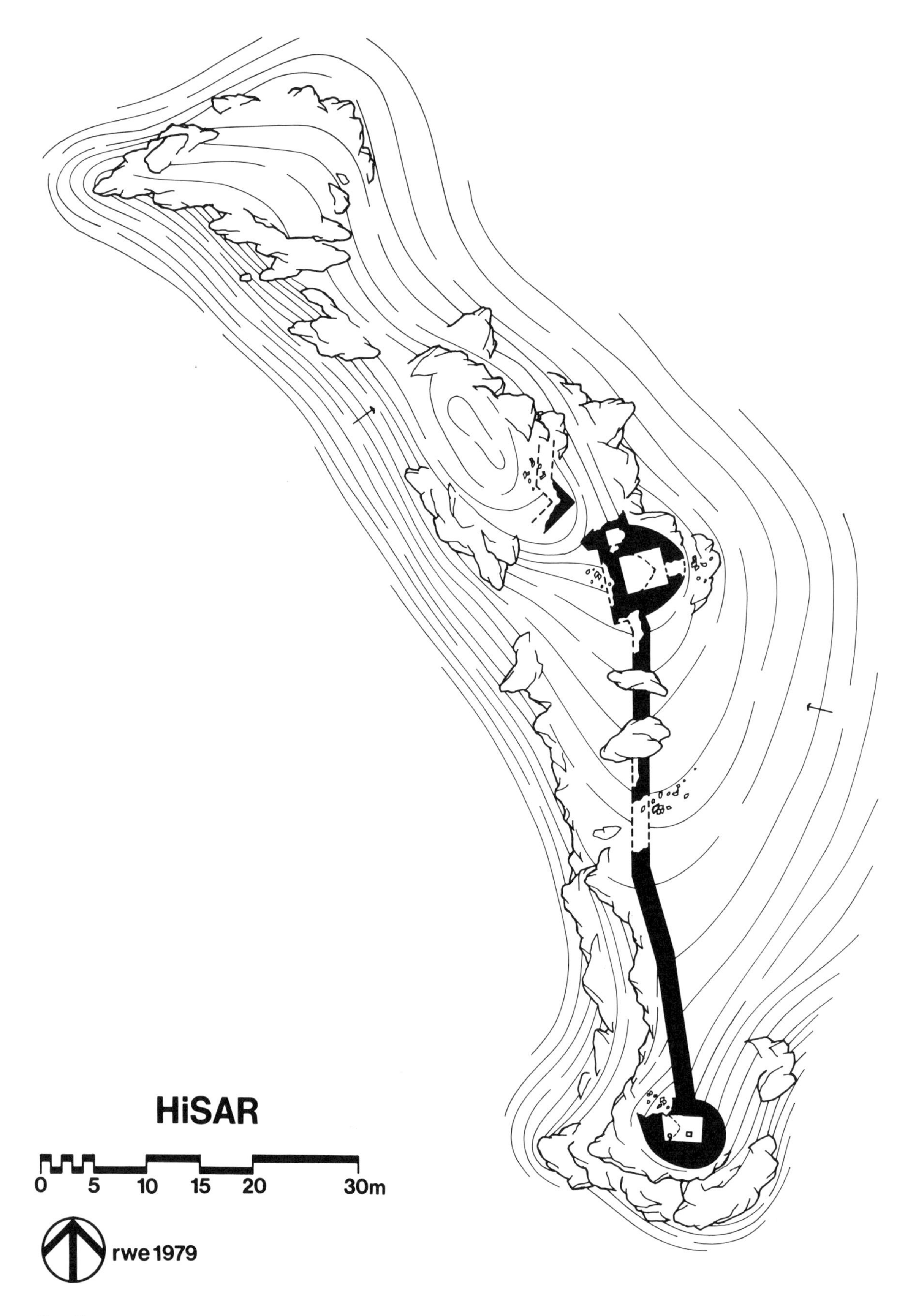

Fig. 40

holes are still visible. The arch of the upper-level window is constructed with stones that are laid radially—a technique seldom used by the medieval Armenians. Today an electric loudspeaker is secured to the top of this wall to call the faithful into a nearby mosque.

Hellenkemper noted that the masonry techniques used at Hasanbeyli are similar to the construction at Amuda, a Crusader fort in eastern Cilicia Pedias.[6] It is quite possible that the fort at Hasanbeyli was built by Crusaders, but at present there is no secure way to determine this.

[1]The plan of Hasanbeyli was executed in June 1981. The contour lines are at intervals of one meter.

[2]Hasanbeyli appears on the following maps: *Cilicie, Gaziantep, Maras, Marash, Malatya*. Cf. *Handbook*, 446.

[3]The modern denizens of Hasanbeyli refer to Savranda Kalesi as "Kaypak Kalesi" or simply the "Baraj." For a discussion of the Amanus pass see U. B. Alkım, "The Road from Samᵓal to Asitawandawa: Contributions to the Historical Geography of the Amanus Region," *Anadolu Araştırmaları* 2.1–2 (1965), 1–41.

[4]Hellenkemper, 120–22; Forstreuter, 65.

[5]Also, Karafrenk is made out of a "black" stone, and Hasanbeyli is not. We have no way of knowing which site in the Anti-Taurus is referred to in the charter of 1271 (Hellenkemper, 121). Undoubtedly further explorations in this region will uncover presently unattested medieval sites.

[6]Hellenkemper, 122.

Hisar

Just off the south flank of the road that runs from Gösne to Arslanköy is Hisar Kalesi.[1] This small garrison fort[2] can easily be reached by driving 7 km southwest from Kızıl Bağ along a fine road of packed gravel. The fort sits on the top of a spinelike outcrop of limestone (fig. 40). To the west there is an excellent view of the upland valley near Arslanköy and Gediği Kalesi. To the southwest the landscape is dotted with fertile valleys and rolling hills. A second trail from the south passes below the fort at the southwest. Around the site there is an abundance of water from mountain streams and artesian wells. This fort rises to an altitude of about 1,050 m. No historical names can be associated with this site.

For the most part the fort consists of two rounded towers and a connecting circuit wall (pl. 103a). The exterior facing stones throughout are a rough type V (pl. 103b). Except for the north tower, the interior masonry consists of types III and IV. Just northwest of the north tower are the shattered remains of a small angular wall.

The north tower is actually a two-storied structure. The upper story has collapsed, and only its foundation is visible. To the north and adjacent to this upper story is a *tiny* apsidal structure, now in ruins. It may have been a storage area; its size seems to preclude the possibility that it was a chapel. Below this upper level there is a chamber with a slightly pointed vault (shown on the plan; pl. 104a). The only apparent entrance into this room today is through a breach in the east wall of the tower. Undoubtedly a lower-level door at the west is buried. The interior walls of this room are constructed with type IV masonry, while the vault is made with type VIII. The color and texture of the limestone in the wall and vault differ greatly, but they do not appear to be the result of separate periods of construction. This room was not a cistern. A finer and stronger type of masonry was employed in the vault because it supported an upper level.

From this north tower the circuit wall runs directly south and terminates at the south tower (pl. 103a). Due to the collapse of its battlements, the top of the south tower is relatively flat today, and it has two openings in the floor to the cistern below. At the east end is a small, square hatch to draw water. In the southwest corner is what appears to be a clay pipe. Undoubtedly rain water collected at this end and drained into the cistern. The cistern is covered by a pointed vault and is completely stuccoed on the interior (pl. 104b). Today one can enter this room through the collapsed northwest corner.

Hisar Kalesi was simply a small troop depot on an important highway. Judging from the plan and the presence of type V masonry, it appears that the fort is an Armenian construction. Near the north tower, a local shepherd found a coin from the reign of King Hetᶜum II. Obviously, the Armenians occupied this site well into the thirteenth century.

[1]The plan for this previously unsurveyed site was executed in June 1979. The contour lines are spaced at intervals of 50 cm.

[2]Hisar appears on the *Mersin (2)* chart. It may also be the Ziaret Tepe Kale on the map of the *Cilician Gates*.

Hotalan

The fortified watch post of Hotalan[1] is one kilometer north of the modern village of Posyağbasan.[2] Hotalan Kalesi is at the south-central edge of the vale of Karsantı and may have had intervisibility with Tamrut to

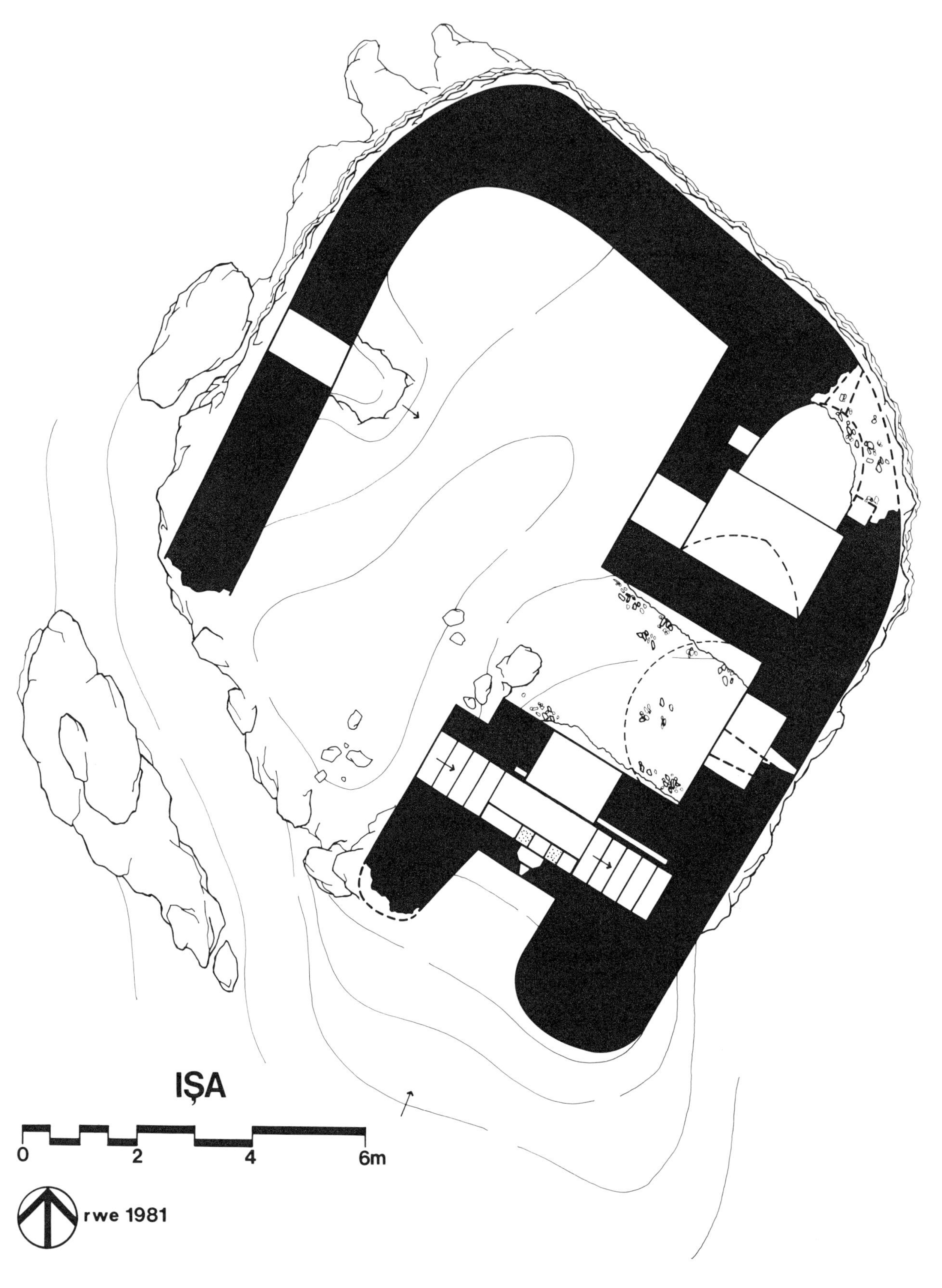

Fig. 42

the west and Meydan to the east. This watch post also has a commanding view to the north across the entire valley. Posyağbasan can be reached by driving first from Adana (via Çatalan) to Etekli. From Etekli a rough dirt road leads to Posyağbasan at the east, a distance of 5 km. Today the forests in the area are being clear-cut to open the land for wheat production. The spring runoff provides a bountiful supply of water. There are no inscriptions or historical references that give the name or date of construction of this site.

This watch post crowns a small sloping hill (pl. 105a). On the exterior Hotalan Kalesi is a simple tower with a rounded east side and a single door in the center of the flat west facade (fig. 41). This structure has two stories. The lower level is covered by a slightly pointed vault and today the upper level (not shown on the plan) is a terrace open to the sky.

The exterior and interior facing stones are not consistent throughout. The inner and outer facing of the *first* level consists of type V. The vault over this level is constructed with type III masonry (pls. 105b, 106b). The stones of the vault are laid radially. On the interior the vault and its walls appear to be stuccoed, but in fact this is only the disintegrating core oozing through the interstices. There is some evidence that the rock floor is scarped. The facing stones of the upper (or second) level consist of a large type III. Only a few stones of the type V are carried into the upper level (pl. 105b), mainly as quoins for the west end. A collapsed section of the wall above the west door reveals that the core was not layered in neat courses but is a jumbled mass of small fieldstones and mortar (pl. 106a). In this exposed section of the west wall it is clear that the core is identical for both types of masonry and that the fort represents one period of construction.

There are no windows in the first level, but there is evidence that two damaged windows opened the second level. One is located at the east; the other is above the west door. Part of the frame for the roundheaded west window is still in situ on the south side (pl. 105b).

On the exterior the west door is covered by a bipartite lintel that rests on narrow jambs (pl. 106a). The interior side of the door is covered by a partially collapsed arch. The square sockets for a crossbar bolt are placed unusually high in the sides of the door.

In the opinion of this writer, the watch tower at Hotalan is probably an Armenian construction, like all of the other medieval structures in this valley. The two exposed corners at the west and their quoins are very unusual in Armenian architecture (not unprecedented), but the rest of the masonry does reflect Armenian theory. At the lower level, where an enemy could inflict harm with a ram, there is a rusticated masonry of relatively good quality. At the second level, where battering rams could not reach, the stones are of lesser quality. This change to a poorer quality masonry on the exterior facing was probably done out of economic necessity.

[1]The plan for this hitherto unattested site was executed in June 1981. The contour lines are placed at intervals of 50 cm.

[2]Posyağbasan appears on the *Ulukişla* map.

Işa

Işa Kalesi[1] is one of a number of garrison forts that guard the vale of Karsantı, the largest Highland valley in the Cilician Taurus.[2] The other forts that secure the valley are Meydan and Yeni Köy at the east and Hotalan and Tamrut at the west.[3] This network of skillfully constructed forts reveals that this vale supported a sizable and prosperous Armenian population. The small garrison fort of Işa is at the extreme west end of the valley, approximately 4 km southwest of the village of Sivişli. The easiest way to reach Sivişli from Adana is to drive on the paved road around the west flank of Lake Adana to Çatalan and then north to Etekli. The journey from Etekli to Sivişli should be undertaken only by a four-wheel-drive vehicle. There is just a moderately difficult footpath from the village to Işa. Sivişli and the fort at Tamrut are visible from Işa Kalesi. This extremely small garrison fort would have provided an early warning of any intruder entering into the valley from the west. The history and date of construction of this site are unknown. Its masonry and architectural features are purely Armenian and consequently identical to the neighboring fort of Tamrut, which is dated by an inscription to A.D. 1233(?).

Near Işa's outcrop at the south and the east is a lush meadow, presently inhabited by a single homesteader. About 0.50 km south of the fort is a natural spring over which a font was constructed. This structure consists of a single arched enclosure that channels water through a number of spouts. The masonry used to build this font consists of relatively smooth ashlar with narrow interstices. It may or may not be contemporary with the fort.

Işa Kalesi covers the top of a pinnacle of limestone (fig. 42; pl. 107a). The only line of approach

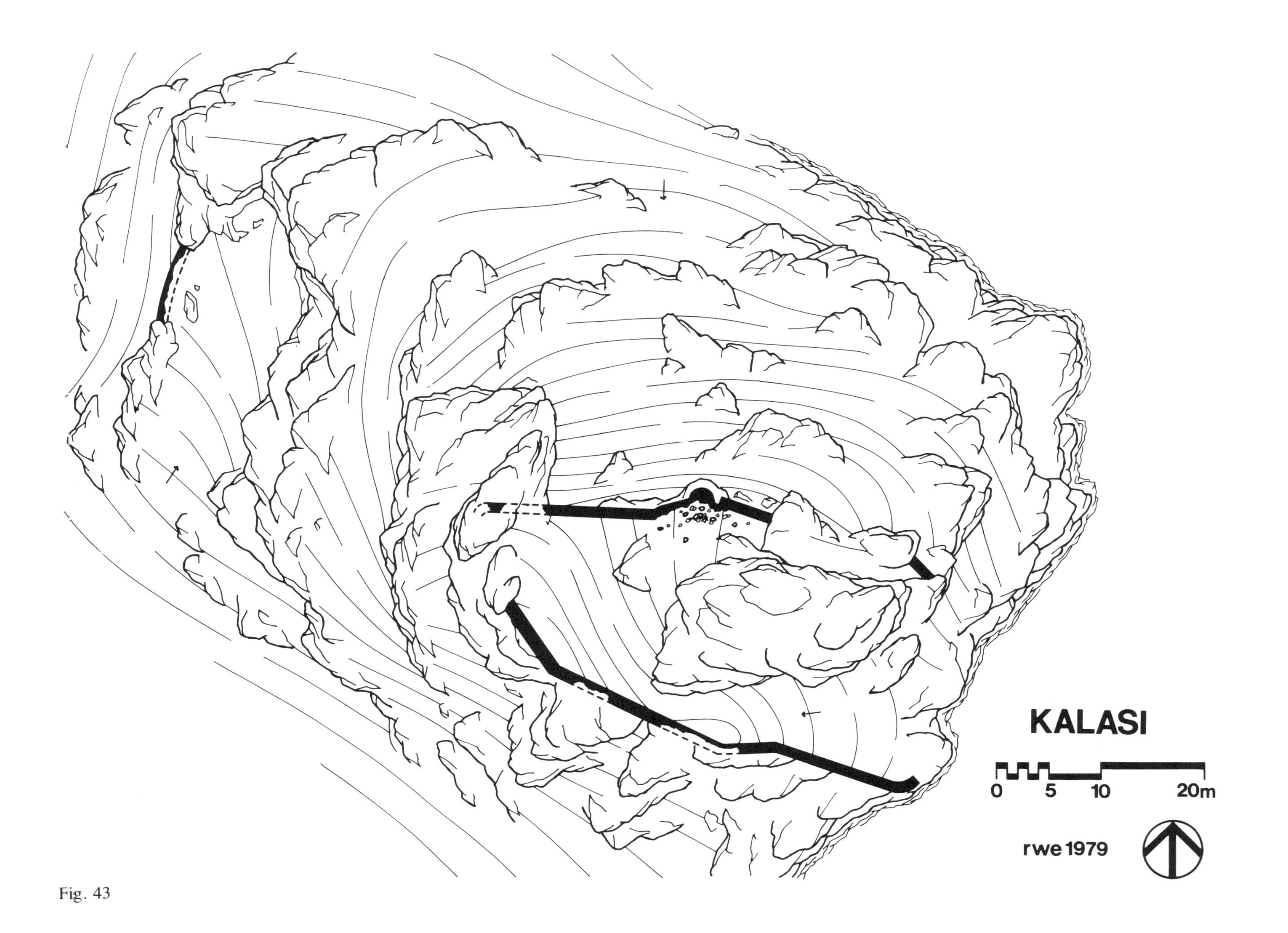

Fig. 43

into the fort is from the southwest. A narrow, winding path leads to the single gate at the south. This gate is flanked by two rounded, projecting bastions (pl. 108a). In keeping with the Armenian principle of avoiding sharp angles and corners, the entire circuit is rounded. At the southwest, the circuit has collapsed under a landslide, but it undoubtedly continued to the gate-tower. The tower to the west of the gate has partially collapsed, but, in its completed state, that salient was substantially smaller than the bastion at the east. It is certain that the east tower was topped by a curving parapet. With the exception of the chapel,[4] the gate complex is the only significant unit in the fort.

The masonry of Işa Kalesi is in keeping with the Armenian tradition (pl. 108a). The exterior facing stones are a well-coursed type V masonry. Around the doors, windows, and machicolations, types VIII and IX are employed. On the interior type IV is common as well as types VIII and IX (pl. 108b). Compared with the exterior, the interior masonry has slightly smaller and cruder stones and thicker margins of mortar.

Işa's gate is part of a vaulted, bent entrance, which requires the approaching party to turn northwest to enter the body of the fort. The actual gate entrance is built on three levels. The door is limited by jambs (not shown on the plan; pl. 108b), which are covered by the typical segmented lintel (now broken). Behind the lintel a slightly pointed vault covers the door passage. A double wooden door, which turned on pivots, was secured by a crossbar bolt.[5] Below the socket of the bolt at the east is a shallow round hole. There is no corresponding opening at the west; its function is unknown. On the exterior the gate is topped at the second level by a machicolation (shown on the plan; pl. 107b), which in turn is surmounted by a single embrasured loophole. As is so common in Armenian forts, the machicolation is sandwiched between the outer arch and diaphragm wall of the gate-door, and the parapet above the jambs. Here the machicolation consists of two square holes that are slightly off-center and partially blocked with debris. Entrance to the narrow platform at the second level of the gatehouse is made by a flight of four stairs (pl. 109a). The jambless opening to the stairs is at the west. From the platform a flight of five stairs leads to the third-level terrace. The second level seems to have been covered by a rather uncustomary vault that consists of a single course of springing stones along each wall and a series of lintel blocks.[6] The second-level platform is rather cramped and requires the archer at the embrasure to straddle the machicolation. From the third-level terrace archers could easily man the parapets along the circuit and perhaps even port-machicolations in the collapsed vault over the bent entrance.

Today there is no evidence that a door and wall were ever attached to the west end of this vaulted bent entrance. There may have been some sort of a wooden barrier. The east wall of the bent entrance is opened by two embrasures (pl. 109b). The lower one is an embrasured loophole with casemate. Like those in Tamrut and other Armenian forts, it is built with a thin, narrow roundheaded slit. Directly above is a splayed window that is centered just to the south of the lower embrasure.

The interior of the fort is covered with debris, and excavations may someday reveal remains of other stone structures. Considering the large forest in the area, it is quite possible that wooden buildings were also adapted to the interior. The only other feature of interest is a small, squareheaded window at the top of the west wall. Below that opening, the dirt from the interior is sliding down the cliff through a puncture in the base of the wall.

[1] The plan for this previously unattested site was executed in June 1981. The contour lines are placed at intervals of 50 cm. In trying to render the gatehouse on a single drawing, some compromises were made. The stairs of the gatehouse (on a northwest-southeast axis) ascend from the first level to the third-level terrace. The latter is not represented, but the second level with its machicolation/embrasure is shown. As a result the jambs of the first-level door directly below and its pivot housings could not be rendered.

[2] The villagers in Sivişli call this fort Işa. This site does not appear on any published charts, nor is it referred to in any modern commentaries. Its medieval name is unknown.

[3] The locals mentioned three other fortifications in this area (see the discussion of Tamrut). Circumstances did not allow me to survey these three sites, which are located north of Karsantı and Sivişli.

[4] For a description of the chapel and its masonry, see Edwards, "Second Report," 143 f.

[5] The dimensions of the sockets for the crossbar bolt are accurately represented on the plan, but in reality those sockets are directly under the upper-level steps.

[6] This is not like the corbeled vault at Meydan Kalesi.

Kalası

Northeast of Andırın and almost directly east of Azgit Kalesi is a very strategic track to Maraş.[1] Today this dirt trail[2] can be traversed in summer by a Murat taxi. The trail is easily located by driving from Azgit to the village of Çurukkoz[3] and then following the only route

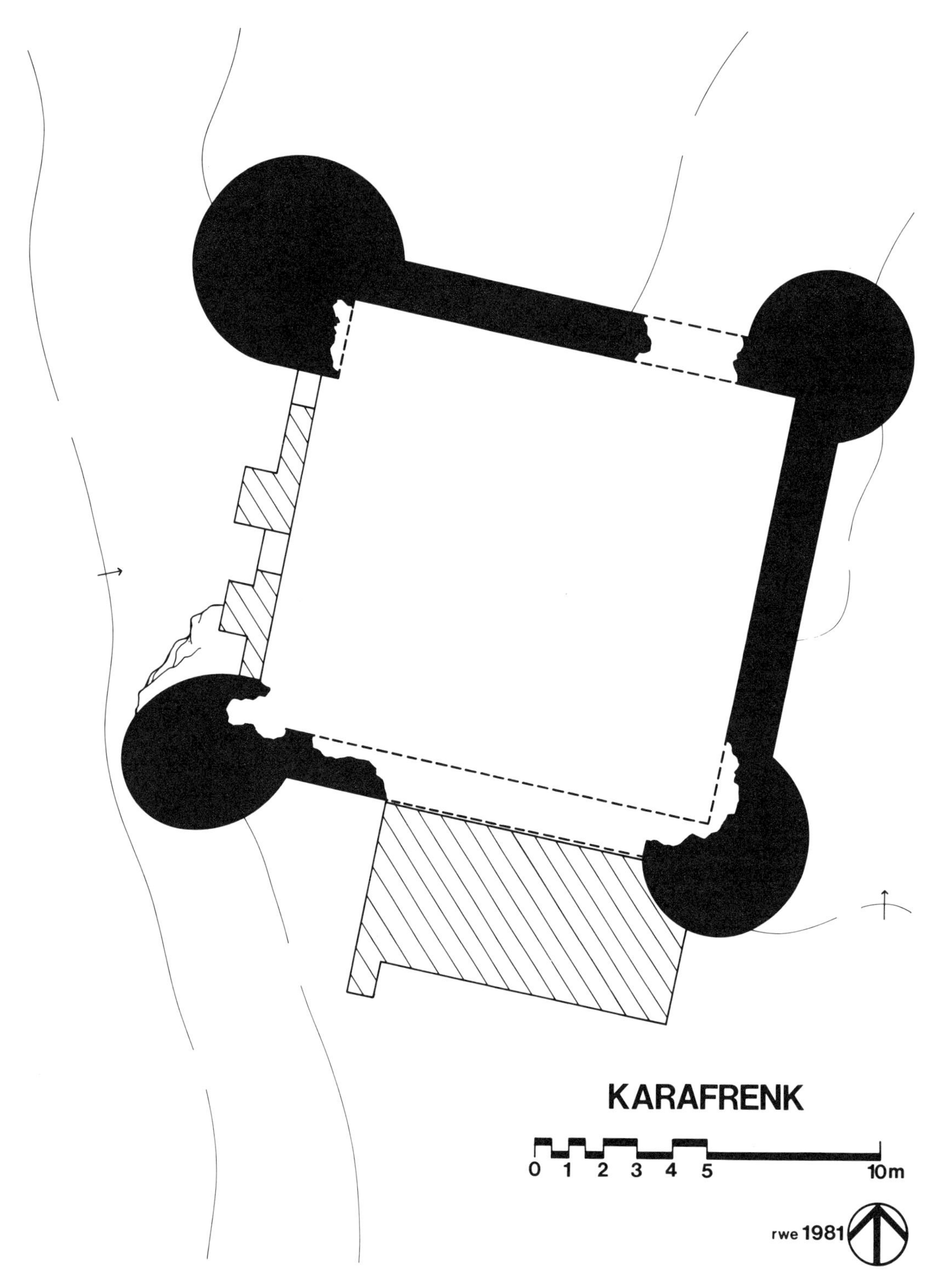

Fig. 44

to Maraş. About 3 km from Çurukkoz there is a huge artesian spring at the village of Karasu Köy. The delicious water, which is icy cold as it gushes from the underground cavern, forms a small stream that flows east along the road. One kilometer east of Karasu Köy is a rock pinnacle known as Kalası Kalesi. This small garrison fort has the simplest of designs, consisting merely of a few walls that close off the summit (fig. 43). Farther to the east (ca. 4 km) is the village of Petler Köy and a similar fortified road station (which I did not survey). From Karasu Köy, a secondary trail (not traversable by any vehicular means) branches to the southeast and eventually turns northeast to rejoin the primary track to Maraş. The secondary trail is guarded by Dibi Kalesi.

Architecturally, there is very little one can say about Kalası Kalesi. The easiest approach to the summit is from the northwest (on the south bank of the stream) where there are the remains of a short wall connecting two rock protrusions (pl. 110a). The only semblance of a tower is on the north side of the summit where a small rounded salient rests atop a jutting rock. The east end of the south wall at the summit has been squared off (pl. 110b); because of the proximity of a cliff, a door in this area seems highly impractical. The interior and exterior facing of the walls is type IV masonry. The locals have found both Byzantine and Armenian coins at this site, and it appears that Kalası, like the neighboring fort of Dibi, was built by the Byzantines and later occupied by the Armenians. Because of its size and lack of internal buildings, Kalası could sustain only a small garrison. We have no historical references to this fort.

[1]The plan of Kalası, a hitherto unsurveyed site, was executed in July 1979. The contour lines are placed at intervals of 50 cm.

[2]For the importance of such a road to the Crusaders, see my discussion of Geben Kalesi in the Catalogue.

[3]Çurukkoz appears on the maps of *Gaziantep* and *Maras*. On the latter the "Kale" east of Çurukkoz is probably Kalası; a second "Kale" to the south is probably Dibi.

Karafrenk

Karafrenk Kalesi[1] is a well-preserved garrison fort situated in a small valley about 7 km south of Bahçe.[2] The fort and the surrounding village are on the elevated east flank of the valley. From the village a stream descends to the west into the plain (pl. 111a). A well-worn road, which can easily accommodate a taxi, links Hasanbeyli at the southwest to Karafrenk, a distance of about 4 km. This road is the north half of a forked divide at the east end of the Amanus pass. Trails continue north from Karafrenk to Bahçe and east to Fevzipaşa, but these extensions are suitable only for pack animals. This garrison fort was obviously an important link in securing the passes. Unfortunately the history and medieval name of this site are unknown.[3]

The plan of Karafrenk Kalesi is a simple castrum with a square circuit and four round corner towers (fig. 44). Unlike many of the garrison forts that have been abandoned and plundered for their stones, the medieval complex of Karafrenk is occupied by the villagers who keep the towers in repair. Because parts of the interconnecting circuit have been removed to accommodate modern houses, the location of the original castrum entrance is impossible to determine. There are two houses that occupy the interior of the fort. One is entered from the south flank (pl. 111b) through a room and porch that extend out significantly from the line of the original circuit. This small house occupies the southeast corner of the fort. Most of the fort is occupied by a home that can be entered from both the west (pl. 112a) and north (pl. 112b). The northeast corner of the fort and part of the area along the interior of the north circuit is an open court. Today the towers stand to a height of about 5 m. The fort is built on a slight incline that descends to the west.

The medieval masonry is a fairly uniform type IV. In the northwest tower (pl. 112a) the stones are especially large and squared off to a greater extent. The few areas of the interior face that are still visible have type IV masonry (pl. 113b). The core of the wall consists of fieldstones bound in a heavy mixture of mortar. I found no evidence to show that the towers ever contained rooms. Behind the northeast tower there is no evidence of an opening, merely the corner formed by the north and east walls.

The northwest tower is the largest of the four bastions. Immediately south of this tower is one of the two modern doors that open the west wall (pl. 113a). All of the west wall, including its two square buttresses, may be of recent construction. Some of the facing on the southwest tower has fallen away (pl. 114a), but enough remains to show that the tower had an enlarged, tapering base. A small section of the original circuit still survives between the southwest tower and the south house. Most of the east face of the southeast tower has fallen away, but a large section of its south face still survives. The original east circuit is still standing (pl. 114b). The north face of the northeast tower has been repaired recently at the base with a crude type of masonry.

Neither the masonry nor the plan of this fort is Armenian. Karafrenk Kalesi has a plan that is similar to the forts at Tumil and Kütüklu, although its masonry is quite different. If Tumil and Kütüklu are Crusader sites, then it is quite possible that Karafrenk is a Crusader construction. The frontier forts of the early Byzantines as well as those of the Arabs have an identical plan. In such a design there is no attempt to cover the single square bailey with a vaulted stone roof. This, of course, is one of the principal features distinguishing it from the Armenian fortified estate houses (for example, Sinap near Lampron and Sinap near Çandır) which are divided horizontally by vaulted stories. Only formal excavations at Karafrenk offer any hope of securely determining the origins of the fort.

[1]The plan of this hitherto unsurveyed site was executed in June 1981. The contour lines are placed at intervals of one meter. On the plan I have not represented the modern construction on the *interior* of the fort, but only the areas where new buildings are adapted to the line of the circuit.

[2]The fort and village of Karafrenk appear on Kiepert's 1890 map of the region (published by Humann and Puchstein); this site is also listed on the *Marash* map. See also Cahen, 145.

[3]Because of the dark-colored ("kara") stones used in the construction of Karafrenk, it is possible that the fort is the Crusader site known as the "Black Tower;" see: Hasanbeyli and Savranda in the Catalogue; Boase, 114; Favre and Mandrot, 35; Forstreuter, 61–63, 235 ff; Hellenkemper, 121 note 4; and Alishan, *Sissouan*, 237 ff. The "frenk" (or Frank) in the name of this site may strengthen this association. There is a possibility that Karafrenk dates from the period of the Arab occupation; al-Kanīsah ("the church") is said to have been built with black stones. See Le Strange, *Palestine*, 477; Ibn Ḥawqal, 163 f, 180, 185; Canard, *Sayf al Daula*, 37, 51, 61.

In fact there is some evidence in our Arab sources to associate Karafrenk with al-Kanīsah as-Sawdāʾ. Ibn Ḫurdāḏbih (*Kitāb al-masālik waʾl-mamālik*, ed. and trans. M. J. de Goeje, *Liber viarum et regnorum*, BGA VI [Leiden, 1889], 100, lines 3 ff) makes it quite plain that al-Kanīsah is part of a *limes:* "The arrangement of Syrian settlements on the frontier is ʿAyn Zarbah, al-Hārūniyyah, al-Kanīsah as-Sawdāʾ and Tall Ğubayr (which is) eight miles from Tarsus." He defines (ibid., 253, lines 4 ff) this frontier by noting: "and what is concerned with both land and sea are the Ṯuġūr called the Syrian, and speaking of that they are: . . . ʿAyn Zarbah, al-Kanīsah, al-Hārūniyyah, Bayās, . . ." Our most complete account comes from the text of al-Balāḏurī (*Kitāb futūḥ al-buldān*, ed. M. J. de Goeje, *Liber expugnationis regionum* [Leiden, 1866], 171, lines 8 ff):

> Al-Kanīsah as-Sawdāʾ was built with black stones; the Greeks built it to last an eternity. It has an ancient fortress which was in need of repair to the extent that it had been ruined (*lahā ḥiṣn qadīm uḫriba fi mā uḫriba*). Then (Hārūn) ar-Rašīd ordered that the city of al-Kanīsah as-Sawdāʾ be repaired and strengthened . . . .
>
> Besides the people of Ṯaġr, ʿAzzūn b. Saʿad told me that the Greeks (*ar-Rūm*) made an unexpected attack on it, and al-Qāsim b. ar-Rašīd was organizing the resistance (inside the city) by using a viscous resin (*dābik;* a kind of naphtha?). Then they (the Greeks) moistened the clothes of (some of) their soldiers and sent a large contingent of them to scale the walls. But they (the Greeks) incurred the wrath of the denizens and soldier-volunteers of Misis (*al-Maṣṣīṣah*), and they (the people of al-Kanīsah) escaped en masse from their fate. They killed a bushel of Greeks; the only survivors were two wounded and mutilated soldiers. Then al-Qāsim dispatched someone to strengthen the city, maintain it and increase its munitions for war. And already al-Muʿtaṣim bi-ʾl lāh had deported to ʿAyn Zarbah and its environs a number of Zoṭṭs who previously used to live in the marshes between Wāsiṭ and al-Baṣrah. And its citizens (those of al-Kanīsah) exploited them (for this purpose).

Since the Zoṭṭs were exiled to Cilicia by 835, this serves as a terminus ante quem for the time of the battle (see: K. Zettersteen, "al-Muʿtaṣim," *EI*, 785; al-Balāḏurī, *Kitâb futûh al-buldân*, pt. 2, trans. F. Murgotten [New York, 1924], 110). In the early 10th century al-Hamaḏānī (*Kitāb al-buldān*, ed. M. J. de Goeje, *Compendium libri*, BGA V [Leiden, 1885], 113, line 21) speaks only of the city of al-Kanīsah as-Sawdāʿ which was built and strengthened by (Hārūn) ar-Rašīd. Al-Iṣṭaḫrī (*Kitāb al-masālik wa ʾl-mamālik*, ed. M. J. de Goeje, *Viae regnorum*, BGA I [Leiden, 1870], 55, lines 13 ff) says that al-Kanīsah is a frontier fort along the Anti-Taurus Mts., and the Ṯuġūr consists "of Malaṭyah, al-Ḥadaṯ, Marʿaš, al-Hārūniyyah, al-Kanīsah, ʿAyn Zarbah, al-Maṣṣīṣah, Adanah, Ṭarsūs. . . ." He follows (ibid., 63, lines 2 ff) with a more specific reference: "al-Hārūniyyah lies on the western flank of the Ğabal Lukkām far removed (*fī baʿda*) from its valleys . . . al-Kanīsah is a fortress and it also has a mosque (*minbar*) and it is a frontier place (*Ṯaġr*) well separated from the sea." It is less than a day's trip from Bayyās to al-Kanīsah and Haruniye (ibid., 68, lines 1 ff). Ibn Ḥawqal (*Kitāb ṣūrat al-arḍ*, ed. J. Kramers, *Liber imaginis terrae*, BGA II.1 [Leiden, 1938[2]], 182, lines 16 ff) adds little new to our knowledge of the site, merely noting "and al-Kanīsah used to be (*kānat*) both a fortress—and there is a mosque inside—and a frontier place (*Ṯaġr*) well separated (*fī maʿzil*) from the coast of the sea. . . ." His use of *kānat* may indicate that this fortress had fallen into Greek hands by 978. On Ibn Ḥawqal's map (Ibn Hauqal, *Configuration de la terre*, intro. and trans. J. Kramers and G. Wiet, I [Paris, 1964], 163 f, pl. 7) al-Kanīsah is listed twice, once as a coastal site and once in the Ğabal Lukkām. The former is actually a church in a village near Haifa (see Le Strange, *Palestine*, 477), and the latter is the Cilician castle. Unfortunately the relative positions of many of the sites on this map are confused. For example, he places Bağras closer to Maraş than to Haruniye, and he actually places the latter northeast of Maraş. More than two centuries later (1225) al-Yāqūt (*Muǧam al-buldān*, s.v. "al-Kanīsah" [or "al-Kanīsah al-Yahūd, called also al-Kanīsah as-Sawdāʾ], ed. F. Wüstenfeld, *Geographisches Wörterbuch*, IV [Leipzig, 1869], 314, lines 9 ff) merely rephrased the accounts of his predecessors on the Ṯaġr: "The Greeks (*ar-Rūm*) built it in the old time(s), and there is a strong ancient fortress (*ḥiṣn manīʿ qadīm*); this site was in need of repair for what had been reduced to ruin. Then ar-Rašīd ordered that it be repaired, and accordingly he restored it and its fortification to what it used to be; he also applied a tax to be collected and gave donations to this place." In 1321 Abūʾl-Fidāʾ (14) notes that al-Kanīsah is part of the Armenian kingdom and is 12 miles from Haruniye. The only fortress that fits all the topographical criteria above and is approximately 12 miles from Haruniye is Karafrenk. Less likely candidates would be Savranda, Hasanbeyli, Çardak, Babaoğlan, Bodrum, and

Kum. Cf. al-Balāḏurī, 264 f; E. Reitmeyer, *Die Städtegründungen der Araber im Islām*, Diss: Heidelberg (Munich, 1912), 78. I can find no evidence to support Cahen's conclusion (148 f) that al-Kanīsah as-Sawdā᾽ is the village of Erzin.

Kız (near Dorak)

In the north corner of the Cilician plain's west lobe there is a fortified estate house known today as Kız.[1] It is 8 km northwest of Dorak Köy (also spelled Durak) and 7 km southwest of Bucak Köy. The easiest way to reach this site from Adana is to take the northbound train from that city (via Yenice). After traveling 55 km by train, the fort is a comfortable two-hour hike from either the Dorak or Bucak station. It is situated on a lofty limestone pinnacle in what is now a semi-arid region where the average summer temperature is over 95°F. Kız Kalesi rises to an altitude of around 600 m. While this site is not more than 12 km from the major road that connects Tarsus and the Cilician Gates, it is difficult today to see that route from Kız.[2] The fort does have a commanding view into the Highlands at the northeast. There is an important road from Adana to Pozantı (via Kütüklu and Milvan) which this fort also guarded. The modern railroad follows most of this route. At the lowest elevations there are a number of scattered homesteads that are supplied with water by small streams and wells. Olives, figs, and melons are the chief crops of the region. The nearest source of flowing water for the fort is a small artesian well at the base of the outcrop. Cisterns in Kız would be a necessity during a siege.

Since we do not know the medieval name of this complex, we are ignorant of its role in history. A few modern explorers have mentioned this site in passing through the area. E. Herzfeld did not discuss the fort, but he did indicate on his map of Cilicia a site called "Kyz" on the east flank of the route from Tarsus to the Cilician Gates.[3] The only published account of Kız is the very brief description and single photograph in an article on Cilicia by J. Gottwald.[4] There is some doubt whether Gottwald ever explored the entire fort since he says that the floors in the "Doppelturm" have disappeared. In 1979 this was not the case. Also, the small, high-placed entrance described by Gottwald is not apparent in the present structure (unless he is referring to one of the windows in A2).

This fortification is unusual in that it has a single circuit wall around the keep-like house (fig. 45; pl. 115a). The baron who resided at Kız may have maintained a small garrison that was quartered in now vanished wooden buildings inside the circuit. The only entrance through the circuit is gate D at the south. This gate is a simple straight-through entrance, and today there is no evidence that it was vaulted. This circuit wall is relatively uncomplicated with only a single rounded salient at point F and a circular bulge at point E. The wall is constructed with type IV masonry on the interior and a relatively crude type V for the exterior facing; in the upper level there are a few sections that have been repaired. The quality of masonry in the circuit wall is inferior to the exterior facing of the keep and may indicate that different builders are responsible (pl. 115b).[5]

On the exterior the keep is hexagonal with a bend of 45 degrees near the center (pls. 115a, 116a). The exterior facing throughout is type VII, except for the upper levels of the east walls of A2–3, B2–3, and C3, where a superbly cut type IV masonry is used (pl. 118a). I suggest that this peculiarity with regard to the use of types IV and VII masonries is due to repairs that were carried out after a destructive earthquake. It is impossible that the areas of type IV represent the damage of a siege since the lower levels of the house (pl. 115b) are a flawless type VII (except for a few areas on the west side where modern treasure hunters have plucked away stones). It is also clear that the type IV was placed after the type VII since the composition of the poured core in the latter is quite different. The core in the type IV has fewer fieldstones.

Aside from the window-doors, there are three distinguishing features on the exterior of this estate house. First, corbels, which once supported brattices and machicolations, almost surround the upper levels of the keep (pls. 116a, 118a). On the west wall of C3 all of the corbels are at the top of that level and were commanded by men in level C4 (now missing); the north wall of C3 has two corbels below the northwest window-door. In the east walls of A3, B3, and C3, all of the corbels are below the doors.

Another prominent exterior feature is a massive chase that runs *vertically* from the top of the southwest corner of B3 (at the junction with A3) to B1 (pls. 117a, 117b). This chase is quite shallow (ca. 5 cm), and it broadens as it moves toward the bottom. At the base of the chase is a small rounded sink with two circular drainage holes in its interior. This sink is only 29 cm wide and 15 cm deep. At the point where the chase meets this basin, two very shallow, crude grooves of about 1.3 m in length run at acute angles to the line of

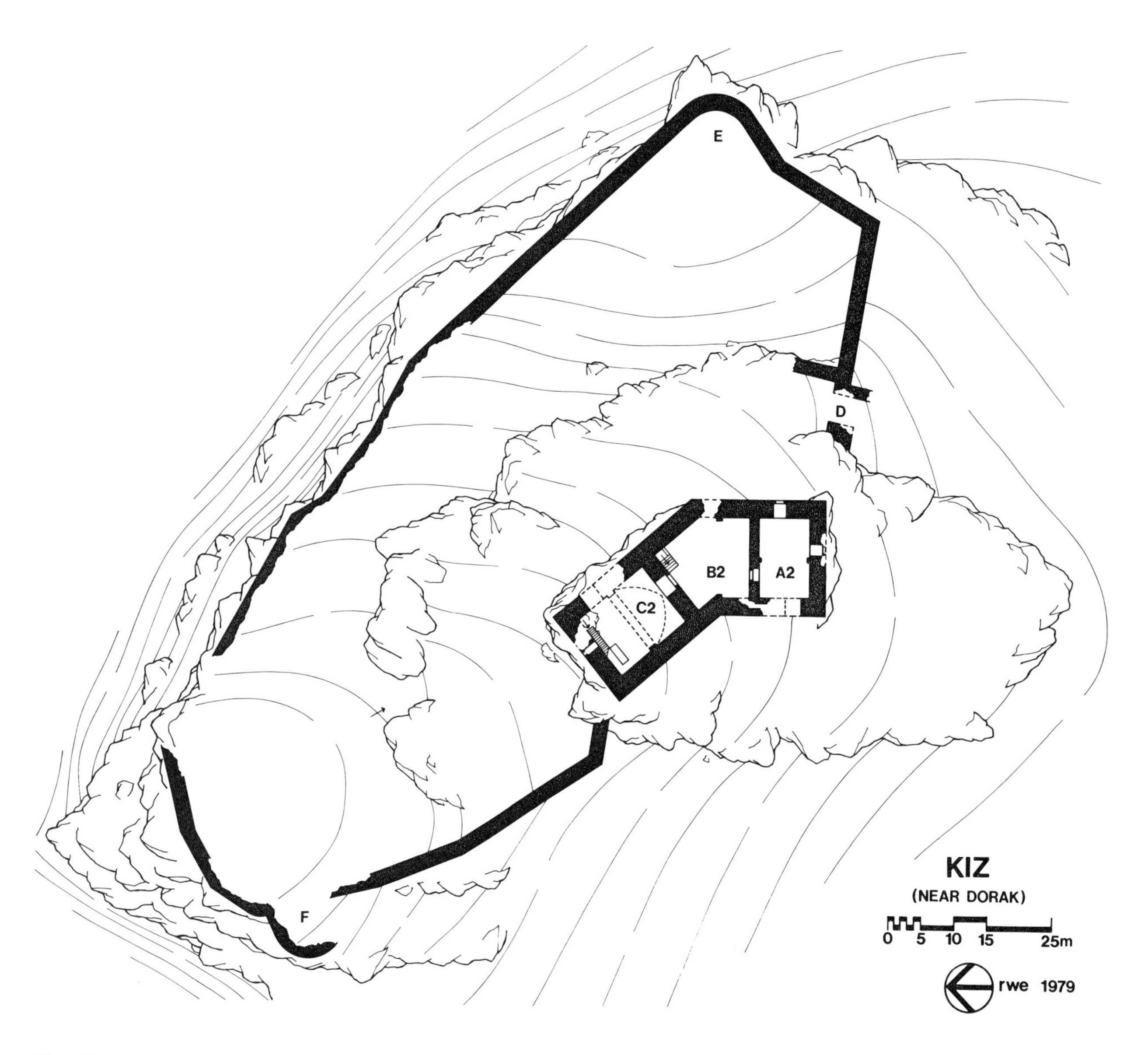

Fig. 45

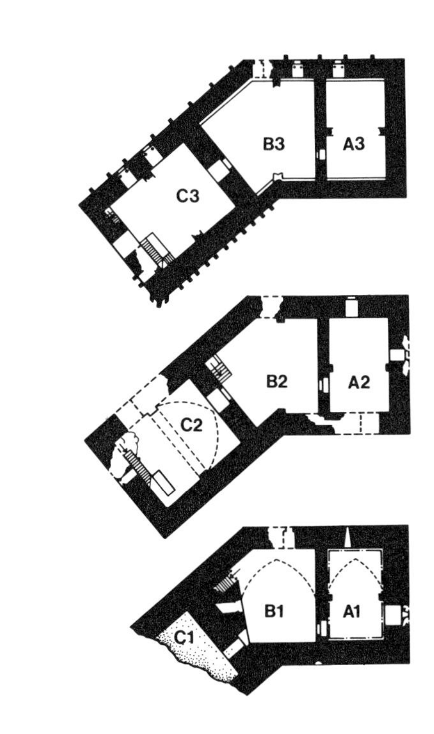
B3
A3
C3
B2
A2
C2
B1
A1
C1

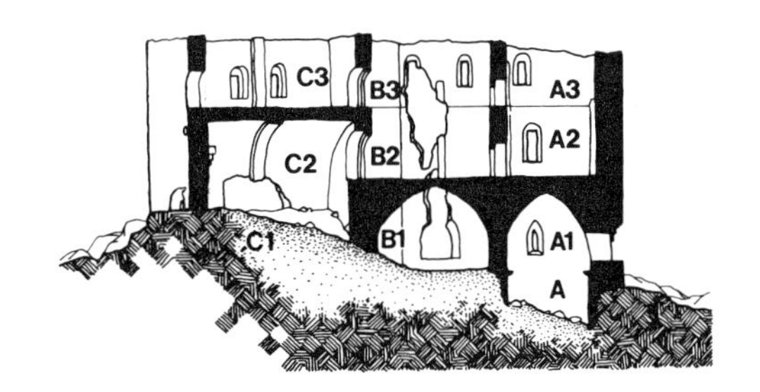
C3
B3
A3
C2
B2
A2
C1
B1
A1
A

the chase, creating the appearance of an arrow. The function of these shorter diagonal marks is unknown. The chase itself was a drainage course for water, which emptied into cistern A and possibly into a cistern below B1. It is likely that some sort of funnel was attached to the exterior of the sink to catch efficiently the water cascading down the face of the wall. This west wall, like the other walls of the keep, is constructed with a slight batter.

The third peculiarity on the exterior is the vertical seam at the junction of rooms B1–3 and C1–3 (pls. 116a, 117a). The two sections obviously represent different building periods. On the interior of the keep we have a clear indication that C1–3 is older. In the center of the north wall of B1 there is a large niche (pl. 118b) that has a height, width, and depth on the interior of 1.1, 0.98, and 2.2 m respectively. The north end of this niche is actually the exterior of the south wall of C1 and is not in alignment with the north wall of B1. The niche could not have served as a door for B1 since it is partially blocked by natural rock. The rest of the north end of the niche is faced with type VII masonry, a style that *never* appears on the interior of the keep. Thus the south wall of C2 was originally conceived as the exterior of a small rectangular building. The complex A–B was added soon after since there is no difference in the tooling marks on the type VII stone in A, B, and C; the core in the type VII of complex C is almost indistinguishable from that in A and B.

The entrance to the keep is not apparent. In the Armenian estate keeps of Cilicia there is always an entrance at the lower level (cf. Anacık and Sinap [near Lampron]). It is possible that there was an entrance in the shattered east wall of C2. The smaller hole in the north wall of this room is probably a breach. Room C2 is the largest single chamber in the complex (pl. 119a). The staircase in the north wall of C2 is built with an extremely crude type VI masonry and a few large ashlar blocks. Types IV and V are the predominant forms of masonry on the rest of the interior. Unlike the rooms to the south, C2 is covered by a vault that is on a northwest-southeast axis. The vault is not stuccoed, but its stones are now coated with a thin mineral deposit from the seepage above. A single transverse arch supports this vault. The only extant window in this room is high in the north wall; it is roundheaded on the interior. Its exterior is simply a small circle—no doubt a protection against intruders. At the south end of C2, two doors are visible (pl. 119b). The upper-level door, which gives access to B2, is in the center of the wall; its sill is about 1.6 m from the floor level of C2. Undoubtedly wooden stairs made access convenient. This door is framed with well-fitted blocks of ashlar and is limited by jambs. All of the doors and windows in the keep are framed by the same well-coursed ashlar. The second door is in the western end of the south wall and is actually below the shattered floor level of C2.[6] The north face of the door has collapsed. It probably led from B1 to the now buried chamber C1. Because of rubble, fill, and the unknown slope of the undulating rock, it is impossible to determine the dimensions of C1, but it was certainly much smaller than C2. That a second lower-level chamber, C, is present should not be ruled out. Only excavation can reveal the true complexity of the keep.

The door from C1 is clearly oriented to the axis of room C1 and may be one of the original openings. In passing through this door, one moves downhill to B1, a pentagonal chamber that is covered by a pointed vault. Aside from the niche and the door already discussed (pl. 118b), in B1's north wall there is a third opening in the form of a staircase to level B2. Post holes at the base of the staircase indicate that it was protected by double doors. In the east wall of B1 is now a shattered roundheaded window. As with chamber C1 there is a possibility of a chamber below (cistern B), although it is not evident in the present debris. At the west end of the south wall in B1 is the door to chamber A1.

Room A1 is rectangular and all of its walls are constructed with type IV masonry. This room is covered by a pointed vault and is opened by two windows. The window in the east wall is a thin, embrasured opening, while the one in the south wall is more broadly splayed and its casemate is clearly designed to accommodate an archer (pl. 120a). Only the rounded head of this embrasured loophole remains today. There are the remnants of a stirrup base in the loophole. The floor in A1 was supported by corbels, two of which are still present in the north and south walls. At the same level of the corbels, there are a number of joist holes which anchored the crossbeams of the wooden floor. Below A1 is cistern A. The walls of this room are carefully stuccoed. As noted earlier, water entered this chamber from the chase on the exterior of the east wall. Water was probably drawn from this cistern through a hatch in the floor of A1.

The stairs from B1 to B2 are now shattered and access is gained only with difficulty (pl. 121a). At the top of the stairs is the only evidence of relief sculpture in the entire keep. On one square block in the northeast corner of B2 is a cross inscribed in a circle (pl. 121b). Its purpose and origin are unknown. The room B2 is hexagonal and has one roundheaded door located in the north wall and another in the south wall. A window may have been located in the east wall, but the present breach has destroyed any trace of it. Attached to the east wall and to the junction of the two western walls are pilasters that rise from B2 to B3 and act as internal buttresses. Similar pilasters also appear in A2, A3, and C3. They are constructed with divergent wedges in the typical Armenian pattern, although strict alternation of single blocks is not followed.[7] The only identifiable mason's mark in the fort appears in B2 near the north door (see below, Appendix 1). The wooden ceiling of B2, which was also the floor of B3, was supported by shelves in the east and west walls and closely placed crossbeams anchored in joist holes.

B3 has four doors: one in the north wall (pl. 121a), one in the south wall, and two in the east wall. The north and south doors are both roundheaded, with the latter having slightly smaller dimensions. The doors in the east wall give access to the breastwork. They are both roundheaded on the interior and squareheaded on the exterior. The northernmost of the two doors is badly shattered. If there was access from level B2 to B3 it was probably by means of wooden stairs.

The only entrances into the rectangular chambers A2 and A3 are through the single doors in the north walls. Chamber A2 appears to have had three other openings, one in each of the walls. Despite the massive break in the west wall of A2 there is evidence of a roundheaded window. The exterior of the window in the south wall is badly damaged, but the interior side is roundheaded, as is the opening in the east wall, which has a squareheaded exterior. On level A3 there is one door in the east wall that is roundheaded on the interior and squareheaded on the exterior. Like the southeastern opening in B3, this door has pivot holes above to accommodate shutters (pl. 116b). The floor of A3, like that in B3, is in part supported by shelves in the east and west walls. Both chambers A3 and B3 were probably covered by a simple tapered roof.

Chamber C3 can be entered by a door from B3 and a stairway from C2 (pls. 122a,122b). In the east wall of C3 the southernmost door is roundheaded both on the exterior and interior (pl. 120b), while the other door to the north is roundheaded on the interior and squareheaded on the exterior. Both of the doors are equipped with twin post holes for swinging shutters. In the center of the north wall there is a large roundheaded door without jambs (pl. 122a). One meter west of the door is a smaller squareheaded window just above the stairway from C2. To the east of the central door in the thickness of the wall is another cantilever stairway that leads to C4. Because of extensive damage it cannot be determined whether C4 was a full-sized room with a ceiling or merely a fighting platform with parapet. Only the foundations of the walls of C4 are standing today.

It is interesting that many of the rooms in the keep are designed to be sealed off from the adjacent rooms, no doubt to inhibit the movement of an enemy who had entered one of the chambers.

[1] A formal survey and plan of this site had never been attempted before I commenced work in summer 1979. The contour lines on the plan are placed at intervals of 50 cm.

[2] Perhaps in the less polluted air of the medieval period, that strategic road was clearly visible.

[3] E. Herzfeld, "Eine Reise durch das westliche Kilikien im Frühjahr 1907," *Petermanns Mitteilungen,* 55 (1909), pl. 3; cf. Frech, 578, 580, fig. 17; Ainsworth, 78; Boase, 162.

[4] Gottwald, "Burgen," 94 f, pl. 3.

[5] As I indicate above, Part I.4, the masonry of the house seems to be quite Armenian, while its exposed corners do not. If this house was built for Crusader occupants they may have added the circuit at a later time.

[6] C2 appears to have had a wooden floor.

[7] F. Gandolfo, *Aisleless Churches and Chapels in Armenia from the IV to the VII Century* (Rome, 1973), 170.

Korykos

The principal guardian of the strategic coastal road between Silifke and Tarsus is Korykos.[1] This fortified port[2] consists of a large land castle and a smaller sea castle (pl. 123a). The latter surrounds the perimeter of an island that was once connected to the land castle by a breakwater. Korykos, which was legendary for its violent winds,[3] became the second port of Armenian Cilicia; commercially it was far less important than Ayas, but strategically it was closer to Cyprus. In 1907 S. Guyer conducted a careful and systematic survey of both forts.[4] His text and plans have served as the bases for later architectural discussions of the site,[5] including my own. My intention here is merely to offer a summary account of the forts and some new observations.

It is possible that the site of Korykos was heavily fortified *prior* to the Arab invasions, but there is no evidence to confirm this. Except for reconstruction during and after the Armenian period of occupation (far more extensive in the island castle than in the land castle), the circuit walls and towers of both units date from the early twelfth century. The erection of the forts can probably be credited to the reign of Alexius I. The emperor's daughter, Anna Comnena, tells us that the

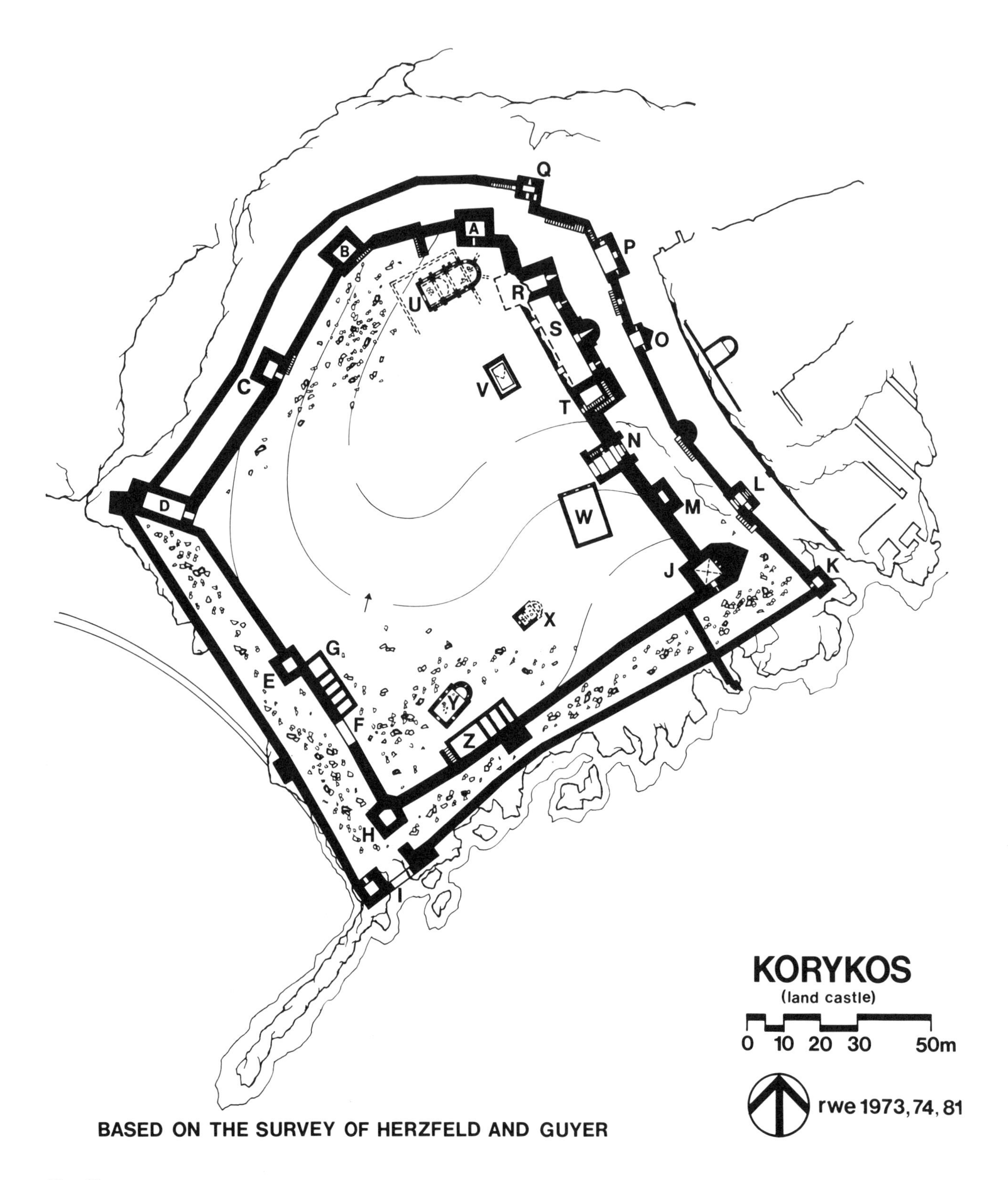

Q
A
B
P
R
U
S
O
V
C
T
N
L
D
W
M
J
K
X
G
E
F
Z
H
I
KORYKOS
(land castle)
0 10 20 30 50m
rwe 1973, 74, 81
BASED ON THE SURVEY OF HERZFELD AND GUYER

Fig. 46

royal eunuch Eustathius was dispatched as an admiral and was directed to fortify Silifke and Korykos.[6] The strategy was to defend these sites from any possible seizure by the Crusader Bohemond. A large garrison was maintained at Korykos under the command of a certain Strategus Strabo. Exactly when the Armenians occupied the Byzantine forts at Korykos is unknown.[7] By 1198/99 the site may be under the control of King Levon I; Simon, the Baron of Korykos, is in attendance at the coronation.[8] Following Vahram's brief tenure as lord of Korykos (1210–12), the Het῾umid Baron Ōšin held the position until the late 1260s.[9] In the fourth quarter of the thirteenth century the Armenian historian Het῾um followed Grigoris as master of the port.[10] Some years later he died tragically in a battle against the Mamluks. In 1318 Het῾um's son, another Ōšin, took three hundred troops from the garrison at Korykos and succeeded (temporarily) in driving out a band of Turks.[11] In 1360 Peter I, the King of Cyprus, assumed control over Korykos when it became clear that the Mamluks were soon to conquer all of Cilicia.[12] Robert of Lusignan was dispatched from Cyprus to administer the port. With Cypriot assistance the residents of Korykos were able to repulse a Karamanid attack in 1367.[13] This fortified port proved to be a profitable toll station until its capture by the Karamanids in 1448.[14]

Because of differences in topography, the plans of the land and sea castles at Korykos are at once similar and yet unique. The land castle (fig. 46), which is built on the relatively flat ground of the rocky shore, is characterized by the almost square shape of a tight double trace with square towers. This is the only fully concentric plan for a fortification in Cilicia,[15] and it reflects the same theories in military science that are embodied in the land walls of Constantinople (cf. Toprak). Since the sea castle is protected by a natural water barrier as well as the formidable shoals of the island (fig. 47), the Byzantines constructed only a single, somewhat geometrical circuit with square towers. This original construction survives only at the south and east and is in sharp contrast to the Armenian reconstruction (with rounded salients) at the northwest. The plan conforms to the topography of the island.

Today the island castle can easily be reached by swimming from the beach resort of Korykos (ca. 400 m) or by renting a small rowboat. The least dangerous approach through the rocky shoals is northwest of tower F (pl. 124a). One of the two posterns that open the circuit is located at this point. The other postern is located directly south of tower J (pl. 123b). The principal opening into the fortress appears to be the bent entrance in tower H.

While walking along the exterior of the circuit it becomes quite apparent that the Armenians refaced the original Byzantine walls at the south with the same type of masonry that they used in their reconstruction of the north and west sides, the distinctive type VII. In towers G and I, types VIII and V masonries are used in conjunction with the type VII (pls. 126a, 127a). The Armenian tendency to vary the height of the separate courses for aesthetic effect is clearly evident. Unlike the Byzantines, who chose to recycle the smooth ashlar from the neighboring city of Korykos, the Armenians seem to have quarried fresh ashlar as well as having reshaped the late antique stones to obtain that distinctive boss on the exterior. In a section of the circuit wall from F to G, almost all the exterior facing stones have been stripped away. It is also apparent that reconstruction with type III masonry occurred *after* the Armenian period (pl. 123b). In some cases, the type III is used with an abundance of mortar, while in other areas there are only sparse traces of mortar in the margins (pl. 127a).

On the interior the masonry shows some variation. In general the south and west walls are faced with type IX (pl. 126b). Much of what remained of the type IX masonry in the north and east walls has fallen away or been surmounted by an almost mortarless type III masonry (pl. 125a). Much of the type IX masonry appears to be spoils recycled from the late antique city. The interior buildings, such as B and E, tend to be constructed with type IV.

The tallest unit of the island castle is the projecting three-story tower F. It was during the Armenian period that this salient was rebuilt on the original Byzantine plan. Today the upper floor has deteriorated badly, and access can only be made with difficulty. In the late 1950s the north face of the lower two stories of F underwent a rather extensive restoration (pl. 124b) that has preserved in situ part of an Armenian inscription.[16] In the nineteenth century a second (now missing) inscription was located in the same area.[17] The latter supposedly mentioned King Levon I as the builder of this site (1206).[18] What can be securely translated from the surviving inscription is that King Het῾um I dedicated his reconstruction at this site in 1251.[19] By the late nineteenth century the entire frame for the second-level north entrance of F (shown on the plan) had vanished; the present reconstruction is quite unlike the original.[20] Adjacent to the doorway, in the thickness of the same wall, is a staircase to the third level. The other three walls of the second level are each opened by two

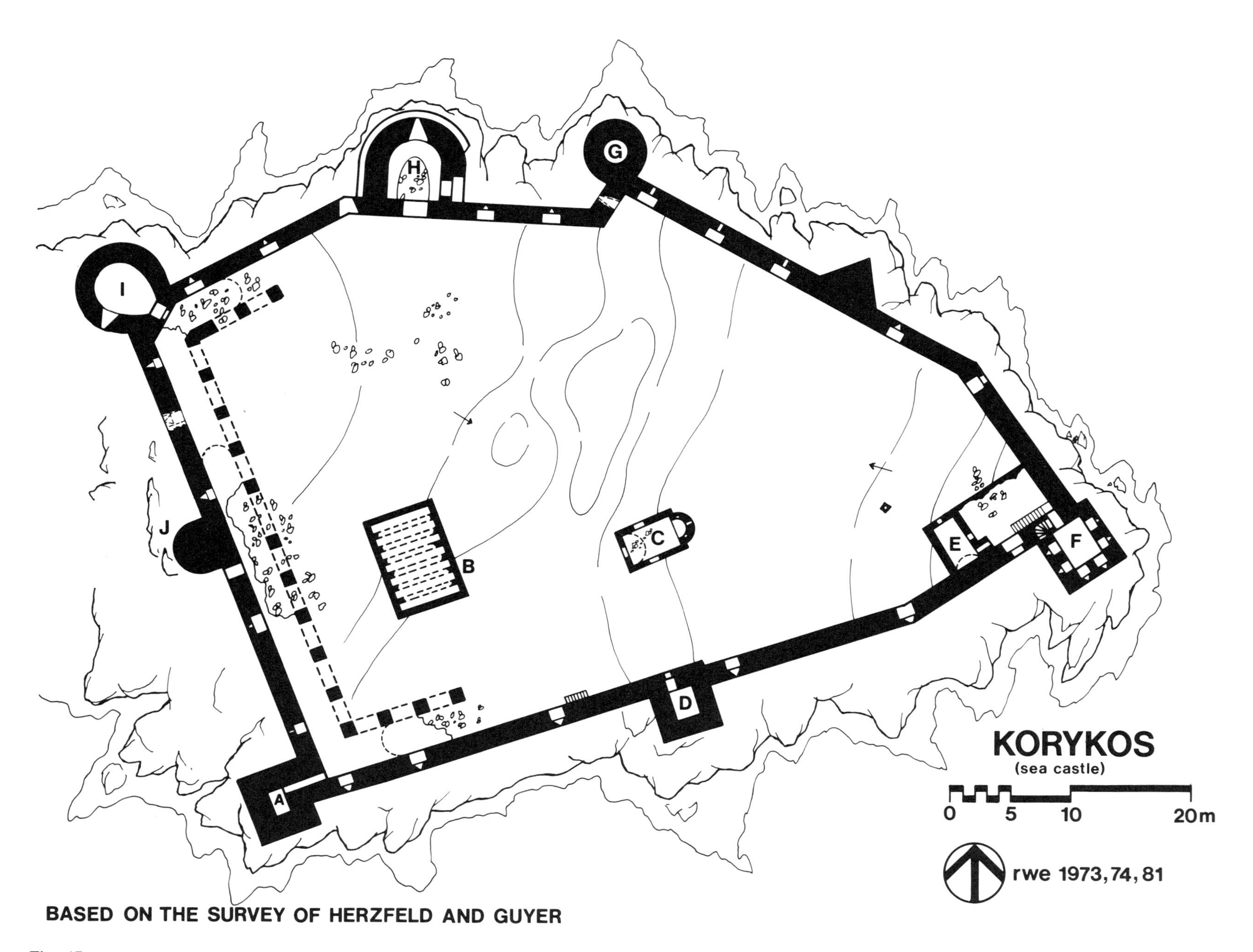
KORYKOS
(sea castle)
0
5
10
20m
rwe 1973,74,81
A
B
C
D
E
F
G
H
I
J
BASED ON THE SURVEY OF HERZFELD AND GUYER

Fig. 47

embrasured loopholes with slightly pointed casemates. Similar casemates appear in the thickness of the circuit wall (pl. 125b). The lower level room immediately north of F is visible only at its foundations. The smaller adjacent room E at the west is fully preserved (pl. 125a). It is covered by a barrel vault and has an entrance at the east. The only other rooms within the fort are the collapsed cistern B and chapel C.[21] There is evidence that most of the interior side of the south circuit was adapted to support wooden buildings. Joist holes are visible at the level of the casemates in the south wall. The other two salients along the south wall, A and D, are not as well preserved, although there is evidence in the second levels of A and D of embrasured loopholes (not shown on the plan). All the loopholes associated with Armenian reconstruction have stirrup bases (pl. 126a).

The west flank of the fortress possesses a very unusual feature. Along the interior side and overlapping at the north and south is an open portico or stoa (pl. 126b). The arcade is formed by a series of carefully placed piers and adjoining arches that in turn support a continuous vault. Most of this slightly pointed vault, which is made with type III masonry and partially stuccoed, is preserved. The west wall is characterized by the rounded bastions I and J (pls. 123b, 126a). As with the south flank, embrasured loopholes with casemates open the west circuit. The lower level of J (an Armenian construction) is solid, while the second level addition is hollow. This upper room has a door at the east and a miniature embrasured window at the west (neither of these Lusignan additions appears on the plan). The postern next to J has a rounded cover that is constructed with well-cut elongated voussoirs. Some of the voussoirs have joggled joints. This is quite unlike the flattened blocks that make up the arch over the east postern. Almost the entire upper half of the circuit between J and I has collapsed.

The only unusual feature in the north half of the fortress is on the interior of tower H (pl. 127b). It appears that the apex of the original apsidal vault from the Armenian period had collapsed. During the Lusignan occupation a second-level parapet was constructed atop this first-level room, but no attempt was made to repair the vault. Either a wooden ceiling was erected, which also acted as a floor for the second level, or the room was simply left open to the sky.

While extensive reconstruction has removed most of the Greek masonry and half of the original plan of the island castle, the land castle is a well-preserved example of Byzantine construction. All of the masonry used in the construction of this fortification is ashlar (pl. 128b) that has been recycled from the late antique-early Byzantine city of Korykos. In one case a triumphal arch has been adapted to function as a gate (pl. 123a). Cornices with dentils are refitted into circuit walls, and elaborate Corinthian capitals have been reused as fill material. In a few areas the cores in these Byzantine walls have not been poured but consist simply of a loose rubble fill or poorly coursed stones wedged together (pl. 128a). Along the heavily defended east flank, headers of recycled column drums are prominent in the exterior facing (pl. 296c). In the few areas where the Armenians reconstructed the east flank (for example, tower S) the headers, which are so common in Byzantine military architecture, are absent; instead, tightly fitted rusticated blocks are prominent, as well as features like the slot machicolation over entrance N. Armenian construction is also evident at the sea-gate I, where a pointed arch (one of the few in the land castle) is so prominent.

Of the three chapels in the land castle, only one, chapel Y, is an Armenian construction.[22] The exterior masonry of chapel Y is a uniform type VI.[23] The only obviously recycled piece of material is the lopsided exterior frame for the apsidal window. On the interior of Y mortar is not prominent in the margins of the regular courses of ashlar blocks; a painted stucco once covered the interior walls. The exterior masonry of Y is quite unlike the facing of the smaller Byzantine chapel X to the northeast, where very crude stones in irregular courses predominate. The interior masonry of X consists of a finer-quality ashlar. The large Byzantine chapel U has a slightly larger and more carefully laid ashlar masonry than in X, but it does not have the regular margins of mortar that appear on the exterior of Y. It appears that the east end of U was reconstructed by the Byzantines with a similar style of masonry; at that time the original apsidal doors and window were blocked, and buildings were adapted to the exterior.

Types III and IV masonries are visible atop the north and east inner circuit walls and probably represent repairs carried out during the Lusignan occupation.

One of the interesting features about the land castle is that there is only one entrance through the outer circuit on the landward side. There was some sort of drawbridge that connected L to the outer flank of the ditch. Also, from the seaward side, a visitor could sail into gate I. Two doors open the inner circuit at F and N. The lower half of door F was blocked with masonry during or after the original construction of the fortress.[24] Some sort of removable scaffold or ladder may have permitted passage through the upper half.

On the other side of the fortress the smaller door at N carries on the exterior side the only surviving piece of Armenian relief sculpture in this fort.[25] This relief consists of a *xač^{c}k^{c}ar;* it has a single cross in the center and is surmounted by stylized rosettes and an ornamented frame. The cross is confined to the upper half of the block. Today the broken lower half shows a few faint traces of Armenian letters. Except for tower D and the wall south of tower J, there are no impediments to free access along the corridor between the outer and inner circuits.

Because the north flank of the fort was once protected by deep marshes (making an assault on that side impossible), there are no bastions on the curving face of the outer north circuit. The only convenient line of access is at the east. Both of the east walls are protected by closely placed bastions and an outer ditch. Traces of battlements can be seen along these walls. The partially scarped ditch is supplemented on both sides with the recycled ashlar from the late antique city. This ditch is now dry, but it is quite possible that an inlet once filled the cavity with seawater. Few towers are placed along the south and west flanks because an assault by sea is difficult (pl. 296b).

Excluding the three chapels, there are only four interior structures in the land castle. Three of these structures (G, V, and Z) are cisterns. V is a freestanding cistern that still preserves most of its pointed vault. The exact function of the rectangular room W is unknown. The highest levels of most of the towers are usually covered by stone vaults. Frequently the rooms at the lower levels have wooden floors/ceilings. The only surviving example of a groined vault in the fortress is the third-story ceiling of the partially collapsed tower J. The impressive complex of R, S, and T may constitute a donjon in that it is heavily fortified and raised above the level of the rest of the circuit.

[1]I first visited Korykos in 1973; in subsequent visits (1974, 1981) I made minor modifications to the plans published by Herzfeld and Guyer. On the plan of the land castle the contour lines are approximated at intervals of 75 cm; for the sea castle the contours are spaced at intervals of 30 cm.

[2]The Armenians refer to this site as Kiwřikos, Kořikos, or Gurigos; in Greek texts it is rendered Kōrykios, Korykos, or Kour(i)kos. The Latin form is Curicus. The Franks call this port: Curcum, Curtum, Curta, Corc, le Courc, le Court, ans Curc; the Arabs use Qurquš, Qurqūs, or Qurqus. Wilbrand von Oldenburg (21) knows it as Cure. Some modern Turks refer to it as Kız Kalesi. Korykos appears on the *Cilicie* map. For general comments on the history and topography of this site see: Carne, II, 48 f; Brocquière, 100; *Handbook,* 189; Texier, 725; Cuinet, 73–75; Schaffer, 63 f; Schultze, 290–96; Ritter, 340 ff; Rey, 350 f; *Les portulans grecs,* ed. A. Delatte (Liège-Paris, 1947), 175 f; C. Irby and J. Mangles, *Travels in Egypt and Nubia, Syria and Asia Minor* (London, 1823), 518 f; Beaufort, 240 ff; V. Langlois, "Voyage en Cilicie, Corycus," *RA* 12 (1855), 129–47; T. MacKay, "Korykos," *PECS,* 464 f; Magie, I, 268, II, 1143, 1168; K. Lehmann-Hartleben, *Die antiken Hafenlagen des Mittelmeeres* (Leipzig, 1923), 260; H. Delehaye, "Les actes inédits de Sainte Charitine, martyre à Corycos en Cilicie," *Analecta Bollandiana* 72 (1954), 9 note 2; Tomaschek, 64 f; Hierocles the Grammarian, *Le synekdèmos d'Hieroklès,* text, intro., and comm. E. Honigmann (Brussels, 1939), 37; Imhoof-Blumer, 462 f.

[3]Choniatēs, 463.

[4]Herzfeld and Guyer, 161–89.

[5]Müller-Wiener, 79 f; Fedden and Thomson, 38 f, 50; Hellenkemper, 242–49. Langlois (*Voyage,* 193–219) and Alishan (*Sissouan,* 393–409) both discuss this site, but they do not deal with the architecture in detail. See also Ēp^{c}rikean, II, 451–56.

[6]The *first* mention in a Greek text of a fortress at Korykos is from Anna Comnena (45 f). Anna refers to Korykos as *once* being a very strong city (πόλις). She makes no reference to a preexisting (or abandoned) garrison fort in her narrative. That the first major period of construction begins in the early 12th century is indirectly supported by Arab geographers, who are *especially* careful to distinguish fortified sites near the frontier. In 864 Ibn Ḫurdāḏbih (*Kitāb al-masālik wa'l-mamālik,* ed. and trans. M. J. de Goeje, *Liber viarum et regnorum,* BGA VI [Leiden, 1889], 117, lines 12–14) says that beyond Tarsus there are several *ruined* Greek cities and one of them is Qurquš (Qurqus). It is not until 1154 that al-Idrīsī (*Kitāb nuzhat al-muštāq,* IV.5, 8, ed. L. Veccia Vaglieri, *Opus Geographicum,* VI [Naples-Leiden, IUON-IsMEO, 1976], 646, lines 3 ff; Le Strange, *Palestine,* 489) refers to Qurqūs as a fortress on the sea.

Unfortunately the imperial spatharios and turmarches on a single Byzantine seal from the second half of the 10th century can be associated only tentatively with Korykos. The seal tells us nothing about the presence of any castles; see G. Zacos, *Byzantine Lead Seals,* II (Bern, 1984), 418. The mid-10th-century *De thematibus* of Constantine VII Porphyrogenitus (ed. and comm. A. Pertusi, Studi e testi 160 [Vatican City, 1952], 77, 147 f) lists ten fortifications (excluding the capital) in the theme of Seleucia, and not one of them can be identified with Korykos (cf. Silifke [note 4] in the Catalogue). Prior to the founding of the theme of Seleucia, a weak case can be made for including Korykos in the Kibyrrhaiote theme. Writing in the early 800s about events in the last years of the 7th century, Nicephorus, the patriarch of Constantinople (ed. K. de Boor [Leipzig, 1880], 40) mentions a certain Apsimarus, who happened to be the commander of the army of Korykosians in the land of the Kibyrrhaiotes. Theophanes (*Chronographia,* ed. K. de Boor, I [Leipzig, 1883; rpr. Hildesheim, 1963], 370), Nicephorus' contemporary, draws on a similar source and refers to Apsimarus as the Kibyrrhaiote colonel of the Korykosians. Neither author uses the toponym Korykos but rather the collective form "Kourikiōtoi." While it is possible that so easterly a port stayed in Greek hands, it is more likely that the Arabs destroyed the Cilician Korykos, and its inhabitants fled en masse to a safe haven west of Seleucia. Unfortunately the evidence for locating a second Korykos in coastal Pamphylia is inconclusive. Cf. Zonaras, ed. M. Pinder et al., CSHB, III (Bonn, 1897), 234; Leo the Grammarian, *Chronographa,* ed. I. Bekker, CSHB (Bonn, 1842), 166; Anonymous, *Chronographia,* ed. A. Bauer (Leipzig, 1909), 64; *De thematibus,* ed. Pertusi, 151 f; Ramsay, 384; H. Antoniadis-Bibicou, *Etudes d'histoire maritime de Byzance à propos du "theme des caravisiens"* (Paris, 1966), 96 note 5; M. Hendy, *Studies in the Byzantine Monetary Economy c.300–1450* (Cambridge, 1985), 652 note 427. H. Turtledove (*The Chronicle of Theophanes,* trans. [Philadelphia, 1982], 68) completely avoids the problem.

By assuming that Korykos Kalesi is an early 12th-century construction, I am at variance with the recent views of A. W. Lawrence ("A Skeletal History of Byzantine Fortification," *The Annual of the British School at Athens* 78 [1983], 177–80) who believes that it was constructed before thc 6th century. My views may also run counter to a forthcoming publication by David Winfield (see C. Foss, "The Defenses of Asia Minor against the Turks," *The Greek Orthodox Theological Review* 27/2–3 [1982], 203 note 14). From the narrative of Foss (ibid., 159) it is not apparent how he (Foss) dates the military construction at Korykos.

An early 12th-century date for the Byzantine reoccupation is also supported by the fragmentary ecclesiastical history of this site. The first bishop of Korykos was probably appointed in the late 4th century. After the Arab invasions a bishop was not reinstated until the late 11th or early 12th century. See: Le Quien, *Oriens Christianus,* II (Paris, 1740; rpr. Graz, 1958), 879–82; V. Laurent, *Le corpus des sceaux de l'empire byzantin,* V.2: *L'église* (Paris, 1965), 373 f; R. Devreesse, *Le patriarcat d'Antioche, depuis la paix de l'église jusqu'à la conquête arabe* (Paris, 1945), 153 f; R. Janin, "Corycos," *Dictionnaire d'histoire et de géographie ecclésiastiques* (Paris, 1955), 925 f.

[7] Korykos may have been captured by the Franks as early as 1109. In the mid 1130s there is sufficient evidence to show that it was held by the Crusaders; see Lilie, 99, 101, 109, 396 notes 42–44. In 1137 John Comnenus recaptured Korykos; cf. Michael Italikos, 250. According to Benjamin Ben Jonah, Baron T'oros II may have occupied this site before 1168; see *The Itinerary of Benjamin of Tudela,* ed., trans., and comm. M. Adler (Oxford, 1907), 15.

[8] See below, Appendix 3 and Christomanos, 148.

[9] *RHC, DocArm,* I, lx, cxviii–ix; above, Part I. 7, note 37; Smbat, G. Dédéyan, 78 note 61, 115. For a genealogy of the lords of Korykos see Toumanoff, 282 f.

[10] Het'um, 469 ff.

[11] Smbat, 660.

[12] Dardel, 31 f, 32 note 1, 51 f; Makhairas, I, 98–102. The two ships sent by Peter I to occupy the harbor of Korykos may not have arrived until 1361; cf. Flemming, 84; and K. Setton, *The Papacy and the Levant (1204–1571),* I, *The Thirteenth and Fourteenth Centuries* (Philadelphia, 1976), 238–40.

[13] Makhairas, I, 174–76, II, 120; *RHC, DocArm,* I, 711, 715 f; Hellenkemper, 244 f; Sevgen, 244 f; Mas Latrie, I, 410 f, 416, II, 42, 77 f; Dardel, 36 note 4; Langlois, *Voyage,* 205 f.

[14] Makhairas, I, 562–64, II, 194.

[15] Cf. J. Rosser, "Excavations at Saranda Kolones, Paphos, Cyprus, 1981–1983," *DOP* 39 (1985), 82.

[16] Langlois (*Inscriptions,* 48) provides this French translation:

Dans l'année des Arméniens 700 ....
par le pieux roi Héthum ....
. . ce château princier a été construit ....
. . le grand prince, (fils d') Héthum ....

Cf. Herzfeld and Guyer, 163, pl. 176. B. Kasbarian-Bricout, *L'Arméno-Cilicie: Royaume oublié* (La Chapelle Montigeon, 1982), pl. 7.

[17] This epigraph appears to have been recycled as a springing stone when the tower was rebuilt; see Alishan, *Sissouan,* 400.

[18] There is some controversy over what the inscription actually says; see Langlois, *Inscriptions,* 48; idem, *Voyage,* 215; Herzfeld and Guyer, 164; Hellenkemper, 248.

[19] King Het'um I is probably responsible for rebuilding the present tower F. What can be credited to King Levon I is unknown.

[20] See Alishan, *Sissouan,* 400.

[21] For a description of the chapel see Edwards, "First Report," 173 ff.

[22] Ibid.

[23] Only the west wall of the chapel appears to be constructed with a masonry that has no mortar in the joints.

[24] Guyer (Herzfeld and Guyer, 175) believes that this portal was built in the classical period as part of a port monument.

[25] Cf. Herzfeld and Guyer, 178, pl. 189.

Kozcağız

Kozcağız Kalesi[1] is a small garrison fort isolated in the Highlands of the Nur Dağları (Anti-Taurus Mountains) at an altitude of about 250 m. This fort is situated atop a hill about 10 km southwest of Islâhiye and about 2 km east of the Kurdish village of Kozcağız.[2] The medieval garrison once stationed here guarded the north flank of a road that now links the Highland villages with Islâhiye. Islâhiye is also situated on the north-south highway that joins Antioch to Gaziantep and Maraş. The villagers of Kozcağız mentioned that an "old" trail (now suitable only for pack animals) continued west to Yarpuz and thus provided a strategic link between Osmaniye and Islâhiye. Undoubtedly this trail was used in the medieval period, which explains the presence of the fort. The sources of water in the immediate vicinity of the site are limited. There is no evidence of springs on the fortified outcrop. In the Highland valleys below, numerous streams and creeks carry the runoff from the mountains farther west into the orchards and wheat fields. The outcrop is relatively easy to ascend, especially at the south. No historical names can be associated with this site.

The polygonal and fairly symmetrical plan of the circuit wall does not conform to the contours of the outcrop (fig. 48). Today much of the circuit at the southeast is missing; this area may have held the original entrance. The presence of a keep at the southwest (A) and the widespread use of types III and IV masonries as an exterior facing indicate that this fort was originally a Byzantine construction. This conclusion is not unexpected, considering that the Byzantine fort at Çardak is the western guardian of the east-west route that links Osmaniye to Islâhiye.

In general most of the exterior facing stones of the circuit consist of type IV masonry. This masonry also appears on the exterior of the north and east walls of keep A (pl. 130b). Quoins are visible in the northeast corner of A. Occasionally, headers are seen in the exterior facing. The stones used for the interior facing of the circuit walls and internal rooms generally consist of type III (pl. 131a). However, a type IV with extremely large stones (and bearing an uncanny resemblance to the masonry at Çardak) is used as the interior facing of the wall flanking tower B (pl. 130a). This wall is thicker than any other section of the circuit and was certainly reconstructed. The type IV masonry shows a remarkable number of variants. The tower in keep A has blocks with less mortar in the interstices and more rock chips than any other type IV facing (pl. 129a). In fact, the stones in this tower have been

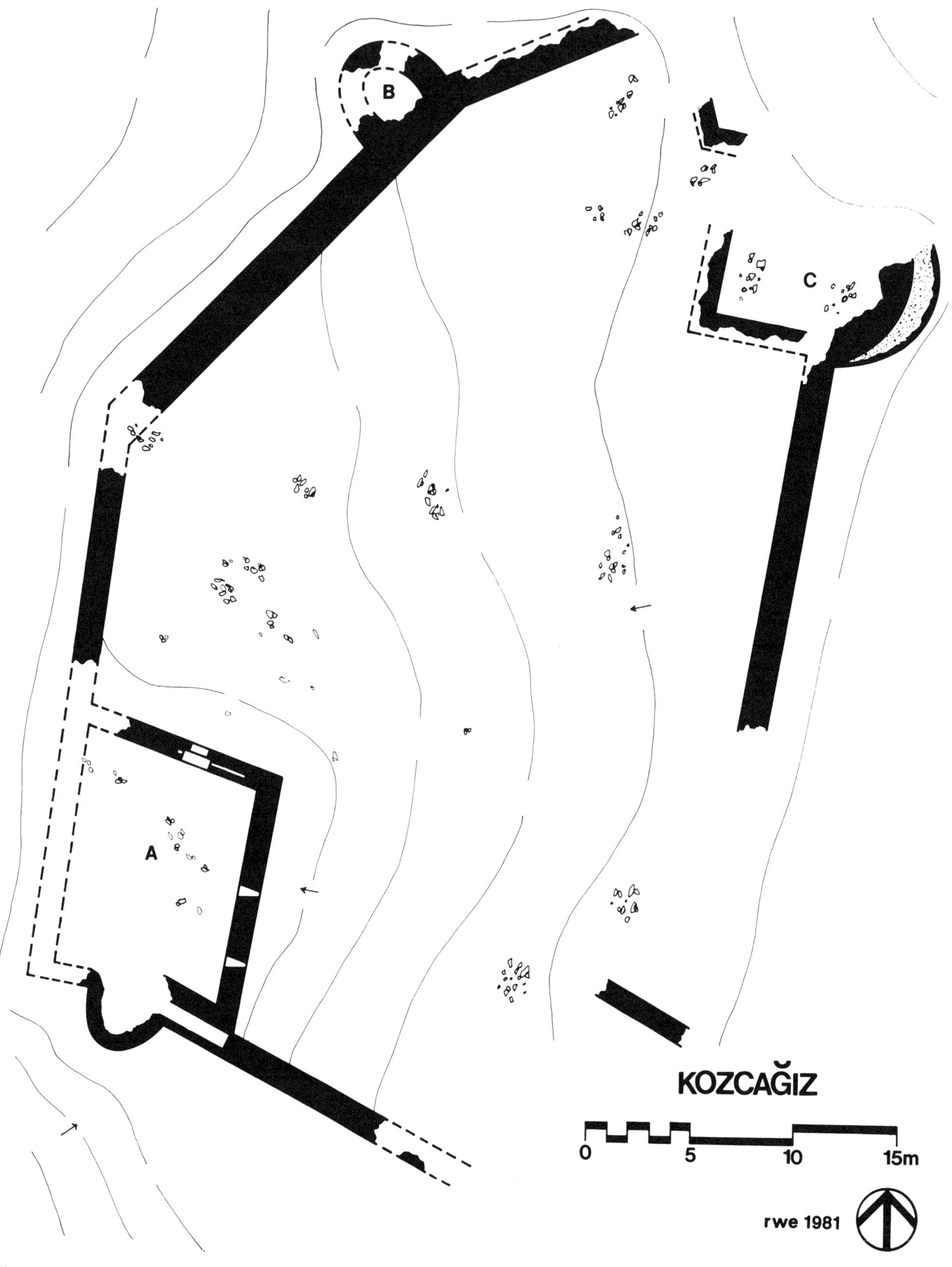

Fig. 48

trimmed to a much higher degree and come very close to resembling type V masonry. Another anomaly appears on the exterior face of tower C where three types of masonries are evident. In the collapsed upper section of the tower it seems that the original east wall of the tower was covered by a new salient wall, which protruded farther to the east. This new wall was separated from the original by a core of unhewn fieldstones and mortar. The masonry of the original salient consists of a combination of types III and IV. The new salient face consists of type VII masonry(pl. 131b).[3] This is undoubtedly an Armenian addition. The variation in masonry types at Kozcağız shows how difficult it is to catalogue the different kinds of stones by appearance. It is certain that the fort has undergone a number of periods of reconstruction, both Byzantine and Armenian. When the Armenians occupied this fort they maintained the existing Byzantine plan.

Today the interior of the fort is covered with rubble and the heavy undergrowth of bushes (pl. 130a). There were probably a number of freestanding structures on the interior (pl. 131a). The rubble is constantly being displaced by the digging of villagers in search of treasure.

The three principal units of Kozcağız Kalesi are donjon A and towers B and C. The donjon or keep holds the high ground in the southwest corner. It has a single door at the north that is protected by jambs and covered on the exterior by a depressed arch of four voussoirs (pl. 130b). The faces of the voussoirs are relatively smooth when compared with the rusticated centers on the large blocks that make up the jambs. On the interior side of the door, square sockets behind the jambs indicate that a crossbar bolt was accommodated (pl. 129b). The interior of the door is covered by a small vault. The east wall of A is opened by two high-placed embrasured windows. The west wall has collapsed, and the only remnant of the south wall is the rounded tower. From the extant remains it is impossible to determine the nature of the roof over the donjon.

Tower B at the northwest is deteriorating rapidly and appears to be a rounded bastion. It is uncertain whether the center of the tower contained a small room. Tower C, which has the enlarged east face, retains the foundation of a rectangular room at the west (pl. 131a). Today the north wall of the room is buried in debris. Since neither the south wall nor west wall is opened by a door, it seems unlikely that the chamber in tower C functioned as a chapel. I found no evidence of a cistern either inside or around the fort.

[1]The plan of this hitherto unattested site was completed in June 1981. The contour lines are separated at intervals of one meter.

[2]The village of Kozcağız appears on two published charts: *Maras* and *Malatya*. See Schaffer, 94–96.

[3]Cf. tower B at Bağras; see Edwards, "Bağras," 419 ff.

Kütüklu

Kütüklu Kalesi[1] is a small garrison fort located about 15 km north of the railroad center at Yenice between Tarsus and Adana.[2] The easiest way to reach Kütüklu is to drive northwest from Adana along the main road to Karaisalı. From this paved road a dirt trail leads to the village of Kütüklu (ca. 3 km) and the fort (pl. 133a). The latter is about one kilometer south of the village on a small hill. Kütüklu was a guardian of the road that linked Milvan Kalesi and Pozantı to Adana and the coast. The closest garrison forts to Kütüklu are the sites at Kız (near Dorak) and Yaka. Kütüklu Kalesi may have had intervisibility with Kız but not with Yaka. About 300 m west of the fort there is a spring of slightly brackish but drinkable water. There appear to be the remains of a cistern (unsurveyed) directly west of the fort. Today the area around the base of the fortified hill supports crops of wheat and melon. No historical names can be associated with this site.

As with most of the exposed garrison forts near the coast, Kütüklu Kalesi is in a relatively poor state of preservation. It is designed like the typical castrum with a square plan and four corner towers (fig. 49). The two southern towers have lost all traces of their facing stones and have deteriorated into somewhat amorphic extensions. It appears that they were once hollow, like the northern towers. The interior of the fort shows no signs of construction, but there is a pit in the center. The only break in the circuit wall is at the south; this may have been the entrance into the fort. There are the remains of a short circuit or revetted fosse to the south of the opening (not shown on the plan). Unfortunately the accumulation of dirt and weeds hides many of the details of this fort. The tower at the southwest stands to less than one meter in height. The salient at the southeast corner consists of an oblong fragment of the tower core. Only a few of the facing stones still protrude on the face of the east wall, leaving a somewhat gnarled appearance. The rounded northeast tower stands to about 5 m in height. Some of the exterior facing stones on the tower and the attached

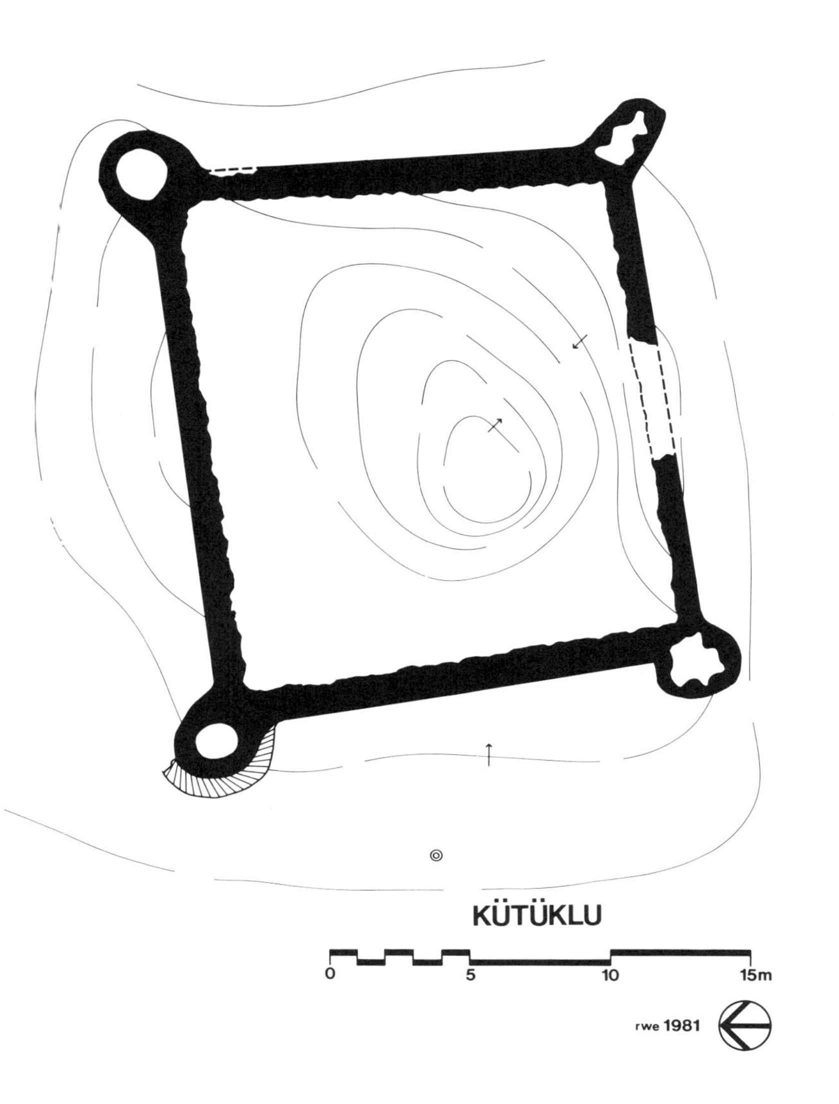
KÜTÜKLU
0
5
10
15m
rwe 1981

Fig. 49

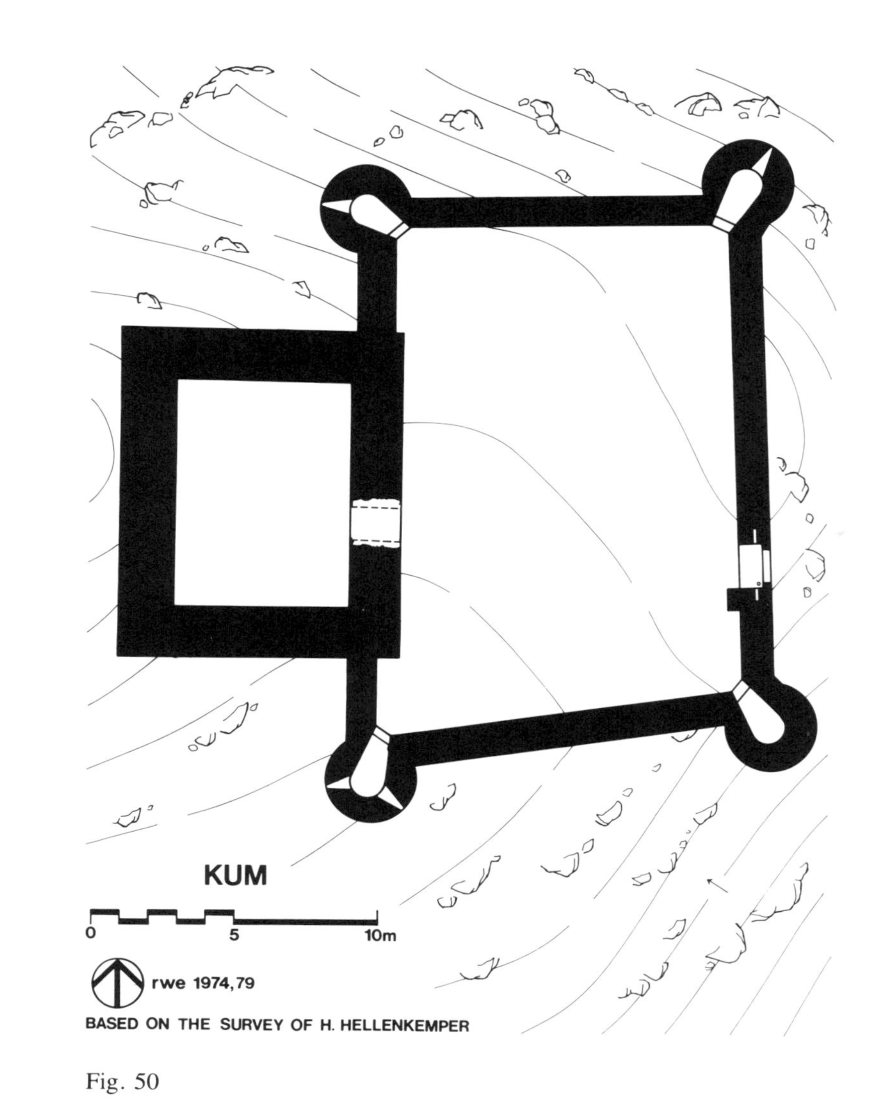
KUM
0
5
10m
rwe 1974,79
BASED ON THE SURVEY OF H. HELLENKEMPER

Fig. 50

wall still survive. The interior of the tower consists of a single circular room that does not appear to have had an interior facing. The north wall of the fort has lost all of its exterior facing stones. The best-preserved tower in the entire complex is at the northwest. This bastion has a talus-like extension at the base of the west side (pl. 132a).

A sufficient number of facing stones still survive on the northwest tower and the adjoining wall at the west to allow us to evaluate the masonry. The masonry and core are quite similar to the forts at Tumil and Yaka (pl. 132b). The exterior facing consists of a type IX that has a rusticated center and broad drafted margins. The core consists of fieldstones and some coursed rubble that are laid in neat parallel courses corresponding to the height of each level of the facing. On the interior of the northwest tower the facing stones consist of type IV. Because of the plan and masonry it seems quite possible that this fort is a Crusader construction.

[1]The plan of this hitherto unsurveyed site was undertaken in early June 1981. The contour lines are separated by intervals of 50 cm.

[2]The village of Kütüklu appears on the maps of *Cilicie* and *Mersin (2)*. The fort of Kütüklu is marked on the chart of the *Cilician Gates*. According to the rules of vowel harmony in modern Turkish, "Kütüklü" would be a more logical spelling. Here I follow the rendering of the modern maps: Kütüklu.

Kum

Kum Kalesi[1] is a small garrison fort that lies on the bank of the Ceyhan River only 2 km north of the neo-Hittite palace of Karatepe.[2] The site can be reached with relative ease by taking the road from Kadirli to Karatepe, a distance of about 24 km. The last 8 km of the road can be treacherous in wet weather. The fort sits near the top of a small limestone outcrop with a commanding view of the valley on the east bank. I could not determine if Kum has intervisibility with any other fortified site. This fort is located at the point where the major tributary of the Ceyhan becomes more navigable; this must also have impressed the earlier neo-Hittite settlers in the region. Kum Kalesi straddles an auxiliary trail that links fortresses at the south to Geben and Göksun (Coxon) in the north. Because of Kum's proximity to Karatepe, it has received some minor attention. In their survey of this district, Bossert and Alkım note that the fort is probably a Crusader construction.[3] J. Thomson also believes that it is a European fortification where "some pretty Frankish knight cultivated sugar cane and raised water buffalo."[4] The only architectural description of this site was published in 1976 by Hellenkemper.[5] My observations were made before and after his publication.

This fort consists simply of a castrum with four corner towers and an attached keep or donjon at the west (fig. 50; pl. 297a). The masonry of both units is quite interesting. Like Hotalan Kalesi, the exterior facing of the lower half of the castrum wall consists of larger and better cut stones than in the upper half. The lower half of the castrum has an oversized, very regular type IV. Occasionally, blocks of the type VI and type VII class are used. In the upper half, type III masonry is much more common. Types III and IV masonries are used for the interior facing. The rectangular keep is constructed with a combination of types V and VII as the exterior facing. In the upper courses the stones become noticeably smaller. For the most part, the corners do not have quoins. In general the interior masonry of the keep consists of types III and IV. The only smooth ashlar used in the construction of this fort is the frame around the east door in the castrum. The cores of both the castrum and donjon consist of fieldstones, some cut blocks, and an abundance of mortar. The mortar is not white in appearance but has a brownish or reddish tint. The masonry is cut from the typical gray and rust-colored limestone. The latter gets its peculiar color from the presence of certain ferrous elements. The stones of different color may come from two separate quarries.

The walls of the castrum, which rise to a height of 6 m, ascend the eastern flank of the outcrop. The only entrance into the castrum is at the east (pl. 297a).[6] The sill of this portal is elevated at least 2 m above the original ground level. This opening is secured by jambs and a crossbar bolt. There is a single post hole in the south corner to accommodate a swinging door. There is also an upper-level door opening into each tower. The small apsidal room of each corner tower has a vaulted ceiling, and three of the towers are opened by embrasured loopholes. On the exterior the loopholes do not have the elongated slit and stirrup base that are common in the Armenian military architecture of Cilicia. The towers cannot be entered from the thin wall walk that connects all four walls. The bases of rectangular merlons are still visible atop the walls and towers. The vaults of the tower rooms supported the floors of open-air shooting platforms. Two of the towers have latrine shoots. Joist holes on the interior of the curtain walls indicate that wooden build-

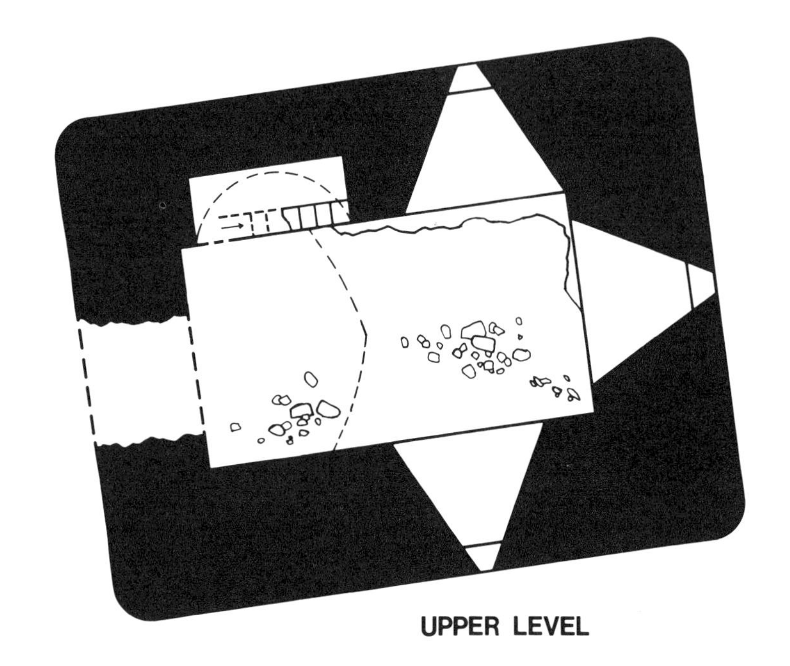

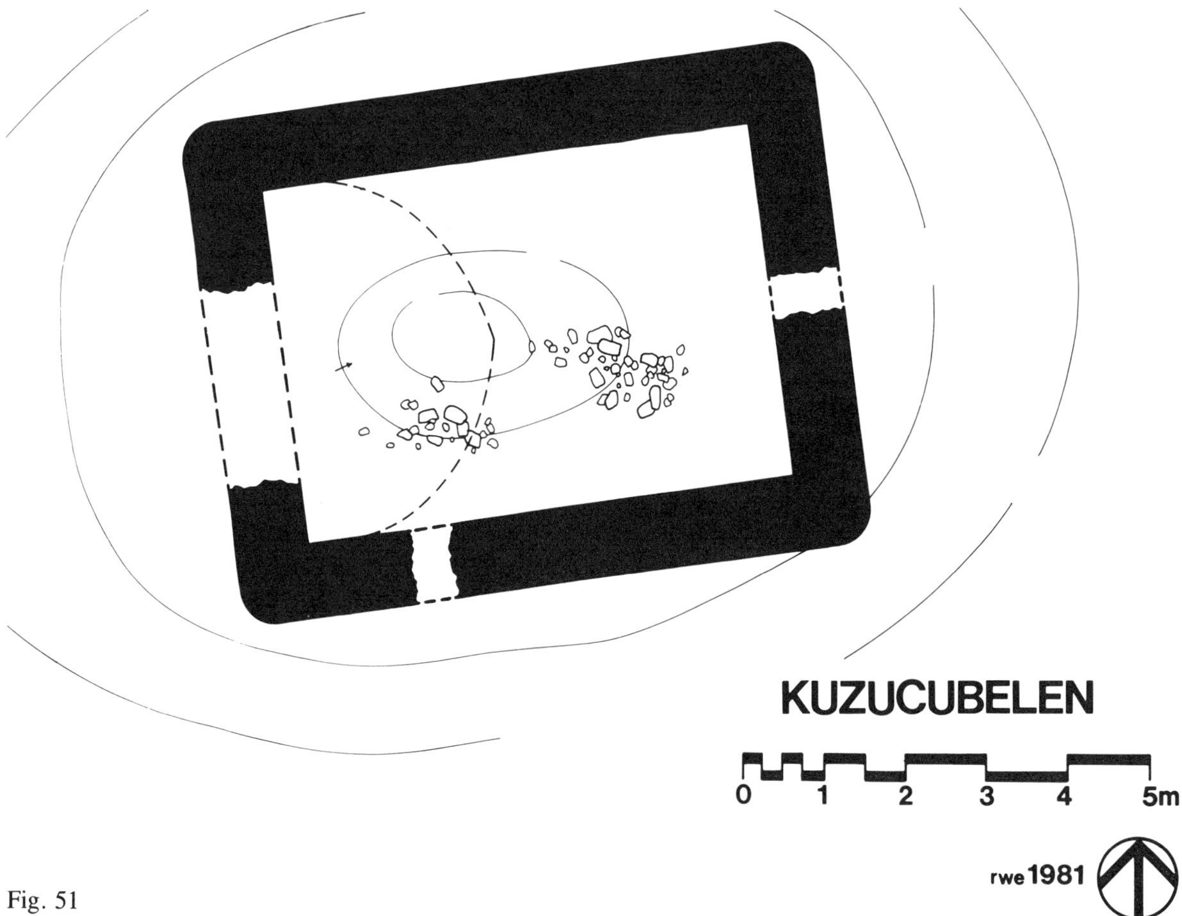

Fig. 51

ings were accommodated.

The donjon, which rises to almost 14 m in height, is securely anchored in the west wall of the castrum and appears to be contemporary. The first level of the donjon is a vaulted cistern.[7] This room is opened just below the springing of the vault in the east wall. Unfortunately the frame for this high opening has been broken away. The second level (not shown on the plan) was covered by a wooden roof, which in turn supported a fighting platform. This second level is opened by a large squareheaded window in the east wall. At the north end of the east wall there is a small embrasured opening. Square merlons surrounded the open-air fighting platform. Only the four merlons at the west were pierced in their centers by single arrow slits.

Hellenkemper believes that Kum Kalesi has a Frankish plan but is constructed by Armenians.[8] He also says it is similar in design to the Armenian constructions at Sinap (near Lampron) and the so-called Tarbas at Sis.[9] However, the Armenians *never* employ the open-air castrum design. The structure at Sinap is a rectangular house with a fully vaulted roof, and the keep-like structure at Sis was also covered by a roof. The fort at Kum was designed as an open-air enclosure for a garrison. It seems likely that Armenian stonecutters were employed at Kum, considering the presence of types V and VII masonries. The architects and masons of this site appear to be non-Armenian, like the plan. If Armenians were involved in the actual construction, then their peculiar architectural features, such as elongated arrow slits, would be present.

Unfortunately there are no inscriptions or references in the medieval texts that testify to the name or builder of this site. There is *no* evidence that the area around Karatepe was given by any Armenian king to a Crusader order. However, from the architectural remains it is possible that this fort is a Crusader site.[10]

[1] I briefly visited this site in April 1974 and June 1979. The contour lines on the plan are spaced at intervals of 50 cm.

[2] Kum Kalesi appears on the following maps: *Adana (2), Kozan, Marash*. See also the map published by Bossert and Alkım. For notes on the topography, see Haçtırın in the Catalogue; Schaffer, 92 f.

[3] Bossert and Alkım, pl. 103.

[4] Fedden and Thomson, 44.

[5] Hellenkemper, 131–34.

[6] Ibid., 132.

[7] I could not accurately determine the shape of the lower-level vault, and thus it does not appear on the plan.

[8] Hellenkemper, 134.

[9] At Sis the square building with corner towers is not the Tarbas; see Edwards, "Second Report," 134.

[10] Although the Byzantines use the castrum plan in association with the donjon, there is no evidence that they ever employed type VII masonry in Cilicia or elsewhere in Anatolia. There is the remote possibility that Armenian stonecutters helped the Greeks to build this fort in the 11th century.

Kuzucubelen

The small fortified watch post at the village of Kuzucubelen[1] not only guards the strategic road from the Mediterranean to Gediği and Arslanköy but has a commanding view into the river valley at the west.[2] At an altitude of about 450 m, this tiny fort rests near the edge of an elevated plateau. Today the areas immediately to the north, east, and south of the site are under cultivation. A number of springs and wells on the plateau provide water for irrigation. Kuzucubelen Kalesi is about 200 m west of the paved road that runs north to Fındıkpınar. Once past the junction to Fındıkpınar the road becomes steadily worse and passable only by jeep until it joins the paved road at Arslanköy. The village of Kuzucubelen, which flanks the road, is the district center for agricultural development. Today some of the villagers call the watch post Taş Kale. No historical names can be associated with this site.

This watch post is a simple two-storied rectangular structure (fig. 51). Except for a breach at the west, the walls stand to their original height (pl. 134a). The vaults over both levels have collapsed. A growing fissure in the northeast corner may soon cause the collapse of the east end. The masonry of the site is typical of Armenian architecture. On the exterior most of the facing consists of a type V. The southwest corner and sections of the east wall have a type VI masonry. A few of the blocks on the facade appear to be type VII masonry. Near the northwest corner of the fort some of the rusticated blocks from the collapsed west section have been recycled into a stone wall that surrounds an adjacent orchard (not shown on the plan). There is no evidence that a medieval circuit encompassed the natural platform to the west of the fort. The masonry on the interior of the watch post consists of types III and IV. The margins of the interior blocks are filled with an unusually large amount of mortar and rock chips. The upper-level vault has crude elongated stones that are laid radially.

Today entrance into the watch post is made through a gaping hole in the west wall (pl. 134a). Whether the actual entrance at the west was at the first (or lower) level cannot be determined. A second- (or upper-) level entrance would require a removable

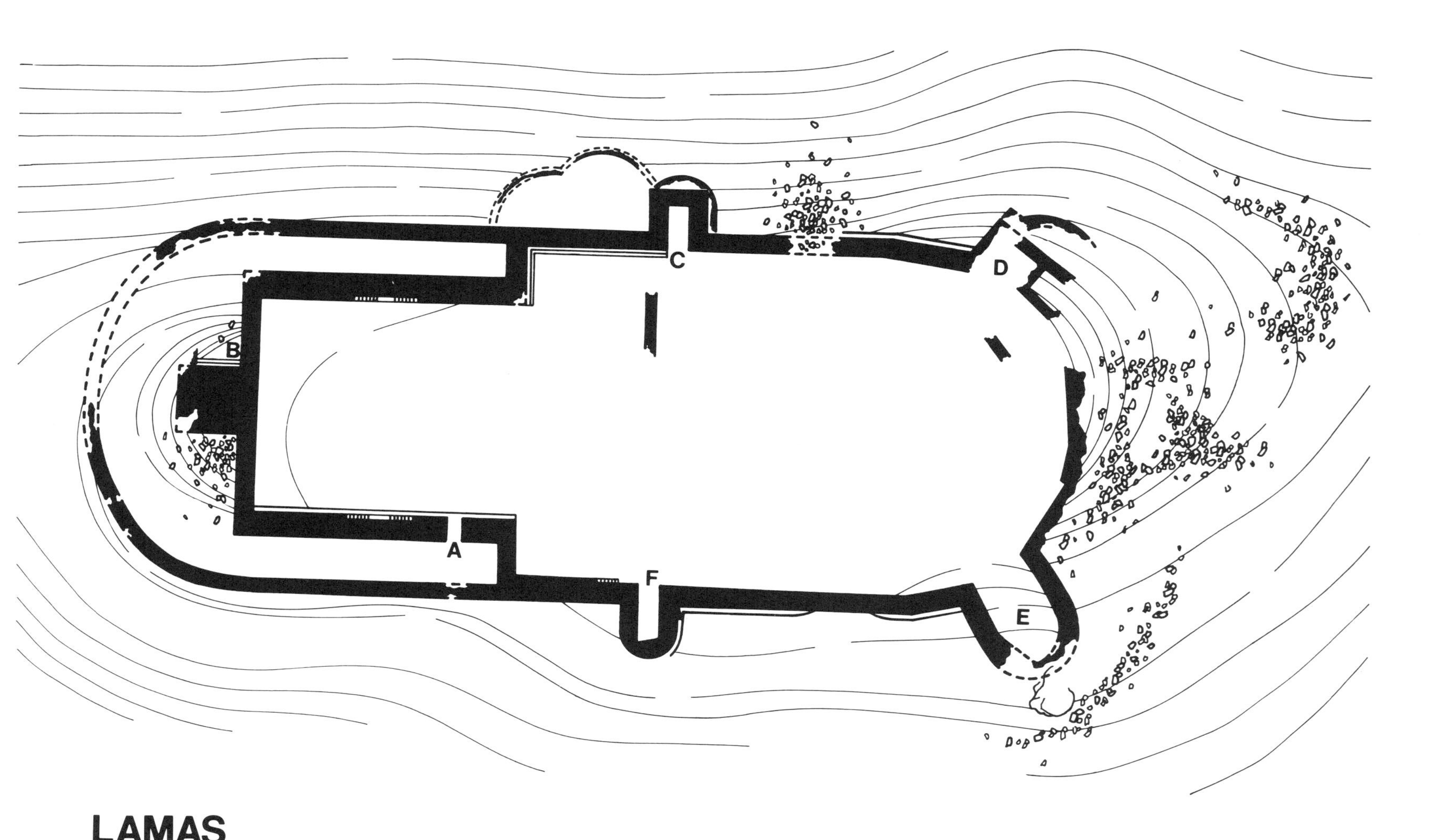

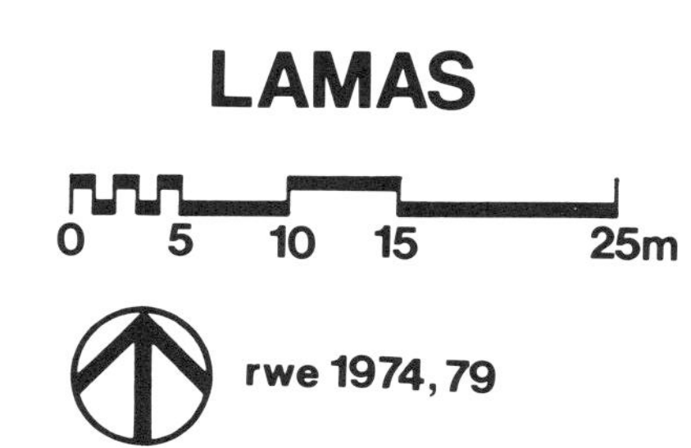

Fig. 52

wooden ladder. The only other openings in the lower level are narrow holes in the east and south walls (pl. 133b). The first level was once covered by a slightly pointed vault. The upper level is opened by three widely splayed embrasured loopholes and a niche (pl. 134b). What is unique about this second level is that its walls are almost twice as thick as the lower level. This expansion of the wall is carried on the springing and lower courses of the first-level vault. That such a construction was inherently weak is seen at the far east where both the lower-level vault and the wall directly below the upper-level embrasure have collapsed. The bases of the three embrasures are about one meter above the floor of the second level. The embrasures are *not* interconnected with any sort of pathway but simply open onto the room. The actual loopholes are rather small in comparison to the size of the embrasures. This design is not as common in Armenian architecture as the straight-sided casemate with embrasured loophole. Today the only apparent entrance into the upper level is at the northwest (pl. 134b), where what may be a narrow flight of corbeled steps leads to a vaulted niche. The vault over the upper level appears to have been depressed and slightly pointed.

[1]The plan of this hitherto unsurveyed site was completed in June 1981. The interval of distance between the contour lines is one meter.

[2]The village of Kuzucubelen appears on the maps of *Mersin (1)* and *Mersin (2)*. See also Heberdey and Wilhelm, 41, map; *Handbook*, 195.

Lamas

Lamas Kalesi[1] lies 300 m west of the small roadside village of Limonlu (pl. 135a). The fort is situated atop a small gently sloping hill near the mouth of the Lamas River. From this site there is a clear view of the Mediterranean Sea. There is no mention of the fortification in ancient sources.[2]

The earliest reference to this fort is in the chronicle of al-Maṣᶜūdī who notes that this Greek post guarded the Byzantine frontier; the Lamas River divided the Greek theme of Seleucia and Arab Cilicia in the tenth century. Much earlier, Greek and Moslem prisoners of war were periodically exchanged here as well as in Tarsus.[3] Since this fort is referred to as a site east of Silifke, it cannot be associated with the Lamos in Isauria.[4] The date of construction of Lamas Kalesi is difficult to determine. In the mid-nineteenth century Victor Langlois published a number of late antique Greek inscriptions in the general region of the fort.[5] Unfortunately they tell us nothing about the construction or the importance of the fortification.

The Armenians occupied this site in the twelfth century. One source reports that Emperor Manuel recaptured Lamas without a struggle in 1158.[6] When the Greek army left, the Armenians simply reoccupied the fort. Vasak, the brother of Bakuran, is listed as the lord of the fortress in the 1160s.[7] Bakuran was the lord of Çandır (Papeṙōn), which is in relatively close proximity, northeast of Lamas. On the Coronation List (below, Appendix 3) Halkam, the brother of Vasak and Bakuran, is the master of Lamas and three other coastal sites.[8] The great distance between Lamas and Anamur indicates that an Armenian baron could have widely dispersed holdings.

The small fort of Lamas Kalesi has a perfectly oblong circuit with two corner bastions at the east (D and E) and one salient near the center of each of the long sides (C and F) (fig. 52). All four towers are hollow; doors do not appear to separate these tower-chambers from the interior of the fort. At the west the oval circuit is actually an outer defense for a square inner wall (pl. 135b). Generally, the exterior walls do not stand to more than 2.5 m in height. The heaviest damage to the masonry is at the east and west ends of the fort. The evidence of internal structures is slight, consisting of wall fragments south of C and D and joist holes in the center of the west wall of the square inner circuit (opposite B; pl. 136a). These holes probably supported the roof of a wooden lean-to. In this same wall is a square niche(?) that was recently filled with stones (the niche does not appear on the plan). Other than the door A at the south there is no visible evidence for other entrances in this fort. The west end seems to have been some sort of strong point. At the north and south the square inner wall has traces of stairs that once provided quick access to the top. The only other extant stairway in the complex is near F. The function of the two shelves west of C and north of B is unknown. A curious, projecting socle is present between towers E and F. The square bastions C and D have the remains of rounded plinths. There is no evidence today that tower D functioned as a cistern.[9]

Most of the masonry of this fort is recycled from an adjacent necropolis and does not fit the paradigms for medieval Cilicia. In many areas the walls have a poured core, but a loose rubble core is also visible in places along the north wall. The inner and outer facing stones consist of square ashlar and occasionally cut fieldstones. There is no consistent attempt to taper the inner sides of the blocks so as to bind them more firmly with the core. Often in the oval circuit and in towers C through F, the largest of the ashlar blocks are confined to the lower courses; these are surmounted by either roughly squared stones or smaller, finely cut ashlar. In places long, massive blocks extend into the thickness of the core. These headers add stability to the construction. In the square inner circuit at the west end, the margins between the exterior facing stones are sealed with strips of mortar. The inner facing stones in this area consist of a mixture of ashlar and fieldstones that have a veneer of mortar (pl. 136a). The interior walls in the east half of the fort have regular courses of ashlar and show slight evidence of mortar in the interstices (pl. 135b). The exterior face of A was recently remortared when a stone irrigation channel was built at its base. The irregularity in the average size of stones in various parts of the fort and the abrupt change in masonry techniques indicate that at least two different building periods are involved as well as a number of different work crews. Since the foundations of most of the walls are buried in debris and soil, only excavation can determine the precise sequence of construction. From surface examination, it appears that salient B and most of the square inner circuit at the west were built at a different period than the rest of the fort. At present there is no evidence to link the rounded plinths of C and D to an earlier building period.

Lamas Kalesi shares some important and very unique characteristics with the only other Byzantine fortifications on the coast, Korykos. The land castle and part of the sea castle at Korykos are built entirely with recycled masonry. Headers are frequently used as well as square and round towers. One section of the circuit appears to be heavily fortified, as if it were a donjon or place of final retreat.

[1]The plan of Lamas was completed in winter 1974. After Hellenkemper published his survey of this site (1976), I returned in 1979 to remeasure the south tower (F). The contours on the plan are placed at intervals of 30 cm.

[2]Langlois (*Voyage*, 233–37) identifies the general site with the ancient μητρόπολις Λαμοτίδος.In the Armenian chronicles the fort's name is usually spelled Lamōs or Lamos (occasionally Lamas). Lāmis is the preferred rendering in Arabic chronicles, while Λαμούσια is known in Byzantine Greek. Travelers in the 19th century spelled the name Lamus. Lamo is found in some Italian Portulans. Lamas or Limonlu appears on the modern maps of the region: *Adana (1), Adana (2), Cilicie, Mersin (1), Mersin (2).* See: *Handbook*, 134, 188, 190; Cuinet, 65, 75–77; Rey, 350 f; Tomaschek, 66; K. Kretschmer, *Die italienischen Portolane aus Mittelalters* (Hildesheim, 1962), 599, 688.

[3]M. Canard, "Cilicia," *EI*², 36; Le Strange, *Caliphate*, 133; A. Philippson, *Das byzantinische Reich als geographische Erscheinung* (Leiden, 1939), 158; C. Huart, "Lamas-Sū," *EI*, 13; idem, "Lamas," *IA*, 10; Langlois, *Voyage*, 105; J. Tischler, *Kleinasiatische Hydronymie* (Wiesbaden, 1977), 87 f; W. Ruge, "Lamos," *RE*, 566; Schaffer, 17, 62 f; Ritter, 347–50; Honigmann, 44, 81 f; Ibn al-Fāqih in E. Brooks, "Arabic Lists of the Byzantine Themes," *JHS* 21 (1901), 75 note 2; S. Runciman, *The Emperor Romanus Lecapenus and His Reign* (Cambridge, 1929), 121, 136, 147 note 1; al-Masʿūdī, *Les prairies d'or*, ed. and trans. C. Barbier de Meynard and Pavet de Courteille, VIII (Paris, 1914), 224 f; al-Ṭabarī, *History*, XXXVIII, *The Return of the Caliphate to Baghdad*, trans. F. Rosenthal (Albany, 1985), 33 note 174, 172 note 835; Vasiliev, 125 note 4, 182 f, 193 f, 243, 254, 281; Canard and Grégoire, 406 f; Canard, *Sayf al Daula*, 68, 83 f; Canard, 282; Ibn Ḫurdāḏbih as translated in H. Gelzer, *Die Genesis der byzantinischen Themenverfassung* (Leipzig, 1899), 84; Silifke (note 4) in the Catalogue; above, Part I.7, note 45.

[4]Hellenkemper, 242 note 2. Alishan (*Sissouan*, 413 f) identified Lamas Kalesi with the Cilician Lamos. See also Ramsay, 380, 382, 455 f; Boase, 170.

[5]Langlois, *Inscriptions*, 36 f; *idem, Rapport*, 12. See also: Heberdey and Wilhelm, 47; Imhoof-Blumer, 465; idem, "Coin-Types of Some Kilikian Cities," *JHS* 18 (1898), 163.

[6]Cinnamus, 180. This site was certainly a Byzantine possession in 1137; see Michael Italikos, 251, esp. note 28.

[7]Smbat, 623; Christomanos, 147.

[8]Smbat, G. Dédéyan, 79 note 66.

[9]Hellenkemper, 241.

Lampron

One of the most famous castles in Armenian history is Lampron.[1] It was the seat of Hetʿumid power in the northwest corner of Armenian Cilicia. This fortress is situated on a pedestal of limestone that projects from the southern tip of the Bulgar Dağı. Lampron[2] is conveniently situated at the intersection of three Highland valleys and has a commanding view to the north and to the south (pl. 140b).[3] This writer knows of only one site, the small fortified estate house at Sinap, that has clear intervisibility with Lampron. The latter is also at a crossroad where trails lead north (via Sinap) to Gülek and across the Bulgar Dağı to Ulukışla. Of the two southern trails from Lampron, one meanders to the west, eventually reaching Çandır, and the other joins the major highway from Tarsus to Pozantı and Cappadocia.[4] The entire length of the Lampron (Namrun)–Tarsus highway (approximately 74 km) is paved, and a bus service (twice daily in 1979) shuttles passengers from the unbearable heat of Tarsus in the summer to the resort of Namrun. In 1969 one visitor noted that the settlement of Namrun at the base of the castle-outcrop had approximately a hundred homes.[5] Ten years later this writer saw over five hundred residences, some with heated swimming pools. Today

Namrun is the largest and one of the most fashionable resorts east of Alanya. As in medieval times, the modern settlement is supplied with fresh mountain water from one of the major tributaries that flow into the Tarsus Irmağı.

It is often remarked that Lampron's great strategic value lay in the guardianship of the Cilician Gates (above Gülek).[6] In 1973 I rode northeast by horse from Namrun to Gülek in two days. In the medieval period it would have taken the same time (if not longer) to move a mounted force to the Cilician Gates. Unless some intervening fortresses are found, it is likely that Gülek did not have intervisibility with Lampron. Thus it is quite possible that an enemy from the north could seize the Cilician Gates before the arrival of troops from Lampron.[7]

In the nineteenth and twentieth centuries many visitors commented on the beauty of Lampron, but only on two occasions have serious architectural studies been conducted. Robinson and Hughes published their study of Lampron Kalesi in 1969, and in the opinion of this writer it is one of the better architectural and historical surveys of an Armenian fort.[8] In 1976 Hellenkemper published a shorter analysis with some new perspectives.[9] It would be of no value to repeat these previous accounts. I will offer here a brief summary of the significant events in Lampron's history as well as a digest of the important architectural features that appear on the new plan of the fort (fig. 53). I will cite the previous studies only when they are relevant to my new observations.

In the third quarter of the eleventh century many Armenian *naxarars* in the regions north and east of Van fled the Seljuk invaders and settled in Cilicia. To one of these Armenian chieftains, Ōšin, the governor of Tarsus, Apłłarip, gave the fief of Lampron.[10] This appointment was confirmed by the Byzantine emperor, who also gave Ōšin the title of sebastos. Prior to this Armenian occupation, the Greeks appear to have had a fortified outpost on the Lampron outcrop.[11] Ōšin's son, Het'um, gave his name to the dynasty that was to reign in Lampron through the fourteenth century. The history of the eventual submission of the Het'umids to the other great Cilician family, the Rubenids, is summarized above (Part I.1). That the Het'umids maintained their independence against the Rubenids until 1201/2 is due to the impregnable nature of Lampron. In 1171, 1176, and 1182 the Rubenids failed to capture the fort by direct assault and long-term siege.[12] Het'um came to the coronation of Levon I (see below, Appendix 3) not as a vassal but as the autonomous lord of Lampron. King Levon finally used the subterfuge of marrying his niece to one of the Het'umid barons in order to draw the Het'umid knights to the wedding feast at Tarsus.[13] During the feast Levon's forces overran Lampron. Levon gave the castle to his mother for her summer residence.

Later, Lampron became the center of one of the serious revolts by an Armenian baron against a Cilician king. In the early 1230s the baron of Lampron, Constantine, made an alliance with the Sultan of Konya, Kayḳubād. In 1245 the former joined Kayḫusraw II and fought against King Het'um I; they held Tarsus until the baron of Papeṙōn (Çandır), the Constable Smbat, drove them out of Cilicia. Constantine was executed in 1250 for his high treason.[14] Lampron Kalesi is seldom referred to after the death of Constantine. In 1309/10 it functioned briefly as the prison for the Lusignan King of Cyprus, Henry II.[15] Sometime in the late fourteenth century a Mamluk garrison occupied the outcrop.[16] In the second half of the twelfth century the nearby monastery of Skevṙay was the home of St. Nersēs of Lampron and the largest library in medieval Cilicia. T'oros Roslin and other great miniature painters received frequent commissions from the court of Lampron. What survives today of this dynastic castle is only a faint testimony to its past greatness.

The outcrop on which Lampron Kalesi is built has an elongated, somewhat bulbous shape (fig. 53).The entire east flank is protected by sheer cliffs that vary in height from 30 to 40 m. The foundation of walls at the southeast end of the summit (H through I) was not part of a protective circuit but simply the surviving traces of buildings. At the extreme northwest end below outwork Q a dry moat has been scarped to sever the fortress from the rest of the outcrop (cf. Bodrum and Anavarza).[18] The moat (not fully visible on the plan) is somewhat trapezoidal in shape in that its north and south sides are splayed, so that the open west end is about 10 m wider than the east. The west flank of the outcrop from the edge of the moat to the southern tip consists of a series of narrow ascending terraces. For the most part the walls of the terraces consist of the natural vertical cliffs that vary in height from 3 to 10 m. Occasional breaks and clefts in the terraces allow for passage from a lower to a higher level (pls. 138a, 138b). Except for the area between E and F, the shelves on the tops of the terraces are usually inclined, giving the appearance of a series of natural taluses. When ascending the outcrop, one does not have the feeling that the terraces are neatly defined wards, but more a complex entrance system built on a scale far larger than those seen at Vahga and Çandır. The defenders of Lampron could always outflank an enemy simply by as-

Fig. 53

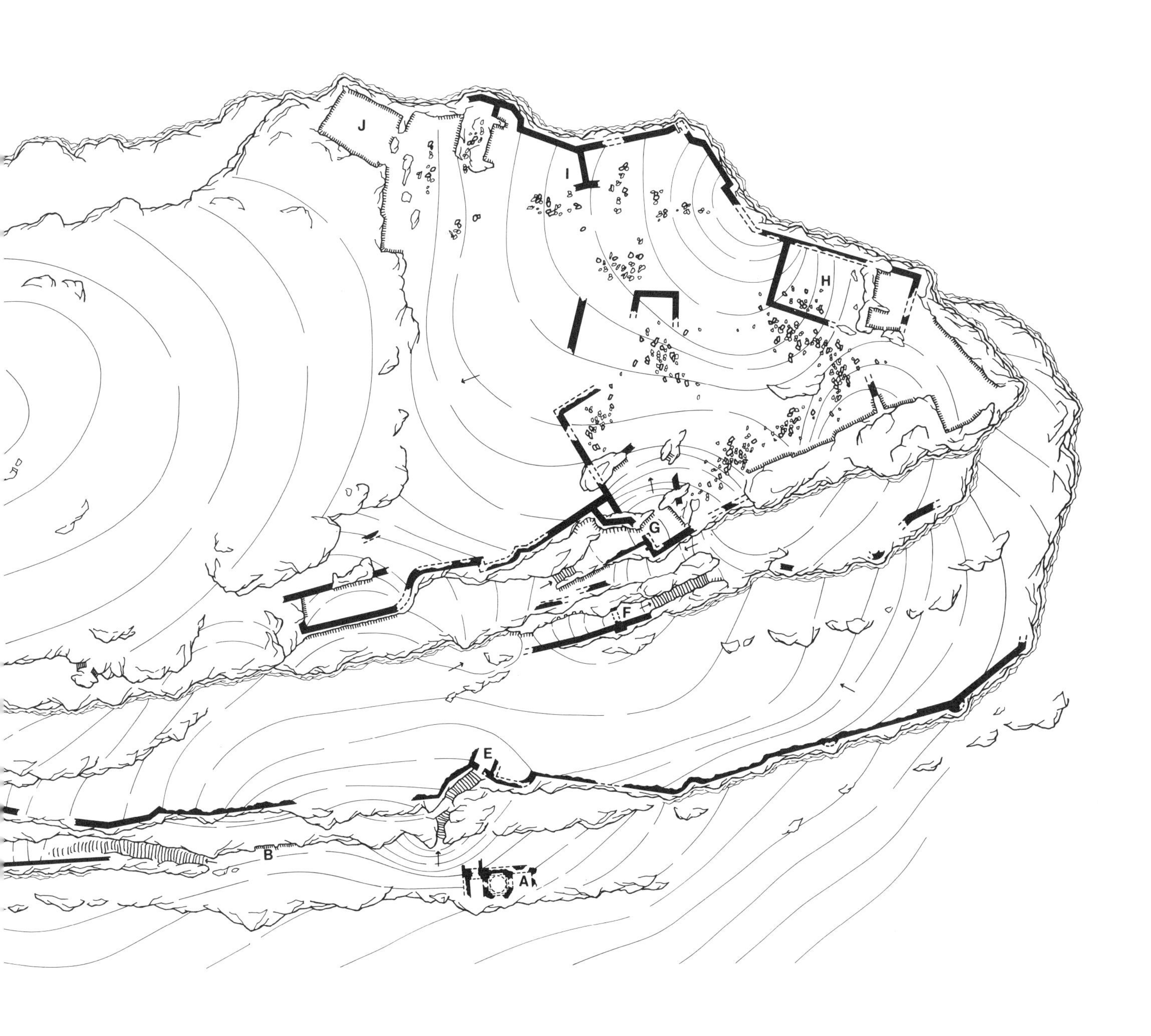

J
I
H
G
F
E
B
A

cending to a higher level. Unfortunately the masoned walls flanking the edges of the terraces have survived only to their foundations, and the extent to which they were lined with embrasured loopholes or battlements cannot be determined. The only evidence of such openings is at outwork Q. From Q to the southern tip of the outcrop is a distance of 420 m; at its widest point (from A to J) the distance across Lampron Kale is 150 m.

When approaching the outcrop from the relatively gentle slope at the southwest, the first medieval construction that comes into view is bathhouse A (pl. 137a). The bathhouse is essentially a rectangular structure divided into three chambers. The central and south chambers are roughly equal in size, while the north chamber appears slightly smaller.[19] A fourth room (or perhaps a porch) may be attached to the east face of the north chamber. Unfortunately the advanced state of decay and heavy undergrowth obscure many of the details. The masonry of A is quite poor in quality, generally consisting of types III and IV. On the interior the walls are stuccoed with a thick lime wash (pl. 137b). Today the central room is covered with the substantial fragments of a cupola. The cupola actually rests on four adjoining arches that crown the slightly recessed walls. The arches are formed by a single course of relatively smooth ashlar. Pendentives rise from the resulting spandrels to make the transition from the square bay to the circular roof. It seems likely that a west door opened this room. Jagged holes in the north and south walls of the central room may once have been portals. Fragments of a pendentive in the collapsed south room show that it too was covered with a cupola. In both the central and south chambers terracotta pipes are visible in the thickness of the walls. These pipes brought hot steam from the subterranean hypocaust. When I surveyed the bathhouse in 1973 the entrance to the hypocaust was blocked, but fortunately Robinson and Hughes measured the three chambers that make up the subterranean level.[20] These chambers were connected by tunnels, and their low flat roofs were supported by closely placed pillars. In 1966 a well immediately east of the bathhouse supplied the site with water. By 1979 this well had completely dried up, leaving the spring at point B (40 m north of bathhouse A) as the only source of water on the outcrop.

Although building A is constructed with rather crude masonry and is the only known bathhouse in a Cilician fortification, it is quite probable that the Armenians are responsible for its construction. In the tenth or eleventh century an almost identical bathhouse was constructed at the fortress of Amberd in Soviet Armenia.[21] It too has three rooms that comprise a long rectangular building; the square bay of the central room and one adjoining room are each covered by a cupola. The other adjoining room was a smaller chamber that held a tub. The hypocaust below has an identical design to one at Lampron. Although the domes over the bathhouse at Amberd are supported by monolithic squinch-joints and the ashlar facing stones at Amberd are far superior in quality to the masonry at Lampron, the near identity of the designs clearly links the latter to Armenian construction.

Just north of bathhouse A are two sets of stairs that lead to two separate entrances (C and E) into an upper terrace. Because of damage, it cannot be determined which was a postern and which was the principal entrance. Over the past century the coursed masonry on the outcrop has been removed and recycled into the constructions of Namrun village. This persistent plundering of the masonry has left today only scant traces of the circuit walls along the west terraces. Unfortunately the nineteenth-century travelers who found a large number of preserved structures on Lampron's outcrop did not properly document their discoveries.[22] The reader should be aware that it is now impossible to determine the sequence of construction or the location of the principal defenses from the fragmentary remains. The steps north of B lead along a narrow terrace-corridor until they reach the shattered remains of gate C.[23] The masonry both above and below the terrace corridor consists of various qualities of type IV. In some areas many of the blocks are more elongated than usual, and their margins have an unusually high number of rock chips. At gate C the approaching party can turn either south to perhaps the remains of another gate at point D or continue north along the shelf to the sloping platform that is situated below apartments L–P and above outwork Q.

Outwork Q essentially consists of an unvaulted square tower and a flanking wall on each side. Q follows the edge of the south wall of the moat and seems to be the result of at least two periods of construction and/or repair. Most of the facing consists of a very regular type III that occasionally has courses of stone like type IV. The circuit wall, which runs south from the square tower and survives only to its foundations, has no traces of loopholes. To the north of the tower, parts of the adjoining circuit stand to almost 2 m in height. The remains of at least five shattered embrasured loopholes are visible today. In the square tower there are two embrasured loopholes and what was probably a door at the south.

On the route to the south through gate D, the path

ascends to another flight of stairs at gate F, which is also the most immediate objective of those passing through gate E. The ascent to gate E from bathhouse A is much more steep than the path to gate C (pl. 136b). Today the masoned steps below have greatly deteriorated, but the revetment wall supporting those steps is still visible. The jambs of the gate are extant only at the foundation level. I saw no evidence of holes to accommodate a crossbar bolt. Immediately south of gate E the wall projects outward as if it was once the foundation of a tower. The masonry of the gate, like other parts of the surrounding terrace wall, is a well-coursed type IV that has a number of elongated blocks. About 2.5 m south of the jambs of E there is a shallow facet on the exterior facing. The point of the facet consists of quoins of relatively smooth ashlar. This use of ashlar with a well-executed type IV is reminiscent of Byzantine construction. The probability that this gate, as well as other sections of the west terrace wall, is from the Byzantine period of occupation is strengthened by the presence of a Greek inscription at gate E.[24] South of E there is evidence of small shallow salients along the circuit.

From E the entire length of the ascending gate-corridor F is visible (pl. 138a). The east flank of F consists simply of the natural vertical cliffs that have been scarped only at the lower level. The west flank is a wall of well-coursed type IV masonry with elongated blocks. Below and west of the masoned wall are what appear to be a series of shelves. Actually these are the steps that result from the quarrying of the limestone outcrop. Above the north half of the corridor, six square holes have been cut into the cliff. Four of the holes are especially deep (shown on the plan), and two are rather shallow. These holes probably supported a wooden canopy over the corridor. If a stone vault were constructed, we should expect to see a continuous shelf in the cliff; such a shelf would anchor the springing of the vault. Today there is no evidence of a tower at either the north or south end of corridor F. At the extreme north end of F there is evidence of a jamb scarped in the cliff (not shown on the plan). A series of small joist holes above and to the north of the jamb probably supported some sort of wooden building that either fronted the north door of F or was incorporated into it. Almost midway in corridor F are the fragments of what may be another jamb or perhaps the springing of an arch (pl. 138b). The southern third of the corridor has the remains of a series of partially scarped and partially masoned steps. The terrace immediately south of the steps is somewhat broad and shows traces of a circuit wall.

At the south end of corridor F a hairpin turn leads the approaching party to the final and most complex gate at G (pl. 139a). Preceding G are the faint traces of partially scarped steps. G is actually a bent entrance, most of which has been scarped out of the natural rock. The west wall and the northwest corner of G show the only traces of what is probably Armenian construction in the entrance. This Armenian contribution is slightly unusual for two reasons. The corner is not rounded, and the lower half of the wall consists not of type VII, like the upper half, but of a superbly dressed type VIII with extremely narrow margins. The type VII has extremely broad margins, and the rusticated face is thinner than usual. My examination of the core showed that the two types of masonry are from the same building period. Just why the different stones do occur and the reasons for a single corbel in the center of the northwest wall are unknown. The salient corner at the northwest would not present any disadvantages, since an enemy cannot effectively use a ram against it in the narrow corridor below, and since the party approaching the outer door of G is substantially above the salient corner. The foundation of a wall extending from the north face of G indicates that a narrow corridor preceded the door. Joist holes outside the gate on the scarped face at the east show that some sort of room fronted the outer door of G. Above this scarpment there is a curious combination of masoned stones that do not fit any of the nine masonry types for Cilicia (see above, Part I.3). The stones in the lower half consist of a smooth, relatively well-cut ashlar that is poorly joined in its courses by rock chips and mortar. The upper half is a mixture of crude fieldstones, mortar, and rough ashlar. I could not determine the number of building periods that are involved in the construction of this wall. In the outer door of G, only the scarped jamb at the east is visible today (pl. 139b). There appears to be provision for a crossbar bolt. If this bent entrance was once covered by a vault, there is no evidence of it now. Except for the west wall and the northwest corner, all the walls of G are scarped.

After passing through the east door of G, the approaching party enters the summit. The summit, which extends from M at the north to the scarped area south of H, is actually divided into three parts (fig. 53). The southern third consists of a sloping, somewhat undulating platform that is covered with a large complex of buildings. The largest part in the center is an elevated, almost triangular hill that today shows hardly any evidence of construction. At the north is a V-shaped terrace that has the rooms L–P at its apex; the sides of this terrace flank the triangular hill.

It appears that most of the southern third of the summit was covered with buildings of stone and wood. Southeast of G a vertical face has been carved in the rock for a length of 60 m. A horizontal chase and a myriad of square joist holes, which once supported adjoining wooden buildings, run along the length of the scarpment. The largest chamber in the fortress is the now collapsed rectangular structure H. There is extensive evidence of scarpment and masonry north and northwest of H. At point I the masoned walls stand to over 2 m in height (pl. 140a). The facing stones consist of a rather small, but well-executed, type IX masonry. Unlike the larger type IX used in apartments L–P, the core of the wall at point I is layered in a neat diagonal pattern at each course level.[25] North of I are the massive scarped walls of chamber J. Sections of this room measure over 3 m in height. Hundreds of tons of coursed rubble are strewn about the lower third of the summit. The original number of structures was obviously far greater than the few traces that are visible today. No remnant of a cistern has yet been found on the summit.

The only vaulted buildings on the summit are the apartments at the extreme north. The other evidence of construction at this end are the two huge scarped rooms at point K (pl. 140b). The walls of both chambers are 3.2 m in height. Today four of the apartments (M–P) survive with their slightly pointed vaults intact. The foundation of a fifth apartment (L) is still visible (pl. 141a). Springing stones on the west wall of N indicate that L was vaulted. The five pointed vaults of the apartments are covered by a single flat roof. Some enterprising Turks have placed two television antennae on the roof. Every year during Ramadan the chief *muhazin* of Namrun erects five tall wooden letters that spell ALLAH. Each letter is fitted with a series of light bulbs, and they are illuminated at night. Unfortunately the children steal the bulbs during the day. When I arrived at the third week of Ramadan, only the first two letters could be seen at night.

The masonry used throughout these apartments is a large, handsome, and very consistent type IX. The only unique feature of the masonry is that the core is double the normal thickness and consists of roughly coursed fieldstones placed in crude horizontal layers. This masonry is used for the interior and exterior facing. Type IX is never used as the exterior facing stone for the circuit wall of an Armenian fort. The elevation and inaccessibility of the summit here removes the apartments and the area of type IX from the possibility of attack.

There may have been a psychological reason for using the very expensive smooth ashlar. In Greater Armenia and in the rest of Cilicia type IX is the facing stone associated with churches. The smooth ashlar is extremely rare in the fortifications of Greater Armenia; the most famous example is in the city circuit of the Bagratid capital of Ani. These walls were constructed by King Smbat II in the late tenth century and were designed to display the wealth and power of the Bagratid dynasty.[26] Perhaps the Hetʿumids also wanted to make a statement about their newly established authority. In Cilicia Pedias this is exactly what Tʿoros I did when he captured Anavarza in the first decade of the twelfth century. There he constructed a very visible church of type IX masonry, and the dedicatory inscription on that building lists the names of his great ancestors. Unfortunately there are no inscriptions or textual references that tell us with certainty when the Hetʿumids built these apartments.[27] Perhaps it is more than coincidental that smooth ashlar (type IX) is used extensively at Çandır (Papeṙōn), another site that is associated with the Hetʿumids.[28] Like Lampron, the smooth ashlar appears at Çandır as a facing stone in areas of the castle that are not exposed to siege. It should be noted that there are resemblances between the Armenian mason's marks at Çandır and Lampron.[29] We do not have enough examples of Hetʿumid construction to conclude that the extensive use of type IX on the secular structures is a Hetʿumid trademark in Cilicia. Just a few kilometers north of Lampron and *certainly* within the confines of Lampron's fief is Sinap Kalesi; here the builders used type VII extensively as an exterior facing stone.

North of the few surviving foundation stones of room L there is the trapezoidal chamber M (pl. 141a). The only door opening into this room was in the now collapsed south wall. The only windw is a high-placed shaft in the west wall. There is evidence of a single hatch in the south half of the vault. A curious step (podium) in the north wall has no apparent function. On the exterior the rounded north face is decorated with three heraldic symbols. I could not get close enough to make out the details of their designs. Fortunately Robinson and Hughes made sketches of them.[30] The only other evidence of a heraldic shield on an Armenian fort in Cilicia is at Fındıklı.

There are three doors with jambs opening into the centrally placed room N. The largest of the three portals is in the west wall. This door is flanked at the north by a high-placed porthole-window. This is the only window in N, and its function is a mystery.[31] Like M and O, the vault over N has a single hatch. The south end of the vault is broken at the point where it rises from the scarped south wall. In comparing the cross section of the vaults over M and N, two features become apparent. The ashlar in the vault over M is 50 percent smaller than the stones in the covering over N. Also, the cores in the walls of N, O, and P are not as neatly layered, nor do they have the abundance of mortar when compared to M.[32] It is quite possible that the vault over N extended south over the flat rock mass that forms the south wall of room N. The scarpment atop the rock, which once accommodated the springing, seems to extend for at least 3 m. Except for the collapsed south end of the vault and a partially broken jamb in the north door, room N is in an excellent state of preservation (pl. 141b).

The north portal from N to O has pivot holes that once supported swinging doors. The small trapezoidal room O has three important features (pl. 142a). Its only window in the north wall is a superbly cut embrasured loophole with casemate. Like a similar opening in the north wall of P, the embrasure is set askew in the casemate. While the masonry of the walls is identical to N and P, the fine ashlar was not carried into the vault. Unlike M, N, and P, the vault of O is built with crude rusticated blocks, fieldstones, and an abundance of mortar. It is quite possible that the original vault over O collapsed and that the present covering is the result of a reconstruction. The third significant element of room O are the well-preserved jambs of its east door (pl. 142b). Here the rather thin jambs are carved as delicate extensions from the final series of blocks that form the door. This identical corbeling technique is used at the other Het῾umid site of Çandır. This is the normal Armenian method for constructing jambs, except that at other sites they are substantially thicker and the arch above them is usually depressed. The delicate, almost pointed arch above the jambs at Lampron and Çandır may be a Het῾umid characteristic. The same technique is used on most of the doors of the apartments at Lampron Kalesi.

P is an elongated apsidal hall in an excellent state of preservation. Only at the south end have some of the exterior facing stones fallen away from the door and the embrasured window above it (pl. 140b). Also, the east half of the south wall seems to have settled slightly over the years and has separated from the vault (pl. 143b). The south half of the west wall of P has a foundation that is scarped from the natural rock. The scarped section is almost 2 m in height. The four windows in the west wall are designed only to let in light. There are holes in the jambs of the windows to anchor iron crossbars. In keeping with Armenian traditions, the vaulted casemates of the windows are slightly pointed and segmented into regular voussoirs of fine ashlar, while the lintels over the jambs are monolithic blocks (pl. 142c). In two of the window-casemates, there are fragments of stone benches. The most interesting feature in P is the construction of the vault at the rounded north end. It is built exactly like the apsidal vaults in Armenian chapels with stones that radiate around a thin oblong core (pl. 143a).[33] The only modern contribution to P is the abundance of spray-painted graffiti on the interior of the apartments.

It is clear that the masoned walls of the fortress represent many periods of construction, with the principal buildings appearing to be Byzantine and Armenian. Because of damage to this unexcavated site and a lack of any historical references to construction, it is impossible to date the architecture accurately.[34]

[1] The plan of this site was begun in spring 1973 and finished during my second visit in summer 1979. I intentionally did not consult any of the published plans of Lampron during my survey in order to see this site in an unbiased light. Due to the limitations of time, many of the contour lines on the plan are estimated. Generally, the contours are placed at intervals of 40 cm. Not all of the scarped surfaces on the fortress-outcrop are indicated.

[2] In Armenian chronicles this site is referred to as Lambrōn, Lambron, or Lambrun. The Arabs altered the spelling with the designation of Tāmrūn. On a French map of the 13th century it is listed as Les Embruns (see Robinson and Hughes, 183 note 2).

On 19th- and 20th-century maps this site is usually listed as Namrun, Namroun, Nemrun, or Lampron. Refer to the following modern maps: *Adana (2), Cilician Gates, Cilicie, Ulukişla.*

[3] In the early 1960s the Turkish government attempted to rename the site Çamlıyayla (or Cumliyaila). The locals still refer to it as Namrun and the castle as Lampron Kalesi. One recent Turkish map ("Cilicia," published by Silifke Turizm ve Tanitma Derneği [Ankara, 1966]) refers to the site as "Namrun (Lampron)." See also Tēr Łazarean, 15 f.

[4] Chauvet and Isambert, 772.

[5] Hellenkemper, 228; cf. Cuinet, 48.

[6] Robinson and Hughes, 183; Boase, 170; Alishan, *Sissouan,* 80 ff; Langlois, "Lampron," 119–22; Ēp῾rikean, II, 85–89; Sümer, 22, 76; *Handbook,* 238, 245; Schaffer, 56 f.

[7] The author of the 10th-century treatise on Skirmishing would probably agree with this assessment; see Dennis, 158, 182 ff.

[8] Robinson and Hughes, 183–207. A good historical survey of the site can be found in Yovhannēsean, 67–79.

[9] Hellenkemper, 228–36.

[10] Der Nersessian, 633; "Lambron," *Haykakan,* 479.

[11] Alishan, *Sissouan,* 73.

[12] Vahram of Edessa, 509 f; Smbat, G. Dédéyan, 54, 56.

[13] Alishan, *Sissouan,* 85; Smbat, G. Dédéyan, 50 note 20; N. Akinean, "Het῾um Hełi, Tēr Lambroni," *HA* 59 (1955), 397 ff. A genealogy of the lords of Lampron can be found in Toumanoff (276 ff) and Christomanos (146).

[14] Kirakos, M. Brosset, 142 f; Langlois, *Cartulaire,* 214; Simon of St. Quentin, *Histoire des Tartares,* ed. and notes J. Richard (Paris, 1965), 70 note 5, 71; Vincent de Beauvais, *Speculum Historiale* (1624; rpr. Graz, 1965), 1282.

[15] *RHC, DocArm,* II, 937; Mas Latrie, I, 325. Prior to this event the fortress was returned to royal control; see Smbat, G. Dédéyan, 112.

[16] In a colophon there is a brief mention of the sacking of

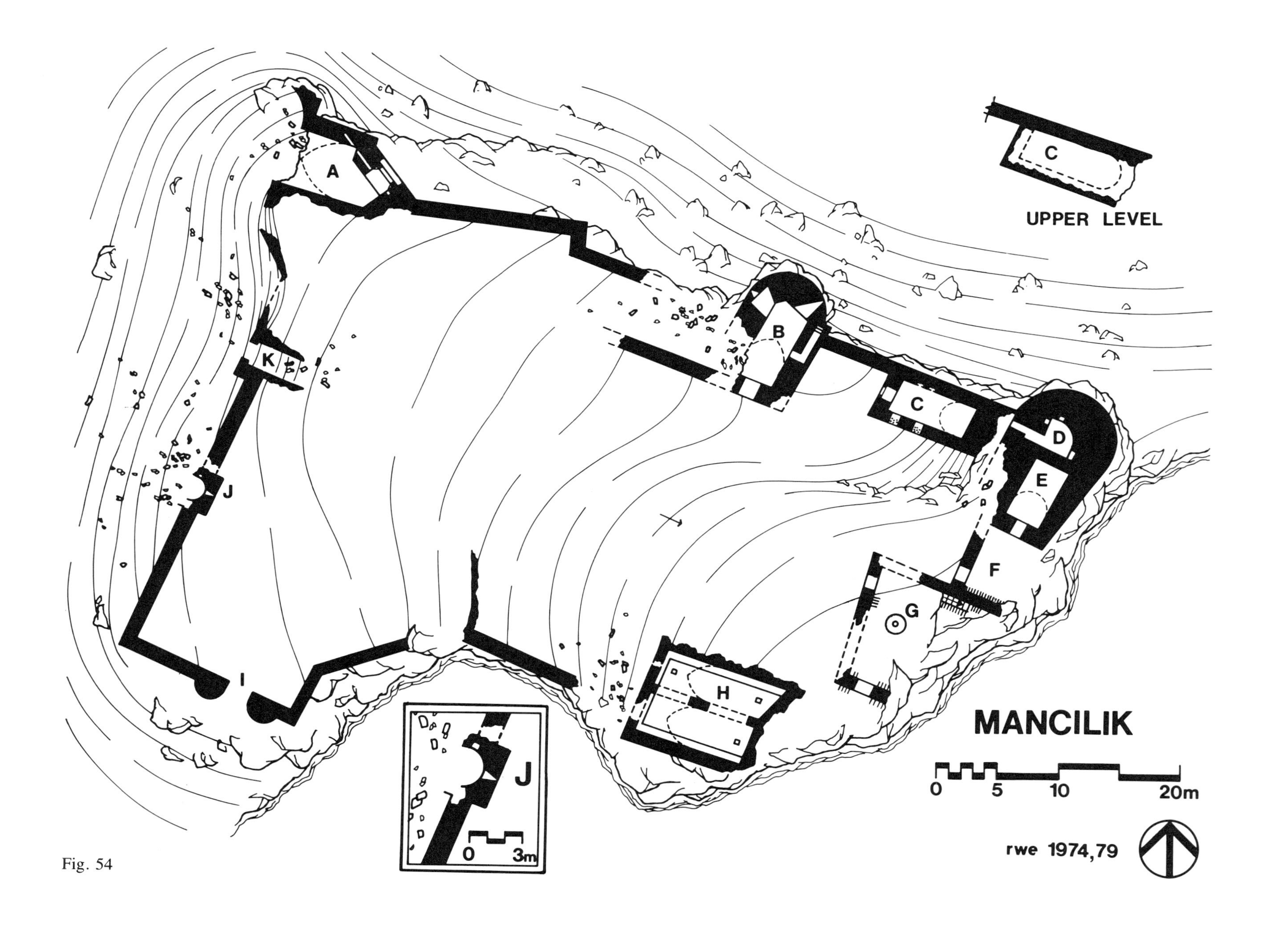

C
UPPER LEVEL
A
K
J
I
B
C
D
E
F
G
H
J
0
3m
MANCILIK
0
5
10
20m
rwe 1974,79

Fig. 54

Lampron in 1376; see Sanjian, 100. The Mamluks refer to this site as Tamrūn; see Gaudefroy-Demombynes, 102.

[17]Robinson and Hughes, 188.

[18]John Thomson (Fedden and Thomson, 38) claims the moat is of Byzantine origin, but there is no evidence to confirm this.

[19]Robinson and Hughes (200, fig. 8) show the north chamber as being substantially larger than the central room. Because of debris and extensive damage, my measurements of this chamber may be inaccurate.

[20]Ibid., 200 f.

[21]*Amberd,* Documenti di Architettura Armena, V (Milan, 1978), 9 f, 28 f, 31, 45. This plan also has parallels in Hellenistic and early Roman bath construction; see A. Farrington, "Roman Imperial Lycian Bath Buildings," *Araştırma Sonuçları Toplantısı,* II (Izmir, 1984), 120–22.

[22]Alishan, *Sissouan,* 89 ff; Langlois, *Voyage,* 360 ff; Langlois, "Lampron," 119 ff.

[23]The elaborate gate and flanking tower that Robinson and Hughes (193) described in 1966 at point C had vanished by the time I conducted my survey in 1973 and again in 1979.

[24]Because of the thick shrubbery and hanging vines I was not able to locate the inscription. Fortunately it is transcribed by Robinson and Hughes (192).

It is impossible to determine if the poorer quality types III and IV masonries of the west terraces are Byzantine in origin. The square tower of Q may be Byzantine or even belong to the Mamluk period. Except for gate G, the lower terrace walls do not appear to be Armenian in origin (cf. Hellenkemper, 235).

[25]Cf. the oratory of the church of T‘oros I at Anavarza; see Edwards, "Second Report," 128 f.

[26]Berkian, 110 ff.

[27]Robinson and Hughes (202) speculate on the dates of the apartments. See my comments on their conclusions above, Part I.6.

[28]See also Kız (near Gösne) in Edwards, "First Report," 172 f and "Second Report," 144 f.

[29]Robinson and Hughes, 203–5. Appendix 1 below contains a selection of mason's marks from Çandır as well as a few from Lampron that are not included in the list of Robinson and Hughes.

[30]Robinson and Hughes, 198.

[31]Ibid., 198 f; E. Rey, *Les colonies franques de Syrie aux XII*^{me} *et XIII*^{me} *siècles* (Paris, 1883), 43.

[32]Cf. Hellenkemper, 233–35.

[33]Edwards, "Second Report," 130.

[34]Hellenkemper, 235.

Mancılık

Mancılık Kalesi,[1] which stands on the west flank of the Nur Dağları (the Anti-Taurus Mountains), is the guardian of the junction of two important routes to the east.[2] At an altitude of 700 m, this fortification surrounds a sloping outcrop that is flanked by streams. Just northeast of the fort, near an artesian well, a peach orchard was cut out of the wilderness. With respect to vegetation, this is one of the most densely forested sites in all Cilicia. Entrance into the fort can be made only with difficulty and from the north. The fort can be located with a guide by hiking east from the village of Rabat (near Dörtyol) for about an hour along the winding course of an ancient stream bed.

There is no solid evidence to associate Mancılık Kalesi with one of the historical sites mentioned in the medieval chronicles. A fragmentary dedicatory inscription in Armenian over gate A tells us that at least the northwest corner of the upper bailey was completed in 1290.[3] Hellenkemper's identification of this site with the Armenian castle of Nłir (Neghir) is uncertain.[4] He also follows Cahen's suggestion and associates the *castrum regis nigrum* in the text of Wilbrand von Oldenburg[5] with Mancılık. According to the text, this fort is located between La Portella and Canamella. The former is north of Iskenderun, while the latter may be identifiable with Hieropolis/Castabala (Bodrum) or Ḥiṣn at-Tīnāt.[6] The problem arises in that Mancılık is not "black" but is actually built with a yellowish limestone. The *castrum regis nigrum* can possibly be identified with Toprak Kalesi, which is built entirely with black basalt and stands on the road from Iskenderun to Bodrum. Even if the identification of Canamella with Bodrum is questionable, Toprak would still be a conspicuous sight when Wilbrand journeyed to Misis (Mamistere/Mopsuestia), his next destination. Because of its location in the mountains, Mancılık is *not* clearly visible from the road, while Toprak, which Wilbrand also calls Thila, is a low-lying guardian of the road to the Amanus from Misis. The only description of Mancılık was published in 1976 by Hellenkemper.[7]

The fort at Mancılık is actually divided into an upper east bailey (fig. 54) and a lower west bailey. Because of shrubbery and landslides, the west bailey was impossible to survey; it does not appear to have complex defenses like the east bailey. The latter is surrounded by a full circuit, except at the east and southeast where walls are unnecessary. Numerous buildings hidden in the thick undergrowth in the center of the east bailey were impossible to survey.

The use of masonry in this Armenian fort is fairly consistent. In general the circuit walls have an exterior facing of types V and VII (pl. 145b). The interior faces of the circuit and the internal buildings consist of large, finely cut types IV, V, and VIII masonries (pl. 144b). The doors and windows are usually constructed with type IX; around gate A the type IX has a bossed face.

The entrance A, which actually gives access to the west bailey, is a tripartite gate with a slot machicolation (pl. 145a). The outer arch of the machicolation has collapsed today, leaving only the springing stones in situ. Below the depressed arch on the inner side of the jambs are pivot housings with holes to accommodate a double door (pl. 144a). The crossbar bolt in this door is unique in that it slides through the north wall of the gatehouse in such a way that part of the wooden bolt is exposed in the angle of the wall. Perhaps this was made to facilitate the replacement of the bolt.[8]

About 35 m east of gatehouse A is tower B. The lower level of this structure is relatively well preserved, except for the collapse of part of the west wall and the floor; the latter is constantly probed by treasure hunters. The chamber is covered by a pointed vault and has four openings. At the south is a straight-sided door that does not have the traditional arched cover of well-fitted voussoirs but simple squared stones placed radially. The interstices of these stones are filled with rock chips and mortar. A similar type of half-round arch is used in the lower-level door of C. In the north apsidal wall of B's lower level is a simple embrasured loophole with a stirrup base. To the west is an asymmetrical embrasured opening with a casemate. In the east wall there is a narrow, vaulted, elbow-shaped passage that leads to a machicolation/toilet; through this hole there is a clear view to the exterior of the circuit. The upper level of tower B (not shown on the plan) has completely collapsed; only the sill of a straight-sided window is present at the north.

Farther to the east is building C which has a barrel-vaulted lower level.[9] In the south wall of this room are two blocked windows, barely visible above ground level (pl. 145b). A single door at the west, which is almost completely buried, gives access into the room (pl. 144b). The lintel, which covers the jambs of the portal, is bored with a single pivot hole to accommodate a swinging door. The interior of the room is built with a large type VIII masonry. Some stones have been pulled out of the north wall, and it is now impossible to determine if this wall once had niches.

The point of junction between room C and tower D–E cannot be determined, but it appears that the upper level of C led into room D. The ground level around tower D–E is much higher than C. Entrance into D is gained by a single narrow passage. The interior of chamber D has no openings other than three small niches. The adjoining chamber E to the south is barrel-vaulted with a single entrance. The upper level of D–E has a few shattered fragments of battlements. From the abrupt changes in the style of the exterior facing, it appears that D–E has undergone at least two periods of construction.

To the south the rooms F and G have badly deteriorated (pl. 297b). Most of the lower part of the south wall of F is scarped. The door in the west wall is covered by a monolithic arch that has keyed springing stones. Adjacent at the south is a scarped stairway in the north wall of G. In the center of G is a subterranean cistern (unmeasured). The type of covering over F and G is unknown; considering that there are no indications of internal supports or vaults, a truss roof seems likely.

At the southwest is building H. This hall, which has a partially scarped foundation, is divided in the center by an arcade of two slightly pointed arches (pls. 146a, 146b). Both halves of this structure are covered by pointed vaults; the north vault is opened by two hatches, and the south vault by one. On the interior of the north and south walls of H there are tall, gently sloping shelves that are made of a combination of scarped rock and cut stones. The two breaches in the west wall of H are highly placed and may once have been the openings for small doors. Because of these and the remains of stucco on the walls, it seems likely that building H is a cistern.

Approximately 35 m west of building H is gate I. This opening is flanked by two solid salients and gives access into the west bailey. To the northwest are the remains of chapel J.[10] In the wall between I and K, the frames of several windows are present (not shown on the plan). A third gate at K is limited by jambs. It appears that the now collapsed vault over K was equipped with a machicolation. The remains of K now stand just over 2 m above ground level. This gate provided a second means of access between the east and west baileys.

[1] I first visited the site of Mancılık in June 1974. The plan was completed during my second visit in August 1979. The contour lines on the plan are approximated at intervals of 50 cm. Because of injuries sustained by one of my cosurveyors in 1979, we were forced to leave the site prematurely. The exact size and position of the circuit wall from H through K is approximated.

[2] This site appears with slightly different spellings on the following maps: *Adana (1), Adana (2), Cilicie*. See Janke, 34.

[3] Heberdey and Wilhelm, 22; Hellenkemper, 105. Unfortunately I was prevented by circumstances from making a new transcription. The recent translation in Hellenkemper reads:

> Diese Burg wurde vollendet im Jahre
> der Armenier 739 [A.D. 1290] während des Katholikats des Herrn
> Stepanos und während der Regierung der Armenier durch Hethum
> Christus Heiland bist der Welt . . .
> würdige mich des Gedenkens in dem schrecklichen [Gericht]
> dem Konstantin, der Herr ist: in Ayaş
> . . . .
> . . . .

[4] Hellenkemper, 107; cf. Alishan, *Sissouan*, 66, 493. Boase (158, 175) believes Nłir to be closer to Partzerpert (Barjrberd). See also Cahen, 148 f. It is likely that Nłir (the Arabic Ḳalʿat Nuḳayr) is located somewhere in the Amanus Mts.; cf. Sanjian, 77, 79.

[5] Wilbrand von Oldenburg, 17.

[6] Ibid., note 98; Cahen, 149 f.

[7]Hellenkemper, 104–8.

[8]Cf. Winter's comments on this problem in Greek military architecture: F. Winter, *Greek Fortifications* (Toronto, 1971), 258–65.

[9]The possibility that the upper level of C functioned as a chapel is discussed in Edwards, "First Report," 167.

[10]Ibid.

Mansurlu

Mansurlu Kalesi[1] is near the northern border of Armenian Cilicia and guards the road that links Kayseri and Sis.[2] The other fortifications on this route are Alafakılar, Bostan, and Bayremker. Mansurlu Kalesi rests on a gently sloping hill near the east flank of the road (pl. 147a). Directly to the southwest of the fort (pl. 147b) there is the village of Mansurlu (often called Küçük Mansurlu); to the northwest is a modern chromium mine. A second road, which leads to Maran (approximately 30 km) and Feke (approximately 50 km), runs in a southeasterly direction from the base of the fortified outcrop. Below this road in the adjoining valley is a second village (often called Büyük Mansurlu). No inscriptions or historical references can be associated with Mansurlu Kalesi. I can find no evidence that this site has intervisibility with any other fortification.

It is most unfortunate that almost nothing remains of this fort today. It was reported that about ten years ago the locals plundered its stones to construct houses and barns. Only small sections of masoned walls still stand in situ (pl. 147b). What is apparent is that the circuit walls of this fort form an almost perfect square. The apex of the hill, in the center of the square, shows no signs today of having been covered by a building. This very simple castrum held a small garrison.

Like Bostan Köy, the masonry of Mansurlu Kalesi consists of crude pieces of phyllite. But in the walls here, there is an inner and outer facing of stones with a poured core. What survives of the facing resembles type IV. Considering the regularity of the plan, it is possible that the fort is a Byzantine construction.

[1]I visited the site of Mansurlu in July 1981. Because of the almost total destruction of the fort, no plan was undertaken.

[2]This site appears on the maps of *Cilicie* and *Everek*. On the latter it is referred to as "Mansurlar." See *Handbook*, 324.

Maran

In the Taurus Mountains between Saimbeyli (Hačən) and Kozan (Sis) a narrow gravel road winds its way west from Feke (Vahga) for about 20 km until it reaches the small fort[1] and hamlet of Maran.[2] This village, which is perched on a slender terrace of rock, has no orchards or ploughed fields in the immediate vicinity, but a yearround spring of the most delicious water does emerge from a cliff below the road. The fortified site, which has no recorded history, once guarded the road that connects Vahga with Mansurlu as well as a secondary route to Bostan. At an altitude of 1,640 m, Maran Kalesi has a commanding view of the entire region (pl. 149b) and clear intervisibility with the great baronial castle of Vahga.

From the village a small, poorly defined trail runs to the northwest for about one kilometer before reaching Maran Kalesi. When approaching this garrison fort from the east, the visible evidence of construction from the ground level is slight. Aside from carefully blocking all potential approaches and integrating the short extensions of circuit walls into the sinuosities of the rock, the engineers tried to camouflage this most visible side of the fort. Because of the more gentle topography at the west, the easiest approach into the fort is through gate A. The west flank has most of the defenses.

Not only the design but also the masonry of this fort is typically Armenian. The doors and windows are constructed with type IX, which occasionally has a boss on the exterior face (pl. 148a). The inner and outer facings for the circuit walls and the facings of most of the buildings are either types V or VII or a combination of both.

The summit of the outcrop defines the peculiar shape of the single slender bailey (fig. 55). Maran Kalesi is opened by a gate in the west flank and one at the east. The best preserved of the two gates is A at the west (pl. 148a). This simple entrance has a square inner chamber and exterior jambs. On the interior side the slightly pointed vault of the former is constructed with three courses of stone and twelve voussoirs in each course; the two central voussoirs serve as slightly oversized keystones. Two of the voussoirs on each side of this vault are segmented. The square chamber has a length of 2.3 m and a width of 1.65 m. From the walls the jambs extend inward 15 cm on each side and have a width of 48 cm. The arch over the jambs is depressed and consists of eight joggle-jointed voussoirs and a single keystone (pl. 148b). The two voussoirs, which serve as the springing stones, have a torus-

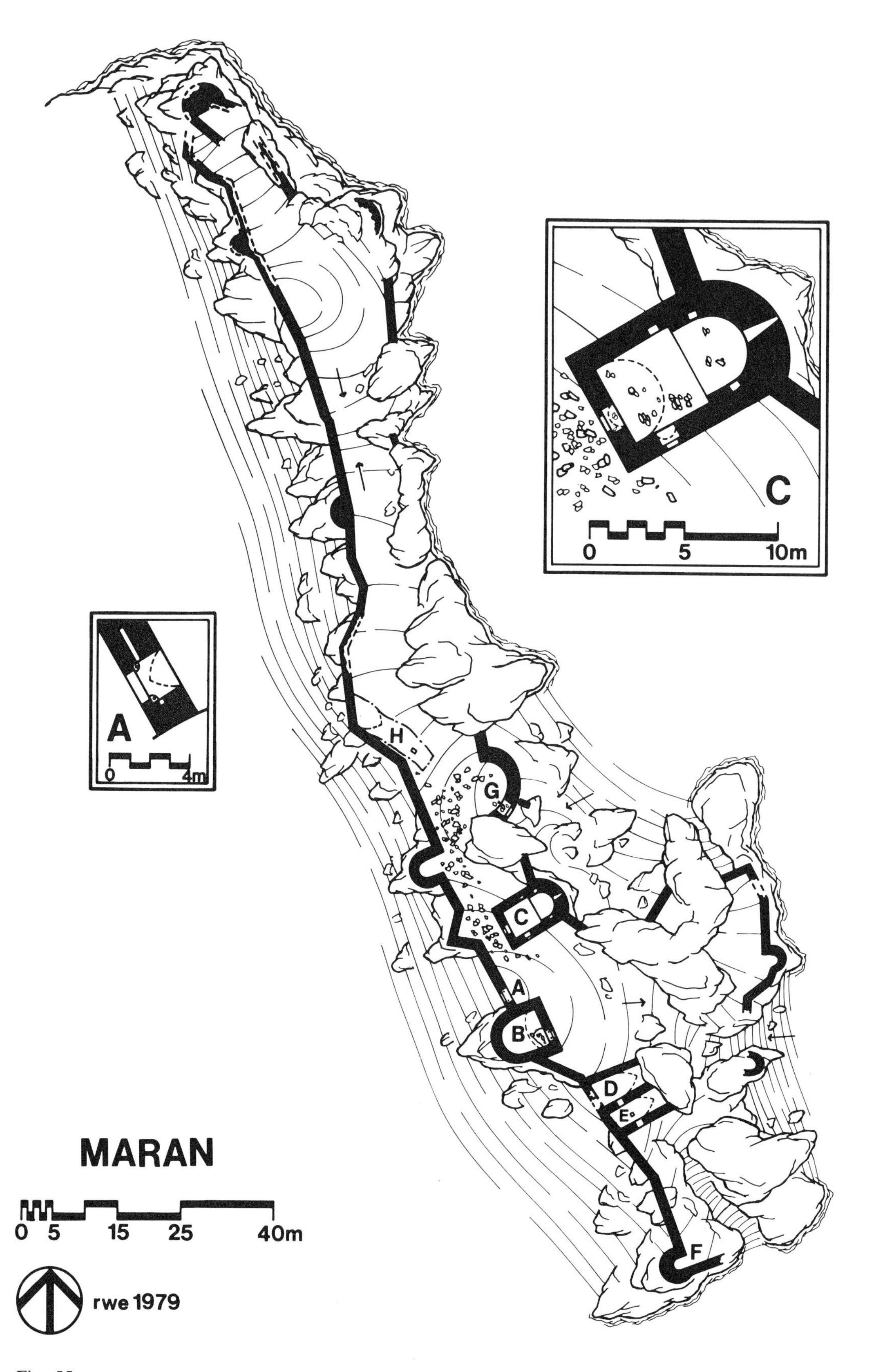

Fig. 55

like extension to protect the corners of the door; the north extension is now broken (cf. Gülek and Meydan). Just behind this arch are two pivot housings with holes to accommodate double doors. The doors were secured by a square crossbar bolt that extends almost 2 m into the north wall but is only 25 cm in width.

Tower B to the south provides flanking fire for gate A. This tower is covered by a depressed barrel vault and has a single door at the east (pl. 149a). Today the southeast corner of the vault has collapsed. The door has a massive monolithic lintel atop the jambs. The lintel is provided with pivot holes for a double door. The interior side of the door is covered with a round vault of seven voussoirs (pl. 150a). This tower has no other openings; it may have functioned as a cistern. The battlements, which once crowned the tower, have collapsed.

Chapel C on the other side of gate A is constructed in the typical Armenian pattern.[3]

South of B are the undercrofts (cisterns?) D and E (pl. 149b). Both are covered by pointed vaults but have a flat roof on the exterior. Their foundations are below ground level. A small portal connects the two chambers. Room E has a door/window at the south, and its vault is opened by a single hatch. The circuit wall at the west continues above the level of the chambers. Above chamber D is a small embrasured opening with a casemate. Just south of this embrasure a small section of the circuit collapsed. The salient F marks the most southern extension of the defenses.

The only other structure of significance is the second gate G, which was built in the roundness of a salient. Today the entire gate is missing, except for part of the south jamb. Northwest of G is the subterranean cistern H. This structure is built in a natural depression; its pointed vault is opened by a single hatch. Most of the north half of the east circuit consists of natural cliffs. At the far north end of the fort there is a shattered salient. From here the outcrop continues to the northwest.

[1]The plan of this hitherto unsurveyed site was completed in August 1979. The contour lines are separated at intervals of approximately one meter.

[2]This site does not appear on any of the maps in the List of Maps, above. The fort is marked on the chart published by Fedden and Thomson (12).

[3]For details on the masonry and design of the chapel, see Edwards, "First Report," 168.

Meydan

Meydan Kalesi[1] is located directly north of Adana in the heart of the Taurus Mountains. At an altitude of 1,955 m it girdles the top of the highest southern peak in the ring of mountains that forms the vale of Karsantı (pl. 158b). From the strategic hamlet of Karsantı, this castle can be reached by hiking south for 7 km over a difficult trail. Nine kilometers south of Meydan Kalesi is the small village of Meydan, which obviously gave its modern name to the castle. There are two possible routes from Adana to Karsantı. The more difficult route is the jeep trail from Çatalan that leads through the Etekli-Tamrut region. The easier approach is at the southwest where a relatively good dirt trail leaves Imamoğlu and passes the east flank of Meydan Mountain and the base of Mazılık Dağı.

From atop this fortress it is easy to see why the site is so important. Meydan Kalesi has a commanding view of the strategic east-west trail from Karsantı to Pozantı, and it guards the route directly north to Kayseri. Other forts completed the defensive network around the vale of Karsantı. To the east is Yeni Köy Kalesi (not to be confused with another Yeni Köy west of Karsantı) which guarded the trail to Imamoğlu. To the west of Meydan Kalesi at least two other forts (Işa and Tamrut) protected the western entrance into the vale. This area and the valleys to the south of Meydan Mountain are extremely fertile agricultural regions, growing a variety of crops. Dozens of streams and rivers move south through this vale and eventually empty into Lake Adana. In the medieval period this region could have supported a sizable population. However, the castle was not a convenient place of refuge, since access is quite difficult. Today Meydan Kalesi is covered by the same pine forest that conceals the limestone cliffs around the mountain. The closest source of flowing water to the fort is a small artesian spring that can be reached in about twenty minutes on foot. Within the complex, four cisterns (E, L, M, and P) supplied water to the residents.

As the medieval name for this site is unknown, so also is its history.[2] No modern travelers have ever discussed this castle; its *approximate* location has been given on a few modern maps.[3]

The circuit walls of the castle are carefully integrated into the saddle-like peak (fig. 56) and are constructed like most of the buildings in the complex with superbly cut examples of types V and VII masonries (pls. 150b, 152a). Occasionally, both types are mixed in the same structure. The interior facings of most of the buildings and circuit consist of a small, well-cut

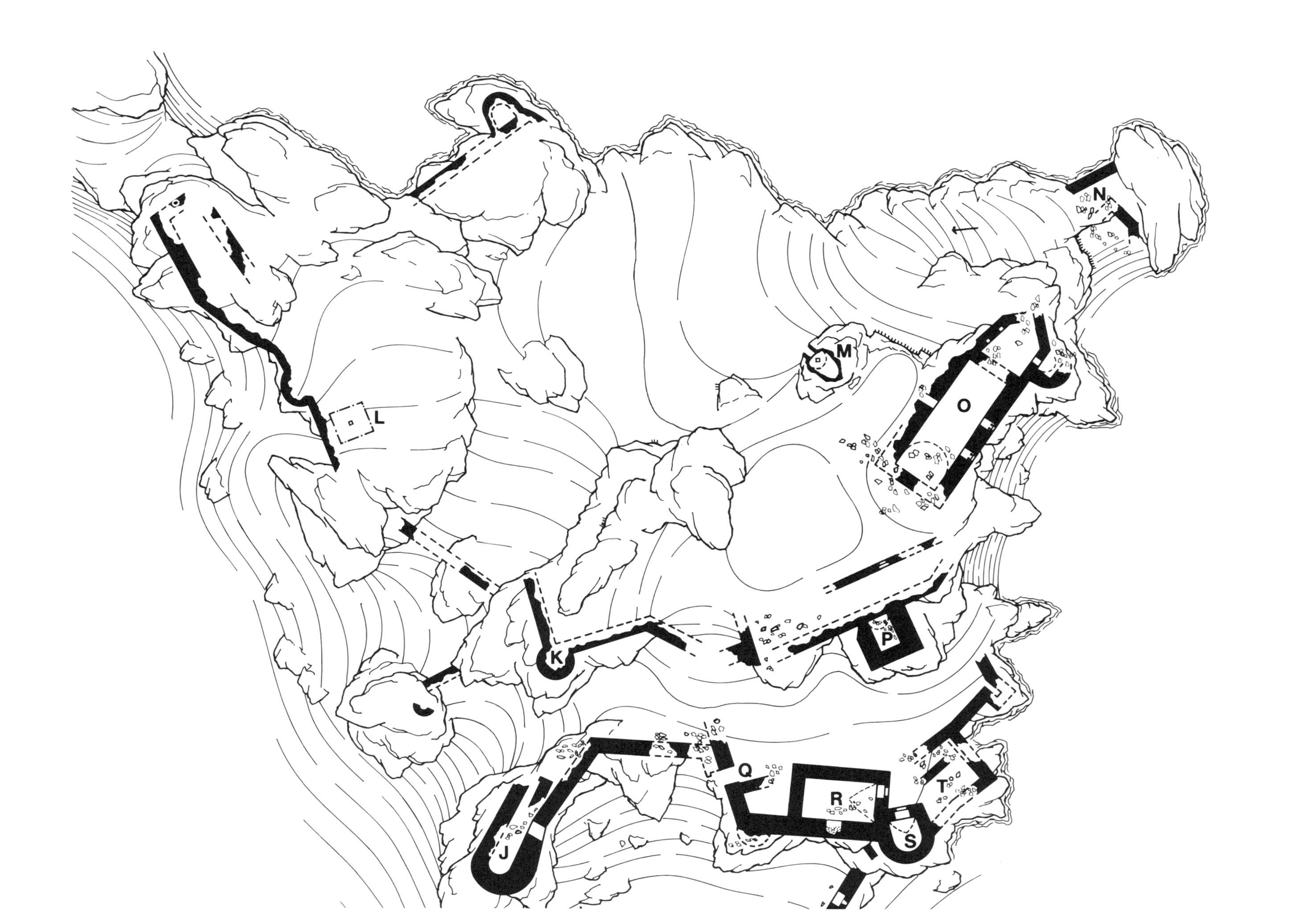

N
M
O
L
P
K
Q
R
T
S
J

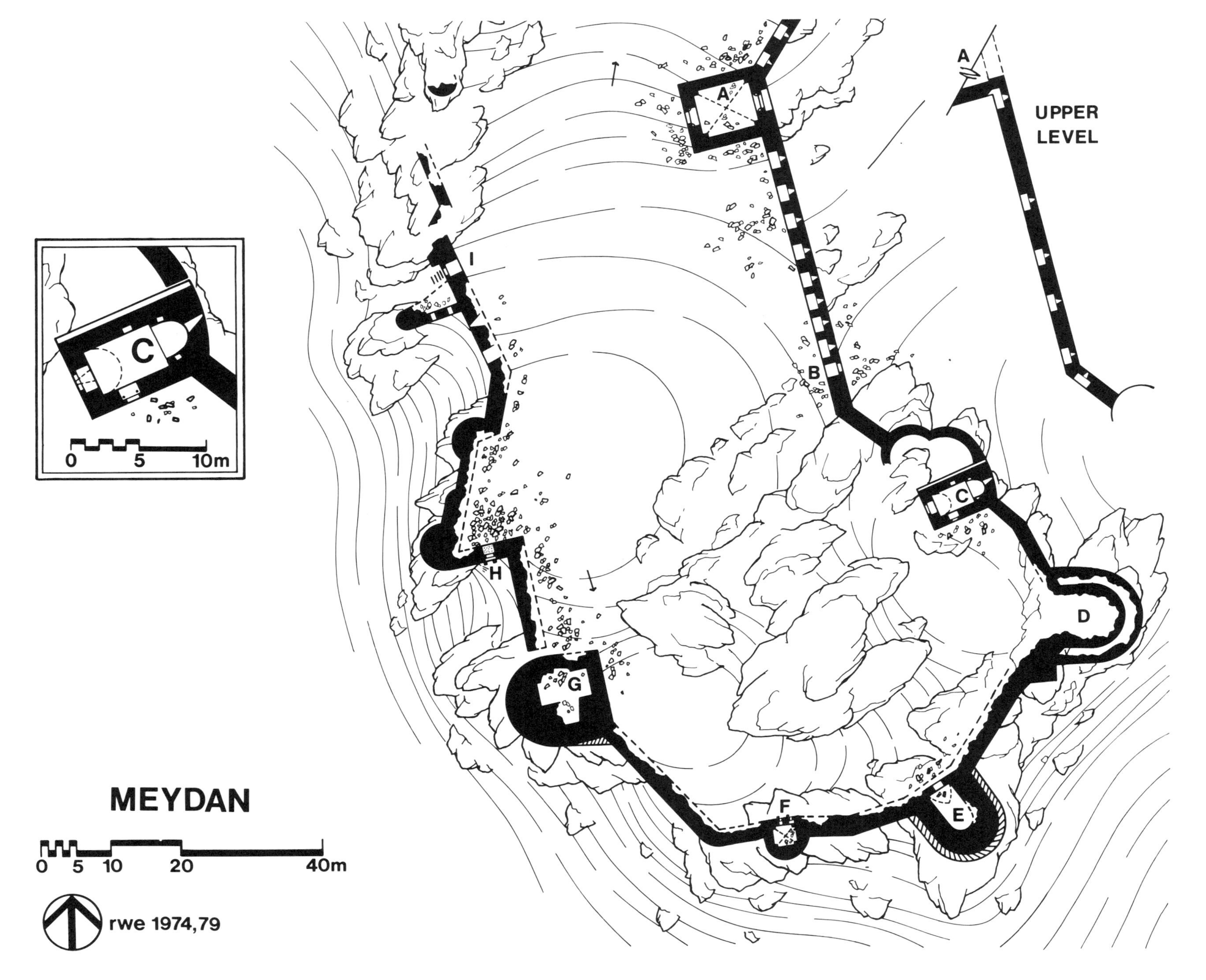

Fig. 56

type V (pl. 299a). Most doors and windows and the interior of the chapel (building C) are built typically with type IX masonry.

When seen from above or in plan, the castle appears to be divided by walls and topography into four units. The lowest point in the saddle of the rock is just south of the center of the complex where gate A and posterns B, H, and I give access into the fort. To the north and south of this somewhat rectangular central bailey, the slopes rise sharply (pl. 297c). At the south there is an almost circular ward running from C to G. There are no masoned walls to separate this unit from the central bailey; only the surging cliff of living rock acts as the north barrier for the south ward. It is clear from the corbels in the north wall of tower F (pl. 155a) that the designers also intended the defenders to fight those approaching from the interior of the castle. North of the central bailey is the largest area in the fort. Here many of the buildings have been integrated into two east-west circuit walls to create ascending wards. The first wall covers the distance from J to T and is opened by gate Q. A second wall from K to P protects the south flank of the highest enclosure in the castle, the north bailey. The area between the two walls is the small north-central bailey. Thus the Armenian concept of a segmented fort, where an advancing enemy has limited mobility, is stressed here.

The trail from Karsantı to the fort takes the approaching party to the east wall of the central bailey and to gate A and postern B. The entire east wall is a masterpiece of construction (pl. 152a). With some sections towering over 9 m in height, it is one of the tallest freestanding walls in any Armenian castle in Cilicia. A number of the type V stones in this wall are over one meter in length. The average length, height, and depth of the blocks in the lower courses of this wall are 74 × 50 × 49 cm. The stones in the upper courses are as neatly coursed but smaller. Generally the type VII stones are confined to the areas around the two gates. Because of earthquakes and vandals, few of the walls in this fort are possessed of their battlements. At the north end of this wall is gate A, one of the most impressive entrances in medieval Cilicia. When approaching from the east, one sees that the slot machicolation over the outer door has collapsed (pl. 150b).[4] Perfectly fitted mortarless stones cover the jambs with a double segmented lintel. The lower of the two lintels has the wedge-shaped blocks so typical of Armenian gates (cf. Yılan). The blocks of the upper lintel have joggled joints; this upper unit is surmounted by a relieving arch. The jambs below are 2.4 m apart. At the point where the jambs and lintels meet, corbeled extensions with rounded faces protect the corners of the door. On the interior side of the outer door of gate A, round pivot housings for a double door are still visible (pl. 151a). Unlike the lintels, the form of the relieving arch is carried over the interior of the door. This outer door opens into the rectangular chamber of gate A. This chamber was once covered by a single groined vault that was supported by a tapered springing stone in each corner of the room (pl. 151b). No doubt this ceiling was opened by machicolations (cf. Vahga). Unlike most Armenian gatehouses, passage A is not a bent entrance. The architects probably believed that such a design was unnecessary since the steep topography outside of the gate makes a direct, rushing attack difficult. Some of the interior masonry of A is a rough type V; most of the interior facing is type IX. The west door of gate A is not as complex as the east door. It consists simply of two arches, the outer (or west) being lower and resting on the jambs. The opening between the jambs is 2.2 m. On the interior side of the jambs there are the same rounded pivot housings that appear on the east door. At the point where the arch and jambs meet there are two rounded protrusions that likewise protect the corners of the wooden doors.

Because the east wall of this central bailey protects the easiest approach to the castle, it is heavily fortified with embrasured openings for archers. At ground level there are three of these openings to the northeast of gate A, where the wall ascends the cliff, and eight from A to B. Each opening consists of a vaulted casemate in the thickness of the wall and a roundheaded embrasured loophole (pl. 153a). On the outside the loopholes are 1.4–1.5 m in length and 10 cm in width. All of these archières have a simple square base (pl. 152a), except for one with a stirrup base (the fourth loophole north of postern B; pl. 152b), which is 1.98 m in length. This anomaly is probably due to different guilds of laborers and not to a separate period of construction, since the masonry throughout is fairly uniform. On the interior of the east wall the open sides of the casemates were once covered by a wooden canopy (pl. 299a). The joist holes that supported this cover are still visible above the casemates. On the upper level of this east wall are similar embrasured openings. At this level there are only six shooting ports; here all of the loopholes are squareheaded (unusual in Armenian construction) and have stirrup bases. Two of the upper-level embrasures are south of postern B. Postern B is a simple vaulted door limited by jambs (pl. 153b).

The east wall of the central bailey ascends to the south ward and abuts against a salient that has two rounded sides (pl. 297c). From there the wall of the south bailey merges into chapel C.[5] From chapel C the partially collapsed wall continues to the southeast until it joins tower D (pl. 154a). This bastion consists of two tiered revetment walls. On both levels the natural rock foundation has been carefully scarped to follow the contours of the tower. Tower D appears to have had no interior rooms. Adjoining the tower to the south is a small square extension of solid masonry. The wall continues to the southwest until tower E rises from its rock plinth. The base of tower E is fitted with a small, carefully placed talus. This rounded bastion enclosed two oblong chambers; the lower chamber appears to be a cistern(?) covered with a pointed vault. The upper chamber (not shown on the plan) is slightly smaller and badly damaged. The nature of its roof is not certain. Both chambers were entered through roundheaded doors at the northwest. The upper-level door is preserved and is carefully built of type IX masonry (a few of the blocks have bossed faces). At the top of this damaged tower on the west flank are the remains of a stirrup base for a loophole (pl. 154b). The circuit wall continues west from tower E to bastion F. The only room visible in F is set far below the upper level of the present circuit wall. This may indicate that some sort of terrace with battlements was accommodated above (pl. 155a). The existing room in tower F was probably covered by a now collapsed groined vault. The entrance was a single door at the northeast. The door is flanked on the west by a window that has half the dimensions of the door. Both openings are limited by jambs and squareheaded on the inside. Two corbels below these openings supported a platform that could be used for fighting or withdrawn to prevent access into tower F. From this tower the wall descends and curves to the northwest until it reaches tower G. Tower G is the largest single mass of masonry in the entire castle. At the south where it joins the circuit wall there is a small talus. The interior of this tower once had a complex of rooms. Today the rooms are a mass of rubble with only a standing foundation of walls.

To the north and just below tower G, the circuit wall descends from the rock to the west side of the central bailey and postern H. This door is tripartite but does not have a slot machicolation. Due to the collapse of the adjoining circuit wall, the interior of the door is completely blocked. The exterior and interior of the door are roundheaded with a flat lintel over the jambs. There are still a few stairs to the southwest of the door. Directly above the door are two corbels that probably supported a brattice. The stairs and postern H could also be protected by flanking fire from the south and west. With an expected attack on the east wall of the central bailey, this door, along with I, would allow the defenders to sally around the south bailey and encircle the attackers. From postern H the wall continues north and is supplemented with small rounded salients. Just south of postern I is an outwork (or elongated tower) that extends at a 90-degree angle from the direction of the circuit wall. This outwork is equipped with two windows. The circuit wall directly to the south has an embrasured loophole in the upper level and what may be a door in the lower level. The outwork creates a very dangerous cul-de-sac to the south and provides flanking fire and concealment for postern I at the north. Postern I is a simple door that is approached by stairs at the west.

From I the circuit ascends to the north and eventually fades into the rising cliffs (pl. 156a). As one walks north from the central bailey the topography forces the approaching party to enter through the now collapsed gate Q into the north-central ward. At the south edge of the ward on a high spur is the elongated tower J (pl. 156b). Its east wall rises from the cliff almost like a natural projection. At the top of that wall is a large roundheaded window that no doubt opened into a now vanished room inside the tower. Today there is still evidence of a vaulted covering. Tower J holds the commanding position on the west flank of the castle. To the south of J a small salient is built right on the face of the rock. Today the wall that connects J and Q is badly damaged. Beyond Q to the east the wall is in excellent condition. Chamber R is the principal structure in this area. This building is a rectangular hall covered by a broken pointed vault (pl. 157b). In its northeast corner is a single roundheaded door that is limited by jambs (pl. 157a). A squareheaded niche (or possibly a blocked window) is located in the center of the south wall. In the southeast corner the junction of R with tower S is a poor one that is characterized by irregular rusticated faces on the blocks of tower S. It appears that hall R was constructed after tower S. Tower S is a large projecting bastion that is faced exclusively on the outside with a well-cut type VII masonry. On the interior of tower S there is an apsidal chamber covered by a pointed vault. The entrance to this chamber is at the north and is barely visible above current ground level. Unlike the door in hall R the opening here has an oversized, joggle-jointed keystone reminiscent of the lintel in gate A. Northeast of tower S the remains of the circuit are in ruins. T was once a rectangular building, but little can be said about it because of heavy destruction in the area. This north-central bailey (that is, the area between J–T and K–P) is relatively small and narrow in shape. If there were interior structures in this bailey, they were probably built of wood since there are no traces of stone foundations today.

The next level to the north is the massive north bailey. At the southeast end of this bailey is the only square salient in this castle and one of the few in all of Armenian Cilicia. On the interior of bastion P is a vaulted cistern. There are provisions for an upper-level door at the north. West of P the circuit wall is badly shattered. There were probably provisions for a gate somewhere between K and P. Tower K is still in a relatively good state of preservation. Below K and to the west there is a small outwork that is now in ruins. Northwest of K the wall moves across and down a large ravine and eventually becomes a revetment as it snakes around the northwest end of the castle. The only significant building in this area is the small square subterranean cistern L. The interior is carefully scarped, and there is some indication of masonry and stucco. A single hatch gives access to this chamber. Except for a small tower and wall at the extreme north end of the north bailey, the Armenian engineers allowed the natural cliff to defend the castle. Only in the northeast corner at point N are walls employed to block a potential access. N itself is an extremely curious structure set between two precipitous cliffs with a sheer drop below. Directly to the southwest of N is a scarped cliff that had some connection to N. Despite the severe damage, there are indications—the presence of springing stones in the north wall—that building N was once a two-story vaulted structure. There is no evidence that a wall once connected N to O. Small scarped walls, holes, and niches on the sides of the rocks are evident throughout the north bailey, and often they are not in alignment with any of the visible masoned walls. No doubt wooden buildings were attached to the scarped areas; this ward probably held a sizable population (cf. Lampron). To the southwest of N is the third cistern in the north bailey, building M. A large part of M is above ground level; its basin is set in a natural sink where the sides have been stuccoed. Like P, its vault is rounded with a single hatch.

The most important structure in the north bailey, which excavation may someday reveal to be the principal part of the most significant unit in the castle, is O. Building O is a rectangular structure in the northeast corner of a large, gently sloping meadow. Today the center of its vault has collapsed. The remaining sections of the vault are built with type IV masonry in the typical Armenian pattern with plates of cut stones. There is no trace of stucco on the vault. The sections of the wall below the springing are built in types V and VII masonries. The south end of this long hall has collapsed. In the west wall are the remains of a single roundheaded door (pl. 158a). In the east wall the two roundheaded windows are limited by jambs. The interiors of these east openings are covered by small corbeled vaults (pl. 155b). This is one of the rare occasions that corbeled vaults are used in any medieval structure in Cilicia. Both windows have two pivot holes to accommodate shutters, and the sides of the jambs have neatly drilled parallel holes to hold crossbars or perhaps the frames for shutters. The north part of hall O is divided from the larger south section by a now collapsed low-level wall. It is likely that the wall was opened by a door. This north unit of building O is pentagonal. On the interior of its west wall are two square niches; the east wall is pierced by a single window. The function of the niches is unknown; it seems highly unlikely that the north end was a chapel. Below the level of the east window is a rounded salient that is now in ruins. Just southwest of building O, I located two well-cut square capitals. The corners of the capitals are decorated with concentric cavetti. On the side of one, a small bird has been carved (pl. 298a), and on the other a rather bulbous cross. Just what the capitals supported is not known. They are close to the collapsed south end of building O and may be part of that building or an adjoining structure.

Between buildings O and P there is a thin wall that is parallel to the north side of P and is pierced by a narrow door. The exact relation of this wall to the two structures is unknown; it appears to have undergone extensive reconstruction. The south face of the wall has types IV and V masonries, and the north face is type VII.

[1] I began my survey of this site in October 1974; the plan was completed during my second visit in June 1979. On the plan the contour lines are spaced at intervals of 50 cm. Because of the limitations of time, numerous post holes and scarped faces in the north bailey are not represented on the plan.

[2] Because of its location, it is possible that Meydan Kalesi is the medieval Kopitaṙ (Copitar); see Alishan, *Sissouan,* 156 ff. Until excavations reveal inscriptions or other pertinent material, the certain identity of this site will remain a mystery. It is less likely that Karsantı stands on the site of the 13th-century settlement of Kostəndnocʿ; see Yovhannēsean, 117, 189–92.

[3] *Cilician Gates* and *Kozan*.

[4] The outer arch (with most of its springing) and the small room over A have disappeared today.

[5] For a detailed description of the chapel, see Edwards, "First Report," 171.

Milvan

Milvan Kalesi[1] is a cloister/fort on the east flank of the Çakıt Suyu. The village of Milvan (also called Kale Köy)[2] can be reached by driving about 19 km northwest of Karaisalı. The mountain and its fortification rise to a height of 1900 m. The fort can be reached by hiking west from Milvan Köy for about three hours (pl. 159a). The trail leads around the north flank of the outcrop and gradually ascends the almost vertical sides of the west face until it reaches the relatively flat summit at the southwest. The vale of Çakıt, into which Milvan Kalesi has a commanding view, is of great strategic importance, since it links Pozantı with the Cilician plain. This north-south route is completely independent of the road that links Tarsus with Pozantı via the Cilician Gates (below Gülek). The northern guardian of the vale of Çakıt is Anahşa Kalesi. Today no vehicular road traverses the Çakıt valley, but it is the route for the railroad linking Adana with the Cappadocian plain. A well-worn trail, which is suitable for pack animals, still carries traffic from Karaisalı to Pozantı.[3] This Çakıt trail must have been popular during the medieval period.[4] Directly north of Milvan Dağı is Kabalak Dağı. The mountain west of Milvan (on the other side of the Çakıt valley) rises to an altitude of 2,140 m and blocks the view into the Cilician Gates.

By reason of its location and name, it is almost certain that Milvan Kalesi is the medieval site of Molivon (Mōlovon).[5] This site was founded by the Byzantines, who recaptured it from the Arabs between 876 and 878, and was later occupied by Armenians.[6] On the Coronation List (see below, Appendix 3) Ažaros is listed as the Armenian (Greek?) Baron. This site probably remained in Armenian hands through the first quarter of the fourteenth century. We know that the bishop of Molivon, a certain Nersēs, attended the church council of Adana in 1314. The bishop of Molivon resided in the nearby monastery of Kamrik Anapat/Gaṙner (Gṙner).[7] The two famous events in Molivon's history both involved violence. King Levon II (1270–89) laid siege to the fort when Baron Levon of Molivon joined Aplłarip and Vardan of Hamus in revolt.[8] In 1299 King Hetʿum II was imprisoned here by his brother Smbat and blinded with a scalding sword.[9]

The ascent up the west flank of Milvan can be treacherous and should be undertaken with great care. The upper third of the west face has a series of natural horizontal shelves that have been widened by careful scarping into a broad footpath. This path is protected by three gates as it gently ascends to the southwest. I will henceforth refer to the area of the path as the lower level. Because of the restrictions of time and the nature of the topography, I was unable to survey the lower or middle levels of Milvan Kalesi.

The first gate of the lower level has collapsed, and only a few of the foundation stones are present. The second gate of the lower level is situated on a sharp natural bend in the path. Most of the gate has collapsed except for the salient corner. In general the exterior facing stones on the remnants of the gate consist of types VII and VIII masonries with a thin poured core. As the trail turns to the south, the path widens and is supported by revetments (pl. 159b). The masonries of the revetment vary from types III, IV, and V. The third gate, which is the largest and best preserved, is perched atop a descending spur of the mountain (pl. 160b). Like the other two gates, all evidence of jambs has vanished. The southwest section of the gate has two types of masonries: the lower two-thirds consists of type VIII and the upper third of type IV. Mortar has been used sparingly.

Once through this gate the path ascends abruptly to the northeast. Between the summit and this third gate there are a large number of buildings and an even larger amount of coursed rubble. This area, which I will refer to as the middle level, testifies to the sizable population that once inhabited the outcrop.

Directly east and above the gate are the lowest sections of the middle level (pl. 160a). The sides of the protruding mass of limestone have been scarped into flat vertical faces to accommodate wooden buildings. The stone buildings, which stand above ground level, are usually constructed with type IV masonry (occasionally type I), and quoins are frequently employed to the highest point of the walls. In the middle level there are two structures (probably cisterns) which are covered by vaults of crudely placed stones. Approximately 55 m above and to the northeast of the third-level gate, there is a deep rectangular pit carved out of rock. This undoubtedly functioned as the foundation for a building. Near the east end of this structure is the base of a monolithic needle. In its undamaged state it probably supported the roof over the building. Near the top of the middle level there is a huge rectangular foundation that is more than 21 m in length (pls. 161a, 161b). The northern half of the rectangle is scarped into the natural rock. Most of the southern half seems to have rested on level ground. About 6 m west of this structure, one of the trails through the middle level passes underneath a roundheaded gate carved in the natural rock.

The summit (or upper level) of the mountain at the northwest is covered by a number of significant

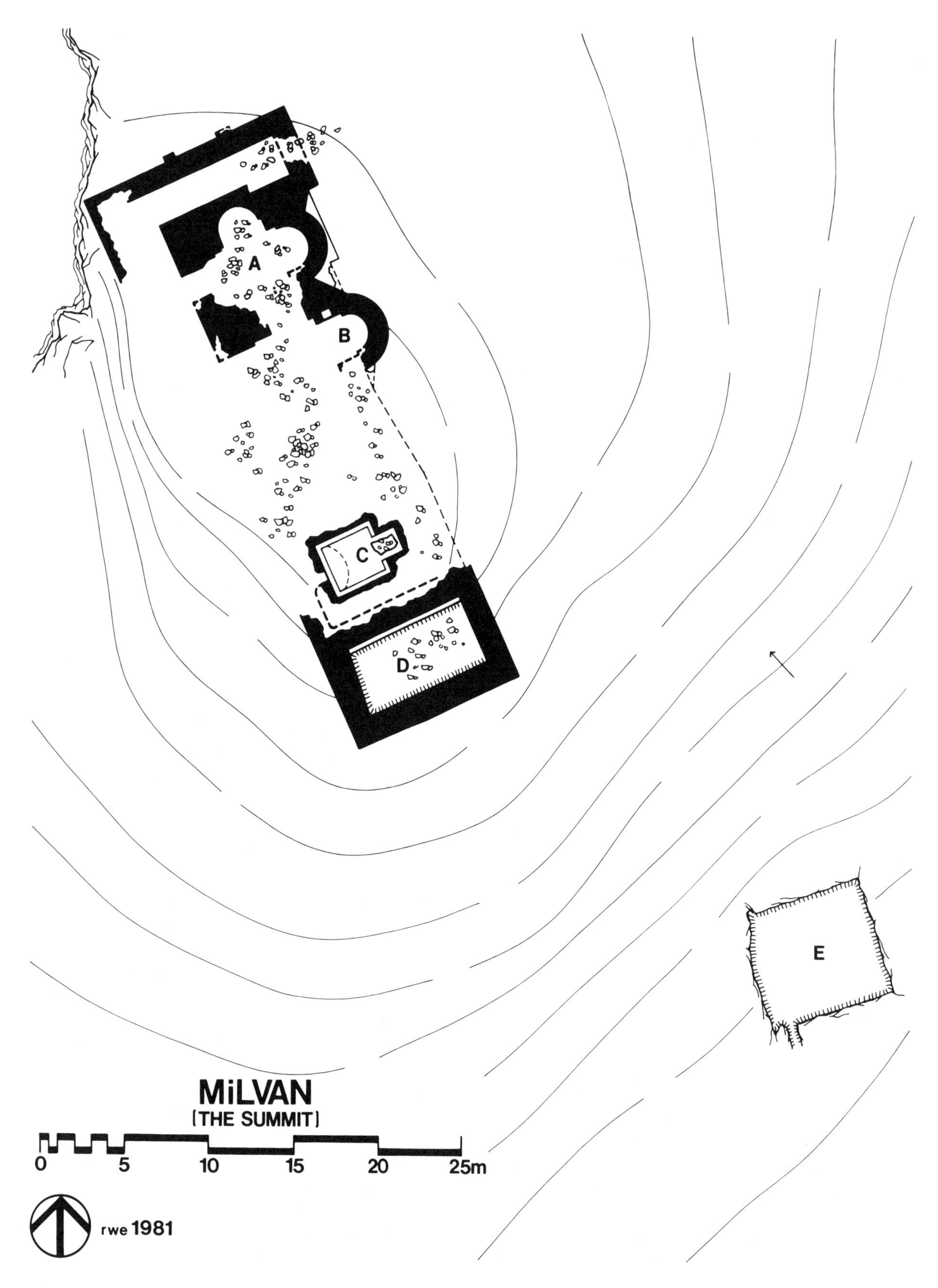

Fig. 57

structures (fig. 57). Most of the summit construction (A–D) rests on a platform (pl. 162a). Building A seems to have been enclosed by some sort of circuit at the north that merges into the raised platform along the east embankment. The west end of the north wall runs to the edge of the cliff. This wall is supported by two small buttresses. The east end of this wall, at the point where it merges into A, has collapsed. The masonry of the north circuit is quite similar to the exterior masonry of A, which is type IV. The masonry on the north wall of A has one curious feature (pl. 162b). At the point where the wall jogs, the stones in two of the courses are laid on their vertical axes. The constructions on the summit have more mortar in the margins of their stones than in the middle- and lower-level buildings. Along the exterior of the east wall, two apses protrude above the level of the platform (pls. 163a, 165a). The southernmost apse, which extends beyond the platform, is part of the adjoining building B. Type IV predominates on the exterior of B, with occasional courses aligned on their vertical axes.

Unfortunately both buildings A and B have suffered severe damage (pl. 165a). All that remains of B is the apse at the east and the wall connecting it to A. The north and east apses of A are the best-preserved units, with each standing to almost 2 m in height (pl. 163b). The most interesting feature on the interior of A is the masonry which shows a curious alternation of brick and type III (pl. 164a). Four courses of brick (where each tile measures on average 18 × 24 × 4.5 cm) are joined into horizontal layers by a white, almost chalky, mortar. The height of the alternating courses of type III masonry is about the same as the bonded units of the four brick courses. At the salient ends of the apses, the type III is replaced with a relatively smooth ashlar (type IX), but the brick courses are allowed to continue around the jutting corners. Since most of the brick has fallen away, the *opus listatum* in the salient northeast corner has an indented appearance. The interior side of the opening at the west end of A is faced with relatively smooth ashlar (pl. 164b). A few small square holes are cut on the exterior faces of this masonry; their function is unknown.

The apse in B is not constructed with brick, but all of the interior facing consists of type III (pl. 165b). This apse is slightly asymmetrical; on its north side there is a square niche. Considering the uniformity in the exterior masonry, I do not think that B represents a later period of construction.

While the function of the corridor between the north circuit and the north wall of A is unknown,[10] it seems certain that A is at least a trefoil and perhaps even a quatrefoil church. Such a design is unprecedented in east Cilicia; in other parts of Anatolia, quatrefoils are built from the fifth through the eleventh centuries.[11] In the opinion of this writer, the multifoiled church at Milvan is not an Armenian construction. A sufficient amount of the apsidal walls survives at the north and east to show that the peculiar features of Armenian ecclesiastical architecture (for example, two apsidal niches and the low-level apsidal window) are absent. The extensive use of brick for the interior masonry,[12] as well as the protruding apse at the east, indicates that church A is probably Byzantine.[13]

Toward the south end of the summit-platform at the point before the topography begins to descend sharply is cistern C and the adjacent room D. C is covered by a depressed vault that is in a relatively good state of preservation, except for a hole at the east (pl. 166a). Unlike the vault over the middle-level cistern, the vault here has been constructed out of smooth uniform blocks of ashlar. The lower half of the vault as well as the walls has been covered with a thick layer of plaster. This plaster has a pinkish tint. Most of the interior walls are constructed with type IV masonry. In the east wall of the cistern there is a large square niche or recess. Just below the springing of the depressed vault (and not in the area of the niche) the north and south walls project inward about 90 cm. The tops of these projecting walls do not form flat shelves but slant downward. Two courses of brick are visible in the masonry of the north and south walls. There is evidence of at least one drainage hole in the vault over C. This structure is opened by a single high-placed door at the west.

Chamber D to the south is constructed partially out of natural rock. The exterior masonry of its west and south walls consists of a roughly coursed ashlar. The interior masonry of D, which is carefully worked around the scarped rock, is a rough type IV (pl. 166b). Most of the stones, which are laid in fairly regular courses and cemented with a white mortar, are small. They measure on the average 28 cm in length, 18 cm in height, and 17 cm in depth. Occasionally, huge monolithic blocks are carefully cut to equal the height of a single course and to span the entire thickness of the wall, acting as headers.[14] In the present remains there are no indications of any doors or windows. The nature of the covering over D is unknown. Southeast of D is the scarped chamber E.

From the surface remains it appears that most of the complex at Milvan is a Byzantine construction. The only evidence of type VII masonry, on the exterior of the second gate of the first level, is probably the result of Armenian reconstruction. The villagers at Kale Köy did show me two Armenian coins from the reign of King Het῾um I that they found at the site. Only excavation and the careful analysis of the finds can determine whether the summit complex was built before or after the Arab invasions. Church A, the apse of what is probably a second church at B, and the large number of crude residential buildings in the middle level may indicate that Milvan Kalesi also functioned as a cloister. Like Kale Gediği, it could have maintained a population of ordained and lay residents as well as serving as a watch post and signal station along a strategic trail. Because Milvan had a resident baron, a small garrison of retainers was undoubtedly maintained. The altitude and inaccessibility of the site would make it impractical for a large garrison. During my surveys I did not locate the monastery of Gaṙner. There is a very remote possibility that the present site of Milvan Kalesi is Gaṙner and that the nearby fortress remains to be found.

[1]The plan of this hitherto unexplored site was completed in June 1981. The contour lines are separated by intervals of one meter. Because of the complexity and size of the outcrop, only the summit appears on the plan.

[2]Kale Köy is listed on the *Ulukişla* map.

[3]Hild, 57 f, 123; Schaffer, 79–83.

[4]The Çakıt (Byzantine: Maurianos) was used occasionally by a Greek army invading Cilicia from Constantinople; see Dennis, 218.

[5]Other Armenian designations are: Mawlowon, Nahangk῾ Dłekin Molewoni and Dašt Mluni; cf. Alishan, *Sissouan*, 150–56. The Franks refer to this site as Mons Leonis. Milvan Kalesi is the Kastron Meluos of the Byzantines; see: Smbat, G. Dédéyan, 77 note 51; Christomanos, 146.

[6]Cf. Yovhannēsean, 120–22; R. Edwards, "The Cilician Tetraconch at Milvan," *Eleventh Annual Byzantine Studies Conference, Abstracts of Papers* (Toronto, 1985), 37 f; Scylitzes, ed. H. Thurn, CFHB (Berlin, 1973), 141 = Cedrenus, II, ed. I. Bekker, CSHB (Bonn, 1839), 213; Theophanes continuatus, ed. I. Bekker, CSHB (Bonn, 1838), 278; Vasiliev, 81; Honigmann, 61; H. Glykatzi-Ahrweiler, "Recherches sur l'administration de l'empire byzantin aux IXe–XIe siècles," *Bulletin de correspondance hellénique* 84 (1960), 46.

[7]*RHC, DocArm,* I, lxx. One of Mawlovon's most famous bishops, John (also called by his baptismal name of Baldwin), was the brother of King Het῾um I. In his capacity as abbot of Gaṙner, John commissioned the monk Constantine to illuminate at least one Gospel between 1263 and 1270. The scriptorium at Gaṙner seems to have attracted considerable attention in the late 13th century. John's ecclesiastical authority also extended into the region controlled from Barjrberd. See: Hakobyan (below, note 9), I (1951), 82, II, 70; S. Der Nersessian, *Armenian Manuscripts in the Freer Gallery of Art* (Washington, D.C., 1963), 55–72.

[8]S. Der Nersessian, "Un Évangile cilicien," *REArm*, n.s. 4 (1967), 103 f.

[9]Smbat, 656; V. Hakobyan, ed., *Manr žamanakagrut῾yunner, XIII–XVIII dd,* II (Erevan, 1956), 170 note 193. Four years prior to the destruction of this fort (1335) a certain Constantine is said to have made at Milvan an Armenian copy of the Proverbs of Solomon; see L. Xač῾ikyan, ed., *XIV dari hayeren jeṙagreri hišatakaranner* (Erevan, 1950), 241 f, 295.

[10]It is remotely possible that the space is part of an ambulatory.

[11]W. Kleinbauer, "Zvart῾nots and the Origins of Christian Architecture in Armenia," *The Art Bulletin* 54 (1972), 256 ff; E. Rosenbaum et al., *A Survey of Coastal Cities in Western Cilicia* (Ankara, 1967), 18 f; H. Buchwald, "Western Asia Minor as a Generator of Architectural Forms in the Byzantine Period," *JÖB* 34 (1984), 206 ff.

[12]The Armenian use of brick in ecclesiastical construction is very limited. It appears only in the semidome of the chapel at Meydan Kalesi and in the barrel vaults of the church of T῾oros I at Anavarza. See Edwards, "First Report," 157–61, 171 and "Second Report," 128–30.

[13]Ibid. The only Armenian apses that protrude on the exterior of churches and chapels in Cilicia are at Korykos and Sis.

[14]Headers are *never* used in Armenian construction, but they are common in the Byzantine fortifications of Cilicia (see above, Part I.3, I.5).

Misis

Next to Tarsus, late antique Misis[1] was considered by the early church fathers to be the second city of Cilicia Pedias.[2] However, the real significance of Misis lay not in its ecclesiastical history, but in its topography. The city is situated on the Ceyhan River (ancient: Pryamus) near the strategic road that links Adana and Tarsus to Gaziantep and the rest of eastern Turkey. Although the Ceyhan is navigable as far east as Kum Kalesi, cargo vessels can sail inland only as far as Misis. In the medieval period the route for trade from Erzurum and Trabzon came south through Göksun (via Geben) before reaching the Italian warehouses in Misis and eventual distribution by barge (or along the land route via Gökvelioğlu) to Ayas. Today, as in the late antique period, a single bridge connects the old city of Misis on the north bank to the Ayas road (pl. 167a). This bridge was probably built during the Roman period and was completely restored by Emperor Justinian.[3] Later it was extensively repaired by the Abbasids, then by the Ottomans. Many of the modern denizens have moved from the heights of the north bank to the more level confines on the south side of the river. Here they can more easily tend their farms at the base of the Cebelinur.[4]

Because the complex history of this site has already been discussed in detail,[5] I will provide here only a summary. The young John Chrysostom frequently preached in Misis, and the teacher of Nestorius,

Bishop Theodorus, was in residence there until his death in 428. In 550 Justinian I convened a synod in the city's great hall. It seems that Misis lay deserted in the first half of the seventh century after Emperor Heraclius ordered the city to be abandoned in advance of the Arab invasions. Muᶜāwiyah is said to have destroyed the empty garrison fort in the city in 651. In 684 Emperor Constantine IV captured Misis, and it remained a Byzantine possession until 703 when ᶜAbd Allāh drove the Greeks out and rebuilt the fortifications on the original plan. He also constructed a mosque and maintained a permanent garrison. In 756 the city circuit, which had partially collapsed in the earthquake of 755, was restored. Later, over eight thousand settlers were moved from Syria and Persia to Misis. The size of the permanent garrison was increased to two thousand soldiers. In 804, the year after a great earthquake damaged the town, the Byzantines carried out punitive raids in the region. Throughout the ninth century Cilicia suffered a series of devastating earthquakes. In 963/64 John Tzimisces failed in his attempt to dislodge the Arab garrison at Misis. In the following year Emperor Nicephorus Phocas besieged the city for almost two months but had to retire briefly when his supplies ran short. Finally, the Greeks succeeded in capturing Misis by siege.[6]

The Greeks held Misis in relative peace until the last decade of the eleventh century when Turkish forces overran the town. The Turks were expelled (1097) by the Crusader forces under Tancred; this action was openly supported by the Rubenid barons. Through most of the first two decades of the twelfth century the city remained a possession of the Franks.[7] Except for a few brief intrigues, the Rubenid Baron Levon I held the town from 1132 to 1137. Levon's capture by Emperor John II led to his deportation to Constantinople and his eventual death. However, the Rubenids had their revenge when Levon's son, Tᶜoros II, captured Misis in 1151. The next year Emperor Manuel dispatched Andronicus Comnenus and twelve thousand troops with orders to capture the town. Outside the walls of Misis the Greek forces suffered an embarrassing defeat that, for all practical purposes, marked the end of any permanent Byzantine rule in Misis. In 1156 the Sultan of Konya suffered a similar defeat at the hands of Tᶜoros II. When the emperor encamped his army at Misis in 1158 (en route to assist the Crusaders) Tᶜoros cleverly acknowledged Byzantine suzerainty over the region while accepting the title of sebastos of Misis, Anavarza, and Vahga. Through most of the twelfth and thirteenth centuries the town remained an Armenian possession.[8] The civilian population of Misis appears to have been quite small;[9] it is likely that a permanent garrison was stationed there to protect the emporia of the Genoese and Venetians. In 1266 the Mamluks briefly captured and plundered the town. In 1275 Baybars captured the city and burnt it to the ground. Even the bridge over the Ceyhan River was destroyed. In 1322 the Armenian forces who had withdrawn to Misis were soundly defeated by the Mamluks. The last Mamluk raid on the city occurred in 1374.

It was reported by the geographer-historian al-Yāqūt that the old city on the north bank originally had a circuit wall with five gates, while Kafarbayyah on the south bank had a similar wall of four gates.[10] In the early thirteenth century Wilbrand von Oldenburg noted that the turreted circuit around the city was in disrepair.[11] Aside from sketchy medieval reports, we know almost nothing about the physical defenses. The superficial comments of the nineteenth-century explorers do not touch on the architecture of the site.[12] In 1879 E. J. Davis published a simple hand-drawn map of the old city on the north bank.[13] This map shows the general rectangular shape of the city outcrop and the approximate location of the circuit and citadel. When I first visited this site in the summer of 1973, substantial fragments of the foundations (north of the river) were still visible at the east. Unfortunately when I returned in 1974 to survey the circuit, much of what was visible in the previous year had been recycled into new constructions in the village below. The few surviving courses of masonry show that the stones were atypical of Armenian medieval construction. In the area of the citadel at the north end of the city, large stones were cut in the classical fashion with *anathyrosis*. It does appear that the course of the circuit followed the almost rectangular shape of the outcrop. On the river side of the outcrop I found evidence only of a circuit at the east end. Today the position of the towers and gates can no longer be determined. The outline of the classical-period stadium is clearly visible below and east of the outcrop. Occasionally, modern travelers visit Misis to view the mosaic floor of a now vanished church/synagogue.[14] Formal excavations at this site should reveal considerable detail about the plan of the late antique-medieval city.

[1]Because of extensive damage and the active plundering of the masonry, I did not survey this site during my visits in 1973, 1974, and 1981.

[2]The toponym Misis is derived from the name of the mythical sage Mopsus. The Greeks often refer to the city as Mopsoues-

tia, Mompsouestia, Mamista, and Manistra. The latter two forms are common in Byzantine texts. The normal Arabic designation is al-Maṣṣīṣah. The Armenians call the city Msis, Mises, Mamestia, and Mamuestia. Today the modern Turkish spelling is Misis. This site appears on the following maps: *Adana (1), Adana (2), Cilicie, Malatya*. See *Handbook*, 715–19.

[3] Procopius, *Buildings*, 5.5.4–7; Seton-Williams, 163 f.

[4] The settlement on the south bank was built originally by Hārūn ar-Rašīd and called the Kafarbayyah. See K. Otto-Dorn, "Islamische Denkmäler Kilikiens," *Jarhbuch für kleinasiatische Forschung* 2.2 (1952), 117 f; al-Balāḏurī, 257.

[5] By far the best account of the history of Misis was published by E. Honigmann, ("Miṣṣīṣ," *EI*, 521–27); see also: idem, "Misis," *IA*, 364–73; W. Ruge, "Mopus(h)estia," *RE*, 243–50; Imhoof-Blumer, 437–78; Ramsay, 207, 341, 381, 385 f, 451; E. Reitemeyer, *Die Städtegründungen der Araber im Islām*, Diss: Heidelberg (Munich, 1912), 77; Haldon and Kennedy, 107, 110 ff; Le Strange, *Palestine*, 26, 37, 62, 78, 82, 505–7; Schaffer, 40, 97 f; Cuinet, 42 f; Schultze, 305–15; V. Langlois, "Voyage en Cilicie, Mopsueste," *RA* 12.2 (1856), 410–20; idem, *Inscriptions*, 4–11; idem, *Rapport*, 47–50; A. Grohmann, "Die arabischen Inschriften der Keramiken aus Misis," *Istanbuler Mitteilungen* 15 (1965), 243–63; Rey, 349 f; Ritter, 96 ff; al-Balāḏurī, 255–59; Magie, I, 273, II, 1148; R. Mouterde, "Inscriptions grecques et latines du musée d'Adana," *Syria* 2 (1921), 280–85; Heberdey and Wilhelm, 11–13; E. Doblhofer, "Eine Grenzregelung in Kilikien," *Jahreshefte des Österreichischen Archäologischen Instituts in Wien, Hauptblatt* 45 (1960), 39–44; idem, "Mopsos, Arzt der Menschen," ibid., *Beiblatt* 46 (1961–63), 6–18; H. von Aulock, "Die Münzprägung der kilikischen Stadt Mopsos," *Archäologischer Anzeiger* (1963), 231 ff; Th. Pekary, "Kaiser Valerians Brückenbau bei Mopsos in Kilikien," *Historia-Augusta-Colloquium 1964/1965, Antiquitas*, ser. 4, vol. 3 (Bonn, 1966), 139–41; Ibn Ḥawqal, 163, 180 f, 184 f, 206; Canard, *Sayf al Daula*, 45, 49 f, 61, 68, 81 f, 214, 299, 381, 392 f.

[6] Canand, *Sayf al Daula*, 173–77, 182–89.

[7] Lüders, 67 note 1; Ralph of Caen, *Gesta Tancredi, RHC, HistOcc*, III, 634–39.

[8] Lüders, 39.

[9] See above, Part I.7. Cf. Brocquière, 94.

[10] "Miṣṣīṣ," *EI*, 526.

[11] Wilbrand von Oldenburg, 17.

[12] Langlois, *Voyage*, 446–63; Alishan, *Sissouan*, 284–90; Davis, 64–69; Barker, 111, 265; Texier, 730.

[13] Davis, 66. Alishan (*Sissouan*, 288) published the same map, but unfortunately the north arrow was reversed. The etching published by Langlois (*Voyage*, 451) shows the ascending circuit wall and towers placed at regular intervals. Budde's general plan of the city (I, 12, pl. 4) shows the outline of a circuit north of the river. This circuit, which appears to have square(?) salient-buttresses, is the one I sighted in the summer of 1973. I did not explore the fragments of the circuit south of the river in the Kafarbayyah. Bossert's short notes (*Türk Arkeoloji Dergisi* 7–10 [1956–59], passim) on his excavations at this site tell us little about the medieval walls. See also L. Budde, "Die Ausgrabungen in Misis-Mopsuhestia," *Deutsch-Türkische Mitteilungen* 17 (1957), 1 ff.

[14] L. Budde, "Die rettende Arche Noes," *Rivista di archeologia cristiana* 32 (1956), 41–58; idem, "Die frühchristlichen Mosaiken von Misis-Mopsuhestia in Kilikien," *Pantheon* 18 (1960), 116–26; idem, I, 9 ff, 101; H. Buschhausen, "Die Deutung des Archemosaiks in der Justinianischen Kirche von Mopsuestia," *JÖB* 21 (1972), 57–71; E. Kitzinger, "Observations on the Samson Floor at Mopsuestia," *DOP* 27 (1973), 135 ff; H. Thümmel, "Das Samson-Mosaik in Misis (Mopsuhestia) und seine Inschriften," *Zeitschrift des Deutschen Palästina-Vereins* 90 (1974), 69–75; M. Avi-Yonah, "The Mosaics of Mopsuestia—Church or Synagogue?," *Qadmoniot* 5.2 (1972), 62–65; R. Stichel, *Die Namen Noes, seines Bruders und seiner Frau* (Göttingen, 1979), 15–19, 98–103.

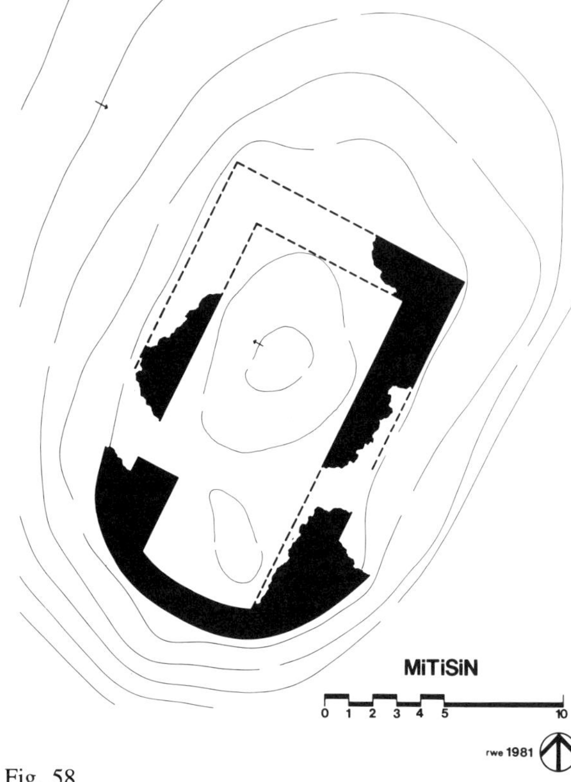

Fig. 58

Mitisin

Mitisin Kalesi[1] is a watch post in one of the more inaccessible regions of the Nur Dağları (the Anti-Taurus Mountains). This site can be reached by driving southeast from Osmaniye on the paved road to Zorkun. The small village of Mitisin Köy is 3 km northeast of Zorkun. The distance from Osmaniye to Mitisin is about 30 km. Although the paved road from Zorkun does not continue east, it is safe to assume that some sort of medieval trail did continue to the Maraş–Antioch road. Mitisin was built to observe traffic on this trail. Even though the area immediately around the watch post is covered with pine trees, there is still a clear view of the Cilician plain and Osmaniye to the northwest. The formation of mountains in this area *seems* to block Çardak Kalesi from view. There may yet be undiscovered forts or watch posts in the area that had intervisibility with Mitisin Kalesi. This medieval site is just north of the village of Mitisin Köy (pl. 168b);[2] its original name is unknown.

The watch post at Mitisin is essentially a four-sided structure (fig. 58) that sits atop a densely forested mound. The southern end of the building is rounded, and the southeast corner extrudes beyond the limit of the east wall (pls. 167b, 168a). Unfortunately the structure has suffered considerable damage, and large sections of the west, north, and east walls are not visible above the present ground level. The interior space is almost rectangular, except for the jog at the southwest where some sort of door seems to have been accommodated. There is a shallow pit in the north half of the fort. There appear to be no internal divisions on the interior of the watch post.

The masonry of the fort consists of a type IV that has an extremely wide core. The roughly cut fieldstones used in the construction are not limestone but serpentine.[3] On the exterior facing more care seems to have been taken to make regular parallel courses. Near the northeast corner, the interstices have been carefully filled with a white sandy limestone mortar. This same mortar is used in the core of the wall. The stones of the interior facing are coursed to a lesser degree and in some cases are hidden in a thick matrix of mortar.

The plan and masonry of this fort are *not* typical of Armenian construction. Considering that the Byzantines have also constructed forts at the nearby sites of Çardak, Savranda, and Kozcağız, it is likely that Mitisin Kalesi is Greek.

[1]The plan of this hitherto unattested site was completed in July 1981. The contour lines are spaced at intervals of one meter.

[2]To my knowledge, the village of Mitisin does not appear on any of the maps in the List of Maps, above.

[3]Serpentine is also used in the nearby Armenian church at Frenk; see Edwards, "Second Report," 124.

Payas

On the west flank of the coastal road that links Iskenderun to Adana and Osmaniye is the fortified complex of Payas.[1] Here two separate and roughly contemporary forts were built by the Ottomans. The larger of the two is situated near a coetaneous bedesten and mosque, while the smaller once guarded the nearby port. Even though these structures postdate the fall of the Armenian kingdom, I have included them in this study because one or both is occasionally mistaken for an Italian or Armenian construction.[2] Both of the forts at Payas have a type of exterior facing stone that is common in Armenian military architecture. Over twenty years ago the Turkish authorities began the careful restoration of the bedesten and mosque. Because a detailed architectural study of both structures and the adjoining fortress is presently under production, I was asked to confine my critical survey to the small fort at the port. Thus I will supply only summary comments on the bedesten fortress.

There is little verifiable information on any pre-Ottoman settlement at Payas.[3] Construction of the present complex was completed in the fourth quarter of the sixteenth century under the patronage of Sokollu Mehmed Paşa.[4] It was intended as a medrese, emporium, and a complementary port to Ayas. Payas remained a thriving trading center until the mid-nineteenth century when considerable damage was inflicted on the site by the troops of Küçük Ali.

The fort protecting the bedesten and mosque has a roughly pentagonal plan and is surrounded by a deep moat whose vertical sides are faced with masonry. The only entrance into this fort is in the center of the east wall (pl. 169a). This opening is preceded by a barbican,[5] which rises from the bottom of the moat, and by a massive salient extension of the entrance itself (pl. 170b). Both the north and south corners of the east wall have a single huge round tower (pl. 169b). At the west ends of the north and south walls stands a massive hexagonal tower. Between the hexagonal and round towers there is a single square salient. The V-shaped west flank of the fort has one massive round tower at its apex (pl. 170a). The third story of this salient has a smaller diameter than the lower levels, creating a circular terrace atop the second level. There are three types of embrasured openings in the circuit and towers; all of these openings appear in conjunction with casemates. The first type is identical to that seen at Ayas, which consists of a loophole and a highly enlarged stirrup-shaped opening. This opening constitutes the lower half of the loophole. The second type consists of a loophole with a round hole in the center. The third type has a stirrup-shaped opening in the center of the loophole (pl. 171b).

Except for the crumbled walls of a few internal buildings, the single bailey of the bedesten fort is flat and empty (pl. 171a). Unlike many of the Cilician forts, the two-story-high circuit walls here are periodically opened at the second level by the casemates of the embrasured loopholes. Although stairs frequently lead from ground level to the wall walk, these casemates could only be entered with wooden ladders.

While the plan and architectural features of the fort are non-Armenian in character, the exterior facing stones are very familiar. Except for the frames around doors and windows, most of the exterior facing is type

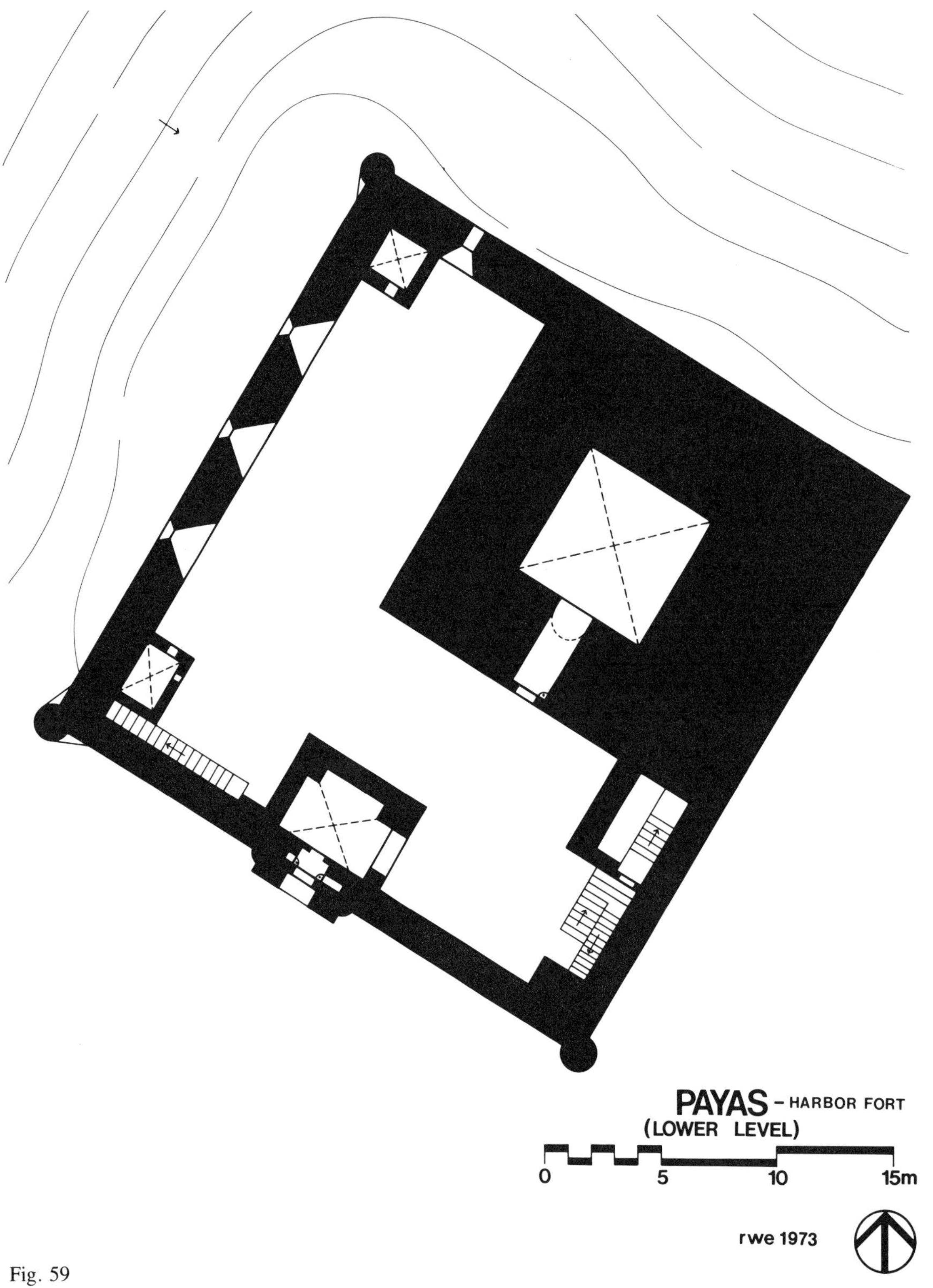

Fig. 59

VII with extremely broad drafted margins (pl. 170b). Since this type of masonry is not used in Ottoman fortresses outside Cilicia, it is likely that local Armenian masons were hired and directed by Turkish architects. The only evidence of brick masonry is in the small ruined mosque on the interior of the fort and in the rounded covers of the casemates (pl. 171b).

An identical kind of type VII masonry is used extensively in the harbor fort of Payas (fig. 59). This structure has a simple square plan and a single, centrally placed entrance at the south (pl. 172a).[6] There is neither a barbican nor a moat at this complex. In the northeast corner a three-story donjon occupies over 30 percent of the total surface area of the fort. On the exterior of the circuit there are a number of curious features. Three of the four corners are protected by miniature salients (pl. 172b). As with the other two towers, the salient at the southwest is fully rounded, but it has in addition two peculiar, broadly splayed attachments at the top (pl. 173a). The north salient has only one such attachment. A buttress was not deemed necessary for the northeast corner because of the presence of a solid donjon. Only the donjon is articulated on the exterior with a single torus molding; this decoration continues around the entire keep (pl. 175b). The circuit attached to the donjon has only four windows (pl. 175a). In the north wall the single window is straight-sided on the exterior and embrasured on the interior. In the west wall the three oversized embrasures have curious reverse splays on the exterior (pl. 173a). This feature also appears on the second-level windows of the donjon (pl. 175b). I see little advantage to this design from a military standpoint; it increases only slightly the enfielding range of the cannons. In fact, it may offer a disadvantage in that enemy fire is more easily deflected into the interior.

The entrance at the south (pl. 173b) resembles its counterpart in the bedesten fort in that it projects from the surface of the wall and has rounded sides. Likewise the exterior door of the entrance is covered by a depressed arch of joggle-jointed voussoirs and is fronted by a tall pointed arch. Below the latter and set within the thickness of the portal are small benches. The dedicatory inscription (or coat of arms?) above the depressed arch has been removed; flanking the outer arch are traces of a lion relief. As with its counterpart in the other fort, this portal leads into a small gatehouse that constitutes a bent entrance (pl. 174a). On its interior the groin-vaulted covering and part of the upper-level walls are constructed with small crude blocks of ashlar (pl. 174b). Generally, the walls and openings are made with large, finely cut ashlar blocks. Since the junctions between the two types of masonry are rather irregular, the crude ashlar may represent a reconstruction. While the east (or interior) door of the gatehouse is jambless, the exterior door has two sets of jambs. Behind the outer jambs are the fragments of two pivot housings and the sockets of a crossbar bolt to secure the double doors.

The interior of the fort consists of an L-shaped courtyard (fig. 59). In the far south corner there is a bipartite stairway. The first flight leads into the rectangular corridor that gives access (via a third stairway) to the second floor of the donjon (pl. 176b). The second flight of stairs reverses direction and leads to a now collapsed wall walk. The only other stairs in the courtyard are west of the south entrance. These stairs surmount a small groin-vaulted room in the west corner and rise to the level of the wall walk. A similar room is in the north corner (pl. 175a).

The three-story donjon has many unusual features (pl. 175b). Its lower floor (the only level shown on the plan) has no windows and only a single roundheaded door that leads to a small vaulted room on the interior. The arch over this door is constructed like its other counterparts at Payas with joggled joints. The vault over the square interior room is constructed of crude stones and shaped into a canopy (pl. 176a). It resembles a groined vault in which the salient junctions are inverted. The walls of this room are made with a relatively well cut ashlar. The lower-level walls of the donjon are unusually thick, as if it were intended as a bunker.

The second level is opened by numerous embrasured windows on all four sides. The four windows in the west wall, unlike most of the openings at this level, are splayed so that the broad end is at the exterior. On the interior these openings consist of simple square holes. The south wall has one such embrasured window and two damaged embrasured loopholes (pl. 178a). These loopholes, which also appear in the north[7] and east walls as well as in the third level (pl. 178b), have a single round hole in the center. They are identical to one of the three types in the bedesten fort.[8] Occasionally, fragments of corbels protrude from the flat top of the second level. The only unusual feature on the exterior of the second level is a chase that is located two courses above the horizontal torus molding. This chase is visible only in the west and south walls. Regarding the windows in these walls (not the embrasured loopholes at the south), the chase runs across their entire widths. The spaces between the chase and the sills are filled with small stones. It seems that the sills for the five windows were intentionally raised after the initial period of construction, perhaps to accommodate the wooden roof that was temporarily placed over the courtyard.

The third level of the donjon consists of a polygonal tower whose diameter is about 50 percent of the width of the second level (pl. 178b). The construction here is reminiscent of the west tower in the bedesten fort (pl. 170a) and provides a continuous wall walk to man the corbels of the second level. At the south end of this tower there is a roundheaded door. The only evidence of reconstruction at the third level are some modern cinder blocks that once formed part of a military watch post. As with the lower two levels, the exterior facing here is a fairly uniform type VII and represents one period of construction for the entire donjon. The only curious feature in the third story is the half-buried gun port at the present ground level. This may represent a mistake that was corrected during the initial period of construction. A curving stairway links the second and third levels.

The interior of the second level consists of a square inner court surrounded by an arcade that forms an almost continuous ambulatory (pls. 177a, 177b, 178a). The embrasured loopholes and windows are approached from the ambulatory. The groined vault, which covers the inner court, is opened by a single hatch in the center. Each side of the square arcade consists of two arches. Above the central spandrel of the arches there is a single square window. The barrel vaults over the ambulatory and central groined vault are constructed like the covering in the bent entrance. In 1968 certain areas on the interior of the second level were refaced with a white ashlar (pl. 176b).

[1]I visited the site of Payas in May 1973 and July 1981. During my first visit only the lower level of the harbor fort was surveyed. The contour lines on the plan are at intervals of 50 cm.

[2]*Blue Guide* (Hachette), *Turkey* (Paris, 1970), 654.

[3]This is possibly the ancient site of Baïae or Baiae; see Alishan, *Sissouan*, 469. Payas is probably the Arab Bayās; also spelled Baiyās or Bāyyās. Hellenkemper (107) identifies Payas with the Canamella of Wilbrand von Oldenburg (17, 20, 53, 57, 72). However, Cahen (149 f) prefers to identify Ḥiṣn at-Tīnāt with Canamella. See also: Heberdey and Wilhelm, 18; Brocquière, 90 note 1; Sümer, 61 ff; Abū'l-Fidā', 28; Taeschner, I, 146 f; Barker, 111, 264; *Handbook*, 146, 318, 722 f; Seton-Williams, 166; W. Ruge, "Baiae," *RE*, 2775; R. Hartmann, "Bayās," *EI*, 684; Humann and Puchstein, 160–63; Le Strange, *Palestine*, 422; Cuinet, 105 f; Tomaschek, 71; Schaffer, 97. Al-Muqaddasi (*Description of Syria*, trans. G. Le Strange [London, 1892], 8 f) lists Payas as an important settlement in a northern district of the province of Syria. Cf. Canard, *Sayf al Daula*, 44, 49–51, 61; Ibn Ḥawqal, 163, 184 f, 188; Janke, 10–12, 53–55.

[4]Langlois, *Voyage*, 471 f; Deschamps, III, 70; Alishan, *Sissouan*, 469 f; J. Kramers, "Ṣoḳolli Muhammad Pasha," *EI*, 474 f; K. Otto-Dorn, "Islamische Denkmäler Kilikiens," *Jahrbuch für kleinasiatische Forschung* 2.2 (1952), 113–16 notes 3–6; F. Akozan, "Türk Külliyeleri," *Vakıflar Dergisi* 8 (1969), 308, pls. 33–34; Ainsworth, 91 f; B. Darkot, "Payas," *IA*, 531 f; K. Erdmann, "Zur türkischen Baukunst seldschukischer und osmanischer Zeit," *Istanbuler Mitteilungen* 8 (1958), 30 f; S. Faroqhi, *Towns and Townsmen of Ottoman Anatolia* (Cambridge, 1984), 29, 76, 86, 95. Payas and Iskenderun were the Turkish harbors for Aleppo; see A. Wood, *A History of the Levant Company* (London, 1964), 76 f.

[5]Fedden and Thomson, 51.

[6]The plan of this harbor fort bears some resemblance to a Turkish medrese; see M. Sözen, *Anadolu Medreseleri* (Istanbul, 1972), 30 ff, 286 ff.

[7]The north wall of the second level has four openings. Three are embrasured loopholes with a round hole in the center. Only the opening at the west has the reversed splay on the outside.

[8]This type of embrasure is frequently used by Ottoman builders outside of Anatolia; see Meir Ben-Dov, *Jerusalem's Fortifications: The City Walls, the Gates and the Temple Mount* (Tel-Aviv, 1983), 95 ff.

Pillar of Jonah

The Pillar of Jonah[1] is located in the Portella, only 50 m from the shores of the Mediterranean and about one kilometer southwest of Sarı Seki (pls. 179a, 179b).[2] Legend has it that here Jonah was released by the whale. The Pillar was a gate complex that opened the route east to Kırıkhan (via Çalan); it also served as a road station. The ancient and medieval road that ran from Iskenderun (Alexandretta) to the Cilician plain passed through a small defile which the Pillar still surmounts. Today that old road is occupied by railroad tracks. The modern north-south highway of asphalt is immediately east of the defile. In some modern accounts the height and magnitude of this defile are exaggerated;[3] it is actually a very small natural cleft. Today the only surviving part of the Pillar of Jonah is a single tower, crowning the west side of the cleft. (pl. 180a). A photograph taken in the 1920s shows a post-medieval building (with two windows) on the summit of the east side of the cleft.[4] Unfortunately the construction of the modern road demolished this structure. In historical times the only known enlargement of the defile occurred during the reign of Justinian when it was widened to accommodate more traffic.

The troops of Alexander the Great and later the Crusaders passed through the Portella. Alexander probably stationed a force here to keep the road open and to prohibit any of the Persian forces from crossing west past Çalan and eventually linking up with Darius, who was moving toward Issus via the Amanus pass.[5] The value of having a permanent garrison here was

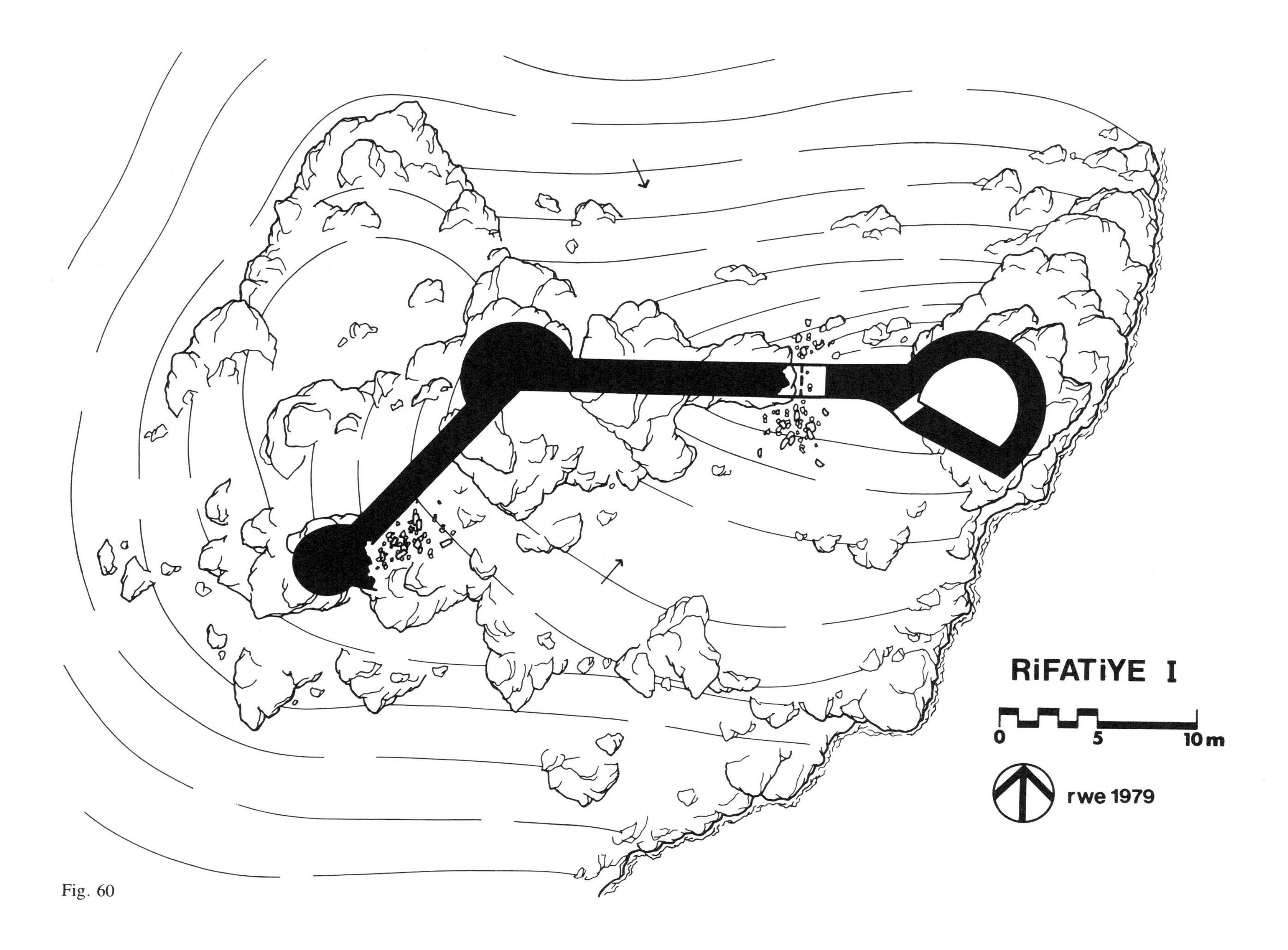

Fig. 60

realized in medieval times when the nearby fort of Sarı Seki was built. The Portella is frequently mentioned as a geographical feature and the location of a Crusader toll station. In 1154 the Templars aided Stephen, the brother of Baron Tᶜoros II, in attacking a Seljuk patrol near this site.[6] Although it was principally controlled by the Templars, there were alternating periods of Greek and Armenian occupation in the twelfth and early thirteenth centuries.[7] Hetᶜum posted one of the three divisions of his army within the Portella in 1266 to counter a Mamluk attack.[8]

There is no reliable literary evidence on this unexcavated site that tells us anything about the content of its architecture or the periods of construction.[9] What is certain is that the surviving square tower is not part of a medieval building. This rapidly deteriorating tower, which measures 4.7 × 3.8 m at the base, is constructed out of perfectly cut blocks of ashlar (pls. 180a, 181a). On the average these stones measure 78 cm in length, 66 cm in height, and 61 cm in depth. There is no poured core and no evidence of mortar on the exterior (pl. 180b). The interior of the tower is solid, consisting of neat courses of ashlar with occasional patches of mortar. While the stones on the exterior are quite smooth, the interior stones are less carefully cut. A few pieces of fallen masonry show evidence that they were cut with *anathyrosis*. This is not a medieval masonry technique but one common in the ancient world.[10] Perhaps the assessment of H. Kähler and T. Boase, that this is actually part of a Roman triumphal arch, is correct.[11] Instead of limestone, the masonry consists of a grayish-white marble; this too is atypical for medieval Cilicia. There is no evidence of further construction in the area immediately east of the Pillar and the modern road (pl. 181b).

[1] I did not execute a plan during my visit in July 1981 because so little remains of this site.

[2] There is some confusion whether "Pillar" should be plural or singular. Since a single tower of stone is present *today,* I will use the singular. The Portella is that small portion of the coast from Alexandretta to Sarı Seki where the western ends of both the Belen and Çalan passes can be approached. By a process of metonymy "la Portella" is frequently used to identify the Pillar of Jonah. Both Cahen (148 f) and Deschamps (III, 70 f) identify the Pillar with the Arab/Turkish Sakaltoutan. This site has been associated convincingly with Kodreigai (see J. Kubitschek, 'Εν Κοδρείγαις ὅροις Κιλίκων, *Numismatische Zeitschrift* 27 [1896], 87–100); cf. Janke, 22 f, 154. The Pillar appears as "Kız Kulesi" on the *Adana (1)* and *Adana (2)* maps. See also: Alishan, *Sissouan,* 497 f; Dussaud, 183, 446; Ainsworth, 92; above, Part I.7, note 7; *Handbook,* 317; M. Sanudo, *Secreta fidelium crucis,* trans. A. Stewart, PPTS (London, 1896), 4; Magie, II, 1153; Tomaschek, 72; E. Honigmann, Σύριαι πύλαι, *RE,* 1727 f; H. Treidler, Πύλαι Κιλίκιαι, *RE,* Supp. 9, 1352–66.

[3] This exaggeration can lead to serious misinterpretations of history; see Engels, 131–32.

Geographical studies prior to the construction of the railroad show that the pass has changed little since the nineteenth century. W. Ainsworth ("Notes on the Comparative Geography of the Cilician and Syrian Gates," *Journal of the Royal Geographic Society of London* 8 [1838], 185) says that "over this narrow pass the road is carried with care, and, although steep, is paved throughout." R. Oberhummer and H. Zimmern (*Durch Syrien und Kleinasien* [Berlin, 1899], 104) merely report the presence of the Pillar of Jonah and the mile-high mountain on its east flank. They do not complain about a difficult passage through the defile.

[4] Jacquot, 147. This structure is probably the "Gendarmeriekaserne" photographed by Janke (fig. 2).

[5] Engels, 48 ff and 131 ff.

[6] Gregory, 171 f.

[7] Langlois, *Cartulaire,* 110.

[8] S. Der Nersessian, "The Armenian Chronicle of the Constable Smbat," *DOP* 13 (1959), 164.

[9] Wilbrand von Oldenburg (16) describes it as a vaulted gate of polished marble. According to the local legend, the gates contained the bones of Alexander the Great.

[10] W. Dinsmoor, *The Architecture of Ancient Greece* (New York, 1975), 173.

[11] H. Kähler, "Triumphbogen," *RE,* 459 f; Boase, 167. The plan and brief description of the site published by Heberdey and Wilhelm (19) show two surviving piers of a gate complex. In Ēpᶜrikean (II, 77) fig. 35 depicts a more complete gate below Sarı Seki; cf. Alishan, *Sissouan,* 477; Alishan, *L'Armeno-Veneto,* pt. 1, opp. 16; Carne, III, opp. 46; Barker, 91, 263.

Rifatiye I

Six kilometers southwest of Çokak is the village of Rifatiye, which gives its name to two garrison forts[1] in the immediate area.[2] Although Rifatiye I and its sister fort to the northwest, Rifatiye II, are very simple in plan, they guard one of the more strategic roads in northeast Cilicia.[3] At Çokak, which is directly north of Andırın, three northern routes merge into a southern road that follows the course of a long, slender valley past Rifatiye and Canbaz and eventually terminates at Kadirli. Rifatiye I rests on an intermediate outcrop in the short range of mountains that parallels the east flank of this road. Visible from the east and south sides of the fort there is a narrow and very difficult track on a north-south axis (pl. 182a). The local residents indicated that this alternate route was recently cut by the Forestry Service. This fort is less than one kilometer southeast of the agricultural village of Rifatiye. A stream supplied by a natural spring flows past this village throughout the summer. A creek, created by a smaller spring, is located in a gorge, about one kilometer east of the fort. The fort itself has no source of flowing water but does have a single cistern. The medieval name and history of this area are unknown.

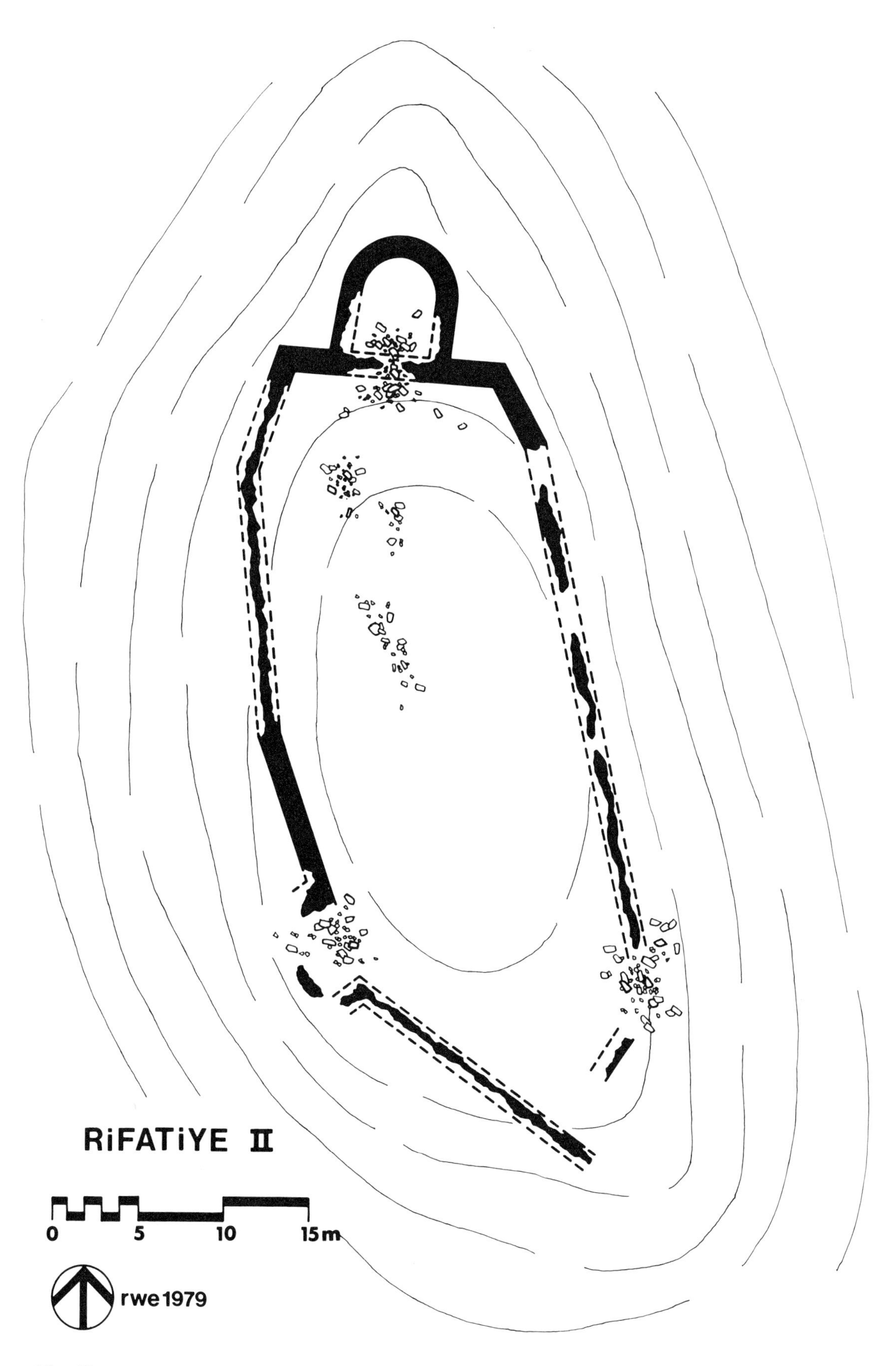

Fig. 61

An important feature of this fort is the masonry. On the interior side the masonry consists of type III and a large type IV (pl. 183a). The bulk of the exterior of the wall has a type V facing (pl. 182b). A few of the lower courses appear to have type IV masonry. Because of the tendency of the Armenians in Cilicia to use *only* types V, VI, and VII as an exterior facing on the circuit walls, it is quite possible that they are responsible for most of the construction here. The lack of exposed corners and the presence of rounded towers would certainly associate the present plan of Rifatiye I with the Armenians. The lower courses of type IV may represent an earlier Byzantine occupation.

Rifatiye I is a simple bent circuit with three towers and a single entrance (fig. 60). This fortified wall, which crosses the top of a limestone mass, is designed to protect the easiest route of access from the northwest. Behind this wall the small bailey is limited at the east and south by sheer cliffs; here walls were deemed unnecessary. At the southwest, where access is easier, there is today no trace of masonry. Just why this area was left undefended is a mystery.[4] Unlike the area to the east, the natural terrain does not offer a defense. The western and central towers on the parapet are both solid. The central bastion actually covers the point of junction where the wall angles to the east (pl. 182b). The height of the tower and walls in the central area is only 1.6 m. The eastern section of this wall and the adjoining east tower are almost double in height due to the plunging topography (pl. 182a). It is at the east end that a single door (now shattered) pierces the wall (pl. 183a). Beyond this door there is the hollow tower that is covered by a semidome. A single squareheaded door leads into the interior of its stuccoed room, which certainly functioned as a cistern.

[1]The plan of Rifatiye I, a hitherto unattested site, was executed in July 1979. The interval of distance between the contour lines is 75 cm.

[2]The village of Rifatiye appears on the *Kozan* map.

[3]If any of the narrower aspects of this road have been given specific names as passes, they are unknown to this author. The Bağdaş and Mazdaç passes are to the west on the road that connects Çem with Göksun. See the discussion of Geben Kalesi (note 4) in the Catalogue and Alkım, map 3.

[4]It is possible that a now vanished wall extended across the south.

Rifatiye II

With regard to the general location and importance of this site see the discussion of Rifatiye I. Rifatiye II[1] is a Byzantine (or late Roman) garrison fort that lies on a gently sloping hill just northwest of Rifatiye Köy and on the west flank of the Kadirli–Çokak road (pl. 183b). It was coguardian of this strategic route. The closest source of water for the fort is the spring near the village.

This fort encircles the top of the hill in a very angular and symmetrical fashion (fig. 61). Because of *severe* damage, it is impossible to determine where the entrance is located. A modern path leads to the shattered remains of what was probably a square tower at the southwest. Today large sections of the walls have fallen down the sides of the hill. In some cases the facing stones have fallen away, leaving an exposed core (pl. 184a). In the areas where gaps were created by the collapse of the walls, the local denizens have plugged them with dry thorn bushes to make the fort a corral for sheep. The considerable amount of rubble within the circuit indicates that buildings of stone were once present. If there was a principal point in this fort it would be the now shattered north tower (pl. 184b). The nature of its covering and the position of any doors or windows are uncertain. Both the interior and exterior masonry of the towers and circuit is type IV. Unlike Rifatiye I, the circuit was not rebuilt by the Armenians. This may indicate that they relied exclusively on Rifatiye I for protecting the road. The Armenians underplayed the importance of this area, implying that their shield of security extended farther to the north (that is, to Geben).

[1]The plan of this hitherto unattested site was executed in July 1979. The distance between the contour lines is 75 cm.

Saimbeyli

At an altitude of 1,300 m, the village of Saimbeyli[1] (Armenian: Hačən) straddles the strategic road that links Feke to Kayseri and the Cappadocian plain.[2] Saimbeyli is built along the slopes of two adjoining valleys (pl. 185b). A natural spur of rock provides a majestic foundation for the fort which is located at the extreme southern end of the outcrop (pl. 185a). This V-shaped spur is at the junction of two major tributaries. The Obruk Çay, which cuts the route for the north-south road, is on the west flank of the fort. Today most of the modern village of Saimbeyli overlooks the Obruk Çay. From the east the Kirkot Çay flows through a steep valley and merges with the Obruk Çay directly south of the fortified outcrop (pl. 185b). Before 1920 the valley of the Kirkot Çay held a substantial portion of the Armenian settlement.[3]

Although Saimbeyli (Hačən) is mentioned frequently in studies on Armenian Cilicia and is the subject of an 857-page monograph, we have no documented information on the foundation of the medieval fort or the nearby monastery of St. James.[4] Because of its isolation, Hačən did not play a prominent role in the politics of the Armenian kingdom. The fortification and garrison were placed here because of the proximity to a major road. After the fall of the Armenian kingdom, the importance and size of Hačən increased dramatically. In the second half of the nineteenth century the Armenian population numbered over twelve thousand.[5] Aside from the monastery of St. James, which is at the far northwest end of Saimbeyli, fragments of the Armenian settlement can be seen in the Kirkot valley and along the top of the outcrop.

There is a relatively easy approach to the fort at the southwest end of the outcrop (pl. 185a). The trail zigzags past the south wall of a two-story brick and concrete structure, which was dedicated in 1912, and turns to the north. Because the height of the natural cliff diminishes at the northwest end of the fort, the medieval masons scarped a vertical face to continue the rock barrier and limit the line of access. The modern trail follows the base of the cliff and abruptly ascends to the terrace northwest of towers A and B (fig. 62; pl. 186a).

The masonry of Saimbeyli Kalesi is entirely in keeping with the architectural traditions of Armenian Cilicia. The vast majority of the exterior facing stones consist of type VII masonry. In some cases the consistently rusticated face of type V is evident. It is also quite interesting that in some of the courses the margins of the type VII have been stuccoed with white mortar and studded with small dark stones. Such a technique normally occurs with the poorer quality rusticated stones of type VI masonry, where its wider margins give the exterior stucco a firm anchorage. The attempt to stucco the relatively thin margins of the type VII may have been done for aesthetic effect. On the interior of the fort there are a number of types of facing stones. At the extreme south in salient E type III facing stone is common; it occurs again along the west wall of F. Type IV is also present in room D (pl. 186b). In the wall north of chapel C the interior facing improves decidedly, for in this area types V and VII seem to predominate. This is also true for the lower half of the north circuit (flanking A and B). But the upper half was refaced at a much later date (fifteenth century?)[6]. This post-medieval masonry (pl. 188a) consists of recycled blocks and fieldstones bound in a thick matrix of mortar and covered with plaster. This crude masonry is also anchored by horizontal wooden headers in the core (pl. 188b). The square area marked as H on my plan stands today only to its foundation. What survives is an extremely crude masonry of fieldstones bound in irregular courses by rock chips and small traces of mortar. The walls of H, which probably are the foundation of a post-medieval church, have a poured core.

This fort consists of a high fortified circuit wall that cuts off and surrounds the tapering end of the spur (fig. 62). There is no evidence of outworks or ditches preceding the north wall. This wall runs from one side of the cliff to the other. This barrier was joined at the east (pl. 187a) by a wall of equal height that extended from the tower-chapel C. The upper portion of this east wall has now collapsed. On the other side it appears that the original west wall, which connected with tower G, was only about 60 percent of the height of the north wall. At the northwest junction the upper section has been squared off, clearly showing the reduced height of the west wall (pl. 188a). The west wall, like the one at the east, rises directly from the cliffs.

The north wall was opened by a single gate in the center that was flanked by two horseshoe-shaped towers (pl. 186a). Today, in the area where the gate once stood, there is merely a large hole in the circuit. In the lower level of tower A (not shown on the plan) the wall is well preserved on the exterior, except for a puncture at the east. Only a section of the upper level of tower B still has facing stones in situ. Each tower has a single apsidal chamber at the first and second levels. In the lower level tower A was probably opened by a single door at the south. It is unlikely that the breach in the tower's east side held a postern or window. As in tower B, only the walls in the lower-level chamber of A have been stuccoed and painted (pl. 188b). The same stucco was applied to the inner face of the north wall of H. The second level of tower A is opened by two embrasured loopholes. The one at the north is undamaged and has a small stirrup base. Because of damage to the upper level of A, only the inner side of the east embrasure is visible today.

The difficulty with the upper level of A, and likewise the same area in tower B, is to determine the means of access. In neither of the upper-level rooms is

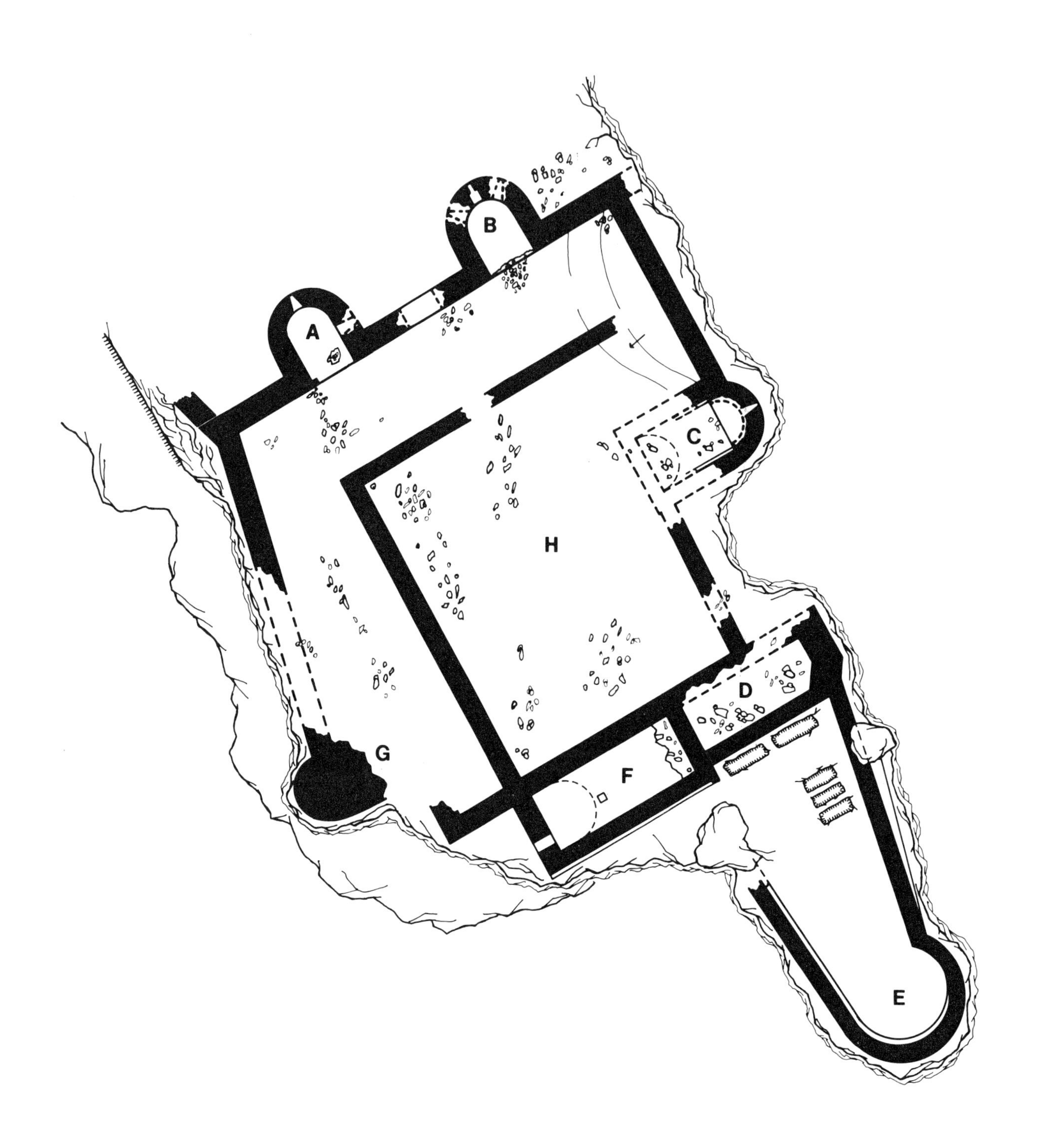

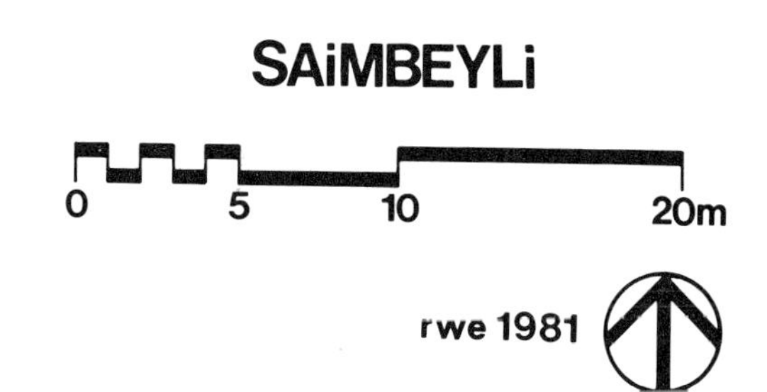

Fig. 62

there evidence of a door at the south. In tower A only a small portion of the open south end (in the west corner) appears to have been walled. This fragment of wall, as well as the squared top of tower A, probably dates from the fifteenth century when this tower became the base of a square campanile.[7] In medieval times ramparts probably topped both of the towers. Because the inner face of the north wall of the fort is so heavily reconstructed, we can only assume that a crenellated staircase once gave access to the ramparts and second level (cf. Yılan and Tamrut). There is the possibility that wooden structures were built directly onto the interior of the north wall in medieval times. The presence of joist holes in the refaced areas indicates that wooden beams were used in the fifteenth century.

Tower B is quite similar to A at the lower level, except the former has a straight-sided window with a miniature casemate on the interior (this opening is depicted on the plan between the two upper-level embrasures). This opening could provide only a small amount of light and ventilation. Both of the embrasured loopholes in the upper level of B are damaged. At the south end of the second level the masonry in the vault consists of long crude slabs laid radially. Toward the north end of this room the apsidal ceiling is constructed with a large type V masonry. This is quite different from the masonry in the upper level of tower A (pl. 188b), where more crude stones in a thick matrix of mortar are common at the north. This difference in masonry types is certainly due to several periods of construction.

What is left of chapel C conforms to the design characteristics of Armenian ecclesiastical architecture in Cilicia (pl. 187a).[8]

South of the chapel there are two adjoining rectangular chambers. Only the foundation of room D is visible today (pl. 186b); its function is unknown. Directly to the west is the vaulted cistern F. Except for a small section at the east, the vault is well preserved (pl. 187b). It is stuccoed like the interior walls and is opened by a single square hatch. The only other evidence of an opening in the cistern is a small round-headed window at the top of the west wall. Although much of the masonry of cistern F is fairly crude, the exterior of its west end is constructed with type VII masonry, since it extends into the space of what should be the circuit. At the far south, a fingerlike projection of rock is surrounded by a single wall, forming the rounded bastion E. The function of a thin ledge on the interior of the circuit is unknown. At the north end of E and just south of room D are the remains of some scarped graves (pl. 186b). These tombs probably date to the fifteenth century, when the fort was converted into a cloister.

Tower G, which is badly damaged today, appears to have been a solid bastion (pl. 185a). Only a few fragments of the circuit connecting G to the north wall are visible today.

[1]The plan of this hitherto unsurveyed site was executed in July 1981. The few contour lines on the relatively flat outcrop are at intervals of 75 cm. Only the upper levels of towers A and B are represented on the plan; the rest of the fort is represented at ground level.

[2]Texier, 583 f; Hogarth and Munro, 657 ff; Hild, 127 ff, maps 11 and 14, pls. 98–99; Alishan, *Sissouan*, 62 ff; Schaffer, 90 f; Sterrett, 239 and map 2. Saimbeyli appears on the following maps: *Central Cilicia, Cilicie, Everek, Marash*. See also: *Handbook*, 69, 91, 343, 337–82, 703; Cuinet, 94 f; Rašīd ad-Dīn, *Die Frankengeschichte des Rašīd ad-Dīn*, intro., trans., and comm. K. Jahn (Vienna, 1977), 44 and notes 78 and 79.

[3]Pōłosean, 106 f, 122 ff; Yovhannēsean, 179–82; Aghassi, 97–101.

[4]Ibid.; Alishan, *Sissouan*, 174–77; King, 240 f; for a discussion of the monastery of St. James (S. Yakob), see Edwards, "Second Report," 125 ff. Concerning the medieval name for this site see below, Appendix 3, note 6. Ramsay (281, 291, 312) believes it to be the late antique Badimon.

[5]The Armenian community here and in Sis occasionally prospered as a result of alliances with the local *derebeys*. See A. Gould, "Lords or Bandits? The Derebeys of Cilicia," *International Journal of Middle East Studies* 7 (1976), 485–506; cf. A. Sanjian, *The Armenian Communities in Syria under Ottoman Domination* (Cambridge, Mass., 1965), 233 ff.

[6]Edwards, "Second Report," 130: f.

[7]Ibid. The square campanile was part of the post-medieval church of the Holy Mother of God (Holy Astuacacin). Before its destruction by the Turks in 1915, it was heavily damaged on two prior occasions: first by a fire and later by an earthquake. For information on this and other Armenian churches in Saimbeyli see: note 3, above; Oskean; G. Galustian, *Marash* (New York, 1934).

[8]Edwards, "Second Report," 130 f.

Sarı Çiçek

The complex at Sarı Çiçek[1] stands on an outcrop of limestone in a small basin that was formed by the mountainous southern barrier of the vale of Karsantı (pl. 189). The closest medieval sites are Meydan, Hotalan, and Tamrut. The summit of Kale Tepe is directly to the north, and Meydan Dağı is at the northeast.[2] The single homesteader who lives near Sarı Çiçek reports that Meydan Köy (9 km south of Meydan Kalesi) can be reached in about three and a half hours by hiking over a difficult trail. The easiest way to reach Sarı Çiçek from Adana is to drive to Etekli via Çatalan on a reasonably good road of asphalt and packed gravel.

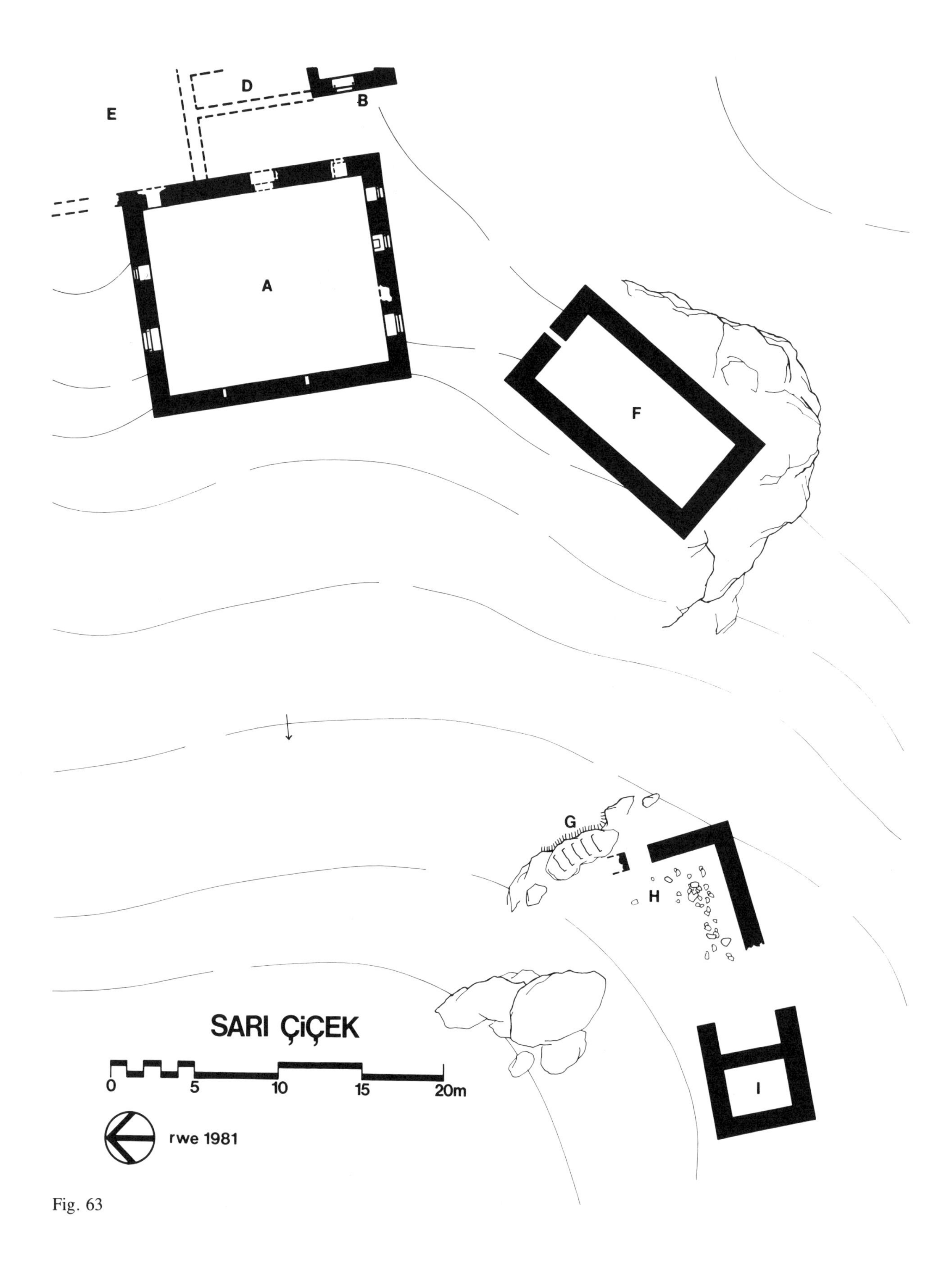

Fig. 63

From Etekli one must hike northeast for about 12 km in the company of an expert guide in order to locate Sarı Çiçek. An unmarked trail leads to this site through a series of narrow mountain valleys; there is an abundance of fresh water, and the lush scenery has not been disturbed by modern settlements.[3] Neither the site nor the name, Sarı Çiçek, which means "yellow flower" in Turkish, is readily identifiable in medieval or modern sources.[4] The masonry and the structural elements of the complex as well as the design of the chapel indicate that Sarı Çiçek is the result of Armenian construction.

What is unique about this complex (fig. 63) is that it is not surrounded by a circuit wall like a garrison fort, nor is it protected by towers and loopholes like the fortified estate houses. There is no single donjon but a series of almost casually placed residential buildings. The first floors of the buildings at Sarı Çiçek are opened by broad windows and doors (pl. 189). Only the exterior facing of fine rusticated stones gives any sense that this site is fortified. These factors, as well as the prominence given to the chapel and the care taken in the construction of the rooms, lead me to speculate that Sarı Çiçek was a monastery or retreat for one of the great Armenian nobles and perhaps the royal family. Defenses were deemed unnecessary because of the relative isolation of the area and because the vale of Karsantı to the north was so heavily fortified.

Generally, the rooms at Sarı Çiçek can be grouped into two separate units: rooms A–E at the northeast and F–I at the southwest (figs. 64, 63). Today the best preserved and largest room in the first group is the great hall A (pl. 189). The first level of this rectangular room is standing, except for the center of the east wall which has collapsed to its foundations. The only remains of a second level are in the center of the west wall. Like chapel B,[5] most of the exterior facing stones of A consist of type VII with occasional blocks of type VIII. Along the exterior of A's east wall the stones are consistently smooth (type VIII), no doubt because this is the west side of the corridor between B/D and A. This corridor may have been covered by a stone vault. In a few cases the boss on the type VII masonry has been shaved relatively flat; this modification of the type VII is rare in Armenian Cilicia. Most of the doors and windows are framed in type IX masonry. On the interior the facing consists of a mixture of types V and VII masonries (pl. 190a). The former is especially large and well cut. An exposed section of the core in the south wall (pl. 191) shows that the mixture of mortar and fieldstones was placed in perfectly parallel beds at each course level. The only unusual feature in the masonry occurs on the interior face of the west wall where most of the stones in a single course near ground level are laid vertically (pl. 190a). In the other courses, vertically laid blocks are rare.

The first level of room A is opened by three doors and five windows. In the south wall the door is at the west (pl. 191). As with the two windows in the south wall (pl. 190b), the interior side of the south door is covered by a vault, while the jambs on the exterior are covered by a flat lintel. However, the exterior of this door differs from the windows in that its lintel is covered by a pointed relieving arch; the resulting tympanum is filled by a recessed monolithic block. In the door and the two windows of the south wall a ledge has been carved into the soffit of the lintel, and two pivot holes have been bored in the corners of the ledge behind the jambs. These pivots anchored some sort of swinging shutters or doors. In addition, the outer half of the soffit of the windows has been bored with two shallow holes (not shown on the plans), no doubt to accommodate vertical bars. The sill of the window east of the south door has a curious square hole on the inner side. Its function is unknown.

The east wall of A is opened by a door near the center and a window at each end (pl. 189). Today only the foundation of the door is visible. Half of the south window is still in situ. It is designed exactly like the two windows in the south wall. The window at the north end of this wall is badly damaged, but it was constructed no doubt on a similar paradigm. The sills of the two east windows are set at a higher level (approximately 60 cm) than the sills of the windows in the north and south walls.

The north wall is opened by a single window and door. The window at the east is identical to the other windows in A. Also, the door is identical to its counterpart at the south, except that the relieving arch on the exterior is not so pointed. The vault over the interior side of the door has collapsed, exposing the back side of the monolithic tympanum block.

Because the ground level at the west was significantly higher than the other areas around A, there are no openings in the first level of the west wall (pls. 189, 190a). On the interior of the west wall there are two long thin niches, whose purposes are unknown. At a height where one would expect a second level, there is a tall rectangular window that is squareheaded on the exterior and roundheaded on the interior (not shown on the plan). There is no indication on the interior that the second level had a floor; thus room A has the distinction of having the highest ceiling of any chamber in medieval Cilicia.

The ground level rises sharply just south of chapel

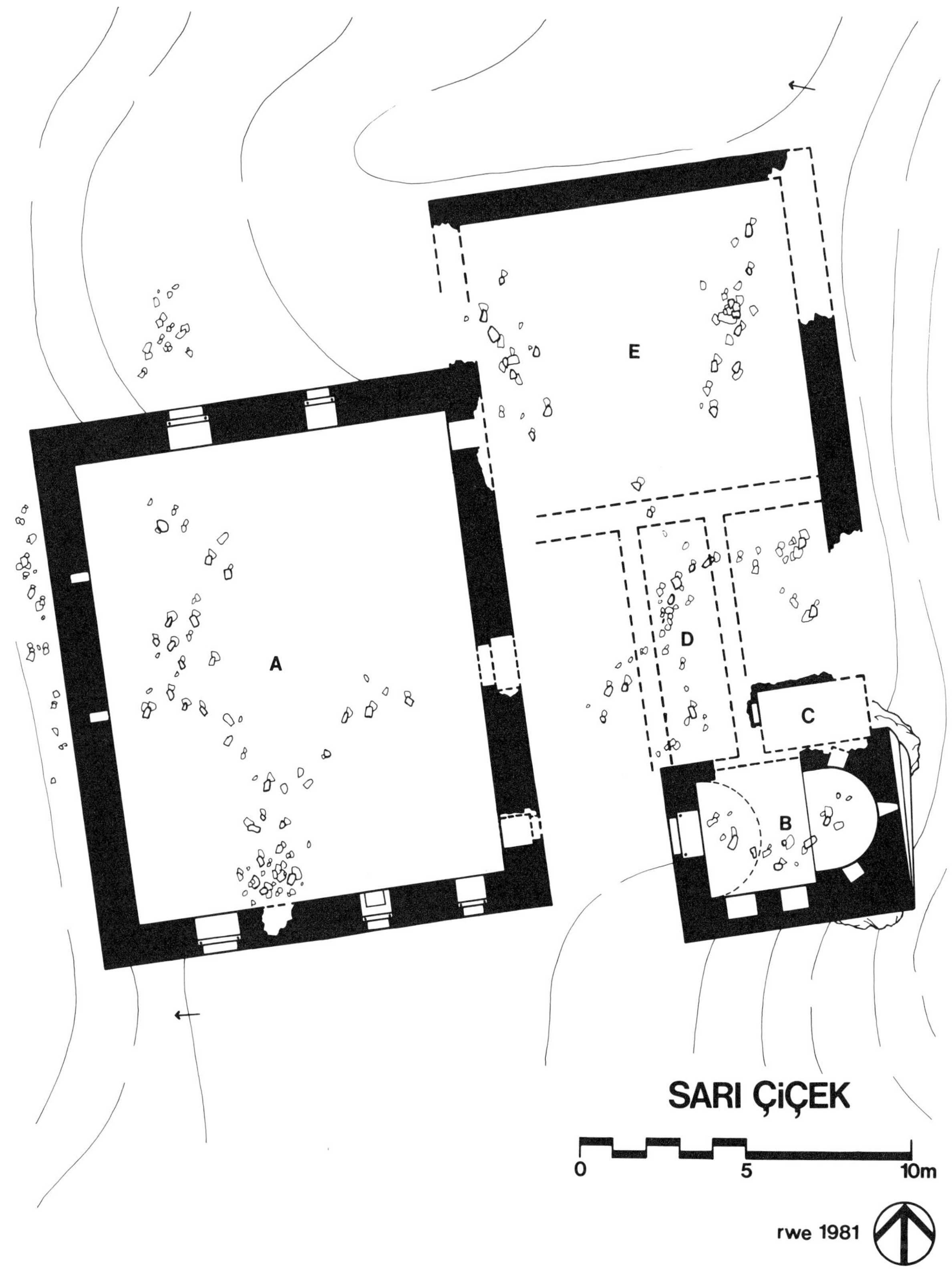

Fig. 64

B and the southeast corner of A (pl. 299b). This slope is covered with debris and is part of a natural incline on which the complex is built. Undoubtedly steps facilitated ascent to the corridor between hall A and chapel B. Since the plan shows only the ground level of A, C, D, and E, the first floor of B (below the chapel) is not represented. The entrance to this first-level room of B is at the south. Today this room is used as a barn by the homesteader, and entrance can be made only with difficulty (pl. 192b). Like the interior of the chapel above, only type IX masonry is used as the interior facing stone in the walls and the vault. This lower room is rectangular and has niches in the west wall and northwest corner. There is a squareheaded door in the north wall that connects to a chamber below C and D. This chamber is now completely blocked with debris. Little can be said about C, D, and E since only traces of their foundations are visible today (pl. 192a). At the west end of C there is a tiny vaulted niche, whose function is unknown. It is significant that the interior and exterior facings of the north wall of E are constructed entirely with type IV masonry. This may represent a repair.

Southwest of the area A–E are rooms F–I (fig. 63). The area between F and B is littered with tons of coursed masonry. Someday excavations will certainly show the foundations of many new buildings. The rectangular room F is buried in the debris of its collapsed walls. The only visible opening in the trace of F's walls is a narrow squareheaded window at the north (pl. 193a). Here most of the masonry consists of types VIII and IX. Farther to the southwest are G, H, and I. Because of extensive destruction, little can be said about these rooms. Scarped elements below the steps of G (pl. 193b) may be evidence that wooden constructions were also present.

[1]The plans of this hitherto unattested site were completed in June 1981. The interval of distance of the contour lines is approximated at one meter. Due to the limitations of time, the relative alignment of rooms H and I was estimated.

[2]Both of these mountains are listed on the *Kozan* map.

[3]The hike from Etekli should be made only by those in good physical condition and those willing to backpack supplies for a stay of two days.

[4]If Meydan Kalesi can be associated with the fortress of Kopitaṙ, then it is possible that Sarı Çiçek is the famous monastery of Drazark. In addition to standing at the base of the fortified Mt. Kopitaṙ, this monastery is said to be a day's journey west of Sis. It was rebuilt early in the 12th century by Baron Tʿoros I. Drazark (Latin: Tres Arces) became famous as a place of burial for the royal families of Cilicia and for the distinguished clerics of the region. The scriptorium and library there were renowned. See: Smbat, G. Dédéyan, 69, 71, 74, 87, 93–95, 123; Matthew of Edessa, 114, 148, 199 f; Samuel of Ani, 449, 458 f, 462, 467; Vahram of Edessa, 499, 508, 510, 525; Tēr Łazarean, 31; Alishan, *Sissouan*, 265–72; Ēpʿrikean, I, 623–25; above, Part I.7, note 34; Kʿēlēšean.

[5]For a description of the chapel and its masonry, see Edwards, "Second Report," 141 f.

Sarı Seki

The fort of Sarı Seki[1] lies in the Portella about 11 km north of Iskenderun and less than one kilometer from the Mediterranean. Perched on a rocky knoll (pl. 179b), it has a clear view of the road into the Cilician plain and is within cannon-shot of the Pillar of Jonah. Unlike the latter, it appears in part to be a medieval construction.

Sarı Seki was probably first built by the Templars in their effort to secure the south end of the Nur Dağları. Between 1135 and 1150 it is likely that this site was occupied once by the Armenian Baron Tʿoros II and probably twice by the Byzantines.[2] Around 1154 Reginald of Antioch took control of the forts in this region and returned them to their original owners. According to Jacquot, the Templars then held Sarı Seki continuously until 1266, when Malik al-Manṣūr captured it by siege and massacred the occupants.[3] Jacquot also identifies this fort with the Gastin (Gaston) of the medieval chronicles, but the present fort at Bağras is more likely to bear that name.[4] Unfortunately there is no evidence to associate Sarı Seki with any specific medieval site. Cahen identifies it with the Castrum Puellorum, but this association is in doubt since Albert of Aix links the latter with a site north of Payas.[5] There is a possibility that Sarı Seki is the medieval Merkez,[6] but this name can also be associated with other forts in the Amanus.

Today Sarı Seki is inside the perimeter of a high-security military compound. I was prohibited from visiting the fort and was forced to photograph the site at a great distance (pl. 179b). Two photos of the fort published by Jacquot in 1931 give some information about the nature of the site. Unlike Armenian forts, its single circuit wall appears to be rectangular and buttressed with only a few towers.[7] The walls are crenellated, and most of the masonry appears to be of poor quality. Cahen and Boase both believe that much of the fort was reconstructed by the Ottomans.[8] Boase reports the presence of a ruined tower of Crusader masonry within the Turkish entrance. The far south tower of Sarı Seki appears to have two types of masonry.[9] The upper two-thirds have small stones bound in a thick matrix of mortar.[10] However, the lower third consists of large rectangular stones, some of which have drafted margins. These stones show less mortar in their interstices than the masonry above. This south tower is rounded and protrudes like the horseshoe-shaped salients in Armenian construction. The circuit attached to this tower has sharp right angles which are atypical in Armenian garrison forts. Considering the history and location of this site, it is likely that Armenian additons are slight and that most of the fort is an Ottoman reconstruction of the Templar plan. Jacquot mentions the presence of ancient walls in the foundations of the fort at the north.[11] He also found evidence that a masoned wall ran from the fort to the sea, cutting off access to the Cilician plan from the south.

[1] Since I was prohibited from visiting this site in July 1981, there is no plan. See map XI in Dussaud.

[2] Der Nersessian, 640. In part this assessment is derived from the general history of the region.

[3] Jacquot, 150.

[4] Ibid., 97; A. W. Lawrence, in Boase, 35 ff.

[5] Cahen, 149.

[6] Boase, 173, 178; Jacquot, 121; Alishan, *Sissouan*, 497, Heberdey and Wilhelm, 18 f; Ainsworth, 92; Langlois, *Voyage*, 472–74.

[7] Jacquot, 121 ff. Cf. Janke, 17–20.

[8] Cahen, 149; Boase, 178. K. Otto-Dorn ("Islamische Denkmäler Kilikiens," *Jahrbuch für kleinasiatische Forschung* 2.2 [1952], 116 f) reports the presence of a Turkish inscription over the entrance; this epigraph probably records the *extensive* reconstruction here by the Ottomans in the mid-16th century. This entrance may have been preceded by an Armenian-type slot machicolation. It is not inconceivable that an Ottoman inscription was inserted in an Armenian-built gate complex. Cf. Janke, 23.

[9] Jacquot, 153.

[10] This type of masonry is similar to that seen in the Templar forts of Bağras and Trapesak; see Edwards, "Bağras," 419 ff.

[11] Jacquot, 150; see also Brocquière, 88 f.

Savranda

Savranda Kalesi[1] rises on a lofty outcrop of limestone at the southern flank of the Amanus pass.[2] At Savranda this strategic route is confined to a narrow canyon that was formed partially by the flow of the Gökpınarsuyu. This stream, which is fed by springs and runoff from the mountains at the south and east, meanders to the northwest until it joins the southern flow of the Ceyhan River. The new road from Osmaniye to Gaziantep does not turn south into the basin of the canyon but continues to the northeast where modern engineers have blazed a highway to Fevzipaşa. There is actually a third route near this Amanus gorge at the site of Bahçe to the north. This Bahçe pass also appears to be one of considerable antiquity.[3] Savranda Kalesi is north of a lesser-known road that runs from Çardak to Yarpuz and Islâhiye. There is a connecting trail between Yarpuz and Savranda. Just what importance this road had in medieval times is unknown. The conifers that surround many of the hills and outcrops adjacent to Savranda are the "forest of Marris" (Armenian: Mari).[4] Beginning in 1977 the Turks began a project to block up the tributaries in the south part of the Amanus gorge. To construct this dam and hydroelectric station the entire pass in the immediate area to the east and north of the castle was dynamited and leveled (pls. 194a, 194b).[5] Soon the lake will be a source of recreation and irrigation for the inhabitants of eastern Cilicia. The level of this reservoir will not reach the castle.

This castle's first mention in medieval history is rather insignificant. It seems that in 1069 a band of Seljuks invaded Cilicia by passing through the Amanus near Savranda.[6] It is safe to infer that the fort was garrisoned by Byzantine soldiers, since Cilicia was under the rule of Constantinople at that time. With the advent of the Crusades there is a period of Frankish occupation at Savranda. It was here in 1101/2 that Tancred of Antioch held Raymond of Saint-Gilles prisoner.[7] For a third of a century thereafter the castle's history is unattested. In 1135 the Rubenid Baron Levon I captured the fort, probably from its Frankish master, the Count of Maraş. A struggle quickly ensued in which Levon and his ally-nephew Joscelin of Edessa faced the forces of Baldwin of Maraş, Raymond of Poitiers, and Fulk, the King of Jerusalem. In late 1136 Levon was captured by Raymond of Poitiers. The Armenians briefly continued to fight, but their internal rivalries eventually resulted in the blinding of Levon's illegitimate son, Constantine. Two months after Levon's capture, Raymond released him after receiving as ransom sixty thousand tahegan, the cities of Adana and Misis, and the fort of Savranda. An unspecified number of Levon's sons were also taken as hostages by Raymond.[8] Sometime between 1172 and 1175 the castle fell into Armenian hands, probably during the period of Mleh. In 1185 Ruben III was ransomed by the Prince of Antioch in exchange for Savranda and two other forts.[9] Two years later the head of the Rubenid dynasty, Baron Levon II (later to become King Levon I), chased a large group of Turkmen out of Cilicia. Levon ended his pursuit at Savranda.[10] This fort is again mentioned in 1198/99 when its Lord, the Het͑umid Baron Smbat, assisted at the cornation of

Levon.[11]

The heirs of Smbat were to rule Savranda until the late thirteenth century. Smbat's son and successor was Geoffrey, who stayed at Savranda until his death in 1261.[12] Smbat II, Ōšin, and Constantine, the three sons of Geoffrey, followed each other in succession to the lordship of Savranda from 1261 to 1274. In 1266 Malik al-Manṣūr invaded Cilicia and defeated a large Armenian force in a place called Marri,[13] which can be identified with Savranda.[14] There is no indication that the Armenians lost control of this fort. Bar Hebraeus reports that an earthquake in April 1269 did considerable damage to Savranda and other Cilician settlements, killing thousands of inhabitants.[15] One of the most significant events in the castle's history occurred in 1271 during the lordship of Constantine. On June 15 Constantine made a treaty with the Teutonic Knights that allowed them to maintain a toll station on the east flank of the Amanus at the "Black Tower,"[16] but not to build other structures outside of (that is, west of) this point.[17] It seems that there was a clearly defined sense of an eastern frontier for the Armenian kingdom of Cilicia; part of this border ran down the spine of the Nur Dağları from Haruniye to Sarı Seki.[18] After the death of Baron Constantine in 1274 his son Ōšin assumed the leadership of Savranda. In 1298 a unit of Egyptian soldiers temporarily captured Savranda. In 1337 the Armenians lost this fort permanently to the Mamluks.[19] Despite the wealth of information on the medieval history of this site, it has never been the subject of an archeological survey.[20]

Savranda Kalesi envelops the entire precipice of an ascending outcrop of limestone. At the southern tip, where the point of access is least severe, massive bastions rise to an impressive height, as if to extend the upward movement of the rock mass (fig. 65; pl. 194a). The approach path below is tactfully positioned to flank this entire circuit until the approaching party arrives at gatehouse A. A thin outwork extends directly east from the corner of the gatehouse to the edge of the cliff in order to block any access other than the intended approach (pl. 194b). At the extreme north end of the outcrop, walls were deemed superfluous since an entry up the vertical cliffs is impossible. As the edge of the summit rises on the southwest, the base of the outcrop broadens. Leaving nothing to chance, the engineers constructed a thin circuit to prevent the foolhardy from sallying up the west face. Since a projection of natural rock runs across the outcrop dividing the upper (or southern) third from the rest of the castle, the masons built a circuit on the same line, creating an upper and a lower bailey. Entrance into this upper bailey was probably at point F.

Gatehouse A on the northeast flank of the outcrop is one of the most complex structures in the castle (pl. 298b). It consists of a square bent entrance that is flanked at the southeast by a bulging tower. The masonry of A consists of three types. The springing of the groined vault over the gatehouse and the frames around the doors and windows are constructed with type IX (pls. 195a, 196a). The walls of A are built with type V. On the exterior the type V is larger and superbly coursed. On the interior the size varies greatly, and more rock chips are used in the interstices. The center of the partially collapsed groined vault is built with type IV masonry (cf. Yılan and Anavarza); these stones are somewhat elongated.

The south wall of the entrance is opened by the gate which consists of an arrière-voussure with a slot machicolation (pl. 298b). Both arches in this entrance are pointed. This type of gate is common in the forts of Armenian Cilicia. At Savranda the lower arch of the arrière-voussure is decorated with alternating colors of stone. It is interesting that the natural pigmentation of the darker stones is not carried onto the interior side of the arch (pl. 195a). Below the arrière-voussure on the interior side of the jambs there are the openings for a crossbar bolt and pivot housings above to accommodate double doors (neither are shown on the plan). The inside of the door is covered by a rounded vault. The tower in the southeast corner is entered through a narrow door near the gate. This door is covered on the north end by a slightly pointed lintel that consists of two tapered stones. The passage leading into the tower is covered by a rounded vault. On the interior the tower is perfectly round. The only window is a narrowly splayed embrasured loophole with a rounded hood. This rounded top is segmented into a number of voussoirs, which is somewhat atypical in Armenian architecture. Its appearance here is probably the result of hasty construction. The wooden beams that once supported the floor for the upper level of this tower have vanished. On the exterior the tower rises on a square plinth (pl. 298b). The east wall of gatehouse A, which is attached to this tower, gradually thickens at the north end. The wall is opened by a single window (pl. 195a). The exterior frame for this window is covered by a thin depressed arch. On the interior side the frame has been enlarged into a niche covered by a similar arch that is slightly pointed (pl. 196a). The base of the niche was probably a bench. The north wall of A is opened by a small squareheaded window. The west wall has a wide door in the center (pl. 195b). The door is covered by two rounded arches; the westernmost is lower and rises

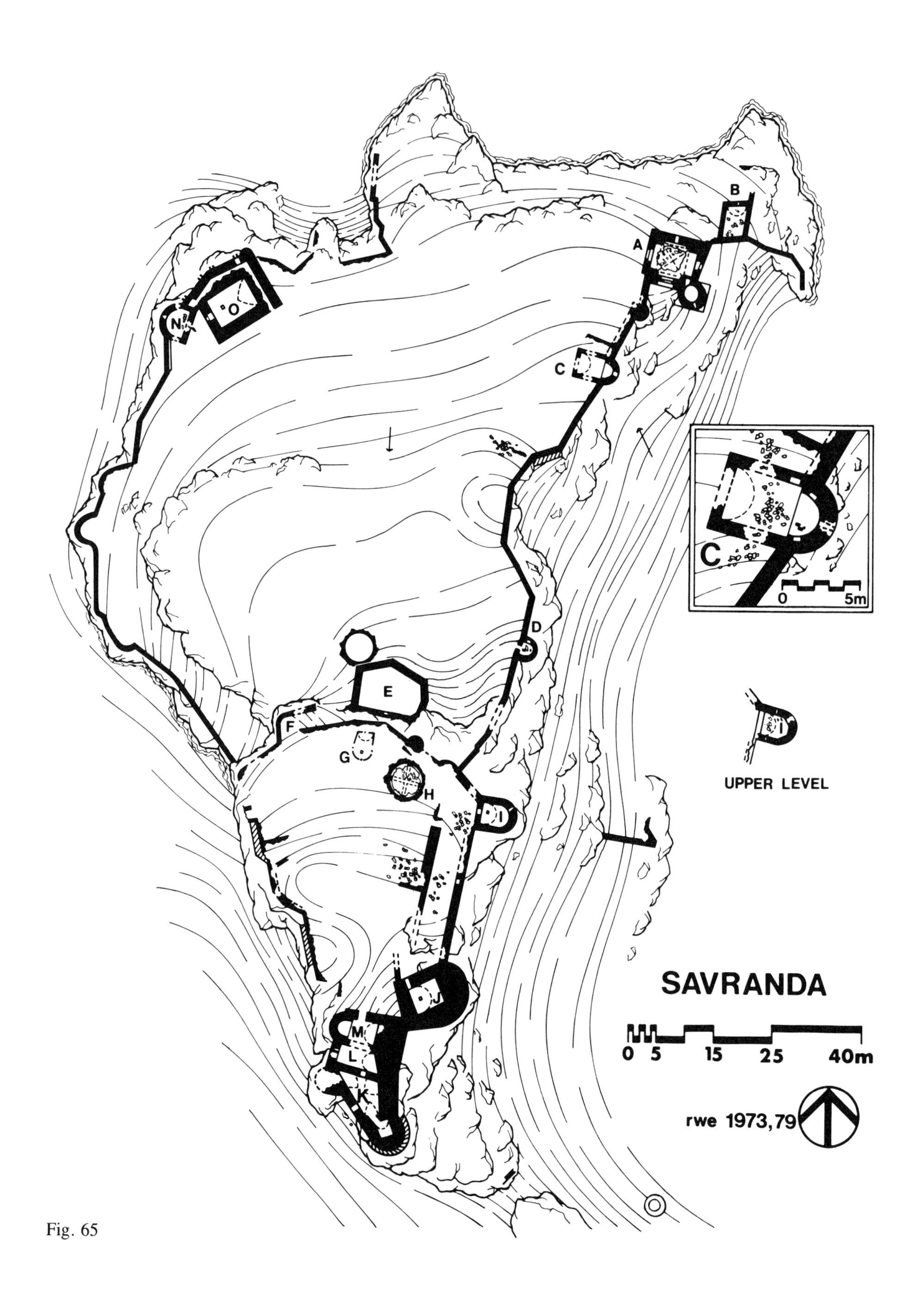

SAVRANDA
A
B
C
D
E
F
G
H
I
J
K
L
M
N
O
0
5m
UPPER LEVEL
0 5 15 25 40m
rwe 1973,79

Fig. 65

on the jambs. Behind the jambs are the holes for a crossbar bolt (not shown on the plan). If an enemy breached the lower ward, gatehouse A could be sealed off from the inside.

Near the interior corner of the south and west walls of A, the local construction workers, who were digging for coins in their spare time, uncovered parts of a wall in the foundation (not shown on the plan). This wall, which is made of type V masonry, seems to run between the two western piers of the groined vault. Its function may be explained by studying the interiors of the north, south, and east walls. In the upper levels of those walls is a deep horizontal chase. This chase functioned as a continuous joist for the floor of an upper level. This floor may have had ports that allowed the defenders above to shoot at an enemy who had breached the door (cf. Vahga). The ceiling for this second level would be extremely low and uncomfortable for long periods of occupation. Other than the floor, the only opening into this level is the straight-sided window (door?) in the north wall. The chase is not present in the west wall, which may indicate that the wall that joins the two western piers served as the western terminus for the upper-level floor. At the lower level this inner wall would have to have a door. The presence of two western walls would create a small corridor and an additional line of defense before passing through the outer west door. The groined vault of A, which covered the second level, is made in the traditional Armenian way, using small plates of stone that abut at the corners.[21] The groined vault is framed on each side by a transverse arch, and it springs from a thin tapering triangular base in each corner. Today most of the vault has collapsed. Because of the design and masonry, it seems quite probable that gatehouse A is Armenian.

Because of the presence of type IV as an exterior facing (pl. 194b), building B and the outwork to the east may be Byzantine constructions. The vault over B has collapsed, and its walls are deteriorating rapidly. B has only one thin opening in the north wall. In a few places the type IV masonry of B is repaired with a finer-quality Armenian ashlar (type V). The Armenians probably built their gatehouse in an area where the Greeks had their entrance. The nature of the original Greek plan is impossible to determine.

That the Armenians used and repaired the Byzantine walls is most evident where the southwest corner of gatehouse A attaches to the east circuit and a small solid tower (pl. 197a). At the base of the circuit wall the first two or three courses consist of large square blocks whose interstices are filled with a thick layer of mortar. Above is type IV masonry with broad mortarless interstices. The upper third of the wall consists of a small, but well-cut, type V. This type V is identical to that used in gatehouse A and is probably Armenian. The base of this wall is a socle built by the Byzantines to support the type IV wall.[22] This is the first case in Cilicia where I have seen the Byzantines use such a large socle. What is significant is that the analysis of mortar samples taken from the type IV section differ completely from samples in the type V above. However, the chemical components of the mortar in the type V of this wall are quite similar to a sample I extracted from the interior of the gatehouse above the south door.[23] The circuit, which extends from A to I (pl. 194a), has large sections with a similar type IV foundation and with type V additions above.

South of gatehouse A, tower C contains an Armenian chapel in its lower level.[24] Its upper level has collapsed, along with the barrel vault over the nave and most of the apsidal semidome. Parts of the nave wall are still standing. South of the chapel, sections of the interior side of the circuit wall have horizontal chases. Undoubtedly these supported the roofs of wooden rooms. About 15 m south of C the Armenians constructed a short talus. Farther south is tower D. The vault over tower D has collapsed; only the remains of a niche are visible in its north wall. The wall that flanks tower D at the south appears to be built with type IV masonry. South of tower D the topography begins to ascend quickly. The rock mass and the north wall of the upper bailey turn the pedestrian toward building E at the west. This hexagonal structure had no doors or windows. If E was covered by a vault, it has completely vanished. The only recognizable feature in E is a small roundheaded niche in the south wall. Since the walls were not stuccoed, it seems unlikely that E was a cistern. Building E is constructed with type IV masonry. Directly northwest of building E is a circular pit whose interior is faced with masonry. As with building E, the function of the pit is unknown. The only salient in the north wall of the upper bailey is a small solid structure south of E. Undoubtedly many other stone structures occupy the lower bailey, but the thick jungle of shrubs and trees makes a careful survey impossible.

It seems likely that the opening at ramp F is the only entrance into the upper (or south) bailey. Sections of the north face of F have fallen away. In the area of F two distinctively different types of masonry are evident in the walls. Type IV is often flanked or found below types V and VII. The latter is probably Armenian; Armenian repairs and construction seem to be extensive throughout the castle and may have been required after the earthquake of 1269. At the north end of the upper bailey there are two skillfully placed cisterns, G and H. Cistern G is fully subterranean; its barrel vault is opened by a small square hatch at the south. To the east the circular cistern H was once covered by a cupola that rested on two intersecting ribs. Today the vault has collapsed, and the entire area is covered in a heavy undergrowth.

Southeast of H a wall parallel to the east circuit forms a narrow corridor. At the north end of this corridor is tower I. On the exterior the walls of I are constructed with type V. Sections of its upper level appear to have been reconstructed hastily with a smaller, poor-quality stone (perhaps Mamluk repairs; cf. the south bailey of Anavarza). On the interior of I, the barrel vault of the upper level has collapsed, and there is a breach in the northeast wall. This upper room is hexagonal; there is one very narrow, slightly splayed window in the north and south walls of the room. These windows are covered by a series of flat monolithic lintels. The sill of the south window inclines sharply upward at the south end. The interior of the lower level of I (pl. 197b) is covered by a barrel vault that terminates in a semidome. Unlike the upper level, the walls are not faceted but apsidal. In medieval times this ground-level room had a lower basement level (probably a cistern); a horizontal chase and joist holes show where the wooden floor was once located. The main apsidal room is pierced by a single, narrowly splayed squareheaded window. Approximately 15 m east of tower I is a fragment of a wall that may once have been part of an outwork.

Proceeding south from tower I, the corridor terminates at tower J (pl. 198a). In the east wall of this corridor (that is, the interior side of the east circuit) is a large roundheaded niche. The circuits south of I and tower J have an exterior facing of type V masonry (pl. 298c); the west wall of the corridor is constructed with type IV. The upper level of tower J is badly shattered, and little can be said of its nature. The lower level is entered through a single squareheaded door at the northwest. On the interior a barrel vault, which is pierced by a small square hatch, encloses the square room. This tower, as well as the adjoining buildings to the south, has exceptionally thick walls, which *may* indicate that the Armenians enlarged the earlier Byzantine/Frankish plan when constructing their type V walls. Only excavations can determine the sequence of construction.

It is quite likely that the extreme southern salient K and its apron-talus are Armenian. About 45 m southeast of K is a cistern on a lower slope; its internal dimensions are unknown. The interior of K is badly damaged, but there are indications that the small square room at the south was covered by a barrel vault and that the larger triangular room of K to the north had a faceted ceiling of three abutting pointed vaults. The south wall of K has a large window, and there is the foundation of a door at the north to connect with L. L, M, and the northwest salient of K have lost much of their exterior facing (pl. 198b). In the center of the west wall of L, there is a thin horizontal chase on the exterior; its function is unknown. On the interior a chase is also present in the south wall. The west wall of L is opened by a roundheaded door and a thin, high roundheaded window. Like L, room M was once covered by a barrel vault. The apsidal west end of M is a tower. The north door of room M has collapsed, but the springing of an arch indicates that it was roundheaded. The west wall of the upper bailey is supported by two taluses.

The west wall of the lower bailey, which is in a better state of preservation than its southern counterpart, is built mainly with type IV masonry. Tower N is the only salient in this wall. Today its roof has collapsed along with sections of its north, south, and east walls. A widely splayed embrasured window at the west is still present. Tower N is built exclusively with type IV masonry (pl. 199a). Room O and the circuit wall flanking O at the north and east are built mainly with type V on the exterior (pl. 196b). Only the lower sections of the circuit north of N are of type IV masonry. Clearly, the Armenians undertook extensive repairs in this area. On the top of the circuit north of O are four sets of twin corbels, designed to support a removable breastwork. The interior of O consists of a square room and a pointed vault opened by a single hatch. A single high-placed door that has accommodations for a crossbar bolt (not shown on the plan) opens room O at the west. The door is covered by an arch of four voussoirs. A thin, poorly constructed circuit of type IV masonry meanders to the east from the north corner of the wall that surrounds O (pl. 199b). Most of the north end of the castle has no circuit wall at all because of the cliff.

[1] I began my plan of this hitherto unsurveyed site in September 1973. The plan was completed on my second visit in August 1979. The interval of distance for the contour lines is approximately one meter.

[2] This site was known to Brocquière (93, esp. note 4) and was located by E. G. Rey at the end of the 19th century; see Deschamps, "Servantikar," 379. The modern names used for this site are somewhat confusing. Neither the locals nor the published charts are consistent. I have chosen "Savranda," the Frankish spelling, since it is frequently used (along with Savuran Kalesi) by the

local Turks. In the 19th century the names Kaypak Kale and Serfendkiar are found. The most common Armenian spelling is Saruandikʿar. In Arab texts the three most frequent spellings are Isfandakār, Sirfandakār, and Sarwandkār. Honigmann (121 note 2) identifies τὸ Σαρβάνδικον ὄρος in Skylitzes (ed. I. Bekker [Bonn, 1839], 684) and Attaleiates (ed. I. Bekker [Bonn, 1853], 138) with Savranda in the northern Amanus. In the Syriac text of Bar Hebraeus the fortress is called Kēfā dhe-Serwand.

This site appears on the following maps: *Adana (2), Gaziantep, Maras, Marash*. See also Fedden and Thomson, 12, 22 f, 35.

[3] See Hasanbeyli (note 3) in the Catalogue.

[4] Deschamps, "Servantikar," 382–84; Le Strange, *Palestine*, 538. Cf. Abū'l-Fidā', 34; William of Tyre, *RHC, HistOcc*, I.2, 755, 789; Cahen, 145–48; Deschamps, III, 68 f; below, note 14.

[5] The old trail leading to the castle on the south side of the Gökpınarsuyu was visible in 1973. On the right flank of the trail, a square bulding was also present. This structure had a door at the south and one high-placed squareheaded window in the east and west walls. A gabled roof once covered the building. Its walls were built with type IV masonry. This building was not visible in 1979.

[6] Deschamps, "Servantikar," 380; Honigmann, 121 note 2.

[7] Matthew of Edessa, 57.

[8] Smbat, 616; Cahen, 355 note 30.

[9] Smbat, 628.

[10] Ibid., 629.

[11] See below, Appendix 3. A genealogy of the lords of Savranda can be found in Toumanoff (288 f) and Christomanos (147).

[12] *RHC, DocArm*, II, 840; Rüdt-Collenberg, table II; Smbat, G. Dédéyan, 107 note 78.

[13] Deschamps, "Servantikar," 386 note 4.

[14] This conclusion finds support in the text of Bar Hebraeus (446). See: Alishan, *Sissouan*, 239, 480; Grigor Aknercʿi, 356; Cahen, 715; above, note 4. Two years prior to this event King Hetʿum I had assembled his troops at Savranda to counter the army of Baybars; see Cahen, 712; Canard, "Le royaume," 226. Al-Maqrīzī (I.1, 234) specifically refers to this site as Sarfand in the events of 1264.

[15] Bar Hebraeus, 448.

[16] Hellenkemper (121) believes this "Zollhaus" to be the site of Hasanbeyli. However, it is quite possible that Karafrenk is this "Black Tower."

[17] *RHC, DocArm*, II, 840; Alishan, *Sissouan*, 239; Forstreuter, 236 f. Cf. Röhricht, 359. Indrikis Sterns is mistaken in assuming that the Teutonic Knights erected "a customshouse in the city of Sarvantikar"; see I. Sterns, "The Teutonic Knights in the Crusader States," *A History of the Crusades*, ed. K. Setton, V (Madison, 1985), 375.

[18] Sites at the south end of the Nur Dağları, such as Bağras, were never a permanent part of the Armenian kingdom; see Edwards, "Bağras," 415–18, 431f.

[19] Deschamps, "Servantikar," 387 note 4; Gaudefroy-Demombynes, 99, 217. Cf. Popper, 17.

[20] Hellenkemper, 114 f.

[21] The other Armenian method is to use V-shaped corner stones of ashlar. In Templar architecture these stones tend to overlap; see Edwards, "Bağras," 424.

[22] There is a possibility that the Franks are responsible for building the type IV walls. Also, the Armenians could have replaced the lower courses with the present large blocks in order to discourage sappers. Since each course is fused to the central core, the lower courses can be changed without disturbing the courses above.

[23] See below, Appendix 2. The type V from the interior of the gatehouse is SAV-2; SAV-1 is from the type V above the type IV (SAV) in the circuit wall. The chemical differences in the mortars between the types IV and V indicate the different recipes were used.

[24] See Edwards, "First Report," 165.

Silifke

Silifke Kalesi[1] crowns a large hill near the delta of the river Göksu (the medieval Calycadnus) in southern Cilicia Tracheia (pl. 200a).[2] Adjacent to and on the east flank of the castle outcrop is the modern town of Silifke and its ancient predecessor, Seleucia. The castle's elongated and somewhat symmetrical outcrop, sculptured by erosion and centuries of terracing, rises to almost 86 m above sea level at the southern entrance to the Göksu canyon. From the river's west bank this fortress commands the strategic coastal road, which runs on a northeast-southwest axis, and the highway linking Silifke to Karaman and the Anatolian plain (pl. 202a). Within the present castle there are no structures that date from the classical or late antique period.

Silifke plays an important role in Armenian and Crusader history, and its strategic importance was recognized in the late antique period.[3] Sometime after the seventh century but before the eleventh, the Byzantines constructed a fort on this site as a bulwark against the Moslems. In the 930s a newly created coastal theme was administered from the town.[4] During the reign of Emperor Alexius I, Eustathius, a eunuch, admiral, and imperial secretary to the emperor, was ordered to sieze Korykos and Seleucia for the purpose of building up their defenses.[5] Korykos is separated from Seleucia "by six stadia."[6] When Eustathius had finished his task (after 1104), Strategus Strabo was left in command of the garrisons at both forts to prevent Crusader advances. The Byzantine emperors were especially sensitive to incursions into this area.[7] John Comnenus personally led his armies against the Rubenid Baron Levon I who was conducting intrigues near Seleucia. The emperor captured Levon in 1138 and conducted him to Constantinople. From this period until the 1180s Seleucia remained in Byzantine hands.[8] At some undocumented time before the arrival of Frederick Barbarossa in 1190, the fort passed into the possession of Baron Levon II (later to become King Levon I), never again to be under Byzantine control.[9]

Initially, Levon gave the hand of his niece, Silifke castle, and the adjoining town to a noble from the court of Sasun, Šahnšah.[10] Three years after the assassination of the latter (1193), the castle returned to royal control. In 1194 the site was given to Constantine, the son of the sebastos Henry. Undoubtedly this is the same Baron Constantine of Seleucie who appears on the Coronation List of 1198/99.[11] Within five years the military situation on Cilicia's western front had deteriorated, and King Levon I invited the Hospitalers to defend his kingdom from the incursions of the Seljuks.

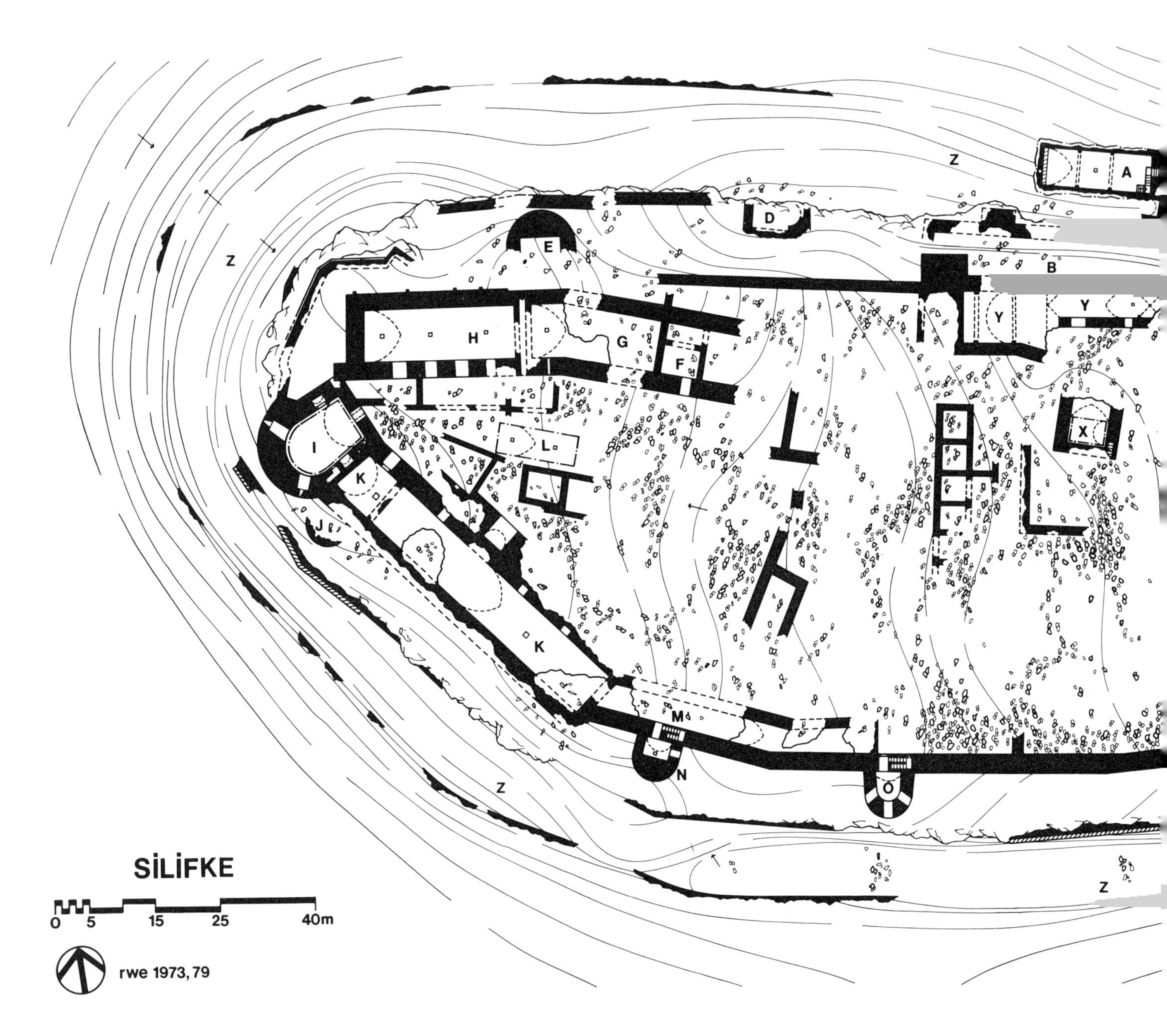

Fig. 66

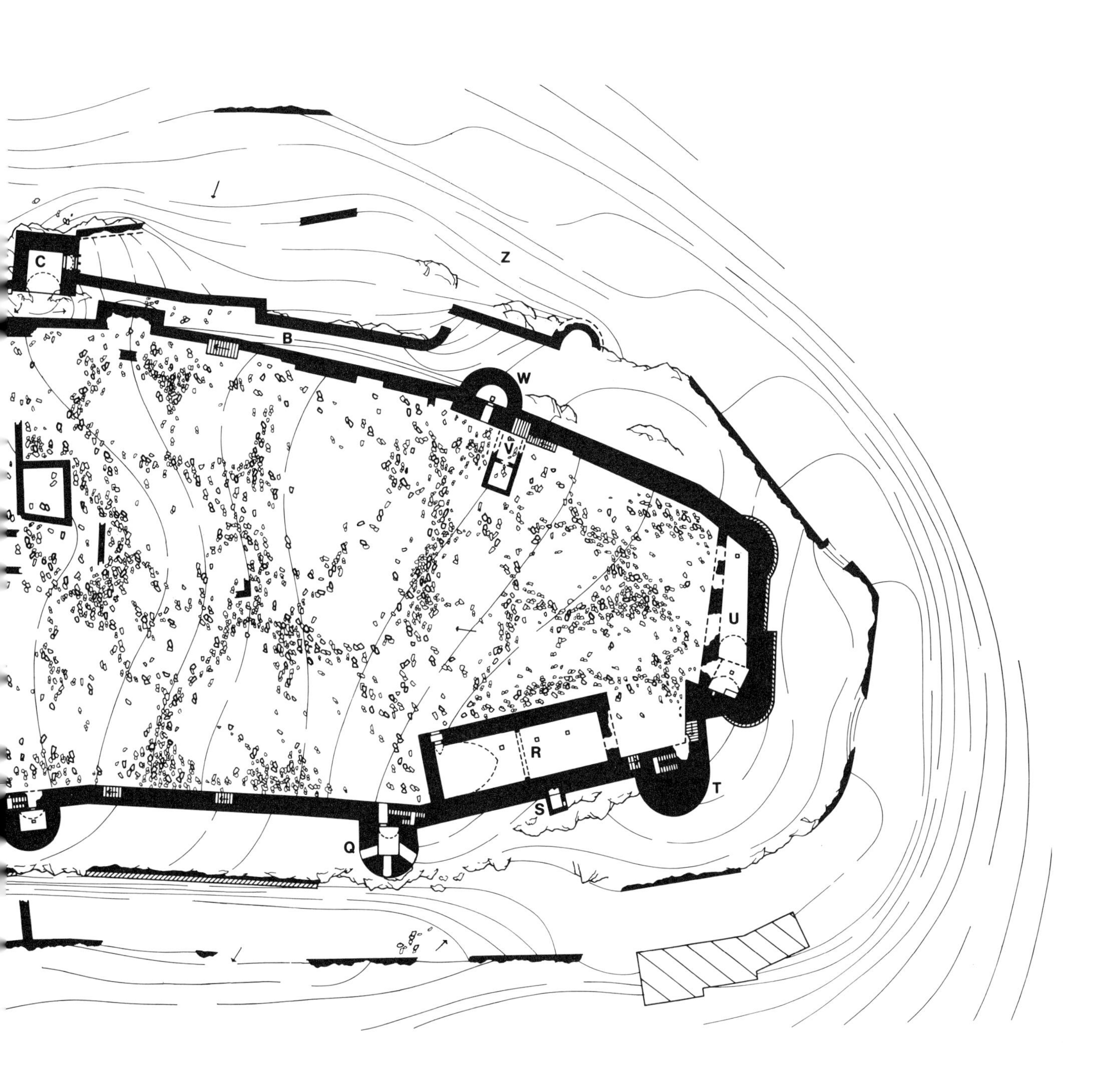

C
Z
B
W
V
U
R
T
S
Q

In 1210 Levon formally granted Silifke castle and the nearby forts of Camardesium and Castellum novum (Norberd) to the Order of the Hospital in exchange for an annual tax and their support of a cavalry unit of four hundred lancers.[12] Thereafter, Garin de Montaigu, Aimery de Pax (the former Castellan of Margat), and Féraud de Barras (1214) assumed in turn command of the castle.[13] During the tenure of the latter, an attack on Silifke castle by Kaykāʾūs I, the Sultan of Konya, was turned back (1216/17). After the death of Levon in 1219 the regent for the Armenian kingdom, Constantine of Çandır, decreed that Levon's designated heiress, Zapēl (the daughter by his second wife, Sipil), was to marry Philip, the fourth son of Bohemond IV of Antioch. Despite his promises, Philip refused to accept Armenian religious practices and showed open disdain for the Armenian nobles. His arrogance eventually brought about his imprisonment in the fortress of Sis (or Barjrberd) and death by poisoning early in 1226. The distraught Zapēl took refuge with her mother in Silifke's fort. Undaunted by his first failure, Constantine arranged for his own son, Hetʿum, to marry Zapēl and demanded that Bertrand de Thessy, the Castellan of Silifke castle, return her at once. The Hospitalers, who would not suffer the humiliation of surrendering the twelve-year-old Zapēl, nor dare to fight the assembled troops of Constantine, simply abandoned their castle (1226).[14] Zapēl's marriage to King Hetʿum I was a success.

In the mid-thirteenth century the castle begins to fade into political obscurity. It is believed that an inscription on the castle records its enlargement in 1236.[15] Actually, two shattered and highly fragmentary Armenian inscriptions were located in the castle in the nineteenth century, one above the gatehouse door and the other on the interior.[16] Neither epigraph gives specific information about the builder, date of execution, or areas of the castle that may have been rebuilt.[17] We know only that in 1248 the Frankish commander of the place was called Guiscard. After this period reliable information on the castle is lacking.

To varying degrees the architecture of Silifke Kalesi has been the subject of a number of modern accounts.[18] It is my intent here to supplement these earlier works with a new and more detailed survey (fig. 66) and to provide new conclusions on the site.

Silifke castle has a design that is atypical for an Armenian fort in Cilicia. It consists of a fortified circuit wall with a continuous lower-level ditch. The size and shape of the castle are suited to the rocky but relatively flat oval summit. The castle's circuit wall and towers do not hug the edge of the summit but are set back. The plateau inside the single bailey slopes gently upward in a northeasterly direction.

Directly below the summit is a dry moat (pl. 300b) that has revetted walls (the lowest points in the moat are designated by "Z" on the plan). The inner revetment wall rises to a much greater height than the outer;[19] *only* at the north (in front of the circuit wall between towers E and W) does this inner wall rise above the edge of the rocky summit and in part surmount it (pl. 200a). The masonry of the outer revetment wall is rather crude in comparison to that of the inner wall of the moat and the castle itself (pl. 300b). The stones of the outer revetment are smaller, crudely cut, and have broad interstices. This is a type IV masonry that is laid in regular courses. In certain areas below towers I and P the inner revetment wall is supplemented by a talus (pl. 300a).[20] This talus, which is made of type IV masonry, is built over the finer-quality ashlar (types VIII and IX) of the moat's interior revetment. The type IV stones represent a later construction, perhaps to prevent the mining of the inner revetment wall. Because of weathering and the need to accommodate a modern paved road, much of the northern half of the moat is missing. The paved road swings into the body of the moat just north of building A (fig. 66) and hugs the summit (in a place where the moat is widest; pl. 200a) to the southwest end, until it terminates at the tourist restaurant. This restaurant is 20 m below tower T. The northern half of the restaurant along with the parking lot rests in the bed of the moat. In the southern half of the ditch there is no evidence of medieval structures, except for two perpendicular walls below tower P (pl. 300a). Below towers N and Q the fill from landslides has raised the level of the dry moat. Many other sections of the inner revetment wall have fallen away. In the northern half of the moat is the rectangular building A. This structure, which is almost subterranean, has a single high-placed entrance at the east with stairs to reach the floor. Two pointed ribs support a vaulted ceiling that is opened by a single hatch in the center. Building A is a cistern that is cleverly placed at the low point in the moat to catch the runoff from the seasonal rains. This runoff, which entered through a conduit at the west end, cascaded down the small interior talus to the floor. Two small, very ruinous buildings whose functions are unknown are attached to the southeast corner of A.

The distance between the circuit wall of the fort and the inner wall of the revetted moat fluctuates greatly. At the south between towers I and Q the inner revetment wall is attached to the almost perpendicular side of the summit (pl. 300b). Here the strip of scarcement between the inner revetment and the circuit wall is relatively thin. As the inner revetment turns to the east end, it moves further away from the main circuit wall. At the north the inner revetment becomes a freestanding wall that creates corridor B, the service passage to and from gate C (pl. 200a). This strong wall is necessary here for two reasons: first, the ascent to the castle is easiest at the north, and consequently it is an appropriate place for an entrance; and second, this north wall, unlike the inner revetment wall at the south, creates a tight enceinte-corridor that could seriously restrict an enemy who had breached gate C.

The entire inner revetment wall (except for the tali and the repair at the south) and the castle's circuit and towers (except for the shattered salient J and postern S) are constructed out of smooth ashlar (pls. 200b, 201b). The size of this ashlar varies greatly.[21] The length and width of an individual block can range from 15 to 100 cm. The typology of this ashlar ranges from VIII to IX, with the latter being in the majority. No one type is consistently used in a single area. John Thomson noted that the stones of the towers and curtains strengthened the cohesion of the walls since their tapering inner faces had less than half the dimensions of their outer faces.[22] But closer observation reveals that this tapering is infrequent and when it does occur, it is slight. This explains why the facing stones at Silifke Kalesi, more than any other site in Cilicia, have fallen away from the core (pl. 300a). This shallow and square type IX is rarely used as a facing stone on Armenian forts in Cilicia.[23] It is apparent from observations in the late antique city of Seleucia below the outcrop (pl. 205b), that the medieval builders of Silifke castle plundered some of their cut masonry from below.[24] No Armenian mason's marks are visible on the ashlar; the few mason's marks that do appear are of an alien typology (see below, Appendix 1). Also, fragments of Greek inscriptions appear arbitrarily throughout the castle (pl. 204a).[25] The sporadic mason's marks may be from the original late antique construction, but it is more likely that they represent the quarrying and shaping of stones in the medieval period. In the case of the latter, the local population of Greeks was probably called upon to assist in construction.

One of the major elements in the fortress is the square bent entrance C (pl. 200a), which is attached to the center of corridor B. This gate sits atop a rock that projects into the body of the dry moat. From the northeast corner of gate C, a badly damaged wall projects eastward across the moat. This wall acts as an outwork to restrict the approach to the gatehouse. When ascending to the gatehouse one is struck by the appearance of three large corbels over the entrance (pl. 299c). Below the corbels is a badly defaced plaque and the inscription mentioned earlier. The square plaque, which may be a coat of arms, appears to have once had a cross inscribed in the center; two rosettes and three curious letters are still visible near the cross. Only the upper half of the inscription is still in situ. It is framed by a raised square border whose inner face is carved with Armenian letters. As they were in the time of Beaufort and Langlois, the letters of the epigraph are badly mutilated and yield only a few translatable words: "Jésus-Christ . . . de qui auras-tu pitié . . . Dans l'année des Arméniens . . . des Arméniens. . . ."[26] When passing into the entrance, two anomalies become apparent. First, the entrance portal, which is covered by a small barrel vault, has accommodations with jambs for a front and back door; the two doors are separated by only 1.5 m. Second, the arch over the front door is not constructed with the typical wedge-shaped voussoirs of the Armenians, but is built with wedges that have a repetitive, square plug-in-gap. This design is seen nowhere else in Cilicia. Once past the entrance portal, one makes a left turn under the barrel-vaulted chamber of gatehouse C. The upper sections of this room are dotted with joistholes which indicate that either the original centering remained in situ (to add strength) or a small second story was supported by crossbeams (perhaps provided with machicolations). The interior of the gatehouse has four identical mason's marks (see below, Appendix 1). When exiting gatehouse C, the visitor must move abruptly uphill, to the east or west flank of corridor B. When one takes the west approach, a stairway, which leads from that corridor to the top of gatehouse C, becomes visible (pl. 200a).

The great enigma of this fort is that there is no apparent entrance through the main circuit wall at any point along or near corridor B. Today visitors use a breach in the northern wall of building Y to enter the castle. Just west of the breach, a small square salient protrudes from the main circuit wall. Like the outwork east of gatehouse C, it is designed to slow the flow of traffic in corridor B. Farther to the west are the remains of what may be the only ecclesiastical building (D) in the castle.[27] Unfortunately D is barely visible above ground level, and only excavations can determine its true function. Just west of D are three fragments of the wall which surmount the inner side of the moat. Posterior to these fragments is the now unattached tower E (pl. 201b). Only the facade of tower E is standing today. From the present remains it appears that a square vaulted chamber was accommodated on its interior.

Southeast of tower E is an area of considerable damage where the junction of that tower to the main circuit wall cannot be determined. Directly south of E are two large undercrofts (H and G; pl. 201a);[28] the smaller building F to the east is badly damaged, but indications are that it was once vaulted. Despite the damage suffered by undercroft G, its pointed vault and one hatch still survive at the west. In the southwest corner of G, a curious, squareheaded shaft (pl. 200b), which slants downward into the interior, may have served as a ventilation duct. Undoubtedly a door once led from G into H. Undercroft H is one of the most impressive buildings in the castle.[29] This long undercroft has a perfectly preserved vault with three hatches. Like other units in this castle, the vault is constructed with type IV masonry, while the walls are made of a finer ashlar (pl. 201a). Four jambless, barrel-vaulted doors open the south wall of H. Indications are that these doors did not lead to the main court but into a smaller (and now shattered) adjoining building. On the exterior of the north wall of H are four upper-level corbels that probably supported a removable breastwork.[30] On the top of the west wall of H the upper courses of stone suddenly project as if to support an adjoining vault of a now missing building or perhaps battlements (pl. 202a). The two or three bossed stones (type VII) in this wall were probably recycled from an Armenian building in the city below.[31] To the west of H and below are fragments of the main circuit wall that emerges from the north side of tower I.

The most intricate single unit in the castle is the three-storied tower I.[32] The only visible entrance, at the northeast, leads directly from the courtyard into the spacious second-floor barrel-vaulted chamber (shown on the plan; pl. 202b). This jambless roundheaded portal is surmounted by a tall squareheaded window. West of the door, a small section of the floor has collapsed. The floor of the second level is also the vaulted ceiling of the first-level cistern. It is likely that a hatch once stood in the collapsed area. As one moves in a counterclockwise direction from the entrance, the first opening on the interior of the second level is a roundheaded niche with jambs. These jambs may have framed shutters. The base of the niche is two courses above the present floor level. The first course of stone at the floor level protrudes about 29 cm from the thickness of the wall. This shelf extends around most of the interior. To the west of the niche is a huge embrasured window with casemate (pl. 203a); its rounded canopy and walls are made of well-coursed ashlar. A similar archière is located at the south. East of this second casemate is an odd, elbow-shaped chamber, perhaps a garder-robe. At the end of this compartment is a curious square socket that measures 50 cm in length and 60 cm in depth. Northeast of the garder-robe is a square niche with a vaulted cover. Unlike the sides of the niche in the north wall, the sides here are not limited by jambs. Finally, in the east corner is a small square niche whose sill has an opening to the cistern below. The upper level of tower I is badly damaged. There may be accommodations for a door at the northeast.[33] Remnants of battlements are still preserved on the west side. In the east corner is a square hatch; it frames a drain that descends through the thickness of the southeast wall into the cistern. Undoubtedly this third level was uncovered, and rain collected in the hatch. The function of a small circular hole next to the square hatch is unknown. A very thin talus (not shown on the plan) skirts the base of tower I at the west.

Below tower I and to the south is the shattered outline of tower J. Standing to no more than six courses in height, it is built entirely with type III masonry. Since the masonry resembles that used by Byzantines in other areas of Cilicia,[34] it is possible that the tower is one of the few vestiges of the prior Byzantine fort.[35]

The undercroft K, which is attached to the southeast wall of tower I, is divided into two unequal units. At the west is a small barrel-vaulted room with a single hatch in the ceiling and a door at the northeast. Strangely, this roundheaded door has jambs on the interior. A second door at the southeast may have led into the larger section of K. This area is the largest extant vaulted unit in the castle. Its pointed vault is broken in two places and is opened by a single hatch. In this chamber's northeast wall there are two square niches at the south and a shattered door at the north. On the outside a small barrel-vaulted chamber is attached to the wall.[36] To the north of K is the rectangular subterranean cistern L. Its pointed vault is opened by two hatches. In the same area are fragments of a number of unidentifiable structures amid the rubble. East of K is the thinner undercroft M, whose pointed vault survives only in places. Farther east, the piles of rubble indicate that other buildings were once attached to the interior of the south circuit wall.

On the other side of undercroft M is tower N. A single jambless door leads from the inside of M to a lower-level barrel-vaulted room in N. A postern(?) in the east wall of this chamber leads to the thin scarcement between the inner revetment and the circuit wall. A stairway, which rises to the unroofed second level of tower N, is positioned in the east wall of the doorway from M to N. Parts of the stairwell are constructed with type VII masonry and even fragments of inscriptions (pl. 204a). At the second level are remnants of battlements and a square hatch in the floor, which connects with the first level.

Farther east is the badly damaged tower O (pl. 300b). There is a single door with jambs which leads to a barrel-vaulted first-level chamber. Like tower I, the barrel vault terminates in a semidome. The rounded sides of the chamber are opened by three huge squareheaded windows. These openings were probably designed to accommodate torsion catapults.[37] The stairway to the second level is similar to the one in tower N, except that an embrasured window pierces the north wall of the stairwell at the point where the steps turn 180 degrees in their ascent to the top.

The first-floor barrel-vaulted room of tower P has no openings to the outside (pl. 300a), except for a hatch in the vault. Like the stairwell in tower O, an embrasured window is placed halfway up the ascent. The door communicating between the courtyard and the lower level of tower P has its jambs on the inside like towers Q and W. The tactical advantages of this inner jamb seem questionable.[38] Because the circuit wall between P and Q rises abruptly in height, two sets of stairs have been placed in the wall walk.

Like tower O, tower Q has three broad, roundheaded windows in the first-level barrel-vaulted room (pl. 300a). There is only a slight indication that these windows were once limited by jambs on the outside. At the top of this tower there are a few remains of battlements. On the exterior of Q at the base is a very shallow talus (not shown on the plan).

Northeast of tower Q is the undercroft R. This structure is barely visible above ground level; it appears that it served as a cistern. It is covered by a pointed vault that is pierced by three hatches. A rib divides the room into two equal halves. In the northwest corner of R is a curious roundheaded window/door(?) that is limited by jambs at each end; a small pointed vault begins south of its inner jamb and crosses the plane of the main vault at a 90-degree angle. It seems likely that a door was in the east wall of R. This cistern is carefully designed to catch the runoff from the higher levels at the west (cf. building A). The abundance of fallen masonry atop R indicates that smaller structures were built there. From the inside of R at current ground level it appears that postern S never penetrated the thickness of the main circuit wall. On the outside the large vaulted door of S is constructed with crude masonry. In 1962 neatly cut ashlar was added to the facade of the circuit (part of a limited restoration), completely covering on the exterior the door's junction with the circuit wall. S may be one of the original openings in the Byzantine castle. Attached to S at the south is a small square parapet of types III and IV masonries with an opening at the southeast. This chamber in its undamaged state probably formed a canopy for the door. Just why the builder of the main circuit wall would maintain this anachronism is a mystery.[39]

Tower T is a solid mass, except for a stairway that leads from a now missing door to the top. Because T is higher than the circuit, two flights of steps link this tower to the adjoining walls. North of T the east end of the castle is rounded off and protected by a talus below. On the interior the asymmetrical undercroft U is badly damaged. The main unit of U is covered by a barrel vault with one hatch; a separate barrel-vaulted chamber at the south is set askew from the axis of the long room. This square compartment has a small niche in its south wall; its vault is opened by one hatch. A now missing door probably connected both of the chambers in U. Battlements crown the wall between U and W. Room V is the only standing building attached to the eastern half of the north circuit. One of the two stairways that ascend the north circuit from the courtyard begins northeast of V. Adjacent to V is tower W, whose single lower-level chamber is opened by a tall roundheaded door at the south and a hatch in the apsidal vault.

In the center of the castle are the remains of a number of buildings. The best preserved of these is the half-buried building X (pl. 204b). A large part of its pointed vault still survives; stairs in the east wall of X lead to the top. Below this building and almost completely hidden is a square, barrel vaulted room (perhaps a cistern). To the north is chamber Y (pl. 205a). The west section of this unit, which is almost twice the width of the east half, is covered by two ribs and a barrel vault. The pointed vault of the east corridor merges into the barrel vault without a transitional rib. At the far east end near the hatch in the ceiling is a broken groined vault. The south wall of Y's east corridor is opened by two well-executed roundheaded doors (pl. 203b). Much of the masonry in this east wing is either the new ashlar of 1962 or the original masonry that was recemented when the new ashlar was installed (Hellenkemper, pl. 70b). These repairs make any assessment of building periods in this area difficult.

Many commentators believe that the Franks merely occupied the Byzantine fort or enlarged upon the Byzantine plan in creating the present castle.[40] I find these theses unacceptable. We know that Eustathius rebuilt the forts at Korykos and Silifke at the same time. Yet no two sites could be more opposite with respect to their plans. Korykos has a continuous, tight double trace with closely spaced square towers,[41] masonry marked by the use of the header-and-stretcher technique, few openings or embrasures in the towers, and a paucity of long undercrofts and pointed vaults. Other commentators have noted that the bent entrance and the round bulging towers at Silifke Kalesi must be Armenian constructions because of their peculiar design and masonry.[42]

In the opinion of this writer, it is impossible to credit any building period to the Armenians with reference to a peculiar style of masonry since no area in the circuit or entrance differs greatly from another. Normally, the bent entrance is a common design in the larger Armenian garrison forts, but the entrance at Silifke has elements that are atypical for the Armenian castles: the sharp protruding corners at the north,[43] uncustomary joints in the voussoirs, and two closely set double doors in the entrance portal. The present fortress, except for the outer revetment wall, the talus in the moat,[44] tower J, postern S, and cistern R,[45] gives no indication of more than one period of Frankish construction. While the Armenians could serve as a source of inspiration for the entrance and rounded towers, the ditch,[46] tower windows, and undercrofts of this fortress have the Frankish hallmark. The Armenians would never use a smooth ashlar as an exterior facing stone on walls subject to direct attack. Only formal excavation can determine the true extent of the Byzantine construction. The defaced Armenian epigraphs at this castle give no date nor evidence of reconstruction. The inscription over the door may be no more than an advertisement of the Armenian occupation after the departure of the Franks.[47]

[1]The plan of this site was begun in winter 1973 and completed on my second visit in late summer 1979. The contour lines are at intervals of 25 cm.

[2]J. Tischler, *Kleinasiatische Hydronymie* (Wiesbaden, 1977), 71. Often this castle is mistakenly labeled as Camardesium; see: Fedden and Thomson, 103–5; Hellenkemper, 249–54. Boase and Lawrence (in Boase, 158, 180) are correct in assuming that Camardesium, while historically associated with Silifke by reason of its proximity, is a distinct fortress. There is sufficient evidence for a separate fortification in Silifke and one in Camardesium; see Langlois, *Cartulaire*, 75, 113 esp. notes 3–5; Smbat, 635–37, 645, 648. The Byzantine and Armenian names for this castle are, respectively, kastron Seleukeias and Selefkia (or Selewkia). The most common Frankish designation is Le Selef; the Arabs employ Salūqiya. Wilbrand von Oldenburg (21) refers to it as Seleph castrum. Cf. Yovhannēsean, 230.

The city of Silifke appears on many modern maps, including *Central Cilicia* and *Cilicie*.

[3]Stephen of Byzantium, *Ethnica*, ed. A. Meineke (Berlin, 1849), 560. In composing this brief historical outline of Silifke Kalesi, I am indebted to the previous research of: Langendorf and Zimmermann, 155–61; W. Ruge, "Seleukia," *RE*, 1203 f; C. Huart, "Selefke," *EI*, 213; M. Tekindağ, "Silifke," *IA*, 643–48; Alishan, *Sissouan*, 328–35; Heberdey and Wilhelm, 100–108; Ramsay, 332, 350, 358, 372, 452; V. Langlois, "Les ruines de Séleucie dans la Cilicie-Trachée," *RA* 15.2 (1859), 748–54; Cuinet, 67–69; Schultze, 223–36; Schaffer, 65; Ritter, 321 ff; Rey, 350 f; A. Delatte, ed., *Les portulans grecs* (Liège-Paris, 1947), 175; Texier, 724 f; Budde, II, 153–63, 224; Sevgen, 292–94; Hellenkemper, 249–54; Yovhannēsean, 218–27; A. Mansel, *Silifke Kilavuzu* (Istanbul, 1943), 3 ff; Honigmann, 43 note 7, 81 f; Oikonomidès, 54 note 35, 241, 247, 350 note 354; Der Nersessian, 644; Smbat, G. Dédéyan, 79 note 62; R. Lilie, *Handel und Politik* (Amsterdam, 1984), 166 f, 618, 622; Lilie, 64, 98, 100 f, 109, 170, 253; Imhoof-Blumer, 481–86; Flemming, 26, 34 f, 84.

[4]An arms factory and mint were founded in Seleucia by the late 7th century; the names of numerous military and ecclesiastical officials assigned there have survived. See: G. Zakos and A. Veglery, *Byzantine Lead Seals*, I, pts. 2 and 3 (Basil, 1972), 727, 1074–76, 1157, 1335–37, 1419, 1433, 1684, 1773; G. Zakos, *Byzantine Lead Seals*, II (Bern, 1984), 144, 336, 338, 344. Initially, Seleucia was included in the massive theme of Anatolikon. Sometime before 732 it was incorporated into the coastal theme of the Kibyrrhaiotes. For information on the 10th-century theme of Seleucia consult: Constantine VII Porphyrogenitus, *De thematibus*, ed. and comm. A. Pertusi, Studi e Testi 160 (Vatican City, 1952), 61, 77, 147 f; Lamas (note 3) in the Catalogue; Canard, *Sayf al Daula*, 62, 67 f, 381.

A bishop was first assigned to Seleucia early in the 4th century; later it became the seat of a metropolitan. The patriarch of Constantinople attached the Seleucian church to his jurisdiction in the 8th century but restored its authority in the 10th century. See: Le Quien, *Oriens Christianus*, II (Paris, 1740; rpr. Graz, 1958), 1011–16; V. Laurent, *Le corpus des sceaux de l'empire byzantin*, V.2: *L'église* (Paris, 1965), 376–79; R. Devreesse, *Le patriarcat d'Antioche, depuis la paix de l'église jusqu'à la conquête arabe* (Paris, 1945), 145 f; E. Honingmann, *Evêques et évêchés monophysites d'Asie antérieure au VIe siècle*, CSCO 127, Subsidia 2 (Louvain, 1915), 84–88.

[5]Anna Comnena, 45 f.

[6]It is clear that Hellenkemper (250) misinterpreted the passage in Anna Comnena (45 f). He says that the fortress of Silifke is six stadia from the city of Seleucia. However, Anna's Greek text is quite specific (. . . τὸ Κούρικον καὶ διὰ τάχους ἀνοικοδομῆσαί τε αὐτὸ καὶ τὸ κάστρον Σελεύκειαν στάδια ἓξ τούτου . . .). Stadia is used in the sense of miles, but her estimate of distance is decidedly too small. See Ramsay, 384.

[7]Kirakos, M. Brosset, 57; Oikonomidès, 54 note 35, 247 note 21; Cinnamus, 16, 179 f; Choniatēs, 21; H. Glykatzi-Ahrweiler, "Les forteresses construites en Asie Mineure face à l'invasion seldjoucide," *Akten des XI Internationalen Byzantinistenkongresses* (Munich, 1960), 185.

[8]Certainly, the picture we have of Seleucia in 1137 is one of a prosperous Byzantine town; see S. Goitein, "A Letter from Seleucia (Cilicia) Dated 21 July 1137," *Speculum* 39 (1964), 298–303. In 1140 the port of Seleucia may briefly have come under Crusader control. At least Raymond of Poitiers grants trading privileges there to the Venetians (*Urkunden zur älteren Handels- und Staatsgeschichte der Republik Venedig*, ed. G. Tafel and G. Thomas [Vienna, 1856], I, 102). It is quite possible that Raymond was merely asserting his claim over the Greek port. Prior to 1140 there is no evidence that the Franks occupied this site. When Emperor John Comnenus and his army passed through Seleucia in the spring of 1137, it was within the contracted borders of Byzantium; see Lilie, 101, 398 note 59. If the city of Nykya can be identified with Seleucia, then Ibn al-Aṯīr (*RHC, HistOrien*, I, 423 f) has completely misrepresented the events there. In his Arab text he states that the army of John Comnenus laid siege to the town and either extorted money from the denizens or occupied the site.

[9]Der Nersessian, 645. For the narratives on the drowning of

Frederick Barbarossa in the Göksu refer to: Choniatēs, 416; William of Tyre, *La continuation de Guillaume de Tyr (1184–1197)*, ed. M. Morgan (Paris, 1982), 97; Langendorf and Zimmermann, 157 note 144. The history of Barbarossa's ties to the Armenian kingdom can be found in E. Eickhoff, *Friedrich Barbarossa im Orient* (Tübingen, 1977), 148–72; and Behā ed-Dīn, *The Life of Saladin*, trans. PPTS (London, 1897) 185–89.

[10]Smbat, 629; Dulaurier, 317 f.

[11]See below, Appendix 3.

[12]Delaville Le Roulx, 115 f, 118 f; Cahen, 614 f; Langlois, *Cartulaire*, 113 f; Langendorf and Zimmermann, 158; Müller-Wiener, 82; *RHC, DocArm*, I, lxxxvi; Wilbrand von Oldenburg, 19.

[13]Delaville Le Roulx, 56, 118 f, 122 f, 164–66. Garin de Montaigu may not have held the formal title of Castellan of Silifke Kalesi.

[14]Smbat, 648. The Hospitalers probably sold it back to the Armenians. See also: J. Delaville Le Roulx, *Les Hospitaliers en Terre Sainte et à Chypre (1100–1310)* (Paris, 1904), 141 f; J. Riley-Smith, *The Knights of St. John in Jerusalem and Cyprus c. 1050–1310* (London, 1967), 164, 432.

[15]Boase, 180; similar opinions were advanced earlier by Hellenkemper (253) and Alishan (*Sissouan*, 331).

[16]The inscriptions were noticed by Beaufort (220 ff); later both epigraphs were reproduced and translated by Langlois (*Inscriptions*, 53 f). Cf. Keil and Wilhelm, pl. 5.

[17]Concerning the possible reference to King Het῾um on the inscription found in the bailey, only the first three letters of what may be his name survive. Langlois (*Inscriptions*, 54) notes that "on ne peut rien tirer de cette inscription." I was unable to make a new transcription of the surviving fragment of the gatehouse inscription; the epigraph on the interior could not be located.

[18]Hellenkemper, 249–54; Keil and Wilhelm, 3 ff; Fedden and Thomson, 103–5; Langlois, *Voyage*, 190; Müller-Wiener, 80 f; Alishan, *Sissouan*, 334; Langendorf and Zimmermann, 161–65; Budde, II, 156 f.

[19]Alishan, *Sissouan*, 333. There are no indications that a parapet stood atop the inner revetment wall at the south, east, or west.

[20]At Crac des Chevaliers the talus is constructed in an identical fashion, although the masonry of the circuit wall is uniform and represents one building period; see Deschamps, I, 38 f.

[21]The very light colored ashlar stones in the photographs are the result of repairs carried out in 1962.

[22]Fedden and Thomson, 104.

[23]Even when used on Armenian churches and chapels in Cilicia, type IX is larger and more rectangular. Compare the fortified walls at Çandır and Lampron

[24]This also explains why occasional pieces of nonindigenous marble are mixed arbitrarily with Cilician limestone.

[25]Langlois, *Inscriptions*, 53, nos. 173 f.

[26]See above, notes 16 and 17.

[27]An etching in Alishan's book (*Sissouan*, 333) indicates that a mosque stood in the castle during the 19th century.

[28]Langendorf and Zimmermann (162) seem to have confused their identification of these buildings.

[29]The masonry, hatches, and other details are almost identical to those in the undercrofts at Crac des Chevaliers; see Fedden and Thomson, 103, fig. 14.

[30]Other than gate C, these are the only corbels in the castle. There is no evidence that the curtain wall and towers were provided with continuous machicolations; cf. Fedden and Thomson, 105.

[31]Occasionally, bossed stones (type VII) appear in other areas of the castle (e.g., on the interior and exterior of tower O). These could represent minor repairs from an Armenian period of occupation.

[32]Langendorf and Zimmermann (162) assert that this tower served the same function as a donjon. This statement is an exaggeration for it is no more heavily fortified than some of the towers at the south. Tower I is not a distinctly defensible unit like a keep. Its broad door at the south has no jambs and is highly vulnerable. This tendency to credit any large tower to the "donjon"-type is repeated by Dunbar and Boal at Azgit; see Boase, 84–91.

[33]The staircase sighted by Langendorf and Zimmermann (162) on the outside of tower I is not visible today.

[34]See above, Part I.3, I.5.

[35]The presence of a Byzantine tower here gives partial support to Hellenkemper's thesis (252, pl. 69b) that the Franks abandoned the original Byzantine circuit wall and made the south front slightly smaller. The presence of postern S (made out of similar type III masonry) may indicate that part of the Byzantine circuit was reused.

[36]The "tower 17" (supposedly attached to K at the south) described by Langendorf and Zimmermann (165) is not visible today. If it existed, it may have had no direct entrance into K.

[37]Broad windows in towers are never seen in the Armenian fortifications of Cilicia. Perhaps torsion weapons were placed on the tops of Armenian towers. Wide, open windows are common in the towers of Hospitaler forts; see Deschamps, I, pls. 89a and 98b.

[38]Fedden and Thomson, 105. These may be the "inner gates" noted by J. Thomson.

[39]There is the possibility that building R was added later when postern S was deemed unnecessary. Archeologically, there is some evidence for this in that the south wall of R is thicker than the rest of the southern circuit wall. Also, the stairway in tower Q seems originally to have extended farther to the east, but was later blocked.

[40]Müller-Wiener (80 f) believes that the present fort is Byzantine. Others believe that the plan of the site is Byzantine: Langendorf and Zimmermann, 165; Keil and Wilhelm, 4; Hellenkemper, 252. Hellenkemper is the only modern writer to locate what he believes to be the foundations of Byzantine buildings on the castle's *interior*. Because of mounds of debris that cover the bailey, I cannot confirm these findings. J. Thomson (Fedden and Thomson, 103) says that it is impossible to determine the original Byzantine plan.

[41]This aspect alone signifies a totally different concept in enfilading.

[42]Robinson and Hughes, 202; Langendorf and Zimmermann, 165.

[43]Armenian bent entrances are always flanked by towers or have a rounded exterior (cf. Tamrut, Anavarza, Savranda, and Yılan).

[44]These may actually represent a second period of Frankish construction.

[45]Ibid.

[46]It is likely that only the foundation for this dry moat dates to the Byzantine period; see Anna Comnena, 46.

[47]The now missing fragment of the unattached inscription on the interior of the fort could have been brought into the castle at any time. My conclusion that the Armenians are not responsible for the present fortress runs counter to the recent study of Clive Foss ("The Defense of Asia Minor against the Turks," *The Greek Orthodox Theological Review* 27/2–3 [1982], 159). Foss' claim, that the square towers from the period of Alexius still survive, is completely without support.

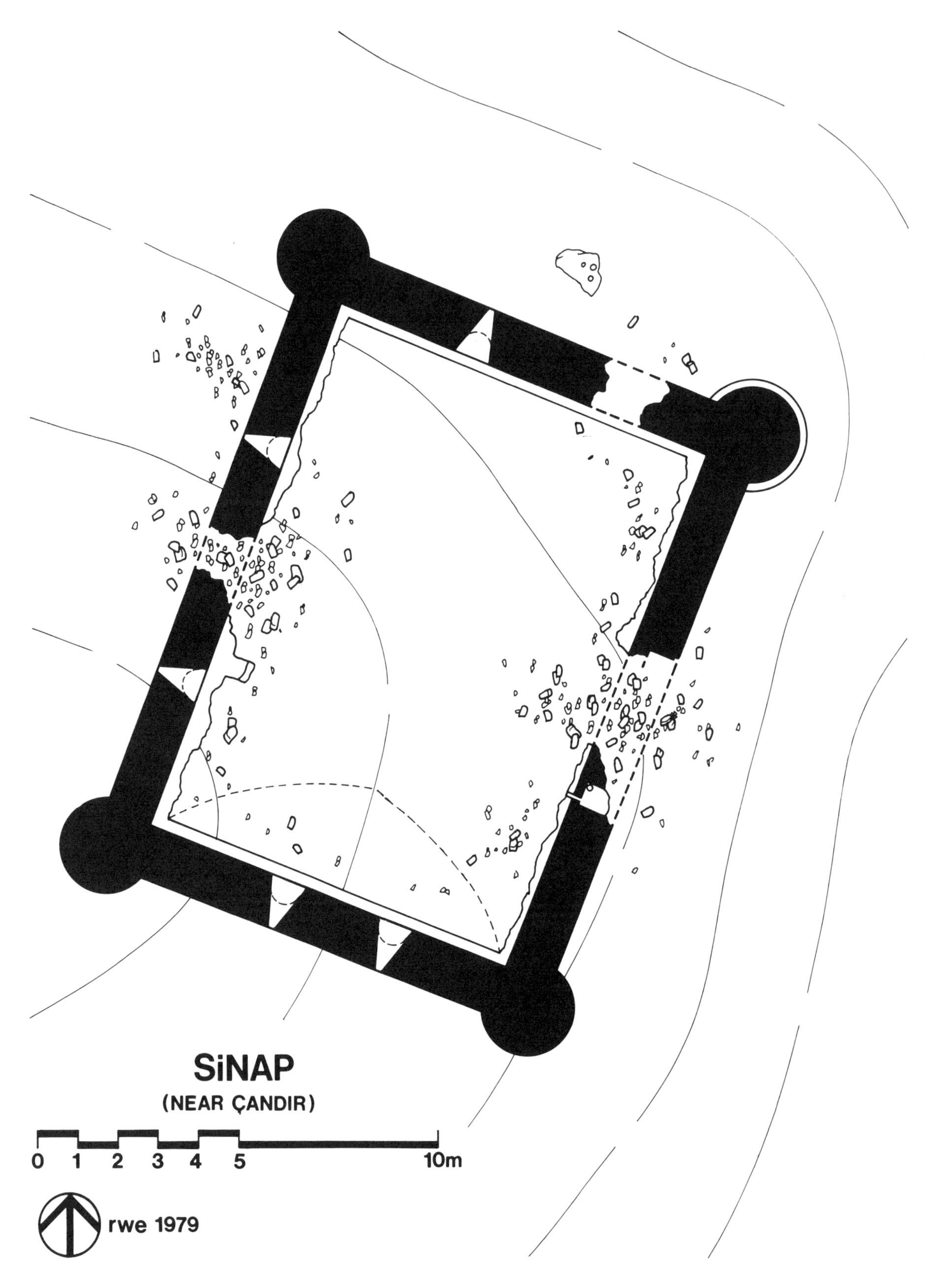

Fig. 67

Sinap (near Çandır)

About 9 km north of Gösne on the west flank of the important route to Çandır is the small fortified estate house of Sinap.[1] It stands near the bottom of a lush agricultural valley that is today watered by dozens of irrigation canals. The medieval name and history of this site are unknown.[2] The modern Turkish name is identical to the designation given to a similar fortification near Lampron.

In plan this fort (fig. 67) is a simple rectangular building with one round salient located in each of the corners (cf. Anacık and Sinap [near Lampron]). Today the fort is a hollow, partially collapsed shell with a massive breach in the west wall and one in the east wall (pl. 301a). There is a similar breach in the north wall. In its predestruction state it probably resembled Sinap near Lampron, which has two vaulted stories and a third terrace level. The two breaches in the east and west walls may be due in part to the intensive irrigation of the land which forced the north and south ends of the fort to settle at a lower level than the center. The fort is being torn into two equal halves.

The exterior masonry is a consistent, well-dressed type VII. Within the fort most of the masonry appears to be type IV.

The towers are identical, except for the northeast bastion which has a projecting shelf (or molding?) at the top. Since the other towers are not quite as well preserved, it is impossible to determine if they had similar shelves. The thin exterior openings of the five upper-level embrasured loopholes all have straight-sided stirrup bases. The entrance was probably in the long east wall; there may be the remains of a jamb at the north end of that east hole. At the south end of this puncture there is a curious square sink set in the thickness of the wall; the sink has a connecting pipe from the interior and a large drain in the base. The function of this sink is unknown. At present the ground level on the interior has risen considerably, but there are still the visible signs of a single, massive, pointed vault that once covered the first floor. Attached to the east and west walls are the shattered remains of the vault. The five extant roundheaded embrasures of the second level were not backed by casemates. In the north and south walls of the second level there are narrow shelves that probably helped to support the second-level floor. Directly northeast of the fort is a rock with three scarped holes; their purpose is unknown.

[1] No previous surveys of this site were ever published. The plan was executed in June 1979. On the plan the contour lines are at intervals of 25 cm. Because of the extensive damage to this fort, it was not possible to measure accurately the thickness of the upper- (or second-) level walls. However, it was possible to record the interior widths of the five second-level embrasures. Thus my plan of Sinap shows all the details of the ground-level room, plus the five upper-level embrasures.

[2] This fort is not listed on any of the principal maps of modern Turkey. Gottwald ("Burgen," 96, pl. 4) *briefly* mentions the presence of this site and gives some erroneous historical information. See also Schaffer, 59.

Sinap (near Lampron)

About 6 km northeast of Lampron on the gentle slope of a partially wooded agricultural valley is the small fortified estate house of Sinap.[1] The typology of this Armenian fort, with its rectangular facade and four small towers in the corners (fig. 68), is identical to the fortified country estates at Anacık and Sinap (near Çandır). Its close proximity to Lampron (pl. 208b) and its position at the junction of two strategic trails from the north (refer to the map, fig. 2) indicate that this fortified residence was also an early warning post for the Hetʿumid stronghold to the south. Unlike Sinap (near Çandır) and Anacık, this fort has three extant stories (pl. 301b). However, the third story, which was probably unroofed, is so badly damaged (only parts of its walls survive) that it is inaccessible and thus not surveyed. Today the closest source of water to this keep-house is a well about 25 m to the southeast. We have no historical references to this site, only occasional sightings by nineteenth- and twentieth-century visitors to Lampron.[2] Directly southwest of Sinap are the remains of a short wall and a considerable amount of coursed rubble. This may be an auxiliary building. There is no evidence of a circuit wall.

In general the exterior masonry is type VII (pls. 206a, 206b, 207a). In the upper half of the second- (or upper-) level exterior walls on the west, north, and east, there are large patches of type IV masonry. This type IV is surmounted by a continuation of the type VII. It appears that the type IV is a later repair, perhaps made after the siege or earthquake. Today the only extensive damage to the towers is in the northwest salient.

The only lower-level entrance is in the south side (pl. 301b). The exterior side of this door is framed with well-fitted blocks of type IX masonry. Directly above this door is the niche that once held the dedicatory inscription (pl. 207b). Three corbels are present above the south door. These probably supported a

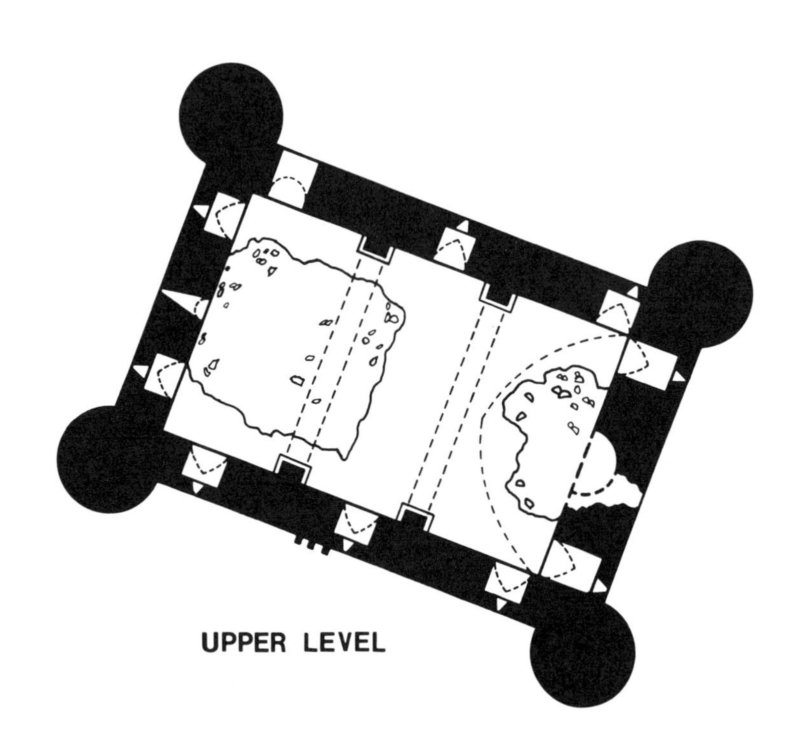

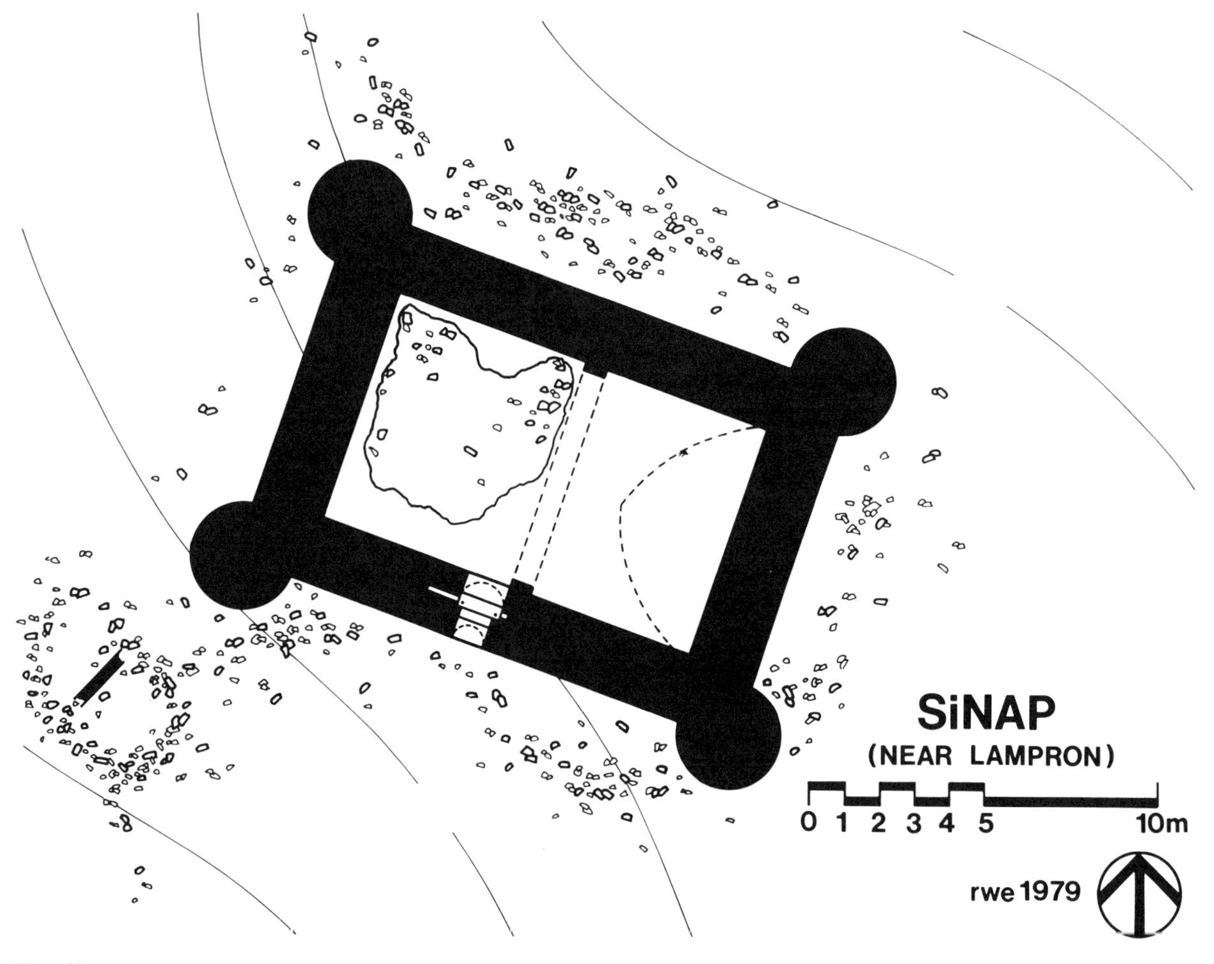

Fig. 68

machicolation that was manned from the roof terrace. The exterior of the upper level is dotted with the slits of nine thin embrasured loopholes with casemates; these have a straight-sided stirrup base (so typical in Armenian forts). Also, visible from the outside are the narrow rounded openings for the single splayed window in the upper parts of the east and west walls of the upper level (pl. 207a).[3] The two windows are considerably above the level of the nine embrasures with casemates. The only other opening in the upper level is a door in the northwest corner (pls. 206a, 206b, 209b). The vault over the door is more round than pointed. On the exterior side there appears to be no accommodation for jambs; the bottom third of the door was blocked on the exterior by facing stones. Perhaps a portable wooden ladder gave access to this level.

On the interior of Sinap types III, IV, VII, and IX masonries are used (pl. 208a, 208b). The latter type is employed in doors, windows, and ribs. The cruder stones are generally confined to the upper sections of the walls. The exception to this rule is the vault over the upper level (pl. 209a) where the type VII of the north and south walls is surmounted first by a well-executed type VIII and then by a slightly cruder form of type VIII masonry. Only in the small sections of the vault at the west is type III masonry used (pl. 209b). This curious placement of type III masonry may be due to the collapse of the original vault. A feature unique to this vault and almost never seen in the vaults of Armenian Cilicia is the small ridge created by the protrusion of the springing level from the face of the foundation wall. The masonries of the east and west walls of the upper level consist of type VII surmounted by type IV.

On the interior the first- (or lower-) level portal is a quadripartite opening with provisions for a double door and a crossbar bolt (pl. 208b). The unique feature about this door is that behind the exterior arch it has two abutting lintels at different levels. Recently, when the lower level of Sinap Kalesi was converted into a barn for three prize bulls (pl. 208a), the rubble from the partial collapse of the vaulted ceilings of both levels was shaped into a dividing wall. This wall is attached to the lower vault's central rib, which springs directly from the ground level. Because of the raised floor level and debris, it is impossible to determine if there was a basement level or a cistern below. Today entrance to the second (or upper) level is made by climbing through the huge hole in the pointed vault over the first level (almost all of the ceiling west of the rib is missing).

Unlike the lower level, this upper level has two ribs that spring from podia in the floor (pl. 209b). The upper third of the east rib has fallen away, leaving the vault intact (pl. 209a). The far east end of the vault has collapsed. The upper portion of the west rib has also collapsed, along with a large section of the ceiling. The high-placed window in the east wall is smaller and more widely splayed than its counterpart in the west wall. The casemates flanking the nine embrasured loopholes are all covered by fairly uniform pointed hoods (pl. 210a). One embrasure in the northeast corner (pl. 210b) and another at the southeast are placed slightly off-center with respect to their casemates, perhaps to accommodate left-handed archers or to enhance enfilading. The east wall of the upper level, which is slightly thicker than the west wall, has a very small apsidal opening in the center. The remote possibility that this functioned as a private chapel should not be ruled out.[4]

[1]Before my survey of this site in late summer 1979, it had not been the subject of any prior architectural study. The contour lines on the plan are at intervals of 25 cm. The "upper level" is actually the second story.

[2]The intervisibility is frequently noted; see: Boase, 181; Schaffer, 57. Sinap may be the "Kala Jik" on the map of Favre and Mandrot.

[3]Because of the presence of the apse, the window in the east wall of the upper level does not appear on the plan.

[4]The lack of niches inside the apse (necessary for the Armenian liturgy) may indicate that it was not a chapel; see Edwards, "First Report." Cf. also Hild and Hellenkemper, 283 f. There is absolutely no evidence to show that Sinap was part of a cloister.

Sis

At that point where the strategic road from Saimbeyli and Feke abruptly descends into the eastern lobe of the Cilician plain, the magnificent fortress-outcrop of Sis[1] rises as a final barrier. This castle has intervisibility with Andıl, Anavarza, and Tumlu. From Byzantine times the unfortified town of Sis has always been located below the east flank of the outcrop where mountain streams provide a constant source of water (pl. 214b).[2] Today the Turks call the modern city Kozan but still refer to the castle as Sis.[3] Since this site served as the capital of Armenian Cilicia and was the seat of the Armenian Katʿołikos, it has attracted numerous visitors and has been the subject of many commentaries and histories.[4] In this discussion I will be concerned only with the castle of Sis.[5] I will preface my remarks on the architecture with a brief historical introduction that is derived from both primary and secondary sources.

Prior to the Byzantine period there is no record of a fortification at this site. Between September 704 and August 705 the small Greek settlement at Sis successfully repulsed an Arab attack with the help of the emperor's brother Heraclius, whose army supposedly killed twelve thousand Arabs.[6] Some years later its Christian population was forced to abandon the town. In the early 800s Sis became part of the fortified frontier for the Abbasids, who planted a small colony on the site; the Caliph al-Mutawakkil reconstructed its Byzantine defenses in the mid-ninth century. Unfortunately we have few specific Arab references to Sis. The fortress must have been of some importance since the Byzantine army of Nicephorus Phocas captured it early in 962.[7] For the next hundred and fifty years the history of Sis lies in obscurity. In 1113 the Rubenid Baron T^coros I captured the town, which also experienced a devastating earthquake in the following year.[8] Emperor John II Comnenus seized it in 1137. At some period before 1172 the Armenians reoccupied Sis and began construction at the site. The usurper Mleh is said to have been assassinated there in 1174.[9] As early as 1177 Sis is referred to in one chronicle as the royal residence.[10] However, ten years later an inscription on the reconstructed donjon of Anavarza describes Anazarbus as the "mother of cities." For military (and consequently political) reasons, the proximity of Sis to the mountains made it a more attractive administrative center; it began to rival Anavarza by the late 1170s as the principal site of Rubenid Cilicia. There is probably no precise date for the official transfer of "the capital" to Sis from Anavarza, but it certainly occurred between 1180 and the 1190s.[11]

During the reigns of King Levon I and his eventual successor, Het^cum I, much ecclesiastical and military construction was undertaken. Unfortunately the only dedicatory inscription still in situ in the castle is a badly mutilated epigraph on tower M. One can still make out "of the Armenians" and what may be the name "Het^cum."[12] Since Het^cum had a reputation as a builder, it is quite possible that he is responsible for enlarging the castle. His wife, Zapēl, is credited with building a hospital there in 1241.[13]

The history of Sis becomes a repetitive account of Mamluk raids from 1266 to 1375.[14] On ten separate occasions the Armenian capital or its immediate environs was successfully sacked by the Egyptians and their allies. In 1348 this misery was compounded by the arrival of the bubonic plague from Europe. The last Armenian king, Levon V, was betrayed and captured in the citadel. Although Sis was the seat of the Armenian patriarchs from 1292 to 1873, the Kat^cołikos resided there permanently only from the seventeenth century.[15] A monastery for the patriarchs was not founded until 1734.

If laid from end to end, the circuit walls of the castle at Sis would measure almost 3 km in length (fig. 69). These barriers carefully follow the irregularities of the cliffs for over 680 m of the exceptionally long and impressive outcrop of limestone (pl. 300c). Because of this adaptation the plan cannot easily be defined in terms of segmented baileys. For this discussion I will divide the castle into the following units. The most complex and amorphic is the central bailey that includes A–B and N–Y (pl. 211a). This unit also has a tall central spur (T–V) and falls sharply at the east and west flanks where the outer circuit walls bow outward. The south-central bailey is a thin, neatly defined unit that is confined to a narrow spur and runs from gate I to tower M (pls. 300c, 215a). The southeast bailey, which includes D–F, is positioned along the sharply falling slope beneath the level of the south-central bailey. The south bailey is set below tower M and includes G and H (pl. 216b). The fifth and largest unit is the north bailey which runs from Y to Z. The easiest approach to the castle is at the southeast end of the outcrop (pl. 214b), where the topography leads the approaching party directly to the vaulted entrance-corridor A (pl. 211a).

The exterior masonry, as with many of the lower east walls, consists of types V and VII. However, the type VII in A tends to be confined to the corners[16] and to the frames of windows and doors (pl. 212a),[17] which is rather unusual in Armenian architecture. The northeast door of A is of extreme importance because it is topped by an Arab inscription (pl. 212b). Approximately 40 percent of the epigraph survives; I was unable to transcribe or translate what remains. I could determine that its letters are identical to the *nasḫī* style of writing that was so common in Egypt during the fourth quarter of the fourteenth century. The careful integration of the epigraph into the northeast wall of A signifies that it can belong only to the permanent period of Mamluk occupation (after 1375).[18] The entire entrance-corridor is probably a Mamluk construction, considering that it has an exposed angle at the north and that the voussoirs of the four lower-level round-headed embrasured loopholes are small and numerous (pl. 213b). These four loopholes in the east wall of A have a simple flat base, while the two embrasured openings with casemates in tower B to the south have stirrup bases (pls. 211a, 211b). The rounded tops of both of the embrasures in B consist of neatly carved monoliths. The latter (pl. 214a) are usual in Armenian architecture, and the variance in styles is certainly due to different periods of construction. A further indication that A was added later to an existing circuit (fig. 69) is the huge gap (with an intervening corbel) between the northwest wall of A and the circuit south of N. When the Mamluks built A they used Armenian masons to cut the facing stones; the addition of A to tower B created a sharp vertical junction (pl. 211b). It is interesting that in the proximity of the northeast door of A, thin metal rods are embedded in the interstices of the types V and VII masonries, but the voussoirs and the few relatively smooth stones around the door do not have these rods. On the interior of the door at the east, there is a square hole to accommodate a crossbar bolt (not shown on the plan) (pl. 213a). Just how this bolt was secured on the other side is not apparent today.

Once past the vaulted corridor and its southwest gate (which is bolted from the outside) there is a second gate of very simple design at the west end of B. This door and the adjoining wall are constructed with type IV masonry. This opening may be part of the original Byzantine entrance at Sis. The foundations of the circuit wall to the south of B (including the square salient C) are built with an identical type of masonry (pl. 215a). Unlike the other types of masonries, the mortar of this type IV has numerous fragments of brick and tile. Another remnant of Byzantine construction may be in salient N. The original tower N with its two doors and ten windows are constructed with type IV masonry and later completely covered on the outside by a new salient-curtain of types V and VI masonries. This second phase of construction is certainly Armenian (pl. 220a).

D, the major entrance into the southeast bailey, is constructed like the typical Armenian tripartite gate (pl. 215b). What is unusual is that the slot machicolation was blocked at a later period with coursed stones. The pointed outer arch of gate D has alternating segmented voussoirs—a style seen at other Armenian castles (cf. Maran). Farther to the south the upper level of tower E has the shattered remains of a small Armenian chapel. (pl. 214b).[19] Except for the partially scarped undercroft E and the adjacent hall to the south, there is no evidence on the surface of other buildings in the southeast bailey.

The gate complex I is the only visible entrance into the south-central bailey. This complex has three units. At the far east is a vaulted chamber (preserved today) that has a round tower protruding from its north wall. This chamber probably functioned as a cistern. The tower is equipped with corbels and certainly provided flanking fire for the two doors at the west. In the center is a room that has only the springing of a round vaulted covering still standing. In the north wall of this room is a door that is limited by jambs and protected by a crossbar bolt (pl. 216a). The arch over the jambs has collapsed. At the south end of the chamber is another door. On the west flank of complex I is a second gate at the north end. Traces of scarped rock in the corner of the door indicate that some sort of wooden portal was accommodated. The corridor behind the west door is covered by a pointed vault.

Once past this double gate a long series of scarped steps ascend the summit of the south-central bailey, passing the foundations of two square rooms at the east. Halfway between complex I and M is the barrel-vaulted cistern J, whose sides are scarped. From this area of the outcrop to the fosse south of M, the entire east flank is supported by a massive revetment wall. The two rooms on the interior of K are difficult to interpret because they are filled with debris. The present ground level covers the ceilings of K and L. On the east flank of K there is a stairway that descends to a vaulted opening and a blocked chamber. The mystery over K's exact function intensifies when entering the vaulted room L via a stairway in its northeast corner (pl. 217a). The stairs leading to L are covered by four monolithic lintels. The slightly pointed vault over L is well preserved and opened by two hatches. A straight wall of a rough type IX masonry marks the southern end of the room. However, at the north, not only is the rounded shape of the exterior wall of K visible, but it is constructed entirely out of type VII masonry. It may be that in an earlier period K was the most southern salient on the summit. Later, new buildings raised and extended the construction to the south. It is possible that the type VII masonry of K's wall was intended as an interior facing, though it would be very rare.[20] It seems likely that room L functioned as a cistern since its lateral walls are stuccoed. Undoubtedly the original east and west walls of M continued north for some distance, creating a second level over K and L. In the center of M there are fragments of a fallen vault and a protruding mass of rock (pl. 217b). This natural rock has not been cut, and just why it is so perfectly preserved in the center of this finely constructed residential quarter is a complete mystery. On the interior of the south wall of M are the fragments of an Armenian epigraph.[21] Substantial portions of the exceptionally large and casemated windows of M are still visible. Directly south of tower M is a massive scarped fosse (pl. 216b). Although the foundation of M has been squared off,[22] this is *not* in contradiction to the Armenian theory about exposed corners (as with corridor A), since the proximity of the cliffs and fosse makes it impossible to attack the base of M (pl. 300c). There is no evidence that M was ever a donjon. Its north end may have been opened and completely accessible to gate I.

The south bailey is carefully defined by cliffs at the east and a circuit wall at the west and south (pl. 216b). Of the three vaulted towers at the south, only H preserves five casemated embrasured loopholes at the upper level. The lower level of H, as with the tower to the north, may have functioned as a cistern. The extreme south tower is fitted with embrasures. The circuit connecting the three south towers has four embrasured loopholes with casemates.

The size and simplicity of the south bailey is in sharp contrast to the massive central bailey. Near the summit of the central bailey is the vaulted cistern U (pl. 218a). Its plastered walls are carefully constructed; the shape of its central rib is reminiscent of the arch over the ground level of the estate house at Belen Keşlik. Not all the buildings in the area of U are so carefully constructed. To the southwest, chamber V (pl. 219a) and the nearby walls are constructed with type III and IV masonries. Just who is responsible for the buildings in this area is difficult to determine. To the north is the complex S with its apsidal east corner. This apse was probably part of a chapel.[23] In this central bailey and to the north there are dozens of small square chambers that served as storage rooms and cisterns. Undoubtedly there were a large number of wooden buildings on the interior. The principal structures of the central bailey are along the east circuit (N [pl. 220a] through Q). The vaulted hall P, which is partially subterranean, is fully independent of the circuit wall. While it is uncertain whether P functioned as a cistern, it is clear that the small semicircular tower directly to the west was used for the storage of water (pl. 218b). This area is the only point on the circuit where corbels are consistently employed. Between towers N and O there is a single and very unusual overhanging machicolation (pl. 219b). Such features normally appear over an entrance.

At the north end of the central bailey, small posterns at R and Y provide access to the city below. The only notable construction at the north end of this ward is the large cistern X. A central arcade of three arches divides the cistern into an east and west half. The west half (pl. 221a), which has a pointed vault opened by a single hatch, is almost entirely built with cut masonry. The east half (pl. 220b), which is covered by a semivault, is scarped into a natural cleft. The tooling marks on the scarped rock are unusually deep and broad.

The huge north bailey is almost devoid of buildings. A modern watchtower (part of the military base in the lower terrace) is located near its southeast corner. At the far north a small postern gate is located at point Z.

Recently, there has been some speculation on the sequence of construction in this castle.[24] However, in the absence of excavation reports, inscriptions, and specific references in the texts, it is impossible to date any of the structures (with the possible exception of gate-corridor A and tower M) on the outcrop. Excluding A and the areas of B, N, and V, most of this fortress is Armenian in respect to its masonry and plan.

[1]The castle at Sis had never been the subject of an accurate instrument survey prior to my own work at this site in summer 1973. The survey was completed after visits in August 1979 and July 1981. The interval of distance of the contour lines on the plan is approximately one meter.

[2]Sis or Kozan appears on the following maps: *Central Cilicia, Cilicie, Kozan, Marash, Malatya.*

[3]In Armenian and Syriac chronicles this site is simply called Sis. The name is derived from the Byzantine Sisin or Sision kastron. The Arabs refer to this settlement as Sīs, Ḥiṣn Sīsiya, and Sīsa. The Franks call the Armenian capital Assisum, Assis, Sis, and Sisa.

[4]V. Büchner, "Sis," *EI,* 453–55; idem, "Sis," *IA,* 708–12; Langlois, *Voyage,* 382 ff; idem, "Voyage à Sis," *JA* 4 (1855), 263–73; Alishan, *Sissouan,* 241 ff; Texier, 583; Ritter, 67–96, 597, 621 ff; Schaffer, 44, 89; Cuinet, 90–92; Favre and Mandrot, 116 ff; *Handbook,* 724 f; E. Lohmann, *Im Kloster zu Sis* (Striegau, 1905), 2 ff; Kʿēlēšean, 226 ff; Müller-Wiener, 77; Fedden and Thomson, 97–100; Mikaeljan, 428 ff; Hellenkemper, 202–13; Rüdt-Collenberg, 21 ff; Barker, opp. i, 111; "Sis," *Haykakan,* 398 f; Mas Latrie, 89; Le Strange, *Palestine,* 538; idem, *Caliphate,* 141; Sevgen, 216–23; *The Bondage and Travels of Johann Schiltberger (1396–1427),* trans. K. Neumann and B. Telfer (rpr. New York, 1970), 86 f, 126, 235; Sanjian, 55; al-Balāḏurī, 262 f; Ḥamd-allāh Mustawfī, *Nuzhat-al-qulūb,* trans. G. Le Strange (Leiden-London, 1919), 258; E. Honigmann, *Le couvent de Barṣaumā et le patriarcat jacobite d'Antioche et de Syrie,* CSCO 146, Subsidia 7 (Louvain, 1954), 45, 51, 68, 74, 167, 174 f; Wilbrand von Oldenburg, 18–20; Kirakos, M. Brosset, 93.

[5]The compound of the patriarchs in the lower terrace is described in Edwards, "First Report," 168–70, and "Second Report," 134 ff. For information on the churches in and around Sis see also: B. Kiwlēsērean, *Patmutʿiwn katʿołikosacʿ Kilikioy* (Antilias, 1939); above, note 4; Oskean; Ēpʿrikean, I, 500.

[6]Theophanes, *Chronographia,* ed. K. de Boor, I (Leipzig, 1883; rpr. Hildesheim, 1963), 372.

[7]Canard, *Sayf al Daula,* 141, 299.

[8]Kirakos, M. Brosset, 57; Matthew of Edessa, 112.

[9]Michael the Syrian, J. B. Chabot, III, 361; Smbat, 625; Smbat, G. Dédéyan, 55 note 36.

[10]Nersēs of Lampron (578) complains that this "metropolis" is without a bishop and suitable churches. Cf. *RHC, DocArm,* I, 301, where the Katʿołikos Grigor IV uses "Sisuan" as a collective term to identify Rubenid territory before 1189.

[11]Wilbrand von Oldenburg (18–20) finds a complete and well-established capital in 1212.

[12] Alishan, *Sissouan*, 247; Langlois, *Inscriptions*, 17 f; idem, *Rapport*, 41 f.

[13] Alishan, *Léon*, 364.

[14] Sanjian, 95, 99, 111, 294; Bar Hebraeus, 446, 453; Dardel, 32 ff, 81–109; Vahram of Edessa, 522; Smbat, G. Dédéyan, 118 f; Gaudefroy-Demombynes, 19, 85, 87, 97, 99 f, 217; al-Maqrīzī, I.2, 55, II.2, 60 f, 116, 131 f, 146, 190, 196, 227 f, 249, 255, 279; Canard, "Le royaume," 231, 238 ff; Yovhannēsean, 49–62.

In the 15th century Sis was a district in the Mamluk province of Aleppo; see Popper, 17.

[15] During the medieval period a monastery at Sis may have been a center for the production of manuscripts; see Sanjian, 92. See also Yovhannēsean, 63–65.

[16] In the corners the type VII stones are only slightly larger than the rest of the facing.

[17] Some of these are actually a type IX with a bossed face (see above, Part I.3).

[18] Hellenkemper, 211.

[19] For a description of the chapel, see Edwards, "First Report," 170.

[20] This does occur at Sis in the west wall of building M and at both Yılan and Çandır; in all three cases it is not used consistently as in K's wall.

[21] See above, note 12.

[22] Hellenkemper, pls. 53a and 53b.

[23] See Edwards, "First Report," 170.

[24] Hellenkemper, 212 f.

Tamrut

Near the southwest corner of the vale of Karsantı at an altitude of 1,420 m is the castle of Tamrut.[1] From this fort[2] there is a commanding view of the regions to the north (pl. 228b) and northwest where the tributaries of the Eğlence Çayı flow from the mountains to their eventual destination in the newly created Adana lake. Also, at the north there is intervisibility with the modern village of Sivişli (approximately 4 km) and the small fort at Işa (approximately 3 km). Tamrut Kalesi does not have intervisibility with Meydan. To the southwest the barrier of mountains separates the vale from the Cilician Gates and Pozantı (pl. 223b). The main purpose of Tamrut Kalesi was to guard two very strategic roads that pass within view of the west and east flanks of the fortress-outcrop. The road at the east, the principal route to Etekli, passes south from Sivişli through the defile cut by the Eğlence Çayı and leads to Adana. The other route, which passes the west flank of Tamrut, also leaves Sivişli and crosses the mountains and Highlands, eventually terminating at Karaisalı. These roads take on added importance because Sivişli is just south of the junction of two major routes to the north. One connects to Karsantı at the northeast; the other, at the northwest, joins an important north-south road that links Pozantı to the Ürgüp region. The Armenian inscription above the gatehouse door of Tamrut Kalesi has not yet yielded any *precise* information about the builder. Considering the size of this complex, it certainly played an important part in the defense of the numerous agricultural communities in this area.[3] Because this region is sparsely settled today, the castle shows no signs of modern vandalism and is in a very good state of preservation.

Tamrut Kalesi crowns the summit of a limestone outcrop. Except for the southwest flank, the sides of the outcrop are almost vertical. From below, the fort's walls are partially hidden by a forest of pine trees. The easiest line of access into the fortress is to approach the outcrop on the north flank and then follow a narrow trail that ascends the west slope (pl. 221b). Eventually the trail leads to the only entrance into the fort at the southwest (pl. 222). The last 50 m of the approach to entrance A consist of an extremely steep and narrow defile, which would deter any enemy from storming the gate. The defenders have a clear line of fire down the pathway from the two towers that flank A. About 9 m south of gatehouse A there is the foundation of a wall that runs across the width of the pathway (not shown on the plan). This was probably some sort of outwork or barbican that restricted entrance into the main gate. Gatehouse A is a typical bent entrance with two flanking towers.

What is not so typical is that the inscription over the gate is fully preserved (pl. 223a). Unless the modern visitor has a scaffold or ladder, it is impossible to view the epigraph at close range. The inscription is about 6 m above the sloping ground level. From the photographs, which were taken at ground level, it was possible to translate parts of the inscription.[4]

(c ankṣ.)	
1. šinecʿoł.a.. cʿankṃ.	the builder ..circuit walls
(tʿ)	
2. .ḅor anvani epł ambrow	.... Lampron (or). by the famous bishopric
3. iyišatak hab.ạgior pr	in the memory of (?) baron
4. ....b.cowṇočʿi. orowtʿọ	?
5. ṇọrogescʿi ẹ̄:tʿv[....]	I will renew
6. ....yiṣ̌escʿowḳʿ	?...We will remember

Directly above the inscription are the remains of three corbels. Only the corbel at the far east is still unbroken. Since the exterior of the gatehouse-door was not provided with a slot machicolation, the brattice supported

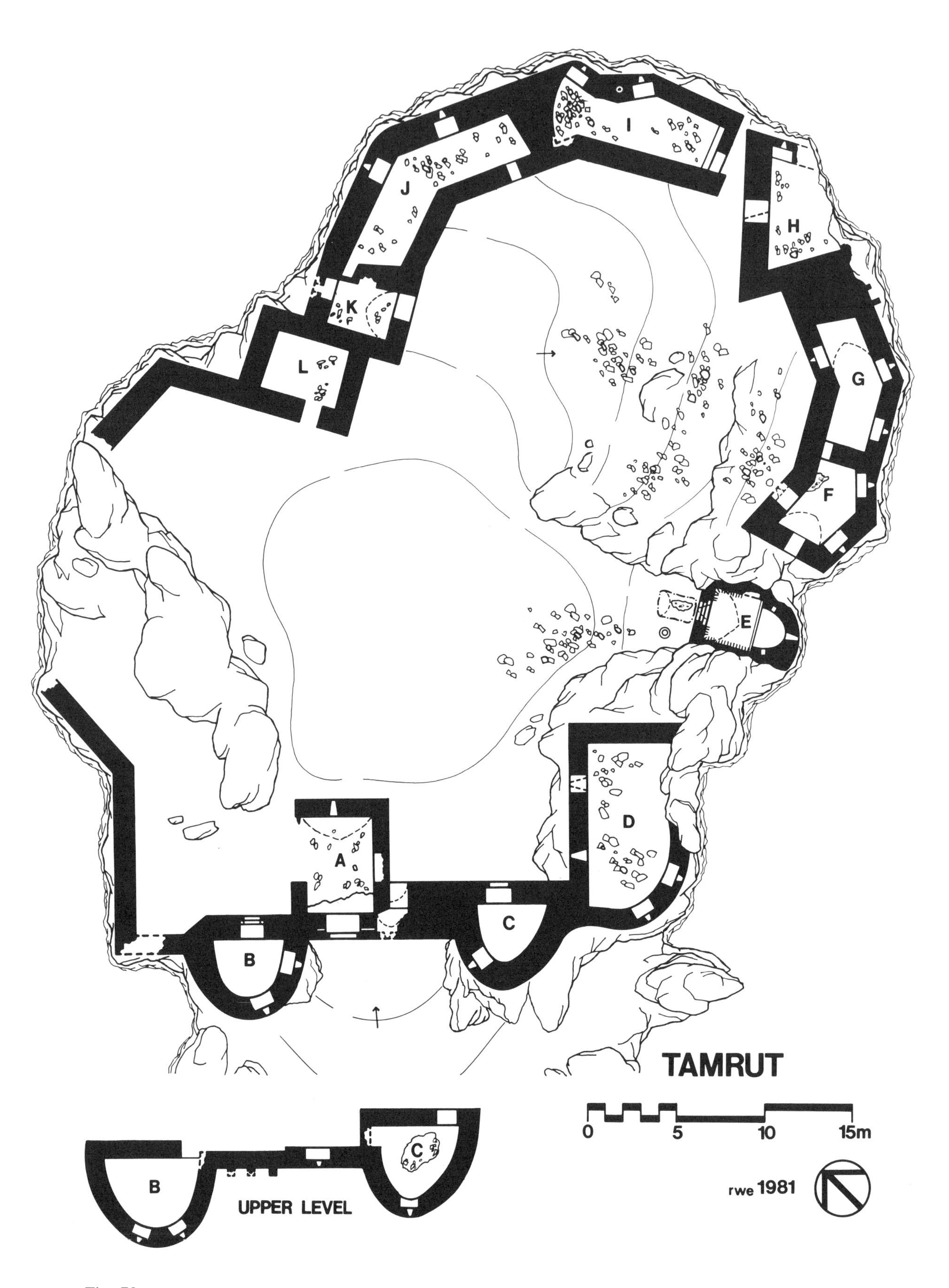

Fig. 70

by these corbels protected the area directly over the gate.

Except for the blocks of smooth ashlar around the frames of the doors and windows, all of the exterior facing stones consist of a harmonious blend of types V and VII.[5] In the areas where the facing has fallen away, the exposed core consists of rough unhewn fieldstones placed in fairly regular layers and cemented by a gray mortar of limestone. Surprisingly, on the interior of the fort *all* of the walls are constructed with the same combination of types V and VII masonries (pl. 224a). Even on the interior of the towers and rooms, the walls have the identical facing (pls. 225b, 226b). The only areas where uniformly smooth ashlar (type IX) appears are in some of the arrow slits and all of the vaults over the casemates and doors. The only crude masonry (types III and IV) occurs in the vaults over the towers and rooms. (pls. 225a, 229b). There is one highly exceptional and very limited stretch of masonry in the rectangular room that surmounts chapel E (this upper level is not shown on the plan; pl. 227b). With the exception of its west wall, the masonry of this unvaulted room consists of the usual type V and type VII. The interior facing of this west wall, which is over the entrance to the first-floor chapel, consists of very rough fieldstones bound in somewhat regular courses with a plethora of rock chips and a small amount of mortar. The exterior facing of the wall is similar, except that the margins are covered with a layer of mortar. This masonry does not fit any of the established paradigms for Armenian Cilicia and probably represents the only evidence in this fortress of a separate period of construction.

The gate complex A is an Armenian-type bent entrance (fig. 70). On the exterior of its south door (pl. 222) there is a depressed arch that is limited by narrow jambs. The sockets for a crossbar bolt, which secured a wooden door, are visible behind the jambs (pl. 224a). The inner side of this south door has a taller and much wider vault. The apex of the vault is slightly pointed. Once past the south door, the visitor must make a 90-degree turn to the left in order to exit from the bent entrance. This broad inner (or northwest) opening of gatehouse A was not fitted with a wooden door. The arch over this opening, as well as the vaulted covering over the gatehouse, has collapsed. The north wall of A is pierced by a single, high-placed embrasured window. This window does not appear to be a loophole. The only other opening in the gatehouse is a niche(?) in the east wall. Once through A, the approaching party does not have unlimited access to the single bailey of the fort (pl. 223b), but must pass through the narrow passage created by the north wall of A and a large mass of protruding rock to the north.

Gatehouse A is protected by the two-storied towers B and C (pl. 222). On the northeast side of B, both the upper-level and lower-level doors are squareheaded on the exterior (pl. 223b). The lower door, which is in the center of the wall, has a segmented lintel. The three segments of the lintel are neatly joined by joggled joints. This lower opening is arched on the interior side, and holes in the sill behind the jambs show that double wooden doors were accommodated. The upper-level door is jambless and squareheaded throughout. Its exterior lintel is smaller and more crude when compared to the lower-level door. The upper-level door is at the east end to facilitate access from the wall walk over the entrance (pl. 224b). The first-level semicircular chamber on the interior of tower B is opened at the south by two embrasured loopholes with casemates. On the exterior (pl. 222) these slits, like the other loopholes, do not have the stirrup bases but do have a circular top. The casemates are constructed in the typical Armenian pattern (pl. 225a). Today a ledge marks the division between the lower and upper levels. This ledge once supported a wooden floor. Likewise, in tower C and in *all* other rooms that have two levels, only the upper story is covered by a vault of stone, while joist holes or stone ledges supported the wooden ceiling over the first floor. All of the wooden planks are missing from the castle today. The vault over tower B is well preserved, except for a small hole near the apex. Two embrasured loopholes with casemates open the second level of tower B at the southwest.

Between B and C at the upper level, most of the wall fronting the wall walk has collapsed (pl. 224b). We can get some idea of the height of the wall from a small section still preserved on the southeast corner of tower B. There must have been some sort of door in the wall to allow the defenders to man the brattice on the overhanging corbels. The foundation of a single embrasured loophole with a casemate is still visible to the southeast of the corbels. Directly below this casemate in the first level of the complex between gatehouse A and tower C is a tiny casemate-chamber that was once fronted by an embrasured loophole. Unfortunately the embrasure has collapsed, leaving a gaping square hole in front of the castle (pl. 222). This first-level casemate was entered from the northeast through a jambless door that was covered by a half barrel vault.

Tower C is the third element of the defenses on the southwest side of the fort. The lower level of the tower is opened by a squareheaded door at the north. This door, which is fitted with jambs, and one embrasured loophole with casemate are the only openings in the first floor (pl. 223b). The upper level is opened at the east by a roundheaded jambless door. One enters this door by walking along a ledge of rock near the northwest wall of room D. On the interior the upper level of tower C is opened by one typical loophole and casemate at the southwest (pl. 225b). The vault over the second floor has a large hole in the apex.

From a purely structural standpoint, it is sound engineering not to place the upper-level casemates directly over the lower-level openings. But at Tamrut Kalesi military considerations also played a part in the positioning of the casemates (pl. 222). At the lower level of towers B and C, the embrasures are positioned to guard the immediate area in front of gatehouse A. At the second level the loopholes are rotated to fire directly down the approach path, taking advantage of the increased trajectory from that height. Today there is no evidence of merlons or other battlements atop the towers or anywhere else along the circuit.

Attached to tower C is room D (pl. 223b). This structure, which is squared at the north and rounded at the south, is separated from chapel E at the east by a mass of limestone. The only entrance into D is a roundheaded jambless door at the northwest (pl. 226a). On the exterior the depressed arch of the door consists of an oversized keystone that is carefully joined into two large springing blocks. On the interior side the door is covered by a semicircular vault (pl. 226b). Directly over the door is the foundation of an embrasured window. To the southwest of this window in the same wall is the foundation for another embrasured opening (pl. 227a). Most of the south wall of D has collapsed, but the foundations of two embrasured loopholes with casemates are still visible there. This rounded south wall does not appear to join with the northeast wall of D, but merges into the intervening rock mass (fig. 70). Some sort of masoned wall may have been adapted to the top of the rock. Unfortunately it cannot be determined if a vault once covered room D. I found no evidence of springing stones in the rubble that covers the interior.

To the east of room D is chapel E, which is sandwiched in a cleft of the rock mass. This chapel is typical of Armenian ecclesiastical construction.[6] Just in front of the west door of the chapel is a small rectangular opening to a subterranean cistern. The cistern, which partially occupies a natural cavity in the limestone outcrop, extends for some distance under the nave of the chapel. The opening into the cistern is actually a collapsed section of its carefully stuccoed vault. Just west of this opening is a drainage hole into the cavity. During my survey this was the only cistern that I located in the fort.

Rooms F–L carefully follow the circular edges of the east half of the outcrop (pls. 228a, 228b). There is only one lower-level entrance into room F (not shown on the plan), and this is at the west. It is a squareheaded jambless door. Large sections of the lower-level walls rise from scarped rock. From the room over chapel E it is possible to gain direct access to a squareheaded second-level door in the west wall of F. This second level has four other openings and is covered by a slightly pointed vault. Except for a section at the northeast the vault is well preserved. In the south wall there are two embrasured loopholes with casemates (pl. 229a). There is one such opening in the north wall, but its exterior loophole has fallen away, leaving the gaping vault of the casemate (pl. 228a). A narrow squareheaded door connects the upper levels of rooms F and G.

The only lower-level entrance into room G (pl. 228a) is a squareheaded door at the west (not shown on the plan). There are four upper-level openings in room G. In the south wall are two embrasured loopholes with casemates (pl. 229b). The west wall has that narrow door that connects to chamber F. The northwest wall of G has a preserved embrasured loophole with casemate. Since F and G are the highest points in the castle, the defenders chose to put the only interior defenses here. It is quite possible that there is a second-level room between G and H, but the means of access into the room is not certain. There had to be some means of entrance into this area to man the corbeled brattice that overhangs the cliff from the edge of the wall.

The triangular room H has suffered considerable damage, and only a part of its upper-level northeast flank survives today (pl. 230b). This surviving section shows the springing of a now missing vault above a well-preserved second-level casemate with embrasured loophole. Just below the casemate in the northwest wall is a line of joist holes that once provided the support of the floor. These holes are especially well preserved and show that entire trunks of trees were embedded in a thick layer of mortar (pl. 230a). This north wall is opened in the lower level by a jambless vaulted door that is surmounted by an embrasured window (both are shown on the plan). The vault over the door is slightly depressed and consists of five voussoirs. The rounded top of the window has a diamond-shaped hole incised in the apex. I have not seen this design used in any other Armenian windows in Cilicia.

The vaults that once covered rooms I–L have all collapsed (pl. 228b). The lower-level southwest wall of chamber I is opened by two straight-sided doors. An embrasured loophole in the lower level of the north wall of I has a very narrow casemate (the lower-level openings are not shown on the plan). In the upper level, I is opened by a door in the southeast wall and two embrasured loopholes with casemates in the north wall. The ashlar stones that make up the vault of each casemate are L-shaped blocks that are reminiscent of the groined vaults in the Armenian gatehouses (cf. Yılan).

Room J, like F, G, and I, bends in the center (pl. 231a). At the lower level its south wall is opened by two straight-sided, squareheaded doors, and the north wall has two embrasured loopholes with casemates (none of these four openings appears on the plan; pl. 231b). In the upper level of the north wall the two casemates are spaced apart at a wider distance. In the west half of this room there is a clear indication of the floor level between the upper and lower casemates, but the division is not so clear in the east half. In the south wall at the east end of the upper level there is a squareheaded window (pl. 231a).

Room K and probably room L were single-story structures (pl. 228b). The springing for a vault is still visible over room K. K is entered through a roundheaded jambless door at the southeast. At the east there is a shallow niche, and in the north wall there is a narrow window that is flanked on the inside by a casemate. L is a simple rectangular room that is opened only by a jambless door at the southwest.

[1]The survey of this hitherto unattested site was undertaken in June 1981. The contour lines are spaced at intervals of 1.5 m. Because of circumstances, the 1981 survey of this fort was cut short, resulting in some shortcomings on the plan. The upper level of chapel E was not surveyed. For rooms F–J, *only* the upper-level openings are shown on the plan (the single exception being the lower-level northwest door of H). The windows and doors of the upper level are more numerous and of greater importance for understanding the defense of the castle. I will mention in my description the presence of any lower-level openings from F through J.

[2]Tamrut Kalesi may appear as Alişekale on the *Ulukişla* chart. The village of Sivişli appears on the map of *Central Cilicia*. For directions to this site see the commentary on Işa in the Catalogue.

[3]Tamrut Kalesi is located in the region that Father Alishan (*Sissouan*, 156 ff) believes to be under the control of Barjrberd (Partzerpert). Because of its size, Tamrut is the most likely candidate to be Barjrberd. For the most recent discussion of the latter see Yovhannēsean, 115–19.

[4]For this translation and transliteration I owe a debt of gratitude to Mr. Virgil Strohmeyer, Jr., of UCLA. Because of the poor quality of the photographs, this translation is highly speculative. From ground level it is clear that some of the letters have weathered extensively. I would suggest that the next visitors to this site make every attempt to take a squeeze.

Virgil Strohmeyer includes the following notes with his translation:

> "In the second line the *ept'amb* could stand for *episkoposowt'eamb* or for *episkopos t'ambrow* which might be a name of a person or place (see Hubschmann's *T'ambarark'*). If this reading is accepted, *row* is possibly a dialectal form of *ṙowben*, but I think the initial *r* argues against this. Finally, if we read *ł* for *t'* we might have the word *Ṯambrow* or Lampron.
>
> The third line is fairly clear except for the name which seems to be in the genitive.
>
> The fourth line is impossible to translate.
>
> The fifth line has a clear *t'v* which may be a phonetic rendering of the more normal *t'ow*.. (number) in classical Armenian."

Prof. Dickran Kouymjian of Fresno State University also commented on the photographs of the inscription, and he reads a date at the end of line 5: "in our era 682." The Armenian year 682 is equivalent to A.D. 1233. My own reading of the epigraph tends to confirm this date, but it should be stressed that this reading is highly speculative.

[5]The two southwest towers, B and C, have a curious change in the size and quality of the exterior facing stones in the lowest courses. These stones are smaller and do not have the sharp rectangular corners of the stones above. A close examination reveals that this facing is not the foundation but is intended merely to cover the rough face of the living rock, which is the foundation.

[6]For a complete description of the chapel are Edwards, "Second Report," 142 f.

Tece

Near the village of Tece,[1] about 4 km north of the coastal road that links Silifke to Mersin, is a structure that combines the features of a garrison fort and a fortified estate house.[2] This site occupies a small part of the summit of a truncated hill, about 15 m above sea level. The primary purpose of Tece Kalesi was to offer a secure haven to the baron of the district and his garrison. Secondarily, it served as a place of refuge for any tenants who were lucky enough to flee inside during an invasion. Tece Kalesi guarded not only the strategic coastal road but also the now abandoned river road that connects the site of Fındıkpınar to the coast. I found no evidence that this site has intervisibility with any other fortification.[3] There are no inscriptions at this unexcavated fort, nor do we have any references to it in our medieval chronicles. Considering that the Armenians often relegated the defense of the coast and borders to their Crusader allies, this site probably belongs to one of the European orders.

Tece Kalesi consists of a single square circuit wall and an unattached, four-story, keep-like residence (fig.

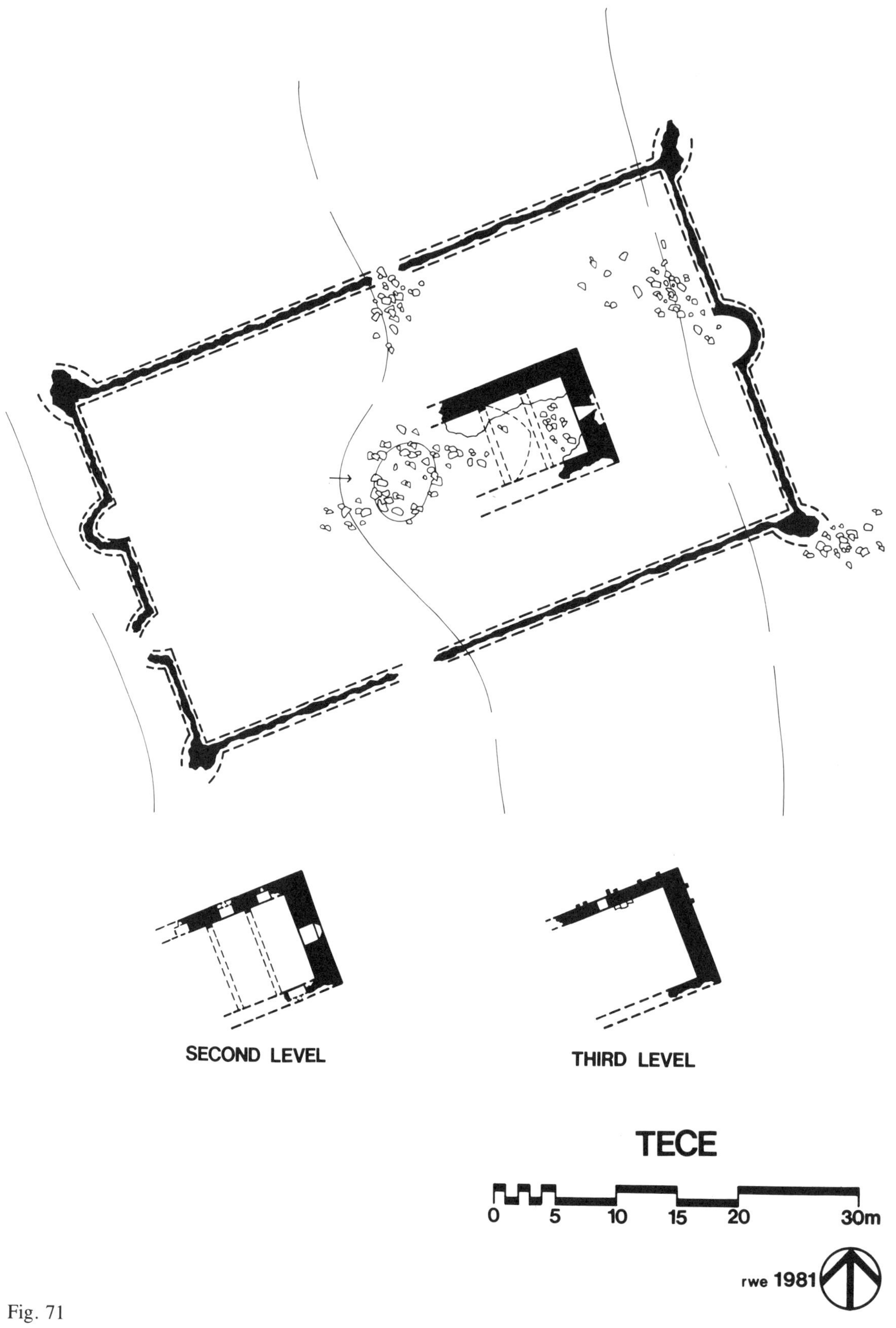

Fig. 71

71, pl. 234a). Today the circuit wall has decayed to the point where the best-preserved section stands to less than 1.2 m in height and shows *no* trace of facing stones. Sections to the north and east are so badly damaged that only small traces of the core are still visible. At the south, enough of the wall has survived to show that the core consists of large uncut fieldstones that are bound in a thin matrix of mortar and rock chips. There is evidence of seven rounded(?) towers projecting from the line of the circuit. Each corner of the square circuit has a solid tower. Since these salients are barely visible, my estimation of their size on the plan may be altered by excavation. Two towers in the west wall and the single tower in the center of the east wall are hollow. The original height of the towers and circuit are impossible to determine accurately, but, considering the quality of the masonry in a comparable circuit at Kız (near Dorak), it is likely that this perimeter wall did not exceed 4 m in height. The only apparent entrances through the circuit wall are at the north and south. Like the fortified estate at Kız (near Dorak), the function of the circuit was to delineate and protect the compound of the estate. Small wooden lean-tos along the interior of the circuit would be adequate to stable horses and shelter a small garrison in this temperate climate. The primary defense for the site was the keep itself.

The keep is constructed with types VII and IX masonries on the exterior and types VIII and IX on the interior (pls. 302a, 232a, 232b). On the average, a stone in the exterior facing measures 69 cm in length and 42 cm in height. These well-cut blocks show few traces of mortar at the junctions. In the lower courses of masonry the rusticated centers are very thin, but 3.6 m above present ground level the rough centers protrude to a greater degree (pl. 232a). There is no structural or military reason for this change, and its presence here must be purely aesthetic. The broad margins and somewhat shallow boss (even in the upper courses) are identical to the type IX masonry in the neighboring forts of Yaka and Tumil, both of which I have credited to Crusader construction. It is *only* in the third level that the rusticated exterior facing stones have pointed inner sides, thus making them type VII. The core in all three levels is consistent throughout, indicating that the keep is the result of one building project. The construction of this site may have been a joint Armenian-Crusader undertaking.[4]

There are other peculiarities about the facing stones that should be noted. At the north end of the east wall, the exterior faces of eight blocks in a single course have been unevenly cut with sixteen small square sockets (pl. 302a). These shallow holes were carved after the construction of the original wall to accommodate some sort of lean-to. About 3 m to the south is a single large hole that was also cut after the original construction.[5] The core in the walls of the keep has much more mortar and fewer fieldstones than the core in the circuit wall (pl. 234a). For the interior masonry, the large well-cut stones are typically confined to the first level and part of the second, while the third level has smaller stones with thick margins of mortar.

What appears to be the first level of the keep may actually have a basement level below. It is common in these keep structures to have a cistern at the lowest level (cf. Kız [near Dorak]). It is certain that the unstuccoed walls of the first floor did not form a cistern (pl. 233b). Today the only opening in the first level is a high-placed roundheaded window in the center of the east wall. This window is embrasured and has lost its exterior facing (pl. 302a). This first story is covered by a slightly pointed vault that is reinforced by two transverse arches (pl. 233a). Most of the vault has now collapsed. There is no evidence for a door at this level. Any large opening in the ground level would be highly vulnerable to direct attack; entrance may have been made with a removable ladder through the second-level east window.

The second level was the major line of defense on the interior of the keep. Today there are the remains of three roundheaded embrasured windows with casemates in the north wall (pl. 234a). The most eastern of these openings is in a relatively good state of preservation, while the one at the far west has almost completely collapsed (pls. 233a, 232a). The central embrasure in the north wall is well preserved on the interior (pl. 232b) but damaged on the exterior. In the small fragment of the south wall there are still the remains of a similar casemate and embrasure. The most interesting feature of the second level is the large window in the east wall (pl. 233b). This opening is limited by beveled jambs and crowned by a rounded top whose well-fitted voussoirs form a slightly pointed apex. Springing stones atop the north pilasters of the second level clearly show that these supports were for transverse arches. On the interior of the east wall there are no indications that the second level was covered by a vault of stone. Perhaps the ribs supported a wooden ceiling.

The only visible opening in the third level is a straight-sided(?) window in the north wall (pl. 232a). On the interior side just above this window to the east are the fragments of three crenellated steps that lead to the fourth-level terrace (pl. 232b). Since no stone steps were constructed below the three fragments, some sort of wooden platform must have been built above the window and below the third step. From the fourth-level terrace, archers could man the brattice that was supported by the corbels at the top of the third level. The nature of the roof atop the third level is unknown. As with Kız (near Dorak), the third level probably had a wooden roof. The fourth level was an open terrace. Since the west end of the keep-estate has *totally* collapsed, its full perimeter cannot be determined.

[1]The plan of Tece, a previously unsurveyed site, was made in June 1981. The interval of distance between the contour lines is 50 cm. The rounded shapes given to the four corner towers of the circuit are hypothetical.

[2]The village of Tece appears on the map of *Mersin (2)*. See also *Handbook,* 196.

[3]The areas to the north and northwest have not been adequately explored and may yield unattested forts.

[4]A strong Armenian influence can also be seen in the slightly pointed vaults over the second-level casemates.

[5]Two holes of a similar size are present in the lower-level north wall.

Toprak

In the southeast corner of Cilicia Pedias on a large somewhat symmetrical mound stands the medieval castle of Toprak.[1] The mount actually consists of an outcrop of dark basalt and the accumulated debris of many periods of human habitation.[2] Considering the proximity of Bronze Age tumuli,[3] excavations at Toprak will probably reveal levels as early as the second millennium B.C. Today there are some fragmentary remains of the medieval village on the west flank of the outcrop. Most of this settlement has been tilled under to accommodate crops of wheat and melon. A constant source of water in the form of the Kara Suyu flows past the outcrop at the south. This stream originates in the mountains near Çardak Kalesi and eventually joins the Ceyhan. At an altitude of about 65 m, this fortress has a commanding view into the junction of five major roads. At the east is the road to Osmaniye and the Amanus pass; directly west, two paved roads lead to Adana. The coastal highway from Iskenderun joins the latter to form a strategic intersection at Toprak. A gravel road leads north from the castle and meets the paved road to Kadirli. The strategic value of the site is enhanced by intervisibility with Tumlu, Anavarza, Amuda, Bodrum, Yılan, and Çardak.[4]

Unlike many Cilician fortifications, the medieval history of this site has not been entirely obscured. One name by which we know this fortress in our Armenian chronicles, Tʿil Hamtun, is not used prior to the twelfth century in any of the extant manuscripts.[5] No name from the classical or late antique world can be securely given to this site. It is speculated (probably correctly)[6] that "Hamtun" comes from the period of Arab occupation, with the name being derived from the founder of the Hamdanid dynasty. It is unlikely that an earlier Arab name for Toprak Kalesi was al-Kanīsah. We are told that when Hārūn ar-Rašīd built the present fortress at Haruniye in 786 he also rebuilt at the same time an older Greek fort with black stones.[7] The Arabs called this fort Kanīsah as-Sawdāʾ ("the Black Church") or simply al-Kanīsah.[8]

This castle is referred to as Tʿil during the initial Crusader occupation and in 1137 when the Byzantine Emperor John II Comnenus captured it during his successful campaign through Cilicia.[9] Baron Tʿoros II captured and garrisoned the fort in 1151.[10] In May 1154 the Byzantine emperor invited Masʿūd I, the Sultan of Konya, to attack Toprak, but the sultan failed to dislodge the Armenian garrison which was assisted by Frankish knights. It was recaptured by Byzantine troops under Manuel Comnenus four years later.[11] Sometime before 1170 it again came under Armenian control. In 1185 Toprak was ceded to Antioch as ransom for the return of Baron Ruben.[12] Bohemond III became the castle's new lord. Within nine years it was back in Armenian hands. On the Coronation List of 1198/99 (see below, Appendix 3) a certain Robert is listed as its lord. It is possible that Robert was a Frank under obedience to the Armenian king.[13] About 1211 Wilbrand mentions the impressive fortified complex of *castrum regis nigrum,* which he also calls Thila.[14] Until 1266 Toprak Kalesi is mentioned frequently as a geographical feature in various chronicles.[15] At that time the Mamluks successfully laid siege and dislodged the Armenian troops. On a number of occasions thereafter it was reoccupied by the Armenians, until it became a permanent Mamluk possession in 1337.[16] The Egyptians held it until 1491 when the Ottomans secured control of Cilicia. Thereafter it was used as a barracks and abandoned.

Toprak Kalesi has been the subject of three serious architectural studies. The first was undertaken by J. Gottwald in the 1930s.[17] His plan of this site is somewhat misleading, but there is valuable information in his architectural descriptions, and his historical notes are generally accurate. Some years later a more precise plan with a one-paragraph architectural description was published by W. Müller-Wiener.[18] In 1976 H. Hellenkemper published a slightly improved plan and a substantial description of the site.[19] Rather than merely repeat what has already been published, I will offer a summary description of Toprak Kalesi and new observations based on my own field surveys.

It should be noted that Gottwald, Müller-Wiener, and Hellenkemper all devised schemes for determining the various building periods at the site and attributed certain sections of the fortress to specific builders. I cannot wholly accept or reject their conclusions, nor can I offer any reliable scheme to replace them. No fortress of a similar size in Cilicia has as many building periods as Toprak. Almost none of the exterior facing stones or the curious variety of architectural features are specifically attributable to the Armenians. In fact, the dominant features of the fortress (such as the three-quarter round tower at the west, the massive talus, the continuous galleries, the rectangular plan, the square and pentagonal salients of the lower bailey, and the peculiar embrasured casemates) have *no* parallels in Armenian military architecture. One of the great difficulties in evaluating this site is that the foundations of the circuit walls and even some of the rooms are buried in debris. Only with the exposure of all levels during excavation and with the careful analysis of mortar, sherds, and other debris can a precise sequence for the construction of the fort be determined. It will become clear in the following commentary that not all of the architectural elements can be attributed to specific builders. From my observations and the history, I believe that there are two *major* building periods. The first major building period, which gave Toprak its rectangular shape, can belong to either the Arabs, Byzantines, or Franks. It appears quite probable that the second major period of construction took place under the Mamluks.[20]

Toprak Kalesi has two specific baileys (fig. 72). The lower bailey at the west can be reached by a modern paved road that ascends the south flank of the outcrop. This road passes below the west wall before entering at gate(?) V. Unfortunately almost nothing is left today of the square salients that may have formed this gate. Most of the east boundary of this bailey consists of the massive talus that is below the west wall of the upper bailey. On the interior of the lower bailey there are joist holes in the circuit wall north of tower Y. These indicate that some wooden buildings were constructed on the interior. At the northeast there is an enclosure that forms the forecourt to the upper bailey. Only a few traces of its foundation survive today. Some sort of gate at point T connected the forecourt with the lower bailey. There is no evidence of construction inside the forecourt.

One of the most interesting features of the lower bailey is the shape and placement of its twelve towers. Three of the towers are square, five are pentagonal, three are semicircular, and one at the extreme south is of uncertain shape. The south half of the lower bailey circuit has only pentagonal towers spaced at close intervals,[21] while the north half has rounded bastions spaced at broad intervals.[22] The northeast tip of the circuit with its short east extension to the forecourt wall has three(?) closely spaced solid(?) bastions.

The masonry of the lower bailey circuit consists of a variety of types.[23] The most common stone is something that closely resembles type IV, except that the edges are more neatly squared than usual (pl. 234b). Another kind, which is used occasionally for repairs to the exterior facing, is type V. Of all of the facing stones used in the fortress, this masonry most closely resembles the Armenian paradigm, and it may date from the period of Armenian occupation (or be the work of Armenian masons). This type V is used to rebuild the southwest tip of tower Y. Some sections (pl. 236a), such as the lower level of the tower southeast of Y, have a relatively smooth-faced ashlar for repair. Entire sections of the lower bailey circuit are built (or repaired) exclusively with a type III that has broad sloppy margins of stucco (pl. 235b).[24] Many fieldstones are used in this masonry as well as recycled blocks from collapsed walls. Sometime in the 1960s the Turkish authorities leveled the top of the lower bailey circuit in the areas around tower Y (pl. 234b) with a layer of cement (supposedly to "restore" and protect the circuit). It is now impossible to tell if Y or any other salients to the north were covered by vaults. A few springing stones indicate that the salient southeast of Y was covered by a vault. Considering the variety of masonry types used in the lower bailey circuit, it is not unwarranted to postulate at least three periods of construction. I can assign none of these periods securely to the two major periods of construction in the upper bailey. The use of widely spaced rounded towers and the closely packed polygonal and square salients represent very different military theories.

A puncture at the extreme south end of the lower bailey circuit exposes the southern tip of the talus and an important piece of information about the construction of the circuit (pl. 235a). This unusually broad talus extends from A past tower I to what appears to be a junction with the south wall of the lower bailey. Most of the talus was not built from a flat ground level but covers an irregular slope. In a few cases the large crude blocks that make up the talus have fallen away

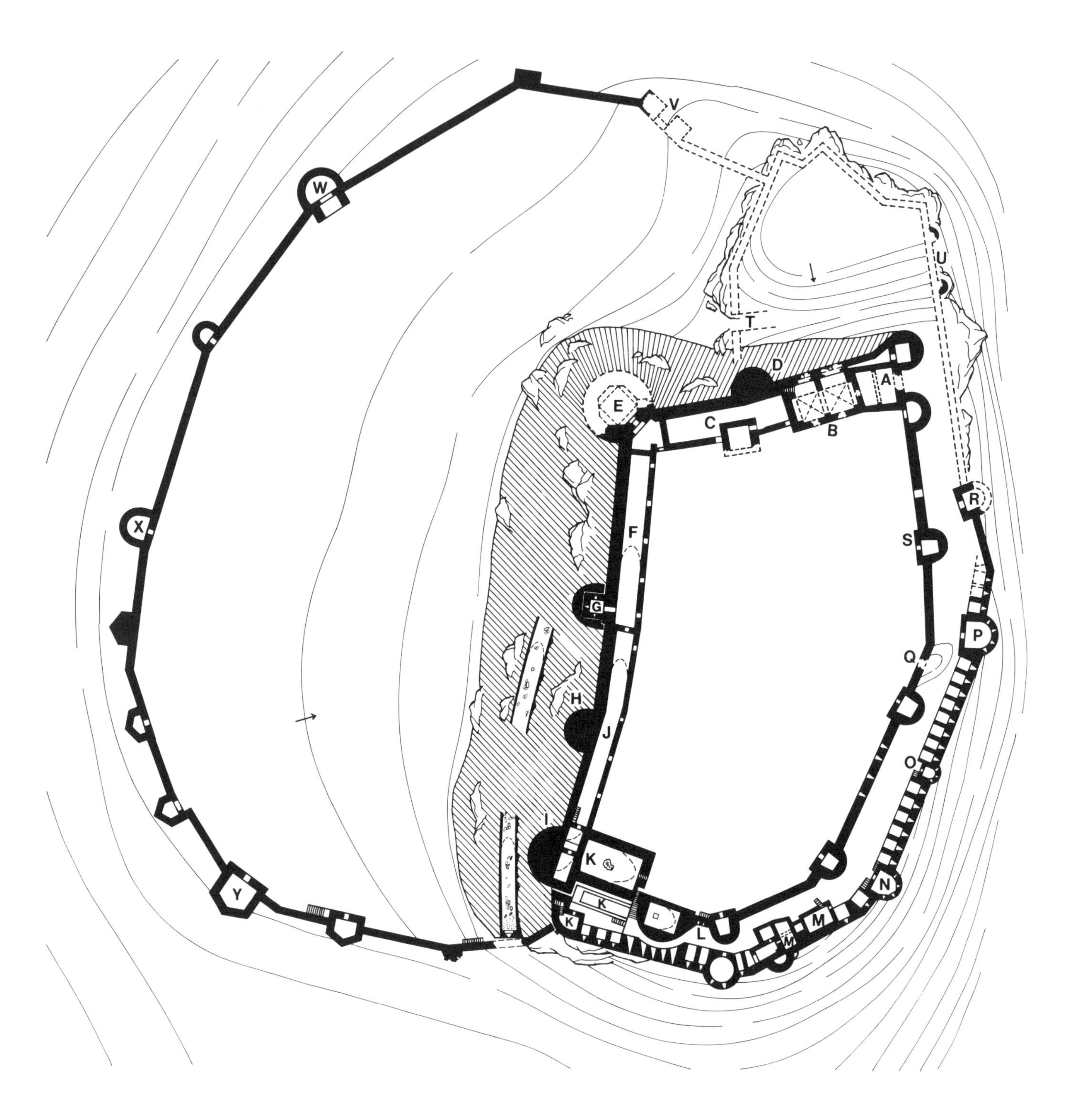

Fig. 72

TOPRAK

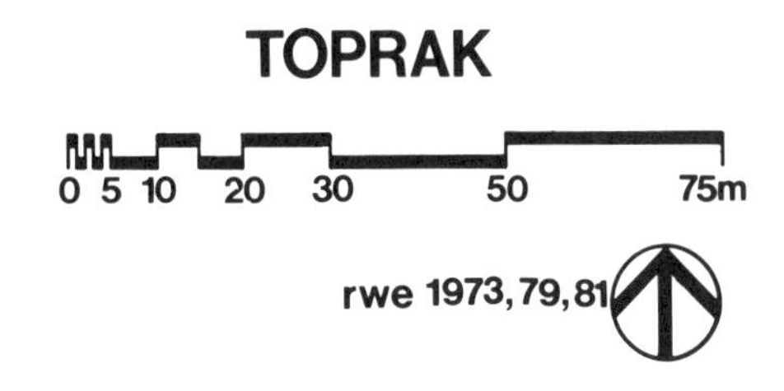

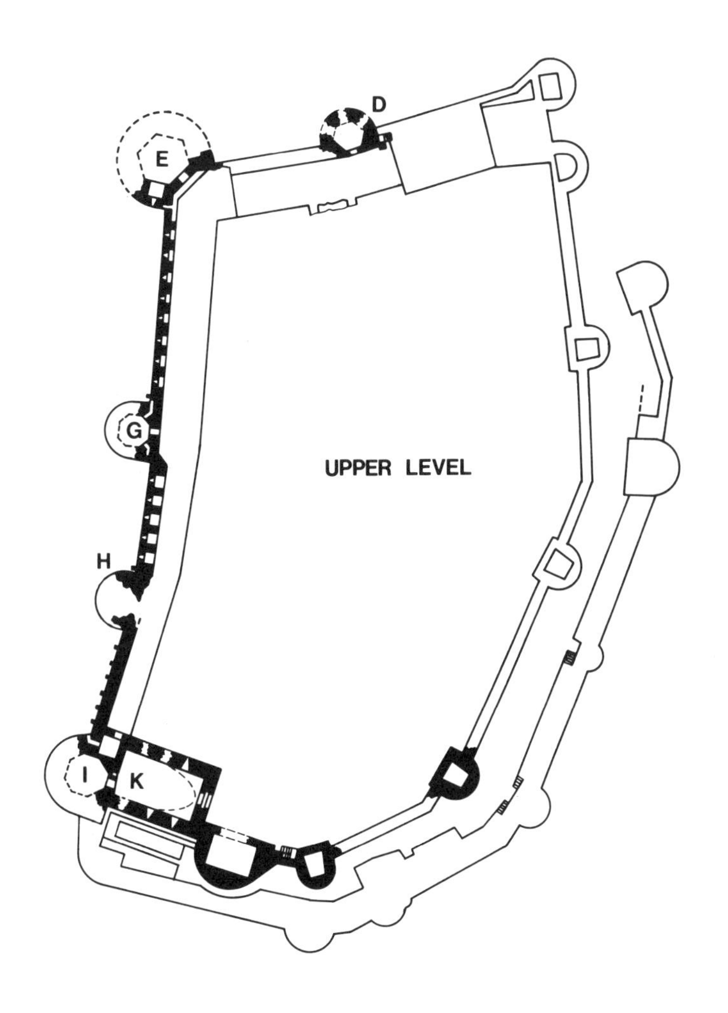

BASED ON THE SURVEYS OF W. MÜLLER-WIENER AND H. HELLENKEMPER

to expose the foundation of dirt and rock. The stones in the talus are not black, as with the basalt in the walls and towers of the castle above, but gray, like the Cilician limestones. This is the only large talus within the boundaries of Armenian Cilicia. It is also unusual because fortified bunkers were built just below the talus-facing between towers E and I. These bunkers were long narrow passages fitted with embrasured loopholes. Because the vaults over these passages have collapsed, their exact number and arrangement is hidden. Some may be preserved, but completely covered. I did locate two sections of what may be a single collapsed passage; one section is southwest of tower G, and the other is west of tower I (pl. 235b). The entire east wall of the latter is still present as well as its south wall, which was exposed by the hole in the lower bailey wall (pl. 235a). The south end of this talus-passage is pierced with the only preserved embrasured loophole in this area of the complex. The top of the loophole is squareheaded on the exterior (the interior side is blocked with debris), and the base is not a flat-bottomed stirrup but is forked. This embrasure was designed to shoot directly south, which *certainly* indicates that the talus was constructed *before* the lower bailey circuit.[25] Except for some areas of repair, the talus seems to represent one period of construction.

Complex A, the vaulted hall flanked by two towers at the east, appears to be the only entrance into the upper bailey. Hellenkemper believes this entrance to be from a later period of construction and that the "Toranlage zwischen den flankierenden Tortürmen greift in veränderter Form das Beispiel der Yılanlı Kale auf,. . . ."[26] However, an entrance between two towers is not the result of any Armenian inspiration but has been widely practiced since the Bronze Age. I see no reason why A cannot be the original entrance into the upper bailey. Hellenkemper postulates that the original entrance into this upper ward was at point L, and Müller-Wiener places it at point Q.[27] In 1973 and in following visits I found no evidence in either location for a gate; the possibility that a postern was positioned at L or Q should not be ruled out. At Q there is not adequate flanking fire to protect a gate. If L or Q were an entrance, it would be most unusual not to have a corridor or gatehouse on the interior side (as with A) to restrict the movement of those entering. A is the only opening that is adequately defensible and restrictive.

There is an important additional reason why A may be the original gatehouse. The preserved north tower of A and what is left of its collapsed south tower have an exterior facing that consists of a small, somewhat irregular type VI (pl. 236b).[28] This type of masonry is most significant because it is used throughout the fortress, usually in the lowest levels of the exterior facing. Along the north wall and in the west towers, the stones of this masonry are larger and in a few cases polygonal; this irregular type VI appears to be the exterior facing of the first major period of construction. Only a few of the architectural features associated with this masonry, such as the white stripes on the towers, are peculiar enough to attribute specifically to either the Abbasids, Hamdanids, Byzantines, or Franks. I have no example where the Byzantines use any form of a type VI in Cilicia or its environs. However, this should not rule out the possibility that the Byzantines are responsible for the first major period of construction.[29] A case can be made to associate the irregular type VI on the exterior of complex A and the rest of the fort with the Franks. At the nearby Crusader fortresses of Amuda and Kum, type VI is used extensively.[30] On the other hand, the single white stripe that occurs in the midriff of the north tower of A may be an important link to the Abbasids. It is very common in eighth-century Arab architecture to use one, two, or three courses of stone to form a single stripe that always differs in color from the rest of the exterior facing on the tower.[31] A single white stripe also occurs on tower I at Toprak, a tower that is faced with the irregular type VI masonry (pl. 243a). It is significant that an identical stripe articulates the round bastion at the only known Arab fort in Cilicia, Haruniye Kalesi.[32] One commentator on Toprak Kalesi has misinterpreted this white stripe on A, believing it to be a "Gesimsband" like those on the Crusader fortress of Crac des Chevaliers.[33] However, the Crusaders do not always articulate their towers, and when it is done it consists of a horizontal molding, step, or chase. These decorations do *not* vary in color from the face of the surrounding wall.[34] The Armenians in Cilicia never decorate their towers in the Arab manner.[35] The possibility that the first period of construction belongs to the Arabs may be strengthened by the fact that the late Omayyad forts were rectangular or square in plan and consistently used rounded towers at regular intervals.[36] Only excavations can shed certain light on the occupants responsible for the first major period of construction.

The interior of gatehouse A is a vaulted hall articulated by one rib. The entire apex of the slightly pointed vault has collapsed. The interior masonry consists of a superbly dressed type IX. The door between A and the double groined vaulted chamber B has collapsed. On the exterior of the south wall the type VI facing of A is not carried onto B, where a rough type V facing appears. On the interior of B large sections of the two groined vaults and the embrasured casemates at the north have collapsed (pl. 237a). The two south doors of B have also collapsed. The interior masonry of B is quite crude, except for the short piers of the groined vaults. At the intersections of the vault the stones are laid in a herringbone pattern, which is quite reminiscent of the Crusader technique.[37] Undercroft C, which is not connected to B by a door, appears to be from the same building period. The south wall of C incorporates a curious square chamber whose south wall has collapsed. This square chamber is opened by a door at the east and shows no signs of ever having been vaulted or having functioned as a cistern.

Much of tower D at the north appears to belong to the first major period of construction. On the exterior most of the facing stones consist of an irregular type VI, but part of the upper level is repaired with a rough type V. This upper level, which is entered via a staircase in the northwest corner of B, is a hexagonal room pierced by three very broad windows. The exterior frames of the openings are shattered, but on the interior side enough survives to show that they were not typical of the embrasured windows in the fort (pl. 237b).[38] It is possible that the windows were exceptionally wide to accommodate *ballistae*. The solid lower level of the tower would provide a strong platform to absorb the recoil of machines. The presence of curving spandrels between the windows indicates that some sort of cupola or faceted vault covered the room. It appears that an open terrace of battlements surmounted this room.

The intricate frames around the upper-level windows of D offer a valuable clue to the origins of the crude type V masonry. The frames, as with the rest of the room interior, are constructed with type IX masonry. The top of each window consists of a depressed arch whose apex is articulated by a deep semicircular niche. The niche is constructed with precisely cut voussoirs that taper to form a flat inner face. The resulting trefoil-crown of the window bears an amazing similarity to the niche decoration in the bathhouse at Saone.[39] This bathhouse was probably constructed during the occupation by the Sultan Baybars. The peculiar architectural features that appear in the new constructions on Crusader fortresses during the Mamluk period are associated at Toprak with only the crude type V facing stones. Thus it seems likely that the widespread use of this type V in the west, south, and east walls of the upper bailey (which constitute the second major period of construction) is from the Mamluk period.

The largest bastion in the fortress is tower E in the northwest corner. As with all the towers in the upper bailey, E is a rounded salient, but it is unusual in that it protrudes for almost three-quarters of its circumference.[40] The subterranean level of this tower (not shown on the plan) has stuccoed walls and probably functioned as a cistern. The ground level and upper level of E have almost completely collapsed. The interior of both levels appears to be polygonal. At ground level an elaborate door connects this tower to the triangular room between undercrofts C and F (pl. 238a). On the interior side this door is framed with neatly cut voussoirs of type IX, while the vault over this room consists of a combination of crude stones in a thick matrix of mortar. Since the vault is adapted somewhat clumsily to the wall surrounding this door, we can assume it is a later addition. Evidence that the upper level of E was reconstructed is visible on the exterior of a surviving fragment of the tower. Most of the exterior facing on tower E is an irregular, somewhat large type VI (pl. 238b), resembling the type VI in the north wall of the fort. However, the upper level of E was clearly rebuilt with a crude type V facing, and one of the embrasured casemates from that period still survives. On the exterior this loophole has a stirrup base and is framed by rusticated blocks with drafted margins.

South of tower E three semicircular bastions flank the west wall along with two long undercrofts on the interior side. This west flank seems to have undergone a number of periods of construction. The only evidence of building in advance of the first major period of construction is the original tower G, which was later covered by a rounded bastion of irregular type VI masonry. The original tower G was probably square on the exterior, although there is no evidence today to prove this. The interior of the tower at ground level is perfectly square and is opened by a single embrasured loophole in the north, west, and south walls (pl. 239b). This present interior room was divided into two floors; joist holes about 40 cm below the sills of the embrasures mark the height of the first-level ceiling. The second level of this room is covered with a groined vault of very crude construction. The roughly coursed ashlar blocks of the vault are not like those in A and B or in Armenian groined vaults. The embrasures themselves are constructed in a very peculiar way. Their rounded heads are made with brick tiles laid in a neat radial pattern. Brick appears nowhere else in the fort, and its use in the original tower G is as a recycled material.[41] The brick survives in a limited quantity and is not used consistently. In the west wall, brick is employed to construct all of the rounded head of the embrasure, while in the north wall the springing level for the rounded top is merely an extension of the fine ashlar blocks of the frame below. The masonry of the walls surrounding the embrasures is difficult to evaluate because the exterior interstices of many blocks have been restuccoed, giving some of the courses the appearance of type VI. It is quite probable that its groined vault is from a later period of construction. Both Müller-Wiener and Hellenkemper believe that the first tower G is a Byzantine construction. The latter asserts that it was built in the twelfth century during the period of John Comnenus (ca. 1137).[42] To accept this conclusion would mean that the addition to G as well as the rest of the castle dates from a later period. However, in the opinion of this writer, the first tower G can either be a pre-seventh-century Byzantine construction or an eighth-century Arab tower. We have numerous examples of small square Byzantine salients that have a single embrasured loophole opening each of the three exposed walls.[43] In the late Omayyad and early Abbasid forts, square towers are extremely rare, but brick is occasionally used to construct embrasures.[44]

The first major period of construction saw the erection of all of the present west towers. If we accept that the exterior masonry of A, D, and the lower three-quarters of tower E is from the same period, then the second tower G and salients H and I are also from the period of the first major construction since they have the identical irregular type VI facing (pl. 243a).[45] In a few places the foundations of the connecting walls on the west flank have an identical masonry.

However, I believe most of the intervening circuit wall between the west towers, as well as the flanking undercroft F and all the upper-level embrasured casemates, to be the result of the second major period of construction. The multiple periods of construction are plainly evident on the exterior of the west circuit between towers G and H. The lowest courses consist of crude ashlar blocks. These are surmounted by six courses of a type IV with the upper third of the wall consisting of a crude type V (pl. 242b). On the interior side of the west circuit the facing stones of undercrofts F and J are quite different (pl. 241a). The facing of F consists of a crude type V that has no mortar extruding from the margins. This undercroft is opened by five roundheaded doors at the east. The interior walls of F (pl. 239a), which are covered by a slightly pointed hatchless vault of type II masonry, are constructed out of a masonry that almost resembles type VII, but is slightly more crude (only the north wall of F appears to be a repair). A continuous horizontal chase at the top of both the east and west walls of F and J probably anchored a wooden ceiling (pl. 240a).[46] The space between the ceiling and vault was used for storage. A similar chase in conjunction with joist holes appears on the exterior of F and J (pl. 241a), which indicates that wooden buildings were attached. The point of division between F and J appears quite sharp because the exterior face of J is constructed with very regular courses of a small rough type IV. The four doors opening the east wall of J are squareheaded. The interior facing of J consists of a small type V (except for the lower three courses of larger blocks; pl 240a); a slightly pointed hatchless vault is constructed out of a similar masonry. At the far south end of J there is a staircase from an earlier period of construction which once led to the upper level of tower I (pl. 240b).

While the masonry surrounding the eight embrasured casemates atop F is comparable to the masonry in the undercroft below (pl. 241a), the five embrasured casemates and the adjoining wall atop J do not have the same kind of masonry as the undercroft below. Although the two sets of embrasures differ in design, it is significant that both types have identical Mamluk-period counterparts in the south of the Levant. The casemates above F are shallow in comparison to the thickness of their adjoining roundheaded embrasured loopholes (pl. 241b). The voussoirs in the conical tops of the embrasures are arranged and beveled in the same manner as those in the embrasures constructed after 1271 at Crac des Chevaliers;[47] the voussoirs taper until they terminate on the inner side of the exterior facing stone that marks the flat top of the loophole. On the exterior all eight of these loopholes are surmounted by smooth ashlar blocks.[48] Except for the lowest courses, all of the exterior facing of the circuit between towers E and G is a uniform but crude type V, which indicates that undercroft F and the casemates above are from the second major period of construction. The five identical embrasured casemates above undercroft J extend far into the thickness of the wall (pl. 242a). Their shallow embrasures are rounded, except for a square notch in the apex. Again, these are identical to constructions at Crac after 1271.[49] On the exterior most of the ashlar blocks that frame the loopholes above J have a rusticated center with drafted margins (pl. 242b).[50] The masonry of the wall surrounding these casemates is that same type V that appears between towers E and G, but only occupies the upper third of the wall between towers G and H. Below the type V in the latter is a type IV that is identical to the exterior facing of undercroft J. The type IV also corresponds in height to the position of undercroft J. Since the masonry and architectural features (for example, squareheaded doors) of J differ so much from the rest of the west circuit, it *must* predate the Mamluk-period construction and possibly represent a Crusader addition to the first major period of construction. The Mamluks probably constructed the embrasured casemates over J immediately before or after undercroft F. This would explain the presence of two styles of Mamluk embrasured casemates and yet the general similarity in the masonry of the upper level of J and undercroft F and its upper level.

The circuit wall between H and I has essentially two periods of construction (pl. 243a). The lower half from the first major period has a masonry identical to tower I (the irregular type VI). The upper half of the wall is that rough equivalent of type V that is associated with the second major period of construction. Near the top of the wall there are six corbels that once held an overhanging brattice. Today almost the entire upper level of the adjoining tower I has collapsed. Complex K attached to the east flank of tower I has at least three periods of construction. All three periods are evident at the extreme south end of K where it joins the south wall of the lower bailey (pl. 243b). Here the slots of a toilet-machicolation are flanked by two quite different types of masonry.

The north half of complex K consists of a two-story hall. The lower- (ground-) level room is a cistern(?) with a partially collapsed vault. The upper level is covered with a slightly pointed vault (pl. 244a). On the interior, the north, south, and east walls as well as the vault are constructed with type IX masonry. The south wall of the room is opened by three embrasured loopholes; the westernmost of the three loopholes is broken today. The east wall is opened by a now collapsed door (pl. 245b) that seems to be fitted with a machicolation. Of the three openings in the north wall, only the one at the far east is preserved today (pl. 245a). The available evidence seems to indicate that this vaulted hall is from a later period of construction, probably Mamluk. The preserved roundheaded window at the east end has those neatly cut voussoirs that terminate on the inner side of a horizontal block, exactly like the embrasured loopholes in the west circuit (pl. 241b) and south wall of K. Also, the vault over the upper level of K is not constructed like the vault of A (which I associated with the first major period of construction) in that the latter is constructed with a rib, while the vault over K has a series of large joist holes that once held wooden crossbeams. The crossbeams were probably part of the centering (or falsework) and were left in situ to add support to the vault.

Perhaps the most conclusive piece of evidence to show that K is a later construction is its interior west wall (pl. 244a), which does *not* have a smooth ashlar masonry but consists of large crude blocks with wide interstices and much mortar. At the south end of the west wall there is a massive door that leads to the upper level of tower I. The top of this door is a depressed arch that bears a remarkable resemblance to the openings on the interior of tower D (pl. 237b; except that the arch over the west door of K does not have a shallow niche in the apex). When the vaulted upper level of K was built, a door was added (or an existing door widened) in the original east wall of tower I. Thus this older wall of crude masonry became the west wall of the vaulted upper level of K. The exterior facing stones of K are a very inconsistent type V; some blocks show almost no mortar in their interstices, while others have broad beds of mortar in the margins.[51] The masonry at Toprak is difficult to evaluate because it is not used with the kind of consistency we see in Armenian forts. Hellenkemper is probably right in assuming that the machicolation above the east door of K is an inspiration from Armenian design (pl. 245b).[52]

The south and east flanks of the upper beiley have a double trace. The inner trace with its rounded towers seems to be built with the same irregular type VI that appears on the north and west walls. Only the lower half of the outer trace has the same type VI. This means that the plan of the double trace dates to the first major period of construction. The upper half of the outer trace (and thus the embrasures) have the crude type V for an exterior facing stone. On the exterior the loopholes are framed with ashlar blocks that have rusticated faces and neatly incised margins. This type of masonry and construction is associated with the major rebuilding of the west front and is probably contemporary with F. This conclusion has further support in that many of the embrasured casemates between M and P fit the Mamluk-period paradigms for the west wall. However, the roundheaded embrasures inside tower N (pl. 244b) do not have the distinctive voussoirs that we see in the embrasures above undercroft F. In N the rounded tops of the embrasures, which consist of a rough conglomerate of fieldstones and mortar, were formed when the vaulted ceiling was poured on the centering. This type of embrasure probably represents a third paradigm from the Mamluk period.

[1]I visited this site in spring 1973 and in August 1979 and 1981. The contour lines on the plan are approximated at intervals of 50 cm.

[2]Because this site is readily visible from most areas of Cilicia

Pedias, it appears on many of the published maps of Cilicia, including *Adana (2), Cilicie, Kozan, Marash*. See also: Heberdey and Wilhelm, 23 f; Tomaschek, 68 f; Schaffer, 94; Janke, 36–43, figs. 3 f, map 1c.

[3] Seton-Williams, 128 ff.

[4] Above Part I.7, note 19; Boase, 184; Favre and Mandrot, 148–50; Chauvet and Isambert, 762 f; Sümer, 13, 25 f; Frech, 577 f.

[5] In Armenian colophons and chronicles this site is commonly referred to as T῾il (Hamtun); the Franks employ Thi, Thila, Tili, or Thil Hamd(o)un. Tell-Champson may be an alternate Frankish spelling. In Arabic the preferred use is Tall Ḥamdūn. Bar Hebraeus calls it Tall Ḥamdōn. Tili is found in Cinnamus. Toprak is the modern Turkish designation.

This site is mentioned in the 11th century as the seat of a Jacobite bishop; see E. Honigmann, *Le couvent de Barṣauma et le patriarcat jacobite d'Antioche et de Syrie*, CSCO 146, Subsidia 7 (Louvain, 1954), 152; Michael the Syrian, J. B. Chabot, III, 187, 503.

[6] Hellenkemper, 140 note 2.

[7] Le Strange, *Caliphate*, 129 f; idem, *Palestine*, 449 f, 477.

[8] If Abū᾽l-Fidā᾽ is right, then Karafrenk is a more likely candidate to be al-Kanīsah (refer to Karafrenk (note 3) in the Catalogue). Unfortunately Tall Ḥamdūn is seldom discussed by Arab geographers. In the first quarter of the 14th century Abū᾽l-Fidā᾽ (29; cf. Le Strange, *Palestine*, 543) falsely claimed that the fortress had been dismantled and was in ruins.

[9] Ibn al-Aṯīr, *RHC, HistOrien*, I, 424. Because the history of Toprak has been recounted in a number of modern publications, I will summarize only the major events. See: Alishan, *Sissouan*, 233–36; Cahen, 147 ff, 325 ff; Fedden and Thomson, 36, 44; Gottwald, "Til," 89–104; Boase, 183 f; Müller-Wiener, 75–77; Hellenkemper, 140–53; Honigmann, 130; Lilie, 70, 109 f, 170, 173, 434 f note 153.

[10] Smbat, 619; Cahen, 389 f.

[11] Gregory, 171–75; Cinnamus, 180; Michael the Syrian, 347; F. Chalandon, *Jean II Comnène (1118–1143) et Manuel I Comnène (1143–1180)* (Paris, 1912), 420, 431 f notes 1–5, 442.

[12] Michael the Syrian, 393 f; Smbat, G. Dédéyan, 58; Cahen, 424.

[13] Christomanos, 151. Cf. Alishan, *Léon*, 242; Cahen, 152, 525, 539; Smbat, G. Dédéyan, 92. King Levon I may have appointed another Frank, Joscelin, to succeed Robert. The Franks probably held the fortress during most of the first half of the 12th century.

[14] Wilbrand von Oldenburg, 16 f, 20 f. I assume that these two sites are identical because Toprak is the only fortress made out of black basalt on Wilbrand's route.

[15] Bohemond IV passed this fortress in search of his son Philip, the husband of Levon's daughter Zapēl. Philip was arrested at Toprak in 1225 and taken to Sis; see Bar Hebraeus, 381; Smbat, G. Dédéyan, 97 note 27; Cahen, 633. In 1265 King Het῾um is known to have celebrated the feast of the Epiphany at Toprak; cf. Canard, "Le royaume," 225 note 39.

[16] Al-Maqrīzī, II.2, 63, 228; Gaudefroy-Demombynes, 99, 101, 218; al-Jazarī, *La chronique de Damas*, trans. of extracts J. Sauvaget (Paris, 1949), 26, 69; Smbat, G. Dédéyan, 114; Alishan, *Sissouan*, 235.

[17] Gottwald, "Til," 89–104.

[18] Müller-Wiener, 75–77.

[19] Hellenkemper, 140–53.

[20] The Egyptian Mamluks probably commissioned Arab (and perhaps Armenian) laborers to reconstruct the fortress.

[21] Because the two southernmost towers ascend the outcrop, both are accommodated with staircases.

[22] Tower X has collapsed, and its foundation is now buried in debris. Tower W has also collapsed, but in 1973 it was visible in outline.

[23] With only a few exceptions, the masonry of the lower bailey was cut from black basalt.

[24] This is not like the small irregular type VI that appears in many of the towers of the upper bailey. To evaluate properly the masonry of this fortress (something that is not attempted here) a separate table of paradigms, quite distinct from the one I offer above, Part I.3, would have to be established.

[25] Cf. Hellenkemper, 151–53.

[26] Ibid., 144, 151–53, pl. 34a.

[27] Ibid., 151; Müller-Wiener, 76.

[28] Hellenkemper, pl. 34a.

[29] Two commentators would credit all or much of the first period of construction to the Byzantines. See Müller-Wiener, 76; Boase, 184.

[30] At both Amuda and Kum it may have been a German order of knights and not specifically "Franks" who employed the type VI.

John Thomson and later Hellenkemper see a strong Crusader influence in the design of Toprak. Fedden and Thomson (36, 49) assume that the Hospitalers completely rebuilt Toprak in the 13th century with Armenian masons. But in my survey I found no evidence to substantiate this claim. The Armenian horseshoe-shaped towers are not evident here. The architect who designed Toprak did not have Crac des Chevaliers in mind. Crac consists of a massive central stronghold that is completely surrounded at a lower level by a circuit with galleries and embrasured loopholes.

[31] The Arabs also use horizontal moldings to decorate towers, but a stripe with a different type of masonry is most common; see Creswell, *Early Muslim Architecture*, 2nd ed, (Oxford, 1969), pls. 92b and c, 94c, 98c. This Arab penchant is displayed dramatically at Bosra (Müller-Wiener, pl. 92) and at Marqab (Margat). In the latter only the Arab tower added to the Crusader fort bears a white stripe; see Müller-Wiener, pl. 52; Deschamps, III, pl. 46a; and T. Boase, *Castles and Churches of the Crusading Kingdom* (London, 1967), 60.

[32] It should be stated that the masonry surrounding the stripe at Haruniye is a relatively smooth ashlar.

[33] Hellenkemper, 152.

[34] Cf. Safita, Crac des Chavaliers, Kolassi, and Rhodes; see Müller-Wiener, pls. 36, 66–71, 143, 158–59.

[35] The only stripes on Armenian towers occur at Fındıklı, Çem, and Savranda. In these cases they consist of smooth courses of ashlar across the rusticated facades.

[36] Creswell, *Early Muslim Architecture*, 522 ff, 578 ff. As with Haruniye, most of the exterior facing of the Omayyad forts is smooth ashlar. Most Omayyad towers are small semicircular bastions. Only one site, Qaṣr Bāyir, has the three-quarter round bulging towers. Gottwald ("Til," 96) believes that the plan and part of the foundation of Toprak date to an early Arab period.

[37] Edwards, "Bağras," 424.

[38] Hellenkemper, 144f.

[39] Deschamps, III, pl. 25b; also cf. Creswell, *The Muslim Architecture of Egypt*, I (Oxford, 1952), 200 f.

[40] Hellenkemper (152) believes, as I do, that the rounded bastion E is of Islamic inspiration. However, Hellenkemper compares it to Kızıl Kule in Alanya, which is not a circuit tower but a freestanding independent keep with five levels; see S. Lloyd and D. Rice, *Alanya (῾Alā᾽iyya)* (London, 1958), 11–16.

[41] The recycled bricks that appear in a few of the embrasures at Tumlu Kalesi have a very different appearance.

[42] Hellenkemper, 151. Cf. C. Foss, "The Defense of Asia Minor against the Turks," *The Greek Orthodox Theological Review* 27/2–3 (1982), 145 ff.

[43] W. Karnapp, "Die Stadtmauer von Resafa in Syrien," *Archäologischer Anzeiger* (1968), 324.

[44] See Creswell, *Early Muslim Architecture*, 529.

[45] As I mentioned earlier, those horizontal stripes around towers A and I are a close link with Arab masonry traditions (see Hellenkemper, pls. 32b, 33a and b).

[46] The chase was formed at the same time as the vault was

constructed by simply placing horizontal beams in the thickness the wall.

[47]Deschamps, I, 163, fig. 24; 260, fig. 53f. This peculiar design has predecessors in the Omayyad period; see Creswell, *Early Muslim Architecture*, 565, pl. 83e.

[48]Hellenkemper, pl. 32b.

[49]Deschamps, I, 162, fig. 22; 259, fig. 53c. The top of the loophole at Toprak is rounded, not flat as at Crac.

[50]Cf. the reconstructed loophole atop tower E.

[51]Reconstruction at the east end of tower I is also evident on the exterior.

[52]Hellenkemper, 152. It should be noted that the Armenians employ slot machicolations only in their gatehouses, rarely on the interior of a fort. Frankly, I do not know what real value a slot machicolation would have at the east end of K.

It is not certain if there was a machicolation over the east door of A.

Trapesak

Trapesak Kalesi[1] surrounds a small oblong outcrop of limestone about 4 km north of Kırıkhan (pl. 247b).[2] This fort guards the strategic road to Antioch and the north approach to the Belen pass. It also secures the secondary east-west route via the Çalan pass.

At best the history of this site is fragmentary.[3] From its inception it appears to have been a Templar fort. It gained its first notoriety around 1171/72 when the renegade Armenian Baron Mleh took possession of the site along with Bağras Kalesi. However, the Templars regained Trapesak in 1175 at the death of Mleh. On 16 September 1188 Saladin captured this fortress after a fourteen-day siege;[4] it was given to ʿAlam al-Dīn Sanjar b. Ḥaidar and briefly fell to the Templars. In 1205 King Levon I failed to take the fort in a surprise attack. In 1237 an attempt by the Crusaders to oust the Saracen (that is, Aleppine) garrison from Trapesak failed. From 1261 to 1267/68 the Armenians held this site (as a gift from the Mongol Khan Hulagu) until the Mamluk occupation. In 1280 the significant event at Trapesak is a very destructive assault by the Mongols. The site was an important staging area for invasions into Cilicia.

Today Trapesak Kalesi is in a deplorable condition. The village of Ala Beyli (also called Beyazil Bostan) has covered much of the original fort (pl. 246a). Many of the medieval walls have been removed and their masonry recycled into homes. In recent years this site has become increasingly popular because a Moslem saint is buried on the summit. A new road to accommodate motor vehicles was terraced into the east face of the outcrop, destroying almost all evidence of circuit walls and buildings. The village is also expanding to the west at the base of the outcrop, where the Crusader aqueduct (quite similar to the one at Bağras) once brought water to the fort.[5] Only at the south end of the site have large traces of the medieval fort survived (pl. 247a). Because of the paucity of the remains and the proximity of the village, a detailed survey was impossible.

The masonry and techniques of construction at Trapesak are almost identical to those at the Templar fort of Bağras.[6] The vast majority of the masonry consists of crudely cut stones whose interstices are filled with rock chips and mortar (pl. 248a). Only in a few places at the base of the outcrop is type VII present (pl. 246b); the same masonry is present only in the "bossed tower" at Bağras. In constructing vaults for the doors and windows the centering was shaped into sharply pointed vaults and the stones were laid in an overlapping herringbone pattern (pl. 248b). These are Templar motifs that are common at Bağras. The type VII probably represents Armenian repairs carried out during the short period of their occupation.

Little can be said about the extant remains at the east end. At the lowest point is an undercroft and adjoining rooms (pl. 247b). From this area a path ascends the gently sloping summit to the northwest. About 15 m north of the undercroft the path crosses the remains of a collapsed wall and tower, both of which probably framed the middle(?) bailey. At a higher level is another wall and the fragments of shattered buildings.

[1]During my visit to this site in July 1979 I was unable to execute a plan. For most of its history Trapesak (as with Bağras, its neighbor to the south) was outside the boundaries of the Armenian kingdom. It is included in this study because some Armenian construction appears to be present here.

[2]The name for this site is also spelled Darbsak, Terbezek, and Trapesac.

[3]Deschamps, III, 361; Cahen, 141, 512, 610, 705, 650 f; *Blue Guide* (Hachette), *Turkey* (Paris, 1970), 673; Sümer, 62, 64; Le Strange, *Palestine*, 436 f; Gaudefroy-Demombynes, 88, 215; R. Hartmann, "Politische geographie des Mamlūkenreichs," *Zeitschrift der Deutschen Morgenländischen Gesellschaft* 70 (1916), 33 note 7; Lüders, 95 note 1; Ibn al-Furāt, 121, 127, 166, 161, 230; Canard, "Le royaume," 229 f, 236; al-Maqrīzī, I.2, 33, 55; Dardel, 14 f note 2; *RHC*, *DocArm*, II, 177 f, 307 f; Dussaud, 162, 183, 435 ff, 436 note 1, 445; Popper, 17.

[4]According to Ibn al-Aṯīr (*RHC*, *HistOrien*, I, 730 f), the active siege lasted from 2–13 September when the garrison finally surrendered. The fort was occupied on the 16th (ibid., 730 f; Abu-Shama, *RHC*, *HistOrien*, IV, 375–77); the Templars were allowed to withdraw unharmed, leaving their property behind and paying Saladin 5,000 dinars. Cf. Imād ad-Dīn al-Iṣfahānī, *Conquête de la Syrie et de la Palestine par Saladin*, trans. H. Massé (Paris, 1972), 141 f.

[5]Deschamps, III, pl. 85b.

[6]Edwards, "Bağras," 419 ff.

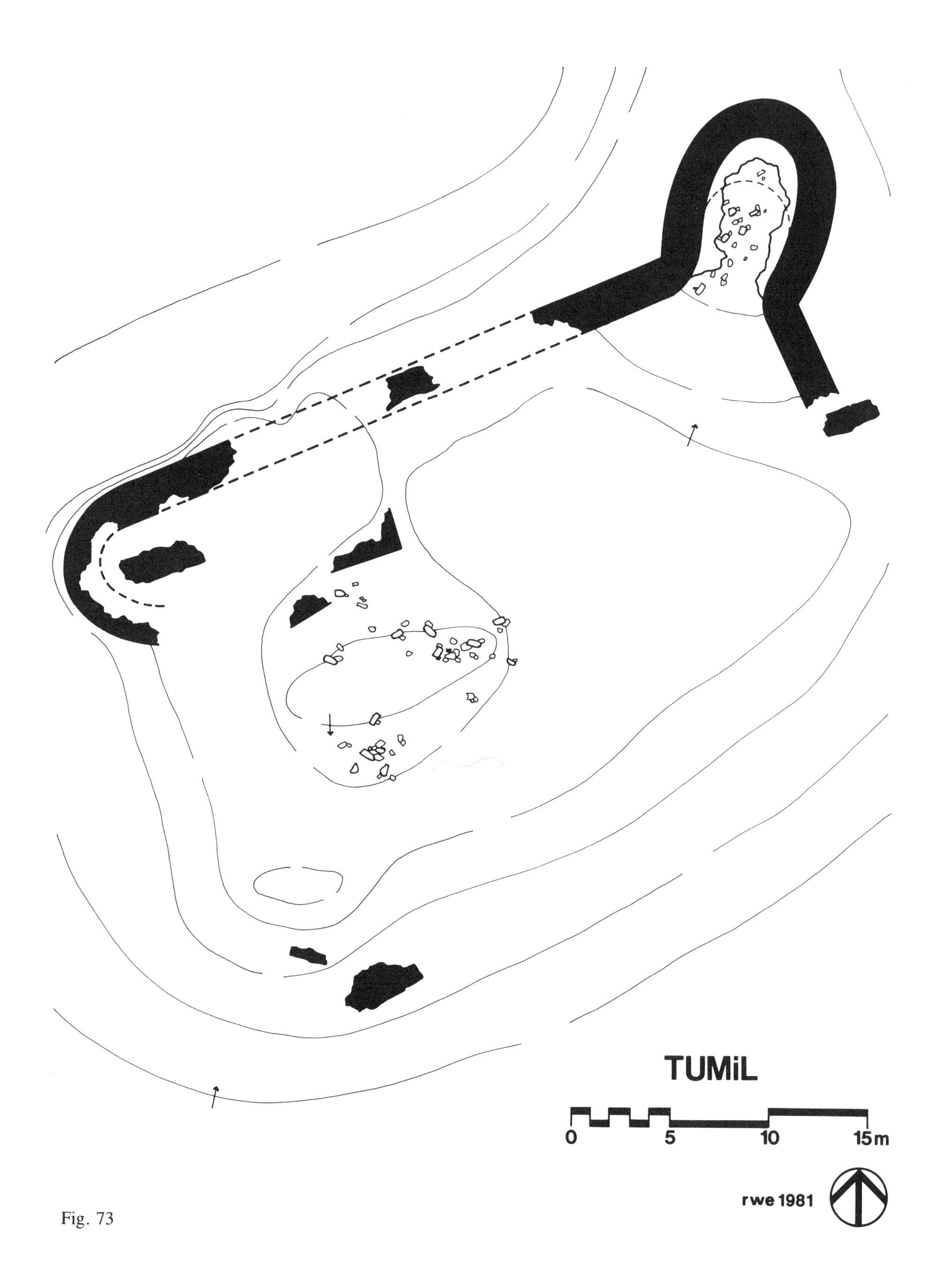

Fig. 73

Tumil

The small garrison fort of Tumil[1] is about 3 km northeast of Mersin and is easily visible from the highway that connects Mersin and Tarsus.[2] Tumil Kalesi surrounds the gently sloping sides of a hill in a suburb of Mersin called Çilek Mahlesi. The fort is just northwest of the suburb and its industrial complex. To the north of the hill, the area is farmed for melons and is heavily irrigated. It appears quite likely that Tumil had intervisibility with Yaka Kalesi to the northeast. Unfortunately air pollution severely limits visibility today. Only a few fragments of the fort now stand in situ (fig. 73). The round tower in the northeast corner rises to the level of the battlements (pl. 250a). The tower at the northwest is preserved to 2 m in height. Its rounded front does not protrude dramatically to the north like the other tower, but is oriented to the west. Most of the circuit wall that connected these two towers has collapsed. In the center of the fort is a shallow pit that is sparsely covered with the debris of coursed masonry. In the north side of the pit there is a large section of a masoned wall that *may* be standing in situ. From the remains it appears that Tumil Kalesi was a simple castrum with irregular corner towers. There are no inscriptions or references in the medieval texts to identify the site.

The northeast tower is in a remarkably good state of preservation. Most of the exterior facing stones (pl. 249b), which consist of an elongated type IX with a rusticated center and broad drafted margins, are preserved. In the sections that have fallen away on the east and west sides of the tower (pl. 250b), it is evident that the core is constructed in neat courses with poorly coursed stones and mortar. The exterior facing is *identical* to the masonry at Yaka, Kütüklu, and much of Tece. What is interesting about this tower is that its entire interior space is defined with a fine facing stone. Unfortunately most of its depressed barrel vault has collapsed. The interior facing stones consist of a uniformly smooth type IX (pl. 249a), and there is no transition or change in masonry style at the springing level. The rather flattened vault and its masonry are reminiscent of the construction at Silifke Kalesi.

The northwest tower has lost its facing stones, but its core is quite similar to that in the northeast tower. Today the apsidal room in the tower is filled with the large fragment of a collapsed wall.

For the reasons that I have outlined above (Part I.5) and in my discussion of Yaka Kalesi in the Catalogue, I believe that the complex at Tumil may be a Crusader construction.

[1] The survey of this hitherto unpublished site was undertaken in July 1981. The contour lines on the plan are placed at intervals of 50 cm.

[2] The village of Tumil appears on the map of *Mersin (2)*. Seton-Williams (171) identifies this site and gives the alternate spellings of Tirmil, Turmil, and Termil. See also *Handbook,* 204.

Tumlu

Tumlu Kalesi[1] is an impressive garrison fort in the heart of Cilicia Pedias.[2] The small village of Tumlu Köy is only 500 m east of the outcrop. At an altitude of 150 m, this fort covers the top of a limestone projection in a relatively smooth region of the plain (pl. 301c). From the heights of this fortress there is clear intervisibility with Yılan, Toprak, Amuda, Anavarza, and Sis. The fortified outcrop is not astride one of the strategic roads or rivers in the plain, but it does flank an auxiliary trail that connects Imamoğlu to the Ceyhan-Kozan highway. The absence of rivers in the immediate vicinity of the outcrop leaves only the three cisterns in the fortress to supply the needs of the garrison. Tumlu Kalesi has a single kidney-shaped bailey that slopes downward to the northwest (fig. 74). It is surprising that this fort is so heavily fortified with massive walls and towers, considering that all sides, except the east flank, are protected by near vertical cliffs. These cliffs vary from 20 to 40 m in height. Only at the north end do the walls become substantially thinner.

Because access into Tumlu Kalesi is difficult, it has only been "sighted" in most nineteenth- and twentieth-century accounts of the Adana region.[3] V. Langlois published a sketch of the fort but incorrectly identified it as Adamodana, a designation more suitable for a site farther east.[4] The first proper survey of Tumlu was conducted in 1962 by G. R. Youngs.[5] This study resulted in a plan and a fine architectural description of the fort. In 1976 H. Hellenkemper published a summary description that was essentially confined to the exterior of the fort.[6] The following account offers a general description of the site and new observations.

There are no specific references in the medieval chronicles or inscriptions that give us the name and date of construction of this site. Hellenkemper attempts to identify Tumlu with the medieval site of Tʿlpałt/Tʿlsap.[7] This association is purely hypothetical and relies on the unwarranted assumption that Smbat, the compiler of the Coronation List, listed *all* the forts

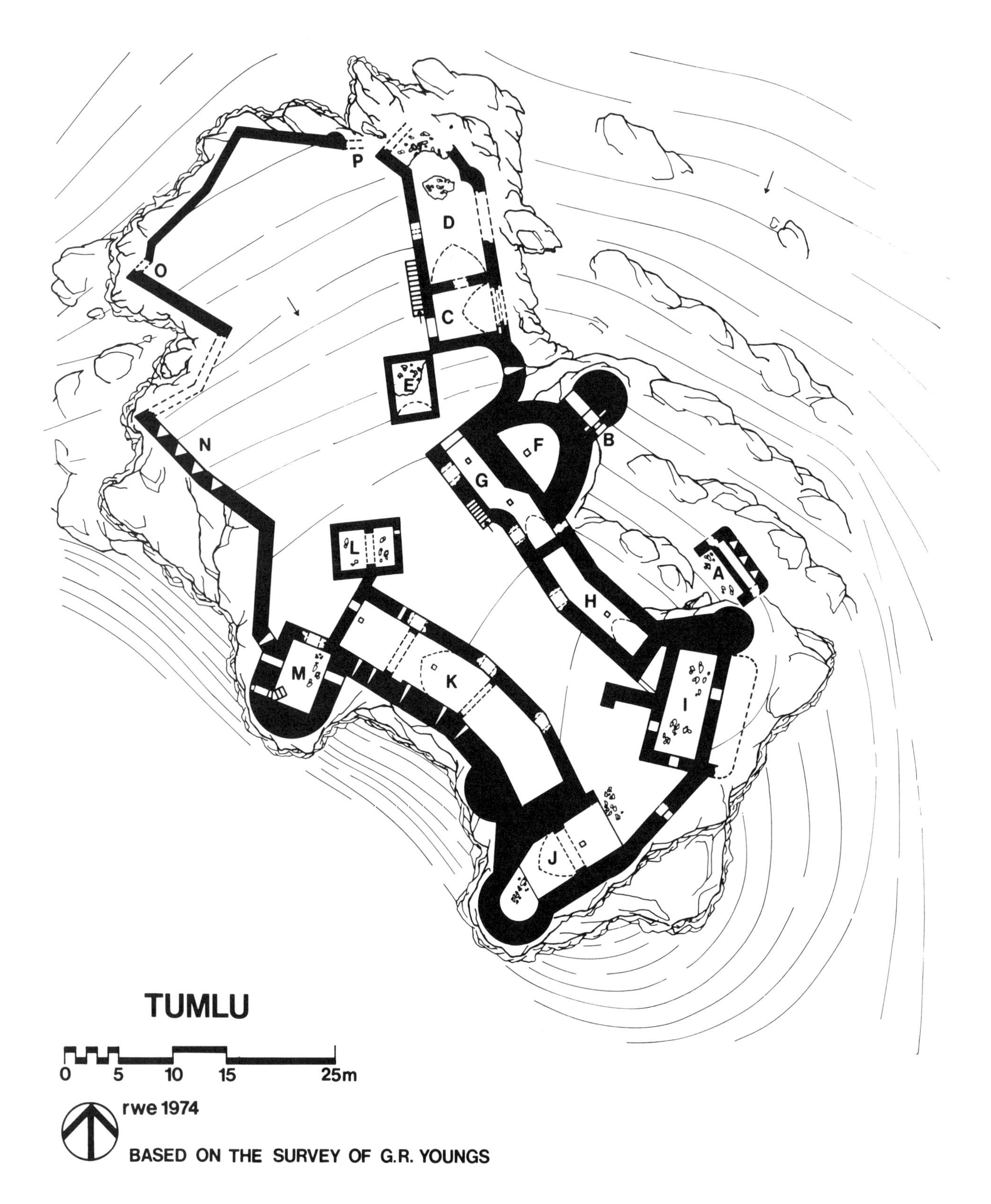

P
D
O
C
E
N
F
B
G
L
A
H
M
K
I
J
TUMLU
0
5
10
15
25m
rwe 1974
BASED ON THE SURVEY OF G.R. YOUNGS

Fig. 74

in the region of Cilicia in a logical geographical order (see below, Appendix 3). From the architectural elements and masonry of the site, it is clear that Tumlu Kalesi is an Armenian construction.

The fort appears to be the result of one major building period, with substantial repairs during the Mamluk occupation. The principal type of exterior facing is a combination of types V and VI. Occasionally, the type V is used alone, as is seen in the lower half of the northeast wall of H (pl. 251b), but normally it is mixed with type VI (pl. 252a). Repairs have been carried out with types III and IV masonries. In a few cases this crude repair has been covered with a thick layer of mortar. Types III and IV masonries (either mixed or used separately) are employed for the interior facing and interior buildings (pl. 254a). Normally, both types have thick margins of mortar when they are used as the interior facing of the circuit wall. Except for the vaults, the masonry of the interior chambers has less mortar in the interstices (pl. 254b). The only brick in this complex appears in cistern E and in five embrasures above undercroft H (the latter are not shown on the plan).

The easiest line of access into the fort is from the northeast (pl. 251a) or the southeast. Natural tiers in the outcrop at the northeast lead to scarped steps below and east of C. These steps do not lead into C but to a path that continues south and eventually joins other approaches from the southeast at gate B. The path from gate B to gatehouse C must have involved the use of some sort of wooden drawbridge (pl. 253a), for today it is almost impossible to climb along the cliff into C. Presently, it is far easier to walk below undercroft H and around the north end of I where the edge of the cliff gives ready access through the collapsed east wall of I.

Outwork A can efficiently control a party approaching from the east or southeast (pl. 251a). There are three embrasures in the east wall of A and traces of one in the north and south walls. A has an open west flank, and there is no evidence of a stone vault. If retreat was necessary during attack, the archers who manned A would be exposed to severe fire when running to B and C. (Like its counterpart in the upper bailey of Yılan Kalesi, A is positioned to guard the most direct line of approach to the entrance. Outworks are not common in Armenian fortifications.

The tower-like barbican at point B consists of a tripartite door with a slot machicolation at the southeast and a short vaulted corridor on the inner side (pls. 252a, 252b). Unfortunately the apex of a slot machicolation has collapsed today, as has most of the pointed vault on the interior side. Behind the jambs of the door are square sockets to accommodate a crossbar bolt. This barbican is actually a lower-level extension of tower F and directs traffic into the main entrance at C (pl. 253a). The salient positioned between C and tower F has a single embrasured window to observe traffic. This window has a pointed top on the interior and is roundheaded on the exterior.

Gatehouse C is a typical Armenian gate-corridor (cf. Meydan Kalesi). The east opening of C has a now collapsed slot machicolation. When V. Langlois visited this site in the mid-nineteenth century, the outer door of C had not collapsed.[8] Today the entire north half of the door is missing, but the south half of the system of arches is clearly visible (pl. 253b). The sill of the door, as well as the lower half of the south jamb, is scarped from the natural rock. A square socket for the crossbar bolt is conveniently cut in the rock. The carefully shaped and rounded protrusion on the *inner* side of the jamb could not function like the jamb-guard in many Armenian doors. The function of this corbeled semicircular block remains a mystery.[9] It appears that most of this outer opening has only one period of construction, and with respect to its masonry it is contemporary with barbican B. As with B, the outer arches of C were designed to accommodate a wooden door. The only repair in the vicinity of the outer door of C is a small section of wall *below* what was once the sill of the door at the north (pl. 253a). It seems that the rock shelf that constitutes the south half of the sill did not extend far enough to the north. So a wall was constructed to fill the gap and provide a foundation for the north jamb. At some point after the original period of construction, the wall below the north jamb collapsed (either due to siege or earthquake) and was rebuilt with type IV masonry.[10] This new wall (and probably the one before it) has a hollow passage that connects with room D. Unfortunately half of the new foundation wall has now collapsed; it is quite likely that the collapsed half contained a small postern gate or loophole. Today the pointed vault over gatehouse C is well preserved. The hole in the north wall of C is a recent puncture and not a collapsed medieval door. The well-preserved arch at the west end of C is a jambless door (pl. 255a). It is common in Armenian gatehouses simply to have the interior side open.

To the north of gatehouse C is the large vaulted hall D (pl. 254a). The only entrance to this room is in the west wall just north of the staircase that ascends to the roof of D. The jambless door is roundheaded and is surmounted by a thin squareheaded window. On the interior of D (pl. 254b) a large section of the east wall has collapsed and the entire north end is missing. The north third of D is on a more westerly axis, and its sharply pointed vault is at a slightly lower level than the vault over the rest of D.

Just touching the tip of the southwest corner of C is cistern E (pls. 255a, 255b). This is the only structure in the fort built exclusively with brick. This parallels the very limited use of brick in Yılan Kalesi. Unlike Yılan, where the tiles are laid in horizontal courses along the rock cleft, all of cistern E from ground level is made with bricks and lime mortar. The cistern is covered by a single barrel vault in which the long brick tiles are laid radially (pl. 256a). The brick in the surrounding walls consists of neatly squared tiles that are laid in a regular horizontal pattern; on the exterior the four corners have been carried to the full height of the vault to create a flat square top. Today a large section of the north half of the brick vault has collapsed. In the upper half of the south wall of the cistern is a terracotta inlet pipe. The walls of the cistern are stuccoed with a thick layer of lime mortar. This cistern at Tumlu is not typical of Armenian construction but may represent the only visible remnant of a late antique settlement on the outcrop.[11]

Tower F is a massive cistern that is opened by a single hatch. It is flanked at the southwest by the angled undercroft G. The two roundheaded doors that open the west wall of G are damaged, but fragments of their jambs are still in situ. A stairway to the flat top of the undercroft is positioned between the two doors. The north end of G is open, and its barrel vault has two hatches. The adjacent undercroft H has only a single door in its west wall and one hatch in its barrel vault. Above H at the second level (not shown on the plan) are six embrasured loopholes. Only the southernmost loophole has collapsed. On the exterior the five remaining loopholes are framed by rusticated blocks that have neatly drafted margins (pl. 251b). The exterior masonry surrounding these blocks and the wall above are the result of a repair/addition. Thus the embrasures in this location may not have been part of the original design of the fortress. The interior sides of these roundheaded openings are quite ordinary, except that the inner half of the semi-cone top is made with a single course of bricks (pl. 256b). These bricks appear to be broken tiles that are laid in a radial fashion and separated by thick beds of mortar. The outer half of the rounded top is cut from a solid block of ashlar. The *limited* use of damaged brick in the five extant embrasures indicates that the material is recycled. What is most interesting about the brick is that it was once completely stuccoed over. On the underside of one embrasure a large fragment of the stucco survives and still shows that it was neatly incised with seven parallel lines. The lines are intended to imitate the actual margins of stone voussoirs. An amateur artist was trying to depict what the rounded top should look like. If these embrasures are repairs carried out by the Armenians, then it would be the only example of Armenian loophole construction in brick. Normally, the Armenians use a monolithic block to construct the entire half cone of the small embrasures. In the opinion of this writer, these five embrasures belong to a Mamluk period of occupation.

Building I has undergone substantial deterioration since Youngs conducted his survey. The foundations for three doors are still visible in the south, east, and west walls of the main rectangular chamber. A wall extending from the west flank of I may have enclosed another room. Attached to the east wall of I was probably some sort of circular extension that had embrasured loopholes (pl. 301c). At this level, windows can be seen in the mid-nineteenth-century etching of Tumlu published by Langlois.[12] In the north wall of I there may be the sill of a door (not shown on the plan). Certainly there must have been a means of access to the battlement-crowned tower on the northeast flank of I. A few small fragments of corbeling are still visible atop this rounded corner of I.

To the south a small courtyard is formed between I and J at the southeast end of the circuit. This circuit is opened by a tall broad window that is roundheaded on the interior and has a lintel on the exterior. Joist holes in the wall as well as post holes in the rock indicate that a wooden building was adapted to the northeast end of undercroft J. This undercroft is covered by a tall pointed vault that is supported by a single rib (pl. 257a). The vault is opened by one hatch and has an open southwest flank that does not cover the hollow salient end of J. Presumably, the tower had a wooden roof.

Attached to the north end of J (but not connected to it by a passageway) is the hall/undercroft K (pl. 258a). The outer wall of K is also the curving south end of the west circuit. The parallel northeast wall of K is opened by three partially damaged, jambless, roundheaded doors. The northwest wall is opened by one door of a similar design. K is covered by a continuous pointed vault that is supported by two ribs. The entire vault is preserved and opened by two square hatches in the north half. The southwest wall (that is, the circuit wall) is opened by five embrasured windows (pl. 258b). All of the embrasures are squareheaded, except for the one at the far north which has a rounded top. The rounded opening is not the result of a separate period of construction and is similar in other respects to the embrasured loopholes in wall N. K is the *only* example of an Armenian undercroft with more than three embrasured windows. It is normal to have the embrasured loopholes on a level above the hall or undercroft (cf. "J" at Yılan).

Attached to the northeast corner of K is room L (pl. 259a). Although this structure stands to only 1.5 m in height, the foundation of a rib in the center indicates that it was once vaulted. There are traces of a lime stucco on the interior walls. L is constructed exclusively with type IV masonry and may be a cistern.

Attached to the northwest corner of K is tower M. The square chamber in this tower is opened by a now collapsed door at the north, one high-placed square-headed window in both the east (pl. 302b) and west walls, and a postern at the southwest. A series of steps descends to the level of the postern. Post holes near its jambs indicate that swinging wooden doors were accommodated. There is no indication that tower M was ever covered by a vault.

The wall directly northwest of M has joist holes which indicate that wooden buildings filled the space between M, K, and L (pl. 259a). Farther to the northwest are the four roundheaded embrasured loopholes of wall N (pls. 257b, 259b). As the slope of the bailey descends to the northeast, a series of joist holes appears below the two northernmost embrasures of N. These holes obviously supported a platform to keep the archers at a constant level with the loopholes. Farther to the north, large sections of the thin circuit have collapsed. There may be the remains of a window at point O. The etching published by Langlois shows that the areas around P had collapsed over a hundred years ago.[13]

[1] I visited Tumlu in September 1974 and made minor revisions to the plan published earlier by G. R. Youngs. On the plan published here the contour lines are at approximate intervals of one meter.

[2] This site is mentioned as Tumla Kale on the chart of the *Cilician Gates;* it also appears on the maps of *Adana (2), Cilicie,* and *Kozan.*

[3] Favre and Mandrot, 148 f; Davis, 69; W. Baker, *Lares and Penates* (London, 1853), 265; Alishan, *Sissouan,* 283; Fedden and Thomson, 12, 37, 100; Ēpʿrikean, II, 56. The latter refers to this fortress as Tʿumla and Tʿumlu.

[4] Langlois, *Voyage,* 444 f; Youngs, 113.

[5] Youngs, 113–18.

[6] Hellenkemper, 188–91. Cf. Yovhannēsean, 210 f; Smbat, G. Dédéyan, 77 note 47.

[7] Hellenkemper, 190.

[8] Refer to the etching in Langlois's *Voyage,* 445.

[9] Youngs suggests that this block is a hinge stone. This is quite possible, although I could find no pivot hole in the block; see Youngs, 118.

[10] The only other evidence of repair in the area of gatehouse C is at the top of the salient between C and F (pl. 253a).

[11] Youngs (118) discovered "Roman" pottery at this site, which is possibly from the 1st or 2nd century A.D.

[12] Langlois, *Voyage,* 445; Youngs, 117.

[13] Langlois, *Voyage,* 445.

Vahga

The castle at Vahga[1] was not only the earliest seat of Armenian power in east Cilicia, but it served as the base from which the Rubenids eventually united all of Cilicia under one king. The site is located about 6 km northeast of the modern village of Feke.[2] The latter, which is located immediately north of the junction of the Asmaca Çayı and the Gök Su,[3] was built in the 1920s when the small population of Vahga Köy (at the base of the fortress) was transferred there. The fortress is of great strategic value because it hovers over the major north-south road that links Kozan to Kayseri. Vahga Kalesi has clear intervisibility with Maran but not with Saimbeyli. Also, when weather conditions permit, a signal could be sent from Vahga to Andıl, which in turn could relay a message to Sis or Bostan (via Alafakılar). About 4 km north of Vahga Kalesi a somewhat rough, but well-worn, trail leads southeast and eventually south to Bucak Kalesi.

Vahga plays its most important role in Armenian history during the twelfth century. According to Hetʿum's chronology, Constantine, the son of Ruben, captured Vahga from its Byzantine owners between 1097 and 1098.[4] Constantine reigned for only two years before his untimely (but natural) death at the castle; he was buried in the nearby monastery of Kastałōn.[5] In the year 1111 Baron Tʿoros I is said to have stored in Vahga the booty from one of his raids in Cappadocia.[6] The major event in its history occurs in 1138 when the Byzantine emperor laid siege to the fort and its Armenian garrison.[7] Unlike the Armenian forts in the plain, which quickly surrendered to the Byzantine army, the garrison in Vahga remained unmoved after three weeks of siege. Finally, it was decided that each side should send its foremost warrior to do personal combat below the castle's walls. The winning party would retain possession of the castle. After an exhaustive combat, the Byzantine champion prevailed and the Armenians surrendered the fort. Shortly thereafter John Comnenus captured the Rubenid chief, Baron Levon I, his wife, and two sons.[8] Levon was to die in captivity, never to see his home again. Surprisingly, within a year and a half of its capture, Muḥammad b. Ġāzī, the Danişmendid emir, captured Vahga and Geben with little resistance from the Byzantine garrisons.[9] Fortunately one of Levon's sons, Tʿoros II, escaped from Constantinople and returned to Cilicia. Collecting his father's vassals, he began to reassert his suzerainty in Cilicia by capturing the fortress of Vahga in 1144 (or early 1145).[10] Like his predecessors, he used Vahga as a base from which to recapture Cilicia Pedias from the Greeks. The fact that Vahga does not appear on the Coronation List of 1198/99 (see below, Appendix 3) may reflect either the incompleteness of the list or simply that King Levon I was lord of his ancestral seat.[11] Vahga briefly reenters recorded Armenian history in 1205, when King Levon I imprisoned his wife there,[12] and in 1275, when the patriarch of Sis fled there to avoid the Mamluk raids.[13] Colophons on certain manuscripts indicate that they were copied in a scriptorium in or near the fortress.

Considering that the area of Vahga was continuously inhabited by Armenians from the late eleventh century until the mid-1920s, there is surprisingly little information on this site. Prior to the formal survey of the castle by Dunbar and Boal, we had only sketchy descriptions of the fort.[14] Langlois avoids this site, and Alishan has less than a page of summary comments.[15] A few years prior to the destruction and abandonment of the Armenian city, a photo was taken of the castle and the adjoining city.[16] Many of its houses appear to have had tile roofs and stuccoed walls. Today only the foundations of the houses survive. In September 1960 Dunbar and Boal carried out an exhaustive and highly competent survey of the fortress.[17] Sixteen years later Hellenkemper published a "résumé" of the Dunbar and Boal study.[18] My work at this site essentially confirmed the findings of Dunbar and Boal. The following is a summary of my own field survey which includes a number of new observations.

From the road that links Feke to Saimbeyli a narrow dirt trail leads northeast to the castle. Except for a few squatters, the vast sloping expanse southwest of the fort, which the town of Vahga once occupied, is abandoned. About 0.75 km from the castle there are the impressive remains of a Byzantine church (pl. 260a). Unfortunately during my visit I was unable to complete a plan of this structure, but I did photograph the church and collect the following notes.

This two-story, aisleless church was covered by a single barrel vault. The east end of the church terminates in a massive protruding apse. In 1974 the entire apsidal semidome and all but a few springing stones of the barrel vault were missing. Most of the south wall is preserved to its original height (pl. 260b), while only a small fragment of the north wall is standing. Substantial fragments of the west wall of the church are still in situ, including the sill of a door. On the exterior of the south jamb of that door there are deep vertical cavettos and flanking tori. The tori have thinly incised edges. The only other evidence of relief decoration is on the exterior of the apse (pl. 260a). Here four horizontal bands, which run around the entire expanse of the apse, define four undecorated registers. The lower two registers and their bands stand to a height of 1.67 m from the present ground level.[19] The first horizontal band, which appears to divide the two lower registers into units of equal height, consists of a deep cavetto flanked by a plain fascia at the bottom and a torus at the top. The horizontal band at the top of the second register from ground level is a broader, more shallow cavetto with a single string course below and a simple squared cornice above. The upper two registers, which constitute almost two-thirds of the height of the apsidal wall, survive only at the north and south. However, what does remain provides valuable clues as to the date of this Greek church. The third register was opened by three roundheaded windows. Part of the sill for the southernmost window still survives in situ. The third horizontal molding gently curves around the tops of the windows. This molding consists of a cavetto flanked by a torus. Similar decorations around the arches of windows are common on the fifth- and sixth-century Byzantine churches in east Cilicia.[20] Another link with that pre-Arab tradition is the course of dentils that makes up the upper half of the fourth (or highest) horizontal band. The lower half consists of a simple congé. Dentils on the top molding of the apse are a ubiquitous feature in the fifth- and sixth-century churches.[21] The springing for the semidome rested directly atop the fourth molding.

The masonry of the church is also quite similar to pre-Arab Byzantine traditions in Cilicia which often have an apse of fine ashlar and nave walls of poor-quality stone. Here at Vahga Köy the exterior facing of the apse consists of huge, superbly dressed blocks of ashlar. A few of these stones are over 85 cm in length and 65 cm in height; some have been keyed into adjoining blocks. On the interior of the apse the facing stones resemble those of the nave wall, a small but well-cut ashlar. Most of the interior facing stones have fallen away, exposing the rough inner side of the exterior facing. Only small fragments of a core survive in the apsidal wall. Typically, in the walls of the nave, the cores are very thin (pl. 261a). In the lower courses of the nave walls, which tend to have much larger blocks than in the upper half, the cores are so thin that the inner and outer facing stones often abut (pl. 260b). The upper half has a slightly wider core of fieldstones and mortar. The largest ashlar blocks in the nave are used for the four roundheaded windows in the upper (or second) level of the south wall.[22] Two similar windows are still visible in the north wall. On the exterior of the south wall of the nave there are the broken fragments of corbels at the level of the window sills (pl. 261a). There is the trace of one large corbel at a lower level. The exact function of the corbels is not certain. Since there are no windows or doors in the lower level of the south wall, the corbels may have anchored some sort of attached building.

In the immediate area of the Byzantine church there are a number of buildings made with similar ashlar or even large well-cut cyclopean stones (pl. 261b). In a collapsed building adjacent to the church, there is a relief of a Byzantine cross on a large block of smooth ashlar. The alpha and omega on the cross are inverted because the block was recycled (probably during an Armenian period of occupation) to function as a door jamb. A similar alpha and omega occur on a cross at the late fifth-century basilica in Kadirli. Although no epigraphs have yet been found at Vahga Köy, it appears that the aisleless church is a pre-Arab Byzantine construction. The church may have been used in the Armenian period from the late eleventh century onward.[23] The few remnants of a pre-Armenian period of construction on the castle outcrop probably came from the Byzantine period.

East of the Byzantine church is the fortress and outcrop known as Vahga Kalesi. Like those of its smaller neighbor to the west, Maran, the circuit walls follow a long serpentine spine of limestone (fig. 75). The entire east flank of the castle is an unapproachable cliff; most of the east circuit wall was intended to function as a mere revetment. There is evidence of only one salient on the east flank (attached to the northeast corner of J). The cliff below the east circuit measures from 18 to 60 m in height. The relatively steep cliffs continue around the north end of the castle but soon spread into a more gentle slope, which provides a convenient point of access at the southwest. The approachable west flank of the castle explains the one-sided defenses (pl. 262a). At its highest point, cistern J, the fortress is about 1,270 m in altitude. It is by no means the highest peak in the immediate area. To the north of the castle some of the mountains are over 2,000 m in height. However, the castle outcrop at Vahga was most suitable for observing the river valley and the north-south road (pl. 265a).

Most of the partially masoned stairs in the approach path at the southwest have vanished, although their scarped foundations are clearly visible. Just before one reaches point A, the area where a now collapsed platform preceded the entrance to the outer gatehouse, it becomes apparent that all the exterior facing stones of the circuit walls and towers consist of a very uniform type VII (pl. 263b). The only exceptions are E and F, which will be discussed below. When looking north into the south door of the outer gatehouse, it seems that a circuit of type VII masonry once extended from the southeast corner of the gatehouse. This wall encircled a small shelf directly west of the vaulted tubular passage. The wall probably formed a small flanking salient whose loopholes could guard the entrance to the outer gatehouse. When this salient wall collapsed, it exposed a door (at point B; pl. 263a) that connected the tubular passage to the interior of the salient. The exterior facing stones around door B are a small, but very uniform, type VIII. However, the type VIII, which was also the interior facing stone of the flanking salient, is actually the result of a second period of construction. Careful observation of the damaged area above door B shows an earlier exterior facing stone (a well-coursed type IV). Thus the construction of the outer gate, the collapsed flanking salient, and the use of type VII as an exterior facing postdate *part* of the tubular passage.

The outer gatehouse, which is a bent entrance, is the result of one period of construction. The only unusual feature on the exterior are the superbly cut voussoirs of the south door (pl. 263b). They are a brownish limestone, while the rest of the gatehouse is built with gray limestone. This intentional alteration in the color of the limestone appears in the other parts of the fortress. This south opening once accommodated double wooden doors. Stone steps are still visible in the sill of the door. On the interior the plan of the gatehouse is an elongated trapezoid, which has two other openings at ground level. At the east there is a door that opens into the vaulted tubular passage. In the west wall there is a single huge embrasured loophole with casemate (pl. 264a). The casemate is slightly atypical of Armenian construction in that its inner half is expanded, giving it a bipartite appearance. The arches over the embrasure and casemate are roundheaded. The unique feature of this gatehouse is in the ceiling (pl. 264b). There is a single groined vault at the north end which has the typical Armenian cross-shaped keystone. To the south the slightly pointed vault over the rest of the gatehouse is opened by three broad machicolations. Today these are blocked by debris; undoubt-

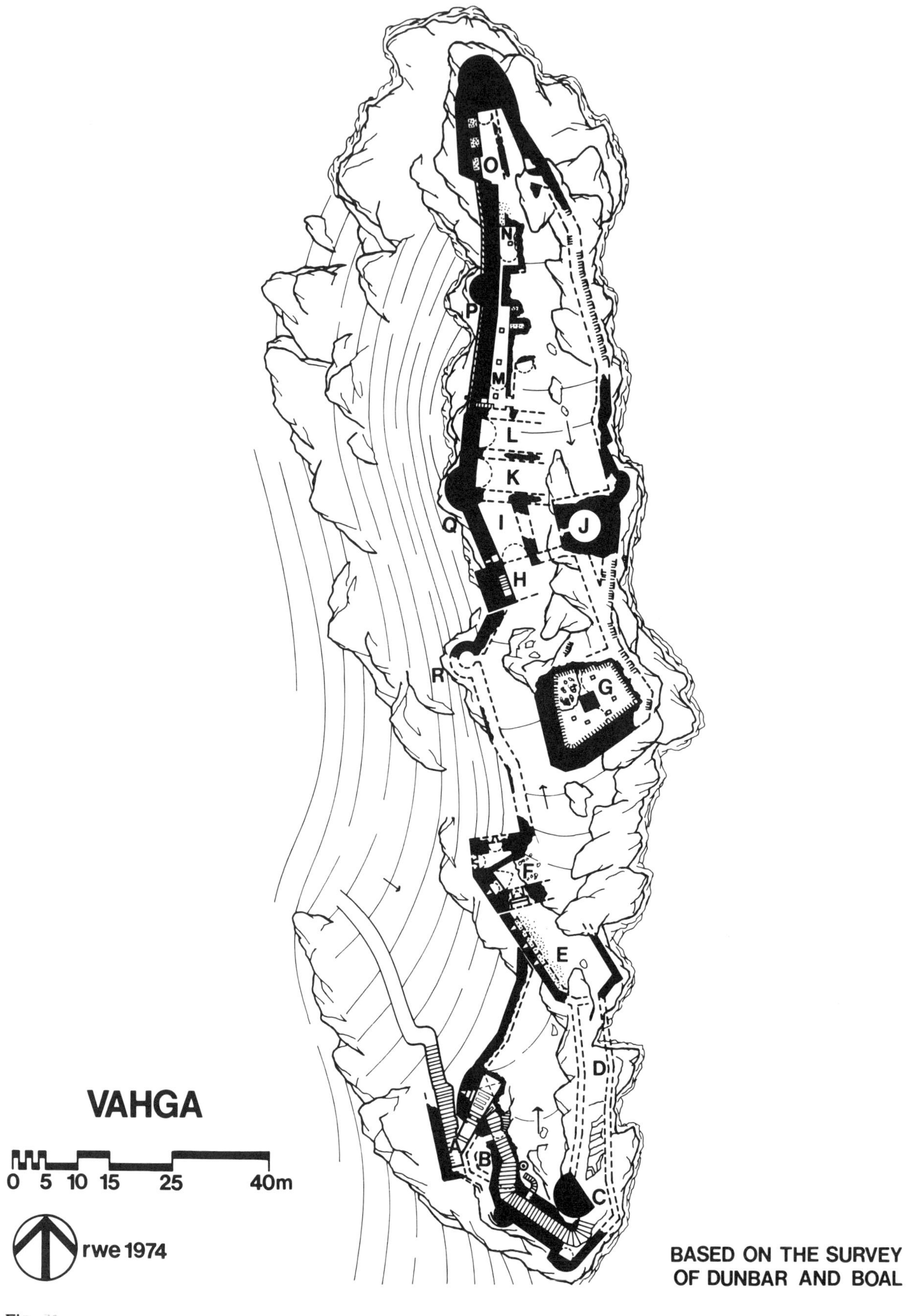
VAHGA
0 5 10 15 25 40m
rwe 1974
BASED ON THE SURVEY
OF DUNBAR AND BOAL
O
N
P
M
L
K
Q
I
J
H
R
G
F
E
D
A
B
C

Fig. 75

edly a now collapsed second level commanded the machicolations. The interior facing stones of the outer gate generally consist of type IX. Stairs in the gatehouse continue through the east door which is guarded by its own machicolation. This door leads directly into the tubular passage which snakes southeast for about 33 m before it abruptly turns to the north.

About 3–8 m south of the outer gatehouse, there is a subterranean cistern whose scarped walls are covered by a barrel vault. Fortunately large sections of the tubular passage over the cistern survive, giving valuable information on the nature of its design. Most of the interior and exterior facing of this entrance passage consists of types VII or VIII (pl. 265a). The continuous barrel vault of the passage was opened by occasional hatches through which the defenders could fire on an approaching party. At the extreme southeast corner of the passage there is a small vaulted chamber; this room was surmounted by a now collapsed second level. At the opposite end of the southeast corner there is the solid, almost semicircular, tower C. The masonry of this bastion is quite unique in that the courses consist of small ashlar blocks that are laid on both their vertical and horizontal axes. This style is quite similar to some of the masonry in the Byzantine church below. In the present gate complex the tower has no apparent function since the tubular passage winds around it along corridor D. The section of the passage north of C has collapsed. Tower C is probably a remnant of Byzantine construction at this site. The masonry and design of the outer gatehouse and the tubular passage are Armenian in character. The two periods of construction above door B may both date from the Armenian occupation. The small enclosure created by the winding tunnel and rock mass at the east could be entered only through an extremely narrow door just east and above the subterranean cistern. This enclosure really does not constitute an inhabitable bailey, but more an open room through which the defenders could pass in order to man the machicolations in the outer gatehouse and vaulted tunnel.

E is a broad corridor, lined with three embrasured loopholes at the west and offering direct access into the south lower-level door of the two-story inner gatehouse F. The three west embrasures in E are now covered with the accumulation of dirt and debris. The walls of E barely protrude above ground level.

Today the entire east wall of F has collapsed (pls. 265b, 266a). The broken lower-level south door of F shows evidence of once having had a slot machicolation.[24] E and F are unique with respect to their masonry. The exterior facing stones of both structures consist of a combination of types IV, VII, and VIII. The latter is in the minority. The smooth ashlar (type IX), which is used to frame the doors and windows as well as the vertical seams in the corner of F, is dark brown in color.[25] The presence of type VII masonry and the selective use of brown limestone indicate that E and F are probably Armenian in origin. What appears to have happened is that F was completely rebuilt following the original pentagonal plan (probably Byzantine) and using recycled materials as well as freshly cut types VII and VIII. The reconstruction appears to have been rather hasty, considering the variety of filling materials used in the margins of the stones. Even after its reconstruction it seems to have been patched. Judging from their masonry and junctions, E was constructed shortly after F. What is most interesting is that the loopholes in E are oriented to the southwest, as if the outer gate complex and the vaulted passage did not exist. When the outer gatehouse was built, a circuit of only type VII masonry attached the outer gatehouse to the lower level of E. In the remains of that connecting circuit are the faint traces of embrasured loopholes (not shown on the plan). After the circuit was constructed, the southernmost loophole in E was no longer functional. Thus E and F may represent the first period of Armenian construction at this site (perhaps in the late eleventh century). After the recapture of the fort by the Armenians in 1144 a decision may have been made to enlarge the entrance into the fort. This period also saw the construction of the circuit from O through R.

Unfortunately only a few minor traces of the circuit wall from F to R survive today. During the Dunbar and Boal survey, substantial portions of the circuit must have been visible, for they noted that the masonry was made of "small, roughly coursed blocks reinforced with horizontal beams."[26] These two authors suspect that the circuit from F to R predates the construction of the inner gatehouse.[27]

Northeast of F is the roughly trapezoidal cistern G (pl. 262b). This massive structure has a scarped foundation. It seems to have been adapted into a natural basin in the limestone. Since the sides of the basin are lowest at the south, the amount of supplemental masonry is much greater at this end (pl. 266b). On the exterior the facing stones consist of a combination of types VII, VIII, and IX. The regular courses are occasionally broken by large ashlar blocks that are fitted into two-course levels. It appears that some of the exterior facing consists of high-quality recycled masonry. The presence of type VII probably indicates that this is an Armenian construction. On the interior the facing stones consist of an attractive well-coursed type IX.

Large patches of stucco still cover the masoned walls. The design of the cistern is simple yet ingenious in that a continuous and slightly asymmetrical barrel vault revolves around a square pillar in the center (pl. 262b). The resulting carousel-like dome was opened by seven square hatches. Two of the hatches disappeared when the northwest corner of the dome collapsed. The cistern was filled by water that drained down through rock-cut channels from the higher limestone mass to the north and northeast. From floor to ceiling the height of cistern G is just over 8 m.

To the northwest of G is the small collapsed room H. Only the west wall of H, which protrudes as a square buttress on the exterior of the circuit, survives today. On the interior side of the west wall, a type VII facing is still visible as well as a fragment of the stairway (pl. 267a). The stairway gives access to the battlements on the top of the circuit. Above K and L only the first course of those battlements is visible. Most of the exterior facing stones of H consist of type VII like the rest of the circuit to the north. However, on the surviving section of the south wall of H there is an interesting variation in that the stones in some courses are exclusively type VIII. The rather thick, but consistently layered, core of fieldstones and mortar shows that H is the result of one period of construction.

Most of the barrel-vaulted chamber I to the north has collapsed. What remains of its interior facing stones is essentially a well-cut type VIII. In the southwest corner of I there is a squareheaded window that is flanked on the inner side by a vaulted casemate. Directly east of I is chamber J (pl. 267b). This structure now stands to less than 4 m in height, but judging from the thickness of its walls it once stood at least two stories. J is not a tower but a trapezoidal room whose west wall was incorporated into a now collapsed east circuit.[28] The east wall of J did not extrude from the line of the circuit. A small rounded salient in the east circuit is attached to the northeast corner of J. The exterior facing of the east circuit is a very uniform type VII. However, the few surviving fragments of the exterior facing of J show the same mixture of masonry types that appears on the facing of G. Thus J is probably Armenian but may not be contemporary with the construction of the type VII circuit. On the interior, J is circular with a narrow entrance in the west flank. The interior masonry is a very uniform type VIII.

To the north and adjacent to I are the vaulted chambers K and L. Unlike I, the slightly pointed barrel vaults of K and L run on an east-west axis. Both vaults as well as most of the north, east, and south walls of K and L have collapsed. In the surviving west walls the facing stones consist of a very uniform type VIII. North of L the topography on the interior descends over a relatively flat slope that is devoid of extensive construction.

In the north wall of L there is a blocked door that gives access to the subterranean chambers M and N. The latter run along the interior side of the northwest circuit. A small portion of the south end of M protrudes above present ground level. In the top of this section the southernmost of the three square hatches, which open the barrel vault of M, is clearly visible. To the west of this hatch there is a small chamber in the thickness of the wall as well as an adjoining stairway. Near the chamber is the shaft of a privy that terminates on the exterior of the circuit wall in a spout. It is clear that other chambers were once constructed above the subterranean rooms M and N. The floor on the interior of undercroft M is cluttered with debris (pl. 268a). Along the entire length of M at the springing level are joist holes that once held crossbeams. A 6-m section of the east wall of M is constructed with a large type IV masonry. This contrasts sharply with the rest of the interior facing of types VII and VIII. Dunbar and Boal believe the type IV to be from an earlier period of construction.[29] This masonry also appears in the east wall of the narrow corridor between M and N. However, the type IV could also represent a later (Armenian) period of construction. The two doors in the northeast corner of M are blocked today with debris. The corridor separating M and N is only 87 cm in width and is covered by a half-barrel vault. The walls of chamber N are constructed almost exclusively with type VII, while type VIII is more frequent in the barrel vault of N (pl. 268b). In the north wall of N there is a now blocked door that leads to another unsurveyed room. A large piece of limestone was allowed to protrude through the east wall of N.

At the extreme north end of the castle there are two long chambers separated by a partially collapsed wall of type VII masonry. The east chamber survives only to its foundations. In the collapsed west chamber the springing for its barrel vault is still visible at the north. In the west wall of the west chamber there are three vaulted niches, all of which are blocked with debris.

The north half of the west circuit is characterized by two important features (pl. 269a). First, the circuit is unusually tall, measuring over 16 m from the surface of the cliff to the top of the battlements.[30] Second, there are no massive horseshoe-shaped towers to provide extensive flanking fire along the circuit, but merely shallow salients (P, Q, H, and R) that act more as buttresses. This plan is a response to the topography. The severity of the cliff and slope in the north half of the west flank would keep the enemy at a safe distance and drastically reduce the need for defenses. The walls are brought to an unusual height not so much to impress an enemy, but more to provide a backing or revetment for the two levels of undercrofts on the interior. The Armenian engineers inset the upper half of the circuit between O and P and between P and Q in order to add stability to the walls. The ledge created by the displacement of the wall was covered by a small talus.

Unfortunately no precise dates can be given to the different periods of Armenian construction at this site. The few fragments of what is suspected to be Byzantine construction (tower C and the circuit between F and R) obviously predate the mid-twelfth century. It is impossible to date the construction of the outer gatehouse at Vahga by comparing it to the upper bailey gatehouse at Yılan.[31] The latter cannot be dated to any specific period in the Armenian kingdom.[32] Also, Yılan is not designed with broad internal machicolations, nor is it built in conjunction with a vaulted tunnel.

[1] My survey of this site was conducted in spring 1974. On the plan contour lines are placed at intervals of 2 m.

[2] Normally, I would transliterate the Armenian name for this site as "Vahka." The alternate spelling is kept here because the Turkish authorities employ "Vahga" (occasionally Vaga or Vaka). Vahga and/or Feke frequently appear on the maps of Cilicia: *Cilicie, Everek, Marash*. On the *Everek* map Feke is not in its proper location at the junction of two rivers but is accidentally placed on the old site of Vahga. See also *Handbook*, 334.

The Armenian name Vahka is probably a variant of the earlier Greek spelling Baka. Bar Hebraeus refers to the site as Bahgai.

[3] The Asmaca Çayı is itself formed by a variety of streams to the west of Feke, including the Yağnık Çayı. Northeast of Feke the Çatak Suyu merges with the Gök Su. From Feke this united Gök Su flows south into the Seyhan.

[4] Het'um, 471; Vahram of Edessa (498) confirms the same event.

[5] Matthew of Edessa, 47 f, 79; Vahram of Edessa, 498. Samuel of Ani (448) gives the same information as Matthew of Edessa. The chronicle of Smbat (610–12) corroborates the information on Constantine's occupation of Vahga and his death, but the chronology is hopelessly muddled. A good historical review of Vahga can be found in Yovhannēsean, 123–34.

[6] Matthew of Edessa, 100.

[7] Choniatēs, 22–25; Cinnamus, 18–20; Smbat, 616; Lilie, 406 note 117; Bar Hebraeus, 266.

[8] Smbat, 616 f; Gregory, 152 f.

[9] Ibid; Michael the Syrian, J. B. Chabot, III, 248; Bar Hebraeus, 266.

[10] Het'um, 474; Vahram of Edessa, 504 f; Smbat, 618. The chronology of the latter is in doubt. We do not know if T'oros captured the fort from a Byzantine or a Turkish garrison. Eventually the Greeks recognized Armenian suzerainty at Vahga; see Smbat 622.

[11] Alishan, *Sissouan*, 172.

[12] Smbat, 642; Smbat, G. Dédéyan, 84 note 84. Vahga's isolation made it an ideal prison; cf. Smbat, G. Dédéyan, 82.

[13] Bar Hebraeus, 453; *RHC, DocArm*, I, 530 note 1.

[14] Schaffer, 91; Cuinet, 96 f.

[15] Alishan, *Sissouan*, 172; Dunbar and Boal, 176 note 5; cf. Tēr Łazarean, 28–30.

[16] Pōłosean, 26 f.

[17] Dunbar and Boal, 175–84.

[18] Hellenkemper, 217–23.

[19] Dirt and accumulated debris appear to cover about half the height of the lowest register.

[20] R. Edwards, "Two New Byzantine Churches in Cilicia," *AS* 32 (1982), 25 ff.

[21] Ibid.

[22] In terms of weight and mass, the largest blocks in the church are the cyclopean stones that make up the salient ends of the apsidal wall.

[23] For information on the Armenian churches and monasteries in the immediate region of Vahga see above, notes 15 and 16, and Oskean.

[24] Dunbar and Boal (179 f) offer a description and section drawing of the various openings in both levels of F.

[25] These cornerstones are not quoins.

[26] Dunbar and Boal, 181.

[27] This may represent one of the few traces of Byzantine construction near the summit. Timber supports occur in other Byzantine forts of Cilicia (cf. Evciler).

[28] The circuit now survives as a revetment along the cliff. Only a small section north of J stands more than 3 m above present ground level.

[29] Dunbar and Boal, 182 f.

[30] The battlements have collapsed, but I estimate them to be 1.6 m in height.

[31] Hellenkemper, 223.

[32] Edwards, "Yılan," 23 ff.

Yaka

Yaka Kalesi[1] is a shattered garrison fort located on the north flank of the strategic coastal road between Mersin and Tarsus.[2] What remains of the fort rises on a small fairly uniform hillock (pl. 270a). At the east there are three towers (fig. 76). The only other remains that still stand in situ are two independent walls in the northwest corner. Despite the fragmentary nature of the fort, it appears that its ground plan was square. There are no inscriptions at the site and no medieval names can be securely associated with it. Today the area in and around the fort is farmed for wheat. The region adjacent to the fortified hill is watered by numerous irrigation projects and also has the distinction of being the worst malaria area in Turkey.

For the modern visitor the most striking feature about the fort are the three dissimilar towers at the east. They are aligned on a common axis, equidistant and hollow. Excavations will undoubtedly uncover adjoining circuit walls. The southernmost tower is a square

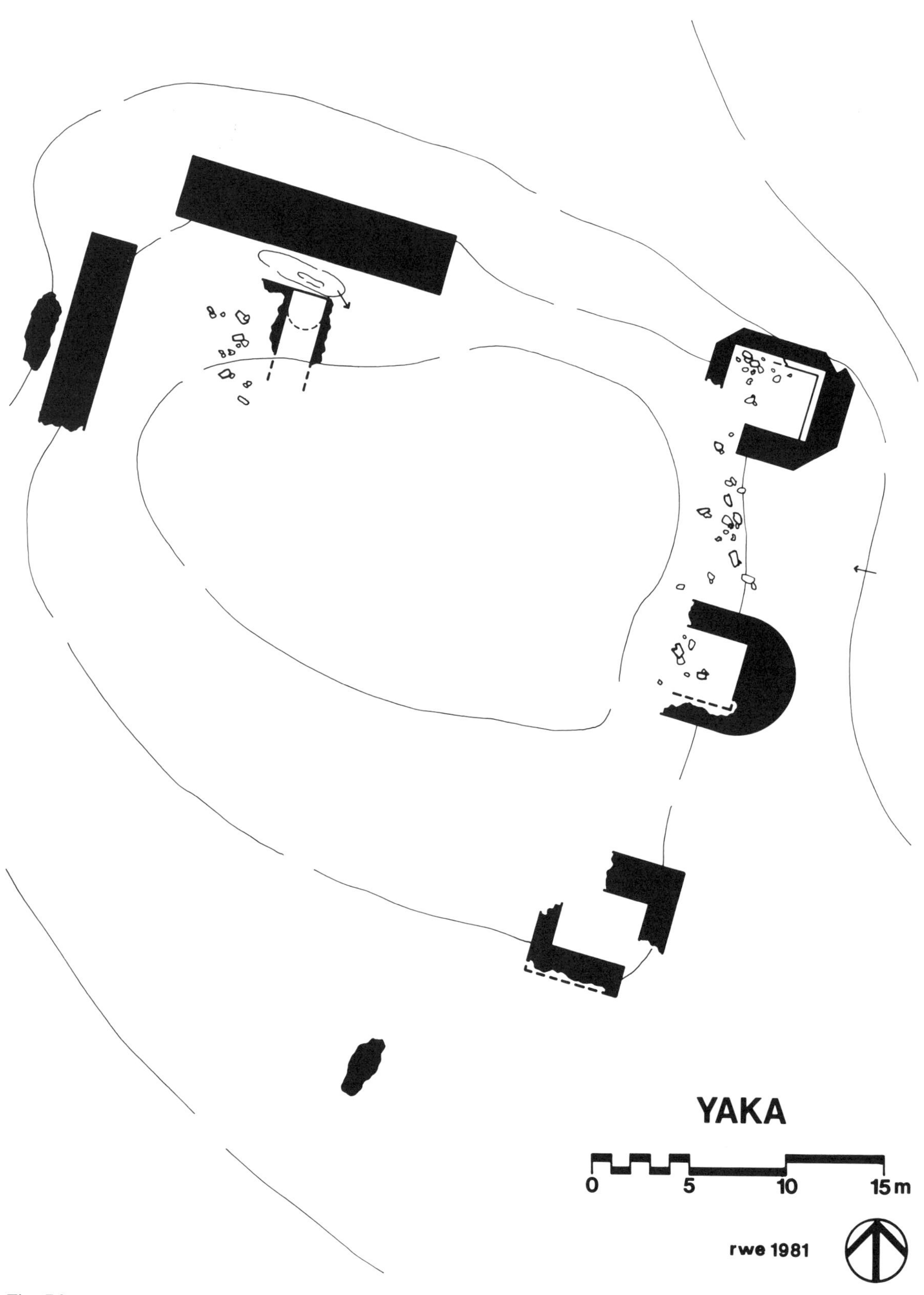

Fig. 76

that has two collapsed corners. The central tower is a rounded bastion that has lost almost all of its facing stones. The exposed core consists of some cut blocks as well as fieldstones. Both are laid in fairly regular courses and bound by numerous rock chips and mortar. The room within this central tower, like those in the other two bastions, is rectangular. The tower in the northeast corner has a polygonal facade (pl. 270b). The exterior facing on the tower consists of an elongated form of type IX masonry that has a rusticated center and broad drafted margins. These margins are much wider than those seen in type VII masonry of Armenian forts. Unfortunately the faces of stone at Yaka are heavily eroded. The interior of the northeast tower had no real facing, but the core was originally laid against a wooden falsework and set in neat courses with an abundance of mortar. There are the remnants of narrow shelves in the north and east walls of the tower room.

At the west end of the north flank there is a single massive wall (pl. 270a). The ends of the wall have been neatly squared off, as if there were once doors on both sides. The exterior facing stones are identical to those on the northeast tower. Unfortunately only the first three courses of the facing are in situ (pl. 271a). What appears to be the exterior of the wall today is actually the neatly layered core of coursed stones. As with the northeast tower, there is no evidence of an interior facing stone. Immediately south of the north wall is a partially buried barrel-vaulted room (pl. 269b). Since this room is significantly below ground level it may once have functioned as a cistern.

At the north end of the west flank there is another freestanding wall that has been squared off at the north (fig. 76). This west wall stands at a 90-degree angle to the axis of the north wall but shows no signs of ever having been joined to it. There must have been an outwork or freestanding tower at the northwest to narrow the point of access between the two walls. The significant feature about the west wall is its masonry. Here the relatively crude stones of the core are laid in irregular courses and show a lack of mortar in the interstices. This masonry is quite distinct from the rest of the fort and may represent a second period of construction.

Because of their proximity, symmetrical plans, and similar (but peculiar) masonry, Tumil and Kütüklu must be studied in the same context with Yaka. These three forts have characteristics that would not be common for Armenian and Byzantine forts.[3] In the opinion of this writer, all three sites appear to be Crusader fortifications. The only known Crusader fort built on the coast, Silifke Kalesi, uses type IX masonry (either freshly cut or recycled). The nearby site of Tece, which employs the rusticated type IX as well as type VII, is in part a Crusader construction, but its plan differs because it housed a district lord as well as a garrison.

[1] The survey of this hitherto unpublished site was undertaken in July 1981. The contour lines on the plan are at intervals of one meter.

[2] Yaka, either the village or fort, does not appear on any of the principal charts of this region. See *Handbook,* 204.

[3] The Byzantines do employ a symmetrical plan but always one in conjunction with some sort of donjon attached to the circuit. Neither the Byzantines nor the Armenians layer their cores with cut stones.

Yeni Köy

Yeni Köy Kalesi[1] is a garrison fort that is located on one of the routes to Meydan. The name Yeni Köy is attached to three different villages in the vale of Karsantı and its environs.[2] The site that concerns us here is at the northeast entrance to that valley. It is of strategic importance since it guards one branch of the major north-south road. This site can be located by hiking east for 1.5 hours from Karsantı through a maze-like narrow defile. Yeni Köy can also be reached from the south by hiking northwest from Mazılık.[3] The fort at Yeni Köy is situated just west of the village on a gently sloping outcrop (pl. 271b). The base of the outcrop has been terraced recently with revetment walls to facilitate farming. Between the village and the fort is a small stream. To the east and south of the village, large sparsely forested hills enclose the arable land. The area is hot and does not have the lush appearance of the great valley to the west. The short range of mountains west of the fort acts as the eastern terminus of the vale of Karsantı (pl. 272a). At the north end of this range the vertical peaks drop, giving access to the vale from the northeast. The name and history of this site are unknown. Today the majority of the locals call it Yeni Köy Kalesi; a few herdsmen refer to the fort as Gire Kale.

The fort consists of a somewhat oval circuit wall with six rounded towers (fig. 77). The wall is carefully adapted to the uneven faces of the limestone crags. There is only one entrance at the east. On the exterior of the fort the walls are constructed with a fairly uniform type V masonry (pl. 271b). Around the entrance there are a few type VII stones (pl. 272b). The frame

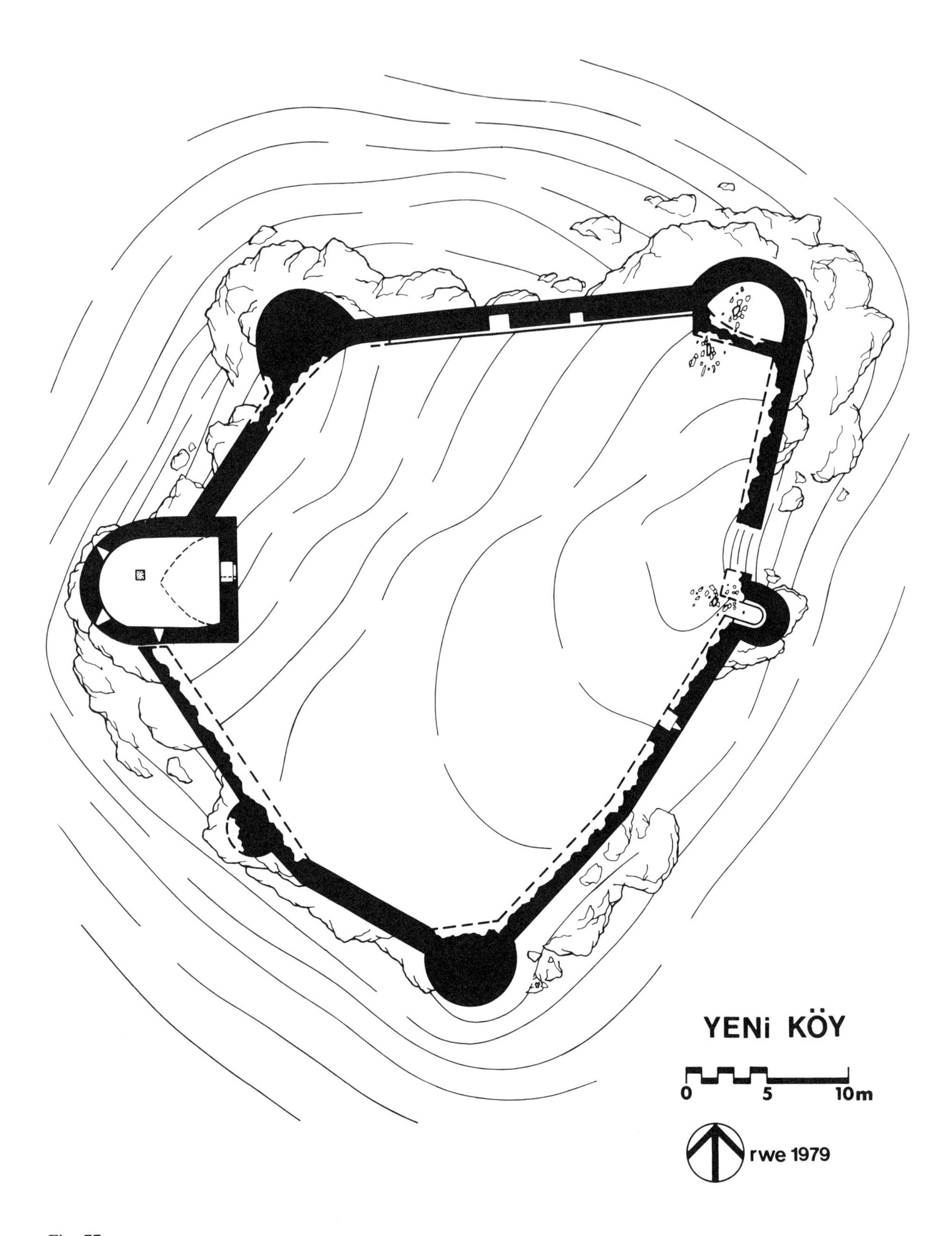

Fig. 77

of this broad door is now shattered, and little can be said about the nature of its construction. The interior facing of the circuit consists of types III and IV masonries. Today much of the circuit is crowned by thick vine-like shrubs that make an appraisal of the battlements difficult.

Adjacent to and directly south of the entrance is a small flanking tower (pl. 272b). This tower rises from a carefully scarped base of limestone. On its interior there is a narrow shelf that may have supported the wooden floor of a second level. To the south of the tower there is evidence of a single crude embrasured loophole and its shattered casemate; both are set low in the wall. At the far south end of the fort, a solid tower is constructed at the lowest point in the outcrop. From here the wall continues to the northwest past another solid tower to the largest bastion in the fort. This west tower consists of a massive chamber covered by a pointed vault. Its door at the east has jambs that are constructed out of huge well-cut blocks (pl. 273a). The covering over the jambs has partially collapsed. A curious corbeled lintel of three stones is set over the inner side of the door. Above the door and to the south is the lower half of a now shattered window. On the interior of the west tower there is a square hatch in the ceiling, which is now blocked by a single boulder (pl. 273b). Of the three embrasured openings in the apsidal wall, only the one in the center is wide enough to accommodate an archer. The two flanking windows were probably used for ventilation. Numerous joist holes are still visible on the interior. From the north wall of this large tower, the circuit continues and merges into a solid bastion before it proceeds east. The north wall of the fort has a thin shelf in the center and two squareheaded niches above. A wooden building was probably attached to this wall. In the northeast corner is the sixth tower which may have had an interior chamber and vault. At present it is badly damaged, and the positions of any doors or windows are impossible to determine.

Today the interior of the fort is divided by new stone walls that serve as corrals for sheep (not shown on the plan). Most of the coins found by the villagers at this site are Armenian; one badly worn and unidentifiable Byzantine coin was discovered. The present fort seems to represent one period of Armenian construction.

[1]This hitherto unattested site was surveyed in June 1979. The contour lines are separated at intervals of 25 cm.

[2]The village of Yeni Köy and its fort are not listed on any of the charts that appear on the List of Maps.

[3]For a description of the location and topography of Mazılık, see R. Edwards, "Two New Byzantine Churches in Cilicia," *AS* 32 (1982).

Yılan

Yılan Kalesi[1] has been photographed more than any other Armenian fortification in Turkey, including the walls of Ani. This notoriety is due to three factors: first, the fort flanks one of the most strategic roads in Anatolia and thus is readily visible; second, its size and complex design make it one of the most impressive military structures in the medieval world; and third, it is one of the best-preserved fortifications in the entire Levant. Most of the comments from nineteenth- and twentieth-century explorers are often superficial observations made from the roadside. It seems that the castle's undeserved reputation for harboring snakes (hence the name "Yılan") and the unbearable humidity of the Cilician summer deterred all but the most hardy from hiking up the fortress-outcrop. Since the late 1930s and prior to my own work at Yılan, five serious studies, which resulted in plans and descriptions of the site, have been published. The first was undertaken by J. Gottwald, followed later by J. Thomson, G. R. Youngs, W. Müller-Wiener, and H. Hellenkemper.[2] Since it would serve no useful purpose to summarize their accounts here, I will offer a brief description of the site and focus on specific units of the castle where I can offer new observations.

Yılan Kalesi girdles the central and the uppermost portions of a limestone outcrop (pls. 274a, 274b) near the city of Ceyhan in Cilicia Pedias. The outcrop was once the northern spur of the Cebelinur Dağı; the Ceyhan River has separated the spur from the rest of the formation, leaving the castle isolated on the northern bank (pls. 276a, 276b). Through most of this century until the mid-1960s, the road connecting Ceyhan and Adana to the Bahçe and Amanus passes proceeded east from Misis on the *south* bank of the Ceyhan River. It seems likely that this route was used throughout most of the Ottoman period.[3] However, the castle at Yılan is designed to counter the most aggressive assault from the north and northwest (fig. 78). In the medieval period the major road from Ceyhan to Misis and Adana must have passed on the north flank of the fort.[4] Yılan would never have been constructed to guard a road south of the river. In the mid-1960s a four-lane superhighway connecting Adana and Osmaniye was built.

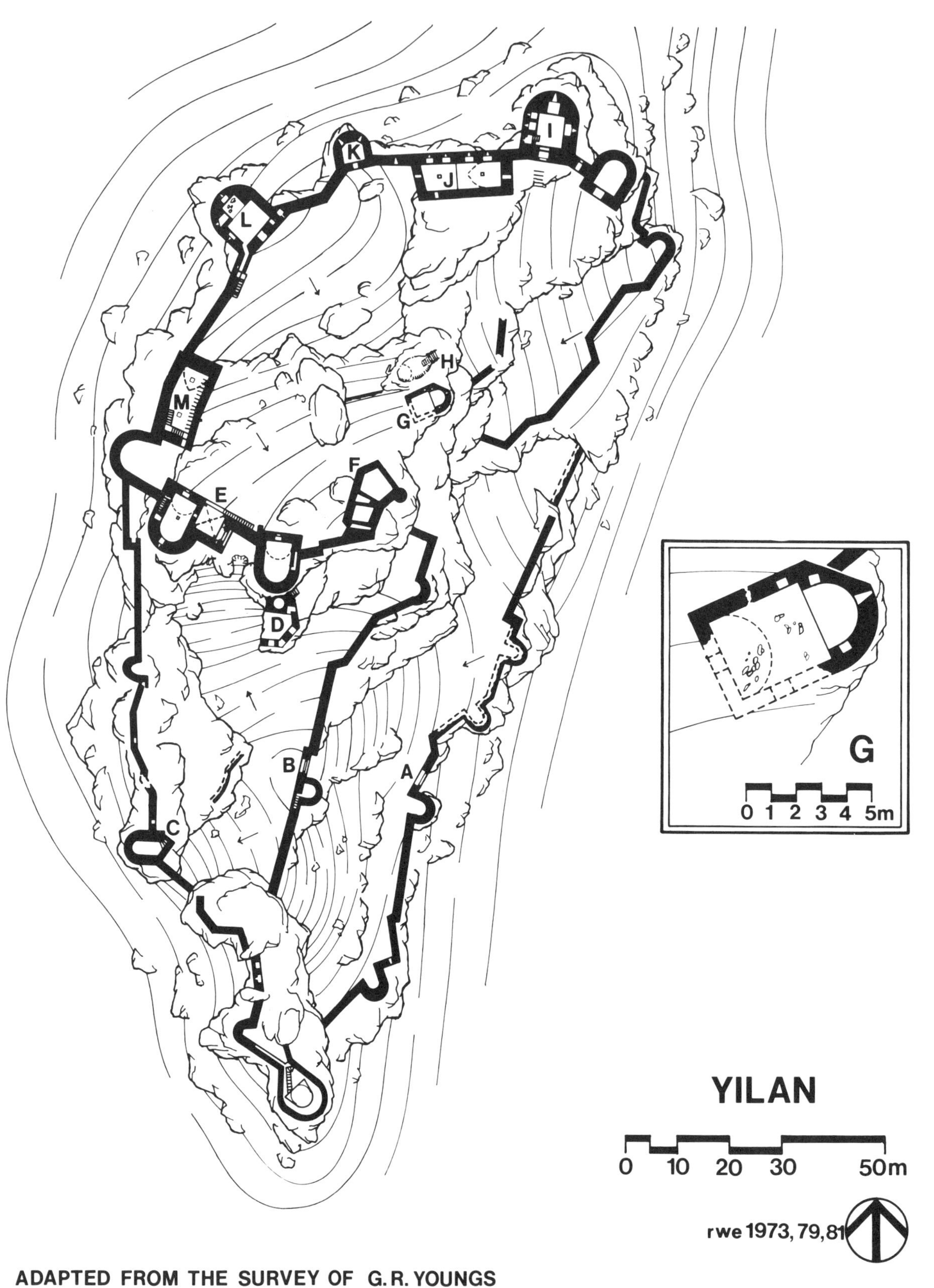
A
B
C
D
E
F
G
H
I
J
K
L
M
G
0 1 2 3 4 5m
YILAN
0 10 20 30 50m
rwe 1973, 79, 81
ADAPTED FROM THE SURVEY OF G.R. YOUNGS

Fig. 78

This new route, which bypasses Misis and follows the north bank of the river until reaching the city of Ceyhan, must follow in part the old medieval road. Today the area around the castle-outcrop is sparsely populated, and the only nearby industry is a quarry. On the bank of the river, southeast of the castle, is Yılan Kalesi Köy. Near the village are the foundations of a number of late antique structures.[5] The approach to the castle is made from the northwest side of the outcrop on an asphalt road that connects to the four-lane highway (pl. 274a). This service road to the castle turns south at the base of the outcrop and terminates at the southern tip.

Despite the visibility of this site[6] and its obvious prominence in medieval Cilicia, there is no way to determine when this castle was built or its medieval name. Hellenkemper's assertion that Yılan is the medieval Govara is purely speculative, for that name can also be associated with other castles.[7]

The walls of Yılan Kalesi, which create three defensible baileys, straddle the ascending ledges of the limestone outcrop (pl. 274b). The lower two baileys (pl. 278a) are designed to protect the southeast flank where the line of approach is easiest. J. Thomson remarks that the walls are so cleverly designed that they push an attacker to the steepest slope, allowing the castle to defend itself.[8] The walls are buttressed and defended by strategically placed towers. Each of the baileys has a single entrance that stands in the lee of a tower. The upper ward, which is the largest and most heavily defended unit, functioned as the home of the resident baron and the garrison. In this highest and northernmost unit are the majority of cisterns and a chapel.[9]

In general, the masonry of this fort is employed in a manner consistent with Armenian traditions. The type IX masonry is confined to the jambs, niches, windows, embrasured openings, and vaults over the doors. The interior facing stones for the lower two baileys are either types III or IV or a combination of both. The interior facing of the upper bailey is decidedly superior, with large sections of walls consisting of types V, VI, and VII masonries (pl. 284b). Type IV occasionally appears as the interior facing in the upper bailey, but usually in conjunction with a better masonry. The vast majority of the exterior facing stones of the lower two baileys is type VI with studding (pl. 275a). In the upper bailey, types V and VII appear alongside type VI (pl. 279a). This mixture of types is not uncommon in Armenian military architecture.[10] It is unwarranted to assume such combinations always represent different building periods or the recycling of materials.[11] The occasional mixture of smooth and rusticated ashlar, such as we see in gate B (pl. 275b) and room D, may simply be the result of masons adding the cut stones of different quarry crews. The only readily identifiable piece of recycled masonry is the base of a loophole that now functions as a quoin in the northwest corner of the chapel (pl. 281b). However, the chapel is unique because it combines so many types of masonries (pl. 282b).[12] If separate building periods are to be identified in the masonries of Yılan Kalesi, it will only be in areas where two very distinctive types of facing abruptly abut and the consistency of the poured core is altered.

Today most of the east wall of the lower bailey is functioning as a revetment (pl. 278a) because the free-standing upper section of this wall has collapsed. The top of this circuit and its four hollow towers barely protrude above ground level. The entrance into the lower bailey, gate A, differs from the other two gates in that it is not a tripartite unit with a slot machicolation (pl. 275a). Gate A simply consists of jambs covered by a now collapsed low-level arch that in turn is flanked on the interior by the higher vault over the door. There is no evidence of a crossbar bolt, but on the interior sides of the jambs at the top there are pivot housings (not shown on the plan). Each housing is bored with a single hole to accommodate one side of a double door.

The path from gate A to gate B winds up the steep and often jagged edges of the outcrop in such a way that an advancing enemy is required to approach at an oblique angle and is thus exposed to prolonged attack from archers. Gate B has a design almost identical to that of the tripartite gate of the upper bailey (pl. 275b). The outer arch of gate B has collapsed, leaving only the springing stones in situ and exposing what was once the concealed machicolation. Above the springing stones there is evidence that a diaphragm wall surmounted the outer arch. The jambs are covered by a segmented lintel, depressed relieving arch, and a diaphragm wall. Like the lintel in gate E (pl. 278b), the three central segments are not wedge-shaped but have flat parallel sides. They are held in position simply with the pressure applied by the relieving arch to the ends of the two terminal blocks of the lintel. What looks like the defiance of gravity is actually sound engineering, for the entire weight of the diaphragm wall is displaced into the ends of the lintel. On the interior side of the jambs, the vault over the door is set at a higher level than the relieving arch. Behind the jambs are the sockets to accommodate a crossbar bolt (not shown on the plan). Immediately south of gate B, the west side of the flanking tower has been closed off.

The walls of this tower enclosure were stuccoed and covered by a vault. This room is the only large cistern in the two lower-level baileys. A small stairway gives access into the tower cistern. The circuit immediately south of gate B has square merlons in an excellent state of preservation and a wall walk that once extended across gate B. The runoff from the winter showers would flow along the wall walk and empty into the cistern.

Once through gate B, the path continues its ascent to the upper bailey at the north. About 30 m south of gate B is a projection of rock (pl. 276a). The west circuit of the central bailey ascends the rock and terminates at the south end of this spur in an almost circular bastion (pl. 276b). Hellenkemper is probably right in assuming that a wooden roof covered this tower.[13] The remnants of joist holes and the lack of springing stones seem to confirm this. Four rectangular merlons are still preserved on the west side of the tower. Based on the present remains, I cannot justify the multiple phases of construction that Hellenkemper postulates for the south tower. The masonry of the merlons is almost identical to the rest of the interior facing in the lower two baileys. The design of the merlons here is quite similar to those south of gate B.

Almost midway between the far southern tower and tower C are two privies. The privy at the south is surmounted by an embrasured loophole. Farther to the north, tower C is fully enclosed by walls and crowned with merlons (pl. 277a). Like the southern tower, it probably had a wooden roof. There is a third latrine in the wall immediately north of tower C. Only a few fragments of merlons stand on the north half of the west wall. Northeast of tower C and directly west of gate B, the foundation for a thin curtain wall is still visible. This wall rises on the low and accessible edge of the rock mass which is crowned by tower C. The obvious function of the wall was to protect the inner side of the west curtain by diverting the forward movement of an enemy rushing through gate B.

The pentagonal chamber D is a curiosity. It is the outwork for the upper bailey, but its most strategic and vulnerable side at the west is wide open and shows no signs of ever having been closed by a wall or wooden barrier. The faceted wall that envelops the south end of D is opened by three almost straight-sided windows that are covered by monolithic lintels on the exterior. The small court created by the encircling wall is bounded at the north by a cylindrical cistern (pl. 277b). A door at the northeast, which is covered by a monolithic lintel with a slightly rounded soffit, gives access from court D to the cistern. The modern visitor should exercise caution since the door also leads to a sheer drop at the east. This cistern is fully stuccoed and covered with a small, partially collapsed cupola. A number of pipes lead into the cistern. The large pipe at the base of the south wall is for drainage.

West of chamber D the path of ascent into the upper bailey is stopped by the rock foundation of the gate complex. Below gatehouse E at the southeast, vertical faces have been scarped on the rock to blunt an enemy advance (pl. 279a). Obviously, some sort of removable ladder was used in medieval times to facilitate ascent. Gatehouse E consists of a bent entrance that is flanked at the west by a horseshoe-shaped tower and linked to a similar bastion at the east. The north side of the bent entrance, which opens the upper bailey, was never fitted with a door or barrier (cf. Tumlu).[14] The east side has a perfectly preserved tripartite door (pl. 278b). The diaphragm walls over the outer arch and jambs rise to their original height. The intervening machicolation is today plugged with several large blocks. On the interior side of the jambs the door is covered by a depressed vault and secured by a crossbar bolt. The two pivot housings, which extend from the soffit of the depressed arch behind the jambs, have been broken away to their bases. There are corresponding post holes in the floor. The rectangular space on the interior of the gatehouse is covered by a single groined vault (pl. 279b). This well-executed vault of smooth ashlar has the typical cross-shaped keystone and the L-shaped stones in the four salient junctions.

The most important features of this gate are the four reliefs above the jambs (pl. 278b).[15] The center block of the segmented lintel shows the traces of a thinly incised cross with four arms, each with triangular ends. Hellenkemper believes that this is a "lateinisches Kreuz."[16] However, such a design was used in Armenian ecclesiastical architecture from the archaic period through the high Middle Ages.[17] The other three reliefs are on the three central voussoirs of the relieving arch. The keystone has a figure in a seated position. The flanking voussoirs each depict a single rampant lion. The voussoir at the north has deteriorated so extensively that only a few traces of the lion are visible. In the corresponding stone at the south, the heraldic lion stands in profile with only his face turned to the front. Standing on his hind legs, he takes on an almost human appearance. His hind paws have been replaced by large feet that are attached to truncated calves. His front legs are curved upward as if he were a boxer. The end of his tail is twisted into a decorative ornament. His face is articulated by a small nose, two eyes, a broad mouth, and two triangular ears. The male figure on the central voussoir is less distinct.[18] The voussoir has a raised frame, and the floor on which the figure is sitting is separated from the base of the block by what appear to be four inverted fleur-de-lis. The sides of this frame do not appear to be columns supporting a stylized *baldacchino*.[19] From the present remains I cannot verify Hellenkemper's description of the facial features. There is no evidence that the figure has hair down to his shoulders. The figure does wear a crown which seems to have a number of points. Today only a slight trace of the extended right arm is visible. The bent left arm seems to hold a sword with the point anchored in the floor.

The central relief is significant because a number of commentators on the architecture of Armenian Cilicia believe the figure to be the kingdom's first king, Levon I (1198/99–1219).[20] This identification is made because they believe that Levon's "reign beheld the most glorious period in the fortunes of Lesser Armenia and it was probably the great age of Cilician castle building, perhaps indicating that the castle was built about the turn of the century."[21] It is assumed that other Cilician fortresses, which are similar to Yılan with respect to their architectural features, may also date from the period of his reign.[22]

However, the iconographic and archeological evidence at Yılan does *not* support this association. Unfortunately no sculptural reliefs of King Levon I have yet been discovered, nor do any illustrated manuscripts from the period of his reign shed light on Levon's appearance or characteristic poses. Our only verifiable and contemporary depictions of Levon come from the huge corpus of coinage that was minted during his reign. Paul Bedoukian notes the following about the seated, full-body portraits of Levon I: "On the obverse of all the silver coins of Levon I (with the exception of the coronation coins), the king is seated on a throne ornamented with lions [sic] heads. He wears a crown and a royal mantle on his shoulders, usually has a cross in his right hand and a fleur de lys in his left. On most of the coins, the king's feet rest on a footstool."[23] On some of the coins his throne becomes so stylized that the carved animal heads on the armrests become circular bulges, almost giving the appearance that the king is sitting side-saddle between a pommel and cantle.[24] In contrast to the numismatic portraits, the relief at Yılan shows a male figure wearing baggy pajamas, and there is no evidence of a long flowing mantle. The relief has no throne, but the figure sits in an oriental fashion on a flat surface with his legs crossed and slightly bowed. The object held by the figure in his left hand is neither a cross nor a fleur-de-lys. On the coinage of Levon I there is no depiction that resembles the relief at Yılan Kalesi.

Bedoukian mentions that most of the trams of Levon I have an unusual design of two rampant lions on the reverse.[25] But this is also a motif that is carried on in the later coinage of Kings Smbat and Ōšin.

On the coinage of Kings Het῾um I (1226–70), Het῾um II (1289–1307), and Levon III (1301–7) there are obverse portraits that are almost identical to the relief at Yılan Kalesi.[26] Armenian kings sitting in an oriental fashion would not be unusual, considering that under Het῾um I and his successors Cilicia frequently became a vassal of its Moslem neighbors to the north and of the Mongols. However, during the independent and highly successful reign of Levon I, we should not expect the Armenian king to be depicted in such a way.

Are we then to assume that Yılan was built and dedicated by an Armenian king who ruled in the late thirteenth or early fourteenth century? Many of these monarchs had short reigns, and their near bankrupt treasuries could never assume the expense of constructing a fort like Yılan. I believe that there is an archeological explanation for this relief. On the interior of the gatehouse we see two distinct types of masonries, and in the southwest corner there is a triangular fragment of the springing for a groined vault (pl. 280a). This fragment is isolated two courses below the present springing of the groined vault and is surrounded by a masonry quite inferior to the masonry in the upper third of the gate interior. The latter consists of type IX masonry. The masonry of the lower two-thirds is type V (excluding the frame and vault of the east door). This distinct contrast of masonry types (not mixture) probably indicates two separate periods of construction.[27] For some reason that is not apparent today, the first groined vault collapsed and had to be replaced at a later period. The collapse was not caused by any external siege since the exterior facing stones are fairly consistent. The only inconsistencies occur in the diaphragm wall over the outer arch, where type IX masonry predominates.

If the groined vault and parts of the arches over the door were replaced in a period after the initial construction, then that cross-legged figure may be the portrait of the Armenian king responsible for the reconstruction. What is certain is that the relief over the gatehouse door cannot be associated with King Levon I. Nor can we assume that he is the original builder, since he was the "first king." The Armenian chronicles report that the first great Rubenid castle builder was Baron T῾oros I, who was active in the second decade of the twelfth century. Since T῾oros I is credited with rebuilding much of Anavarza[28] and occupying most of the Cilician plain, Yılan Kalesi could possibly date from his reign. Only excavations and a thorough scientific analysis of the pottery, coins, and building materials at Yılan can offer a more precise answer about the chronology of this fortress.

The tower immediately adjacent to the groined vaulted entrance at the west consists of two barrel-vaulted stories. The lower level has a single door and window at the north and an embrasured window at the southwest (only the lower-level door is shown on the plan). The entrance at this level has two pivot holes in the soffit which indicates that two swinging wooden doors were accommodated. The upper level, which can only be reached from the roof of the gatehouse by a roundheaded door, has an embrasured loophole with casemate at the southwest and an entrance to a stairway in the west wall (pl. 280b). The stairs ascend to the roof. The casemate is typical of Armenian construction, and the loophole has the traditional stirrup base on the exterior.

The interior of the horseshoe-shaped tower directly east of gatehouse E consists of a single tall room covered by a barrel vault. Although no joist holes are present, some sort of a wooden floor must have divided the upper half of the room (not shown on the plan) from the lower. In the upper half there is an embrasured loophole with a casemate at the south and a door/window at the north. In the lower level there is a closet in the thickness of the east wall[29] and a door at the north. Like the door in the west tower, there are pivot holes in the soffit. Only here there is evidence of corresponding post holes in the sill.

To the northeast of the tower there are several small rooms, the largest of which is the pentagonal chamber F (pl. 281a). Farther to the northeast, at the top of the summit, are chapel G and cistern H (pl. 282a). A gap in the limestone summit is partially closed off by a retaining wall west of the cistern. Because the chapel has been adapted to the edges of the cliff and the sinuosities of the outcrop, the south side of the apse is rounded on the exterior and its north side is beveled flat (pl. 284a). At the base of this north half, a short wall of two courses in height joins the rock mass to the flat side of the apse.

Historically, the most important construction in the castle is the subterranean cistern H, directly north of the chapel. The cistern is adapted into a natural depression in the rock, and much of the rock face on the interior has been carefully scarped (pl. 283a). There is a single stepped entrance at the northeast which leads through a narrow jambless door. The six steps, which descend from the door to the floor of the cistern, are each one meter in width. The steps as well as most of the frame around the door are constructed with large relatively smooth ashlar blocks. The north and south walls of the cistern, which taper outward at a very steep pitch, serve as the longitudinal support for the depressed barrel-vaulted covering (pl. 283b). The apex of the vault is opened by a hole at the east end and is only 1.2 m above the springing course. The vault is made of coursed fieldstones that are laid in neat regular courses and sealed with a white sandy mortar. The interior facing of the vault has been carefully stuccoed. The significant feature about the cistern is the liberal use of brick to supplement and cover the natural rock of the lower walls. The north wall and the apsidal west end has significantly more brick than the south wall to make up for the lack of solid rock. Even though the south wall is a consistent mass of rock, the upper third has been cut back and replaced with a course of brick to support the springing. At the east end of the south wall the brick has fallen away, exposing the springing of the vault. The brick is laid in fairly regular horizontal courses. The thickness of the brick plates varies from 29 to 86 mm. There is no alteration of brick and ashlar. Each course of brick is separated from the next by a thin bed of mortar. The mortar is unlike anything I have ever seen in Byzantine or Armenian Cilicia. In appearance it is like a muddy brown adobe that has not been mixed with small stones or sherds. Widespread traces of plaster indicate that all the brick was covered with a stucco face. There is no evidence of brick stamps. Just who is responsible for the construction of the cistern is a mystery. The Armenians do not have a tradition of building with brick either in Greater Armenia or in Cilicia. In a *few* Cilician sites the Armenians recycled Byzantine brick in small limited areas.[30] It is possible that the Armenians are recycling brick to construct the cistern, but it would not be in keeping with Armenian traditions. When the Armenians use natural basins of rock for cisterns, they never cover the scarped rock with masonry. Cistern H may be the only remnant of Byzantine construction at this site. Since the springing for the cistern's vault is adapted to the irregular breaks in the top of the encircling brick wall, the cover is probably a later (Armenian?) construction.[31]

Northeast of chapel G, two short sections of circuit wall restrict access to the summit as the slope descends sharply to tower I. Tower I and the neighboring tower to the southeast are almost isolated from the rest of the upper bailey because of the intervening outcrop in which J is situated and because the wall walk above J does not continue east. Between the two northeastern towers there is a small postern covered on the interior by a monolithic lintel (pl. 284b). This gate has jambs on the exterior and sockets to accommodate a crossbar bolt. Because of the steep cliff below, a party exiting from the gate would have to use a ladder. The small horseshoe tower east of the gate has a single windowless room that is covered by a barrel vault and opened by a door at the south. It seems quite certain that the top of this tower was surrounded by battlements since the sill for a door or straight-sided window is visible atop the junction of the tower and the wall of the postern. With respect to its mass and height, tower I is probably one of the largest salients in the fort. Its upper chamber (depicted on the plan) is entered at the south by a flight of now damaged steps. The thin jambs on the exterior side of the door are protected at the top with bulbous projections. These would prevent an enemy from prying away the corners of the wooden door with a crowbar. This upper chamber is covered by a slightly pointed vault. The walls of its rectangular room are opened by four embrasured loopholes with casemates and a closet in the southwest corner. There is a large square hatch in the floor of this room, which is the only opening into the lower-level cistern. Water passes through the hatch into the cistern from a drainage pipe in the ceiling of the upper-level chamber. The roof of the tower, which is reached through an opening in the casemate of the north embrasure, is now devoid of battlements.

West of tower I is the undercroft J (pl. 302c). This partially subterranean room is entered through a small door at the west end of the south wall. The lintel of this opening has two pivot holes which indicate that double wooden doors were accommodated. Because this door is at such a low level in the wall, this room could *not* have been a cistern (pl. 286a). The east and west walls of the room are opened by high-placed windows, and the slightly pointed vault is pierced by two hatches. The walls of the room are adapted to a natural cleft in the rock. The limestone, which protrudes into the space of the room, has not been cut away. From the flat roof atop this chamber the defenders could enter the four embrasured loopholes with casemates in the thickness of the wall (fig. 78; pls. 285a, 302c). The flat roof is actually tiered with a single step so that the eastern casemates are at a higher level than the two western ones. This step was made necessary because the east half of the vault over J was raised about 75 cm above the level of the west half. Above the casemates the wall walk is protected by merlons that are each pierced by an embrasured loophole and covered with a pyramidal crown.

Because tower K is set at a much lower level than J, the architects did not connect the two units with a stairway in the wall walk but blocked access to the casemates above with a high wall at the west end of J's roof. Today squareheaded battlements crown the top of K. The upper chamber of K (not shown on the plan) is opened by a high-placed door at the south. The lower-level room (shown on the plan), which also has a single door at the south, is hexagonal and opened by two embrasured loopholes at the north.

As the curtain wall descends to the southwest, stairs are constructed atop the wall walk to give access to tower L. On the south flank of L, a staircase gives access from the upper level to the ground. The lower-level room in the tower (not shown on the plan) is opened by a single door at the south and a window at the west. One can communicate from the lower to the upper level by two openings in the pointed vault. The squareheaded door in the lower level has two pivot holes in the soffit of the lintel. The upper story (shown on the plan) was a partially vaulted roof terrace (pl. 285b). Squareheaded merlons still crown the tower, except at the northwest. At the northwest and west there are three embrasured loopholes with casemates. The one at the northwest seems to have been flanked at the southeast by a small vaulted chamber. Springing stones for the vault of this room are still in situ. Near the southwest corner of tower L where the staircase joins the wall walk there is a typical latrine. It consists of an open niche with a rectangular hole in the floor and a slanted chute below.

About 25 m southwest of L, a long vaulted cistern is attached to the circuit wall (pl. 286b). The vault of cistern M is slightly pointed and opened by two hatches. The top of the north wall has collapsed, leaving that end of the vault unsupported. Most of the lower half of the south wall consists of scarped rock that is stuccoed like the walls. The only door is a high-placed portal at the southeast. Immediately south of the cistern, an open salient joins the west wall of the upper bailey to the gatehouse complex at the south and to the west wall of the central bailey.

[1] The first plan to show the fortress in proper scale was completed by G. R. Youngs. My own plan of Yılan Kalesi was begun in 1973 and modified in two subsequent visits in 1979 and 1981. On that plan the contours are placed at intervals of 50 cm. Not all of the openings are shown in the two horseshoe-shaped towers

flanking gatehouse E. What appear to be four embrasures with casemates in the north wall of J are actually openings in the terrace level above J.

[2]Gottwald, "Burgen," 83–93; Fedden and Thomson, 100–102; Youngs, 125–34; Müller-Wiener, 77–79; Hellenkemper, 169–87. Cf. Frech, 577, fig. 16.

[3]The presence of an Ottoman caravansary on the south bank (east of Misis) and the paucity of remains on the north bank would certainly support this conclusion. See: Taeschner, I, 145 f; Barker, 265 f.

[4]Unfortunately Wilbrand von Oldenburg (16 f), who provides our only itinerary of Armenian Cilicia, does not mention the presence of Yılan castle on the road, though he mentions Toprak and Misis. Hellenkemper's assertion (184 f), that Wilbrand's "patrimonio beati Pauli" is Yılan, is without support. Wilbrand could be referring to Gökvelioğlu or some other site in the region.

[5]Seton-Williams, 121, 173. Yılan Kalesi appears on the following modern maps: *Adana (1), Adana (2), Cilicie, Malatya*. In the late 1870s the village of Yılan was inhabited by "Nogai Tartars" who had been settled there after the Crimean War; see H. Barkley, *A Ride through Asia Minor and Armenia* (London, 1891), 193.

[6]Yılan has intervisibility with Tumlu, Anavarza, Misis, and Amuda (perhaps Bodrum and Toprak as well). For general observations on the site see: Ainsworth, 89; Sevgen, 337–39; Schaffer, 41; R. Normand, "La création du musée d'Adana," *Syria* 2 (1921), 200, pls. 22 f; Davis, 73–76; Deschamps, I, pl. 6b.

[7]Hellenkemper (185) cites the chronicle attributed to the Constable Smbat (Smbat, 667) to support his case that the Mamluks rode in view of Govara to attack Misis in 1322. But too many questions are left unanswered. It is clearly stated in that chronicle that the Mamluks moved from Ayas to Misis along the banks of the Ceyhan River. Since the most direct route would be to the northwest, Yılan Kalesi would be avoided. The site of Govara in 1322 is more likely to be Gökvelioğlu. A number of Armenian forts in Cilicia have names almost identical to "Govara." See: *RHC, DocArm*, I, 818; above, Part I.7, note 16.

[8]Fedden and Thomson, 101.

[9]For a description of the chapel see Edwards, "First Report," 170 f.

[10]Refer to the discussion above, Part I.3.

[11]Hellenkemper, 175, 183 f.

[12]Normally, Armenian chapels are built with greater care. Here at Yılan a lack of money to pay professional masons probably left only volunteers to build the chapel.

[13]Hellenkemper, 174 f.

[14]Ibid., pl. 38b.

[15]Gottwald, 88; Hellenkemper, pls. 39a and b; Youngs, pl. 23a; Edwards, "Yılan," 29.

[16]Hellenkemper, 177.

[17]The design of crosses in Armenian Cilicia shows great variation. See Bedoukian, 78 f, and cf. fig. 40 in Edwards, "Second Report."

[18]Hellenkemper's 1968 photo is quite clear and shows that most of the significant detail was missing at that time. By 1979 the figure's right arm seems to have suffered damage, and dark-colored lichens covered the left half of the relief. Cf. B. Kasbarian-Bricout, *L'Arméno-Cilicie: Royaume oublié* (La Chapelle Montligeon, 1982), pls. 25–27.

[19]Hellenkemper, 177 f.

[20]Dunbar and Boal, 184; Hellenkemper, 184.

[21]Youngs, 130.

[22]Dunbar and Boal, 184.

[23]Bedoukian, 56.

[24]Ibid., cf. pls. 2–14.

[25]Ibid., 57.

[26]Ibid., 60, 62; for Hetʿum I see pl. 30; for Hetʿum II see pl. 36; for Levon III, see pl. 40.

[27]Hellenkemper (178) *implies* that the fragment of springing represents a mistake that was corrected during the original period of construction. But this does not explain the sharp change in the style of masonry. Also, the fragment of springing is not cut with the same smooth face as the springing on the vault above.

[28]Edwards, "First Report." The exterior masonry of the south bailey at Anavarza is quite similar to the masonry in the upper bailey at Yılan. See also Edwards, "Yılan," 30–32.

[29]Youngs calls this space a "garderobe."

[30]Edwards, "First Report," 171.

[31]The mortar in the cistern's vault is similar in appearance to the mortar in the chapel.

Appendix 1

Mason's Marks

The precise purpose of the mason's marks in Cilicia is unknown.[1] The following symbols rarely have a consistent axial placement. The encircled number after each symbol indicates the repetition of that mark in a given area.

Çandır

Building J.
On the rib above the springing stone of the northwest corbel: ꓘ8 ①

On a southwest voussoir of the pointed arch: ↖8 ①

On a northeast voussoir of the pointed arch: Ÿ ①

Building E (lower level).
Room 1: ⩙ ①; ✡ ②

Room 2: ꓘ8 ⑦; ✡ ⑨; ┌┘ ②

Room 3: B⊣ ①; ✡ ②; X ①

Room 4: B⊣ ①

Haruniye

On the exterior side of gate D: ↿ ①; ϟ ②;

ʌ ⑥; ≡ ④; ≢ ②; ≷ ②; + ③

On the interior side of gate D: ⊣ ①; V ①;

ʌ ① ᒐ ①; ≛ ②; + ①; ↖ ①

Kız (near Dorak) Near the north door of B2: ↑ ①

Kız (near Gösne)

These mason's marks occur on the chapel, and they are sometimes difficult to distinguish from the graffiti (see Edwards, "First Report" and "Second Report"):

N ①; ⋈ ④; ≥ ④; ⫰ ①; X ①;

Ħ ①; ˥ ①

Lampron

The mason's marks given here are intended to supplement the catalogue published by Robinson and Hughes (203–6):
Room P, second embrasure from the south: ⴳ ①;
W ①; ↰ ①

Room P, embrasure at the far north: ↘ ①

Maran

On the interior side of the jamb of the door in tower B: + ①

On the interior of gate A: ⊣ ①

Silifke

The mason's marks given here are intended to supplement the seven symbols published by Langendorf and Zimmermann (163):
In the gatehouse: ⊔ ④

In undercroft K: ч ①; ↑ ②

In tower P: ↓ ①

Sinap (near Çandır)

The following types of mason's marks occur on the exterior facing only; they were too numerous to count.

> ; L ; /| ; K ; ⅂ ; X ; Δ ; ⇖

Sis

On the exterior of gate A in the castle, 4 m to the right of the inscription: ◫ ①

[1] Other lists of Armenian mason's marks can be found in: Langlois, *Inscriptions*, 31; *Architettura medievale armena* (Rome, 1968), 38; V. Grigoryan, *Hayastani vał mijnadaryan kentronagmbetᶜ pᶜokᶜr hušar annerə* (Erevan, 1982), 142; *Corpus Inscriptionum Armenicarum*, II, ed. S. Barxudaryan (Erevan, 1960), 92; B. Bagatti, *Gli scavi di Nazaret*, II, *Dal secolo XII ad oggi*, Studium Biblicum Franciscanum, Collectio Maior 17 (Jerusalem, 1984), fig. 32, nos. 22, 23. Cf. D. Pringle, "Some Approaches to the Study of Crusader Masonry Marks in Palestine," *Levant* 13 (1981), 173–99.

Appendix 2

Chemical Analysis of Mortar

The twelve mortar samples in this study are from the walls of five medieval forts in or near Cilicia. These samples were extracted carefully from the interior sides of the facing stones to reduce the possibility of contamination by weathering. The samples were analyzed in February 1980 at the Lawrence Berkeley Laboratory by Dr. Frank Asaro and Mrs. M. Sturz. I am most appreciative for their advice and dedication to this project. The method of analysis was x-ray fluorescence.

When I presented the samples to the laboratory for analysis, I announced that I was searching for distinct recipes (that is, peculiar combinations of chemical ingredients) that were consistently used by the Armenians and Byzantines in making mortar.[1] However, since the number of samples is very limited, and because they were not selected at random but from types of exterior facings that were associated, for example, with Byzantine (types III and IV) and Armenian (types V and VII) architectural features (see above, Part I.2, I.3, I.5), I did not expect irrefutable results but only basic correlations from which to continue the work. In order that Dr. Asaro and Mrs. Sturz show not the slightest bias toward my goals, I took the precaution of labeling samples from the same site with different names.

The results, which were not what I expected, did yield valuable information. There was *no* evidence that the Byzantines and Armenians consistently followed throughout Cilicia peculiar recipes with respect to the making of mortar.[2] What is apparent is that the mortar associated with the exterior facings of types V and VII masonries is *always* different chemically from the mortar in the exterior facings of types III and IV masonries at the *same* site, even if the walls of III/IV and V/VII are adjoining (for example, Savranda).[3] This must indicate that the two groups were constructed at different times, since efficiency would demand that even Byzantine work crews (who theoretically could employ slightly different types of masonry) have a central location for the complex process of making mortar. One should expect minor variations from mortar of the same period. It is possible from the results of this study to say that the Byzantines and Armenians have separate formulae for making mortar at each site.

[1] I thought that this distinction between Byzantine and Armenian formulae might be associated with the use of gypsum or limestone as the primary ingredient. See A Lucas, *Ancient Egyptian Materials and Industries* (London, 1962); Zvi Goffer, *Archaeological Chemistry* (New York, 1980), 103–7.

[2] The high amount of sulfate (S) in SAV (5.5 percent, which is a maximum of ~23 percent gypsum by weight) is the only indication of a sizable gypsum component in this survey. The silicon values (Si) may indicate the relative amounts of sand (or other siliceous materials) that appear in the samples. The magnesium (Mg) may be indicative of the type of limestone used in the fabrication of quicklime.

[3] A specific breakdown of the chemical differences in the mortar samples appears on the following table. A discussion of the significance of the samples can be found in the descriptions of the relevant sites (see the Catalogue) and in Edwards, "Bağras." The abbreviations on the table correspond to the following sites: BAB = Babaoğlan; SIS = Sis; HIS = Hisar; SAV = Savranda; and BAG = Bağras.

Mortar Analysis, 1979, Cilician Field Survey

Location	*Types of Masonry*	*Approximate Elemental Abundances (%)*					
		Ca	Mg	Al	Si	Fe	S
BAB	IV	40.	0.7	0.4	2.7	0.6	<.5
BAB	mosaic	37.	1.2	0.2	4.4	0.7	<.5
BAB	V	30.	3.9	0.8	6.6	1.5	<.5
SIS	V	26.	1.3	2.7	13.0	1.7	<.5
SIS	VII	25.	2.1	2.3	10.0	1.5	<.5
HIS	IV	23.	2.4	2.9	8.4	<.2	<.5
SAV	IV	28.	9.4	0.6	16.0	2.9	5.5
SAV-1	V	29.	6.3	0.3	6.2	1.8	<.5
SAV-2	V	28.	3.4	<.2	6.9	5.5	<.5
BAG	III	39.	0.2	1.3	2.4	1.0	<.5
BAG-1	VII	16.	13.0	0.3	13.0	3.7	<.5
BAG-2	VII	13.	15.0	0.4	14.0	3.9	<.5

Appendix 3

The Coronation List

The most complete list of the nobles in attendance at the coronation of King Levon I (1198/99) is preserved in the manuscript attributed to the Constable Smbat. In this compilation forty-six nobles and fifty-nine separate sites are mentioned. For the convenience of the reader, I reproduce below the most recent translation of the Coronation List.[1] If the location of a given site is *securely* known,[2] it is followed by an asterisk. If a known site is discussed in the Catalogue, its modern name (in parentheses) follows the asterisk.

Le prince de Baghrās,* Adam.[3]
Le prince de Čker, Hostius.
Le prince de Hamus, Arewgoyn.[4]
Le prince de Sarvandikᶜar* (Savranda), Smbat.
Le prince de Harun* (Haruniye), Lewon.
Le prince de Simanayklay, Siruhi.
Le prince d'Anē, Henri.
Le prince de Kutaf, le connétable Aplłarip.
Le prince d'Ǝnkuzut, Baudoin.
Le prince de Tᶜornika, Estève.
Les princes de Berdus, Lewon et Grigor.
Le prince de Kančᶜ, Ašot.
Le prince de Fawṙnaws,* Aplłarip.[5]
Le prince de Kapan* (Geben), Tancrède.
Le prince de Čanči, Kostandin.
Le prince de Šołakan, Geoffroy.
Le prince de Mazot Xačᶜ, Simon.[6]
Le prince de Tᶜil* (Toprak), Robert.
Le prince de Tᶜlsap, Tᶜoros.
Le prince de Vaner, le maréchal Vasil.
Le prince de Barjrberd, Gēorg.[7]
Le prince de Kopitaṙ, Kostandin.
Le prince de Mawlovon* (Milvan), Ažaros.
Le prince de Kuklak* (Gülek), Smbat.
Le prince de Lambrun* (Lampron), Hetᶜum.
Le prince de Lulwa, Šahinšah.[8]
Le prince de Papeṙawn* (Çandır), Bakuran.
Le prince d'Askuṙas, Vasak.
Le prince de Manaš, Hetᶜum.
Le prince de Berdak, Mixayl.
Le prince de Prakana, Tigran.
Le prince de Siwil, Awšin.
Le prince de Kiwṙikos* (Korykos), Simon.
Le prince de Séleucie* (Silifke) et de Punar, Kostancᶜ.
Le prince de Sinit et de Kovas, Ṙomanos.
Le prince de Vēt* et de Vēṙəsk,* Nikifawṙ.[9]
Le prince de Lavzat* et de Timitupawlis, Xrsawfawṙ.[10]
Le prince de Maniawn,* de Lamaws* (Lamas), de Žermanik et d'Anamur,* Halkam.[11]
Le prince de Norberd et de Komardias, le sébaste Henri.
Le prince d'Andawšc et de Kupa, Baudoin.
Le prince de Małva,* de Sik* et de Palapawl, Keṙsak.[12]
Le prince de Manovłat* et d'Alar,* Mixayl.[13]
Les princes de Lakrawēn, Kostandin et Nikifawṙ.
Le prince de Kalawnawṙaws, d'Ayžutap,* de Sainte-Sophie et de Nałlawn, Kervaṙd.[14]

In recent years many scholars have displayed the unfortunate tendency of trying to assign as many medieval names as possible to modern sites. These assignations are frequently made despite the absence of inscriptions, specific references in the texts, and any affinity between the medieval and modern names. To a large degree these correlations are based on the assumption that Smbat's list is a compilation of all the fortresses in a neat geographical order.[15] Two general patterns are apparent in this list: first, from Bağras (Baghrās) to Čanči, the sites seem to move along the spine of the Anti-Taurus Mountains (Nur Dağları) from south to north; and second, from Toprak (Tᶜil) to Anamur, the forts are arranged on an east-west axis. The modern reader must remember that this arrangement is not perfect. For example, Partzerpert (Barjrberd) is followed on the list (in the westerly direction) by Kopitaṙ. Yet the available evidence in other sources seems to indicate that Kopitaṙ was actually east of Partzerpert (Barjrberd).[16]

Smbat's list is merely the collected names of the honored guests who were in attendance at the coronation of King Levon I. It is not a record of the vassals under the suzerainty of the Armenian king, nor is it an indication of the physical size of his kingdom.[17] Considering the large number of unidentified sites that appear to be Armenian constructions,[18] it is likely that the majority of barons on the list owe allegiance to King Levon. G. Dédéyan has stressed the importance of foreign names on the Coronation List (nine of Frankish extraction and eight of Greek).[19] However, it should be remembered that during the barony and subsequent kingship of Levon I, there was a vigorous attempt to westernize the Rubenid court, including the introduction of non-Armenian names.[20] It is quite possible that some of the seventeen foreign names actually belong to Armenians. It is interesting that the three largest fortifications in Armenian Cilicia—Anavarza, Sis, and Vahga[21]—do not appear on the list. This may be a reflection of the incompleteness of the document or simply of the fact that all three sites were directly administered by Levon himself. Of equal importance is the absence of any mention of a baron for the cities of Adana, Misis, Tarsus, and Ayas. Again, this may reflect direct royal administration over the cities or perhaps their lack of importance. Misis, Sis, Anavarza, and Tarsus do have archbishops in attendance at the coronation. In a few cases one baron may be the lord over a number of sites; there is no evidence in our extant texts that a baron had to reside in his castle.

[1] Smbat, G. Dédéyan, 75–80 notes 29–72.

[2] I am less ready to assign medieval names to the fortresses than my predecessors. Ibid.; Alishan, *Léon,* 173 ff, cf. idem, *Sissouan,* 63 ff; Boase, 146–48; Aghassi, 18, 22, and map; Hellenkemper, 13–17.

[3] For a description of this site see Edwards, "Bağras."

[4] Refer to my discussion of Çardak in the Catalogue and to Smbat, G. Dédéyan, 48 note 11.

[5] Fawṙnaws is the modern Fırnıs; see the following maps: *Cilicie, Marash.* Cf. M. Sykes, *Dar-ul-Islam* (London, 1904), map opp. ix, 73; Fındıklı (note 4) in the Catalogue.

[6] I believe that the association of Mazot Xačʿ with the town of Hadjin/Hačən (officially called Saimbeyli after 1935) is tenuous. Hild (in Hild and Restle, 233) compares the Armenian and Turkish words for cross, Xačʿ and Haç respectively, to conclude that their phonetic similarity has a historical connection. Hild assumes that sometime after 1375 the Armenians residing in Mazot-Xačʿ dropped the first half of the town's hyphenated name and adopted (through the influence of their Altaic neighbors) the simplified form Haç (or Haçin [sic]). The only problem with this theory is that the indigenous Armenians did not pronounce the name H à c h i n (Haçin), but H a d j i n (Hačən/ Հաճըն). There is no evidence of a sound shift from չ to ճ . The modern toponym Hačən does not appear in the texts of medieval Cilicia. See Alishan, *Sissouan,* 217.

Pōłosean (130 ff) discusses some of the folk etymology for this site. One legend has it that when donkeys carrying the relics of St. John the Evangelist across the Taurus Mts. to Constantinople refused to continue, the emperor (John I Tzimisces, 969–976) declared that this place was pleasant and ordered that a chapel be built there to St. James. This pleasing (*hačeli*) place becomes Hačən, and the chapel may have become the site for the St. James Monastery (see Edwards, "Second Report," 125–28). Another legend has it that the hills resemble the back of a camel, which the Arabs call *hağīn.* Neither the folk etymology nor the location of this site would lead one to conclude that Haç ("cross") is the source for the toponym. Cf. "Hačən" in *Haykakan* (84 f), where it is conjectured that Hačən results from the combination of two toponyms: Harkʿan and Ačʿe.

[7] As I have mentioned in an earlier publication ("Second Report," 142 note 80), Tamrut and Meydan are the largest fortresses in the region where Barjrberd should be located (see Meydan and Tamrut in the Catalogue). At this time Tamrut is a more likely candidate to be Barjrberd by reason of its location at the west end of the vale of Karsantı.

[8] Smbat, G. Dédéyan, 77 f note 54; Hild, 53 ff. The location of Lulwa (Loulon) is fairly certain. See also Hild and Restle, 223 f.

[9] Smbat, G. Dédéyan, 79 note 64.

[10] Ibid., note 65.

[11] Smbat, G. Dédéyan, 79 note 66. For further information on Anamur see: J. Russell, "Anemurium—Eine römische Kleinstadt in Kleinasien," *Antike Welt* 7 (1976), 3–20 (esp. bibliography on p. 20); idem, "Excavations at Anemurium," *Türk Arkeoloji Dergisi* 13–25 (1964–80), passim.

[12] Smbat, G. Dédéyan, 80 note 69.

[13] Ibid., note 70.

[14] Ibid., note 72.

[15] Hellenkemper, 14. Only a few attempts have been made to identify medieval monasteries with modern locales. Cf. M. Čevahirčyan, "Kilikiayi Akner vankʿi teładrutʿyunə," *Patma-banasirakan Handes* 3 (1982), 126–34; above, Part I.7, note 34.

[16] Smbat, G. Dédéyan, 71 note 98, 77 note 49.

[17] Certainly, Hetʿum of Lampron did not attend the coronation as a vassal of the Rubenids. Among the list of clergy in attendance were archbishops from as far away as Jerusalem; ibid., 74.

[18] See Part I.2.

[19] Smbat, G. Dédéyan, 32.

[20] Alishan, *Léon,* 120 ff. Armenians frequently adopted or were given Frankish and Greek names. Ritʿa is called both Xenia as queen and Marie as a nun; see Rüdt-Collenberg, 48.

[21] Another fortified site, Amuda, is not listed. Its absence in this manuscript is curious since Amuda was built long before Levon I was born and is mentioned elsewhere in Armenian texts (see Amuda [note 9] in the Catalogue).

Indices

1. Toponyms (excluding countries and districts)

The main entry for each fortification is in **boldface** type.

2. Personal Names
(includes only historical figures associated with Cilicia)

Plates

1a. Ak, looking south at the interior of A

1b. Ak, looking north at the exterior of C

2a. Ak, looking southwest at the interior of southwest wall

2b. Alafakılar, looking southeast at exterior of the northwest corner

3a. Amuda, looking northeast at H

3b. Amuda, looking north at the interior of the second level of H

4a. Amuda, looking northeast at the west wall of A

4b. Amuda, looking northeast at B

5a. Amuda, looking southeast at F

5b. Amuda, looking northwest at the interior of the northwest corner of F

6a. Anahşa, looking southeast at the exterior of E and F

6b. Anahşa, looking up into the machicolation of A

7a. Anahşa, looking southeast at the exterior of A

7b. Anahşa, looking south at C

8a. Anahşa, looking south at D from the west side of C

8b. Anahşa, looking northwest at the exterior of H

9a. Anavarza, looking northwest at the circuit west of A

9b. Anavarza, looking northeast at the exterior of the wall between A and B

10a. Anavarza, looking southeast at B

10b. Anavarza, looking southeast at C and D

11a. Anavarza, looking southwest at the interior of D

11b. Anavarza, looking up at the machicolation of G

12. Anavarza, looking northeast at the exterior of H

13a. Anavarza, looking southeast at an embrasure in the south unit of H

13b. Anavarza, looking southwest from the interior of the west chamber in the north unit of H

14a. Anavarza, looking up into the second level of the south unit of H

14b. Anavarza, looking southeast at the groined vault over the west chamber in the central unit of H

15a. Anavarza, looking northwest into the east chamber of the north unit of H

15b. Anavarza, looking south at the groined vault over the east chamber in the north unit of H

16a. Anavarza, looking southwest into the east chamber of the north unit of H

16b. Anavarza, looking northwest across the fosse (north of H)

17a. Anavarza, looking at a relief in the north end of passage J

17b. Anavarza, looking at a relief in the north end of passage J

18a. Anavarza, looking southwest at the interior of the tholos north of U

18b. Anavarza, looking northwest from K

19a. Andıl, looking southeast from Andıl Köy to the Kozan Baraj

19b. Andıl, looking northeast at the lower-level doors of the building near Andıl Köy

20a. Andıl, looking southeast at the exterior of Andıl Kalesi

20b. Andıl, looking south from H to K (lower right)

21a. Andıl, looking northwest at the exterior of Andıl Kalesi

21b. Andıl, looking southwest at the exterior of Andıl Kalesi

22a. Andıl, looking east at the interior of A

22b. Andıl, looking northeast at the window in G

23a. Andıl, looking west from H (the author atop the south wall of D)

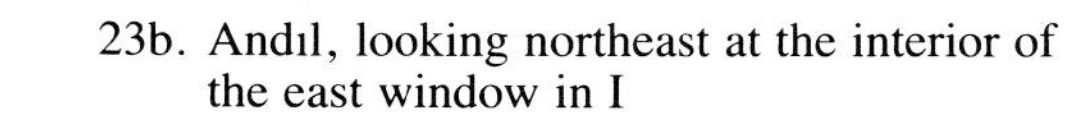

23b. Andıl, looking northeast at the interior of the east window in I

24a. Arslanköy, looking northeast at the fort and outcrop

24b. Arslanköy, looking east at the remains of a door(?)

25a. Ayas, looking northwest at the seaward side of the land castle

25b. Ayas, looking northeast at A and B (land castle)

26a. Ayas, looking southeast at A (land castle)

26b. Ayas, looking east from E (land castle) to the island castle

27a. Ayas, looking southeast at B (land castle)

27b. Ayas, looking east at the exterior of D (land castle)

28a. Ayas, looking northwest at the interior of the central and east casemates of D (land castle)

28b. Ayas, looking west at the interior of the westernmost casemate of D (land castle)

29a. Ayas, looking east at E (land castle)

29b. Ayas, looking west at a building east of C (land castle)

30a. Babaoğlan, looking northeast into the apse of A

30b. Babaoğlan, looking south at the exterior of the circuit from B to F

31a. Babaoğlan, looking southwest at tower D (far left)

31b. Babaoğlan, looking east at the interior of D

32a. Babaoğlan, looking west at D and the adjoining gate

32b. Babaoğlan, looking southwest at the exterior of the northeast wall of ward B

33a. Babaoğlan, looking southeast at the interior of cistern F

33b. Başnalar, looking west at the fort and outcrop

34a. Başnalar, looking northwest at the exterior of the eastern wall

34b. Başnalar, looking northeast at the north end of the fort

35a. Başnalar, looking southwest at the exterior of the north wall

35b. Başnalar, looking northwest at the west tower

36a. Bayremker, looking north from the fort

36b. Bayremker, looking south at the interior of the southeast wall

37a. Bayremker, looking northwest at the exterior of the south wall

37b. Bayremker, looking southwest at the collapsed east face

38a. Belen Keşlik, looking southwest at the fort

38b. Belen Keşlik, looking northwest at the entrance

39a. Belen Keşlik, looking east at the interior of the lower level

39b. Belen Keşlik, looking west at the interior of the lower level

40a. Belen Keşlik, looking west at the interior of the upper level

40b. Belen Keşlik, looking southwest into the interior of the east embrasure of the south wall (upper level)

41a. Belen Keşlik, looking east at the interior of the upper level

41b. Bostan, looking south from Bostan Köy to Bostan Kalesi

42a. Bostan, looking northwest at the gatehouse complex

42b. Bostan, looking northeast at the gatehouse complex

43a. Bostan, looking southwest at the exterior of the east door of the gatehouse

43b. Bostan, looking northeast at the interior of the east door of the gatehouse

44a. Bostan, looking north at the interior of the gatehouse

44b. Bostan, looking north at the remains of buildings on the southeast side of the summit

45a. Bostan, looking northwest at the interior of room A (Bostan Köy)

45b. Bostan, looking northwest at the inscription in B (Boston Köy)

46a. Bostan, looking west at room D (Bostan Köy)

46b. Bostan, looking east at the stone base of a mill near the northeast end of Bostan Köy

47a. Çalan, looking northeast at the fort and outcrop

47b. Çalan, looking west from G to the lower bailey

48a. Çalan, looking east from the lower bailey to the upper bailey

48b. Çalan, looking northeast from E to the exterior of G

49a. Çalan, looking northwest at the upper-level interior side of F

49b. Çalan, looking west at the lower-level interior side of F

50a. Çalan, looking south from F at G and E

50b. Çalan, looking east at the exterior of the north wall of H

51a. Çalan, looking northwest at the interior of H

51b. Çalan, looking east into the interior of H

52a. Çalan, looking southwest at the east end of the north wall of H

52b. Çalan, looking west from the fortress to the Iskenderun road

53a. Çandır, looking north from platform A up the first flight of steps

53b. Çandır, looking northeast at the interior of the gate that divides the steps between A and B

54a. Çandır, looking north from B down the second flight of steps

54b. Çandır, looking southwest at the interior of J

55a. Çandır, looking southwest at the interior of E2

55b. Çandır, looking west at the interior of E2

56a. Çandır, looking northwest at the interior of the upper- and lower-level west walls of E3

56b. Çandır, looking north at the interior of the upper- and lower-level north walls of E3

57a. Çardak, looking east at the interior of the south wall

57b. Çardak, looking east at D from C

58a. Çardak, looking west at the exterior of C

58b. Çardak, looking west at the interior of H

59a. Çardak, looking east at E from the upper-level stairs east of D

59b. Çem, looking southwest at the exterior of the first-level entrance to A

60a. Çem, looking west at the east flank of the fortified outcrop

60b. Çem, looking southwest at the exterior of A

61a. Çem, looking northeast at the interior of the lower-level of A

61b. Çem, looking southeast at the interior of the upper-level door/window over the lower-level entrance of A

62a. Çem, looking northeast at the interior of the upper level of A

62b. Çem, looking north at the south flank of A

63a. Çem, looking northeast at the exterior of C

63b. Çem, looking southeast at the interior of C

64a. Çem, looking south from E to D

64b. Çem, looking east at the interior of the embrasure in D

65a. Çem, looking southwest from F to A

65b. Dibi, looking south at the exterior of the fort

66a. Dibi, looking north at the exterior of the fort

66b. Dibi, looking northwest at the exterior of the southwest corner of the upper (east) bailey

67a. Dibi, looking east along the northern wall of the lower (west) bailey

67b. Dibi, looking south at the jamb(?) of the door between the upper and lower baileys

68a. Evciler, looking southwest at the exterior of the donjon

68b. Evciler, looking northeast at the exterior of the donjon

69a. Evciler, looking west at the interior of the donjon; wood visible in the core of the wall at the right

69b. Evciler, looking northeast at the interior of the donjon

70a. Evciler, looking south from the donjon across the bailey

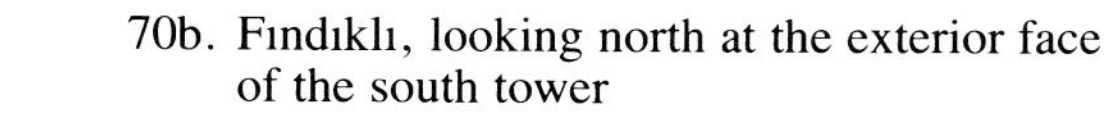

70b. Fındıklı, looking north at the exterior face of the south tower

71a. Fındıklı, looking southeast at the collapsed interior side of the south tower

71b. Fındıkpınar, looking east at the fortified outcrop

72a. Fındıkpınar, looking north from the fort
to the medieval settlement below

72b. Fındıkpınar, looking southwest at the east tower

73a. Fındıkpınar, looking southeast across the south tower

73b. Fındıkpınar, looking southwest at the exterior of the east wall (south of the east tower)

74a. Geben, looking northwest at the central circuit wall

74b. Geben, looking north at the exterior of complex D (left) and I–J (right)

75a. Geben, looking northeast into D1, D2, and D3

75b. Geben, looking southwest into the lower level of D1

76a. Geben, looking northeast into D3

76b. Geben, looking southeast at the lower-level junction of D1 and D3

77a. Geben, looking northwest into D2

77b. Geben, looking west at the interior of the small west room in D2

78a. Geben, looking southeast at the east window (north of the central rib) in I

78b. Geben, looking north into J

79a. Geben, looking west at the north walls of G and H and at the west wall of I

79b. Geben, looking northeast into the interior of K

80a. Geben, looking south from J to K

80b. Gediği, looking west at the entrance between A and B

81a. Gediği, looking southwest from atop the complex to the neighboring outcrop

81b. Gökvelioğlu, looking southwest at the northeast corner of the fortified outcrop

82a. Gökvelioğlu, looking south at G and F

82b. Gökvelioğlu, looking southwest at E

83a. Gökvelioğlu, looking southeast at the east end of B (far right) and the north flank of C

83b. Gökvelioğlu, looking south at the collapsed second level of the north flank of C

84a. Gökvelioğlu, looking northeast at the interior of the north wall of C

84b. Gökvelioğlu, looking east at the interior of C

85a. Gökvelioğlu, looking north at the junction of C and the south wall of B

85b. Gökvelioğlu, looking northeast at the west entrance to D (south wall)

86a. Gökvelioğlu, looking east through D

86b. Gökvelioğlu, looking southeast at the east wall of D

87a. Gökvelioğlu, looking northeast at K

87b. Gökvelioğlu, looking southwest at P and Q

88a. Gösne, looking northwest at the exterior of the west tower

88b. Gösne, looking southeast at the entrance to the east building

89a. Gösne, looking east at the exterior of the west tower (left) and the south flank of the east building (right)

89b. Gösne, looking south at the interior of the southeast door and flanking window in the west tower

90a. Gösne, looking southeast at the interior of the east building

90b. Gösne, looking northwest at the interior of the east building

91a. Gülek, looking northwest at the exterior of A

91b. Gülek, looking southeast at the interior of A

92a. Gülek, looking up through the machicolation of A

92b. Gülek, looking east at B

93a. Gülek, looking west at the exterior of D and E

93b. Gülek, looking south at the interior door of E

94a. Gülek, looking northwest at the exterior of F

94b. Haçtırın, looking southeast at the fortified outcrop

95a. Haçtırın, looking southeast at the exterior of A and B

95b. Haçtırın, looking north at the interior of A and B

96a. Haruniye, looking north at the exterior of E (right)

96b. Haruniye, looking northeast at the exterior of the north circuit

97a. Haruniye, looking northwest at the exterior of the salients flanking Λ

97b. Haruniye, looking north at the exterior of A

98a. Haruniye, looking northwest at the interior of an embrasure northwest of A

98b. Haruniye, looking northeast at the west end of E

99a. Haruniye, looking west through C

99b. Haruniye, looking east through C at D

100a. Haruniye, looking west at the exterior of D

100b. Haruniye, looking south at the junction of D and E

101a. Hasanbeyli, looking southeast at the fortified outcrop from the north end of the village

101b. Hasanbeyli, looking east from the fort to the Fevzipaşa road

102a. Hasanbeyli, looking northwest at the fort

102b. Hasanbeyli, looking east at the fort

103a. Hisar, looking northwest at the fortified outcrop

103b. Hisar, looking southwest at the north end of the fort

104a. Hisar, looking west at the interior of the north tower

104b. Hisar, looking east at the interior of the south tower

105a. Hotalan, looking north at the fortified outcrop

105b. Hotalan, looking east at the entrance to the fort

106a. Hotalan, looking northeast at the north jamb of the entrance

106b. Hotalan, looking northwest at the interior of the lower level

107a. Işa, looking northwest at the fortified outcrop

107b. Işa, looking up at the exterior of the machicolation

108a. Işa, looking northeast at the entrance to the fort

108b. Işa, looking southwest at the interior of the gate-door

109a. Işa, looking northwest at the platform over the machicolation

109b. Işa, looking east at the embrasures directly northeast of the gate-entrance

110a. Kalası, looking southeast at the exterior of the extreme northwest wall

110b. Kalası, looking south at the interior of the south circuit

111a. Karafrenk, looking west from the fort

111b. Karafrenk, looking north at the south face of the fort

112a. Karafrenk, looking southeast at the northwest tower

112b. Karafrenk, looking southwest at the northwest tower

113a. Karafrenk, looking northeast at the west wall

113b. Karafrenk, looking south along the interior of the east wall

114a. Karafrenk, looking northeast at the southwest corner

114b. Karafrenk, looking west at the northeast tower

115a. Kız (near Dorak), looking west at the keep from inside the bailey

115b. Kız (near Dorak), looking west at the base of the south wall of the keep

116a. Kız (near Dorak), looking northeast at the west face of the keep

116b. Kız (near Dorak), looking northeast into the northeast corner of A3

117a. Kız (near Dorak), looking east at the west face of the keep near the junction of B and C

117b. Kız (near Dorak), looking east at the base of the vertical chase on the west face

118a. Kız (near Dorak), looking northwest at the east face of the keep

118b. Kız (near Dorak), looking north into the niche in the north wall of B1

119a. Kız (near Dorak), looking northwest into the interior of C2

119b. Kız (near Dorak), looking southeast into the interior of C2
(the door between C1 and B1 at lower right)

120a. Kız (near Dorak), looking south at the interior of the south window in A1

120b. Kız (near Dorak), looking northeast at the interior of the southeast window/door in C3

121a. Kız (near Dorak), looking northwest at the south doors of C2 and C3

121b. Kız (near Dorak), looking northeast at a cross-relief in the northeast corner of B2

122a. Kız (near Dorak), looking northwest at the interior of the north wall of C3

122b. Kız (near Dorak), looking south at the interior of the south wall of C3

123a. Korykos, land castle, looking south at E and G (sea castle at upper right)

123b. Korykos, sea castle, looking east at J

124a. Korykos, sea castle, looking southeast at F (postern at right)

124b. Korykos, sea castle, looking east at the second-level entrance to F

125a. Korykos, sea castle, looking east at E and F

125b. Korykos, sea castle, looking south into the casemate east of D

126a. Korykos, sea castle, looking northeast at the exterior of I

126b. Korykos, sea castle, looking northwest from A to I
(interior of the colonnade)

127a. Korykos, sea castle, looking east from H to G

127b. Korykos, sea castle, looking northwest at the interior of H

128a. Korykos, land castle, looking southwest at D

128b. Korykos, land castle, looking northeast at the stairs between P and Q

129a. Kozcağız, looking north at the south tower of A

129b. Kozcağız, looking northwest at the interior of A's door

130a. Kozcağız, looking north from A across the interior of the fort

130b. Kozcağız, looking south at the north wall of A

131a. Kozcağız, looking west into the room adjoining C

131b. Kozcağız, looking northwest at the exterior of C

132a. Kütüklu, looking south at the exterior of the northwest tower

132b. Kütuklu, looking northeast at the junction of the northwest tower and the west wall

133a. Kütüklu, looking north along the east wall of the fort to the village of Kütüklu

133b. Kuzucubelen, looking southwest at the interior of the fort

134a. Kuzucubelen, looking northeast at the fort

134b. Kuzucubelen, looking northwest at the interior of the upper level

135a. Lamas, looking east across the interior of the fort

135b. Lamas, looking northwest from E across the interior

136a. Lamas, looking northwest at the interior side of the wall flanking B

136b. Lampron, looking east at E

137a. Lampron, looking northwest at the exterior of A

137b. Lampron, looking northwest at the interior of the central dome of A

138a. Lampron, looking northeast at F and G

138b. Lampron, looking northwest into F (lower left) and at the path to G

139a. Lampron, looking east at the exterior of G

139b. Lampron, looking northwest at the interior of G

140a. Lampron, looking north at I

140b. Lampron, looking northwest at K (foreground) and N–P (background)

141a. Lampron, looking north at L, M, and N

141b. Lampron, looking northwest at the interior of N

142a. Lampron, looking northwest
at the interior of O

142b. Lampron, looking northeast
at the interior of O

142c. Lampron, looking east at the southernmost
embrasured window in the east wall of P

143a. Lampron, looking northwest at the interior of P

143b. Lampron, looking south at the interior of P

144a. Mancılık, looking east at the interior of A

144b. Mancılık, looking northwest at the interior of C

145a. Mancılık, looking west at the north jamb of the outer door of A

145b. Mancılık, looking northeast at the exterior of C

146a. Mancılık, looking northeast at the interior of the south chamber of H

146b. Mancılık, looking southwest at the interior of the north chamber of H

147a. Mansurlu, looking north at the fortified outcrop (upper right)

147b. Mansurlu, looking southwest at a fragment of the wall (Küçük Mansurlu to the left)

148a. Maran, looking east at the exterior of A and B

148b. Maran, looking southwest at the interior of A

149a. Maran, looking west at the exterior of B's door

149b. Maran, looking southeast at D, E, and F

150a. Maran, looking northeast at the interior of B's door

150b. Meydan, looking west at the exterior of the outer door of A

151a. Meydan, looking southeast at the interior of the outer door of A

151b. Meydan, looking west at the interior of the inner door of A

152a. Meydan, looking west at the exterior of the wall between A and B

152b. Meydan, looking west at the base of the fourth loophole north of B

153a. Meydan, looking east at the interior of a loophole in the wall between A and B

153b. Meydan, looking northeast at the interior of B

154a. Meydan, looking southeast at D

154b. Meydan, looking west from E to F (loophole base at bottom left)

155a. Meydan, looking southwest at the exterior of tower F

155b. Meydan, looking southeast at the interior of a window in the east wall of O

156a. Meydan, looking north at the north bailey

156b. Meydan, looking southwest at J

157a. Meydan, looking southwest at the exterior of the entrance to R

157b. Meydan, looking northwest at the interior of R

158a. Meydan, looking southwest at the interior of the west wall of O

158b. Meydan, looking northeast from O to the village of Karsantı

159a. Milvan, looking west from Milvan Köy to Milvan Dağı

159b. Milvan, looking west at the second gate (far left) from the third gate

160a. Milvan, looking west at buildings above the third gate

160b. Milvan, looking east at the third gate

161a. Milvan, looking south into the east half of the foundation of a rectangular structure in the middle level

161b. Milvan, looking southwest into the west half of the same structure as in pl. 161a (gate at right)

162a. Milvan, looking southeast at the north end of the summit complex

162b. Milvan, looking southwest at the exterior of the north wall of A

163a. Milvan, looking northwest from C to A and B

163b. Milvan, looking northwest into the east apse of A

164a. Milvan, looking northwest at the junction of the north and east apses of A

164b. Milvan, looking northwest at the interior of the west entrance(?) into A

165a. Milvan, looking west at the exterior of B

165b. Milvan, looking southeast at the interior of B

166a. Milvan, looking east at the interior of C

166b. Milvan, looking southeast at the interior of D

167a. Misis, looking west at the bridge and fortified outcrop

167b. Mitisin, looking west at the interior of the south end

168a. Mitisin, looking southeast at the exterior of the southwest flank

168b. Mitisin, looking south from the fort to Mitisin Köy

169a. Payas, bedesten fort, looking southeast at the gate complex (barbican to the left)

169b. Payas, bedesten fort, looking south at the south half of the east wall

170a. Payas, bedesten fort, looking south at the exterior of the central-west tower

170b. Payas, bedesten fort, looking southwest at the exterior of the gatehouse

171a. Payas, bedesten fort, looking north across the interior of the fort

171b. Payas, bedesten fort, looking southwest at the interior of the upper level of the central-west tower

172a. Payas, harbor fort, looking north

172b. Payas, harbor fort, looking northwest

173a. Payas, harbor fort, looking south at the exterior of the north wall

173b. Payas, harbor fort, looking north at the south door of the gatehouse

174a. Payas, harbor fort, looking southwest at the interior of the south door of the gatehouse

174b. Payas, harbor fort, looking northeast at the interior of the east door of the gatehouse

175a. Payas, harbor fort, looking north at the interior court

175b. Payas, harbor fort, looking east at the keep

176a. Payas, harbor fort, looking northeast at the interior of the first level of the keep

176b. Payas, harbor fort, looking southwest at the interior of the corridor-entrance to second and third levels

177a. Payas, harbor fort, looking north at the interior of the second level

177b. Payas, harbor fort, looking east through the north flank of the second-level ambulatory

178a. Payas, harbor fort, looking south at the interior of the second level

178b. Payas, harbor fort, looking north at the exterior of the third level (entrance at left)

179a. Pillar of Jonah, looking north at the Pillar and pass

179b. Pillar of Jonah, looking northeast from the Pillar to Sarı Seki (shown by arrow)

180a. Pillar of Jonah, looking east at the Pillar

180b. Pillar of Jonah, looking south at the Pillar

181a. Pillar of Jonah, looking southeast at the base of the Pillar

181b. Pillar of Jonah, looking southeast from the Pillar into the east flank of the pass

182a. Rifatiye I, looking southeast at the exterior of the circuit (east tower at upper left)

182b. Rifatiye I, looking south at the exterior of the circuit (west flank)

183a. Rifatiye I, looking northeast at the interior side of the circuit (east tower at upper right)

183b. Rifatiye II, looking northwest at the fortified outcrop

184a. Rifatiye II, looking at the core of the circuit

184b. Rifatiye II, looking south at the exterior of the north tower

185a. Saimbeyli, looking northeast at the west flank of the fortress-outcrop

185b. Saimbeyli, looking southwest from the top of A

186a. Saimbeyli, looking southeast at the exterior of A and B

186b. Saimbeyli, looking northwest from E across D and F

187a. Saimbeyli, looking north at C

187b. Saimbeyli, looking southwest at the interior of F

188a. Saimbeyli, looking north at the interior side of the wall directly east of A

188b. Saimbeyli, looking north into the interior of the upper level of A (lower level at bottom)

189. Sarı Çiçek, looking southeast at A (foreground) and B (background)

190a. Sarı Çiçek, looking west at the interior of A

190b. Sarı, Çiçek, looking south at the interior side of the east window in the south wall of A

191. Sarı Çiçek, looking southwest at the interior of A

192a. Sarı Çiçek, looking south at B, C, D, E, and A

192b. Sarı Çiçek, looking north at the interior of the lower level of B

193a. Sarı Çiçek, looking southwest at the north window of F

193b. Sarı Çiçek, looking northwest at G

194a. Savranda, looking north from I to A and B

194b. Savranda, looking east from A along the exterior of the outwork

195a. Savranda, looking southeast at the interior of A

195b. Savranda, looking northwest at the interior of A

196a. Savranda, looking northeast at the interior of A

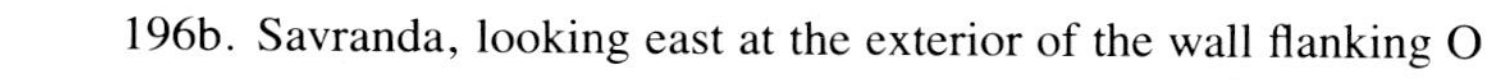

196b. Savranda, looking east at the exterior of the wall flanking O

197a. Savranda, looking southwest from the outer door of A to the curtain wall

197b. Savranda, looking east at the interior of the lower level of I

198a. Savranda, looking south at the exterior of the north wall of J

198b. Savranda, looking southeast at the exterior of M and L

199a. Savranda, looking north at the junction between O (right) and N (left)

199b. Savranda, looking east at the exterior of the circuit east of O

200a. Silifke, looking southeast into B and at the west flank of C

200b. Silifke, looking northeast at the southwest opening in G

201a. Silifke, looking north at the exterior of the south wall of H (G at the far right)

201b. Silifke, looking northwest at the interior of E

202a. Silifke, looking northeast at the exterior of the west wall of H

202b. Silifke, looking north at the interior of the second level of I

203a. Silifke, looking northwest at the interior of the northwest embrasure on the second level of I

203b. Silifke, looking southeast at the interior of the southeast door of Y

204a. Silifke, looking at the interior of the upper level of the staircase in N

204b. Silifke, looking south at X

205a. Silifke, looking northeast at the exterior of the south wall of Y

205b. Silifke, looking northwest at a cistern within the late antique city of Seleucia

206a. Sinap (near Lampron), looking south at the fort

206b. Sinap (near Lampron), looking south at the exterior of the upper-level north door

207a. Sinap (near Lampron), looking west at the fort

207b. Sinap (near Lampron), looking north and up at the exterior of the south wall above the door

208a. Sinap (near Lampron), looking southeast at the interior of the first level

208b. Sinap (near Lampron), looking south at the interior of the first-level south door (Lampron in background)

209a. Sinap (near Lampron), looking southeast at the interior of the upper level

209b. Sinap (near Lampron), looking northwest at the interior of the upper level

210a. Sinap (near Lampron), looking north at the interior of the north-central casemate (upper level)

210b. Sinap (near Lampron), looking northeast at the interior of the northeast casemate (upper level)

211a. Sis, looking north at the exterior of B, A, and N (center), with S–W above

211b. Sis, looking west at the exterior junction between A and B

212a. Sis, looking southwest at the exterior of A

212b. Sis, looking southwest at the exterior of the east door of A

213a. Sis, looking northeast at the interior of A

213b. Sis, looking southeast at the interior of an embrasure in A

214a. Sis, looking east at the interior of an embrasured casemate in B

214b. Sis, looking northeast from M toward E and F

215a. Sis, looking southwest from N to C–F (below), with I–M at upper right

215b. Sis, looking south at the exterior of D

216a. Sis, looking southwest at the east gate of complex I

216b. Sis, looking south at H from M

217a. Sis, looking northeast at the interior of L

217b. Sis, looking east at the interior of M

218a. Sis, looking north at the interior of U

218b. Sis, looking north at the top of the tower east of P (foreground) and at Q

219a. Sis, looking west at V

219b. Sis, looking north at the exterior of the circuit between N and O (machicolation in center)

220a. Sis, looking east at the exterior of N

220b. Sis, looking southeast into the east chamber of X

221a. Sis, looking southeast into the west chamber of X

221b. Tamrut, looking east at the fortress-outcrop

222. Tamrut, looking north at the exterior of B, A, and C

223a. Tamrut, looking northeast at the inscription over A

223b. Tamrut, looking southwest at D, C, A, and B

224a. Tamrut, looking south at the interior side of the outer door of A

224b. Tamrut, looking northwest from the upper level of C to B

225a. Tamrut, looking south at the upper level and ceiling of B

225b. Tamrut, looking west at the upper-level embrasured loophole in C

226a. Tamrut, looking southwest at the exterior of the door in D

226b. Tamrut, looking northwest at the interior of the door in D

227a. Tamrut, looking southwest at the interior of D

227b. Tamrut, looking northwest at the interior of the rectangular room atop E

228a. Tamrut, looking south at F, G, and H

228b. Tamrut, looking north at I, J, K, and L

229a. Tamrut, looking southeast at an upper-level embrasure in F

229b. Tamrut, looking south into the upper level of G

230a. Tamrut, looking northwest at the interior of the lower level of H

230b. Tamrut, looking north at the interior of the upper level of H

231a. Tamrut, looking northeast at the exterior of J

231b. Tamrut, looking northwest at the interior of J

232a. Tece, looking east at the exterior of the north wall of the keep

232b. Tece, looking northwest at the interior of the keep

233a. Tece, looking north at the interior of the keep

233b. Tece, looking east at the interior of the keep (first level)

234a. Tece, looking north at the keep

234b. Toprak, looking southeast at the exterior of Y

235a. Toprak, looking north at the south end of the talus (southwest of K)

235b. Toprak, looking southeast at the interior of a collapsed corridor in the talus below I

236a. Toprak, looking west at the exterior of the salient southeast of Y

236b. Toprak, looking northwest at the exterior of the north tower of A

237a. Toprak, looking northwest at the interior of B

237b. Toprak, looking north at the interior of a window in the upper level of D

238a. Toprak, looking northwest at the interior of the chamber directly southeast of E (main level)

238b. Toprak, looking north from G to E

239a. Toprak, looking north at the interior of F

239b. Toprak, looking northwest at the interior of G (main level)

240a. Toprak, looking south at the interior of J

240b. Toprak, looking south at the interior of J (south end)

241a. Toprak, looking west at the junction of F and J

241b. Toprak, looking northwest into the upper-level embrasures between E and G

242a. Toprak, looking west into an upper-level embrasure between G and H

242b. Toprak, looking east at the exterior of the circuit between G and H

243a. Toprak, looking south from H to I

243b. Toprak, looking northeast at the exterior of K (lower bailey wall at far left)

244a. Toprak, looking northwest at the interior of the upper level of K

244b. Toprak, looking east at the interior of N

245a. Toprak, looking north at an embrasured window in the upper level north wall of K

245b. Toprak, looking north at the exterior of the east door in the upper level of K

246a. Trapesak, looking west at the fortified outcrop

246b. Trapesak, looking southwest at a fragment of the lower east circuit

247a. Trapesak, looking northwest at the exterior of the south undercroft

247b. Trapesak, looking southeast over the top of the south undercroft

248a. Trapesak, looking south at the exterior of the south undercroft (left)

248b. Trapesak, looking east at the interior of the south undercroft

249a. Tumil, looking northeast at the northeast tower

249b. Tumil, looking southeast at the northeast tower

250a. Tumil, looking northeast from the northwest tower

250b. Tumil, looking north at the east flank of the northeast tower

251a. Tumlu, looking southwest at the fortress-outcrop

251b. Tumlu, looking south at the exterior of H

252a. Tumlu, looking north at the exterior of B and F

252b. Tumlu, looking southeast at the interior side of B

253a. Tumlu, looking northwest at the exterior of C

253b. Tumlu, looking southeast at the interior of the east door of C

254a. Tumlu, looking northeast at the west wall of D

254b. Tumlu, looking northwest at the interior of D

255a. Tumlu, looking northeast at E and C

255b. Tumlu, looking northwest at E (lower left) and O

256a. Tumlu, looking south at the interior of E

256b. Tumlu, looking northeast at the interior of the second embrasure from the north in the upper level of H

257a. Tumlu, looking south at J

257b. Tumlu, looking south into the far east embrasure of N

258a. Tumlu, looking northwest at the interior of K

258b. Tumlu, looking south at the northwest end of K

259a. Tumlu, looking southwest at L and M

259b. Tumlu, looking southwest at N

260a. Vahga, Byzantine church, looking northeast at the exterior of the apse

260b. Vahga, Byzantine church, looking southwest at the interior of the south wall of the nave

261a. Vahga, Byzantine church, looking northeast at the exterior of the south wall of the nave

261b. Vahga, Byzantine church, a relief near the church

262a. Vahga, looking southeast at the fortress-outcrop

262b. Vahga, looking south from J at G and F

263a. Vahga, looking east at B

263b. Vahga, looking northeast at the exterior of the south door of A

264a. Vahga, looking northwest at the interior of the west embrasure in A

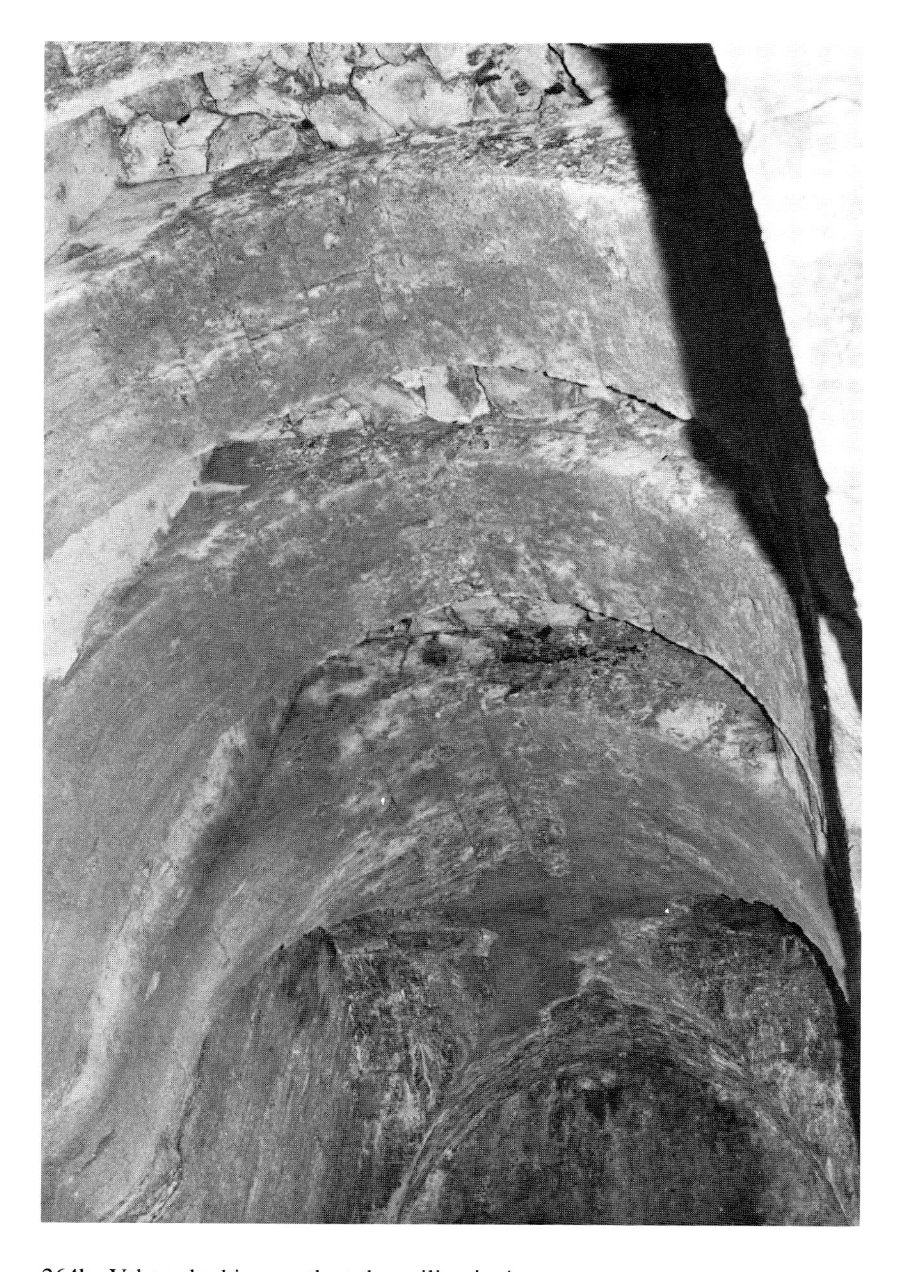

264b. Vahga, looking north at the ceiling in A

265a. Vahga, looking south from E at B and C

265b. Vahga, locking northwest at the south face of F

266a. Vahga, looking southwest at the east face of F

266b. Vahga, looking north at G

267a. Vahga, looking northwest into H and I

267b. Vahga, looking southeast at J (the walls of K, I, and G at lower right)

268a. Vahga, looking north through M

268b. Vahga, looking northeast into N

269a. Vahga, looking south at P

269b. Yaka, looking west at the vault immediately south of the north wall

270a. Yaka, looking north at the fort

270b. Yaka, looking northeast at the northeast tower

271a. Yaka, looking southwest at the base of the north wall

271b. Yeni Köy, looking southeast at the fortified outcrop

272a. Yeni Köy, looking northwest across the bailey

272b. Yeni Köy, looking southwest at the main entrance

273a. Yeni Köy, looking west at the door into the west tower

273b. Yeni Köy, looking west at the interior of the west tower

274a. Yılan, looking northeast at the west flank of the fortress-outcrop

274b. Yılan, looking north at the east flank of the fortress-outcrop

275a. Yılan, looking west at the exterior of A

275b. Yılan, looking southwest at the exterior of B

276a. Yılan, looking southwest from E across the central bailey

276b. Yılan, looking south at the far south tower

277a. Yılan, looking north from the far south tower (tower C at left)

277b. Yılan, looking north at the interior of D

278a. Yılan, looking northwest across the lower bailey

278b. Yılan, looking northwest at the exterior of the east door of E

279a. Yılan, looking north at E

279b. Yılan, looking at the vaulted ceiling over E

280a. Yılan, looking west at the corner of the vault over E

280b. Yılan, looking southwest at the interior of the west tower flanking E

281a. Yılan, looking south from G to F

281b. Yılan, looking east at the northwest corner of G

282a. Yılan, looking east at G (right) and H

282b. Yılan, looking north at G

283a. Yılan, looking northeast at the interior of H

283b. Yılan, looking southwest at the interior of H

284a. Yılan, looking southwest at the exterior of G's apse

284b. Yılan, looking northeast at the postern and tower east of I

285a. Yılan, looking northwest at the eastern casemates over J (left)

285b. Yılan, looking northwest across the top of L

286a. Yılan, looking east at the interior of J

286b. Yılan, looking north at the interior of M

287a. Anacık, looking southeast at the exterior

287b. Anavarza, looking north at the exterior of the south wall of the south bailey

287c. Anavarza, looking northeast at the exterior of H, I, T, S, and R

288a. Anacık, looking north at the entrance in the south wall

288b. Anavarza, looking west at the exterior of tower C

288c. Anavarza, looking south at H

289a. Anavarza, looking north at G

289b. Anavarza, looking southwest into the south end of the central bailey

289c. Anavarza, looking southeast at the exterior of tower M

290a. Ayas, looking west at the watchtower of Süleyman I

290b. Azgit, looking southeast at the exterior of gate A

290c. Azgit, looking west at the interior of postern F

291a. Ayas, looking northeast at the exterior of room A on the island castle

291b. Bodrum, looking southeast at the fortified outcrop

291c. Bodrum, looking west at the fortified outcrop

292a. Bucak, looking south at the interior of the south tower and gate

292b. Çandır, looking west at the entrance complex A–B

292c. Çandır, looking west at apartment E

293a. Çandır, looking southeast (from below) at E

293b. Çandır, looking northwest into the far northwest corner of E1 (lower level)

293c. Çandır, looking north at the interior of the north wall in the lower level of E1

294a. Çandır, looking northwest at the window in the north wall of the upper level of E1

294b. Çandır, looking south into the lower-level entrance of the stairway in E3

294c. Çandır, looking southwest at the central window in the upper-level west wall of E3

295a. Çandır, looking southwest into the lower level of E3

295b. Çem, looking northwest at the tympanum in the outer door of gatehouse A

295c. Fındıklı, looking northeast at the fortified outcrop

296a. Gökvelioğlu, looking northeast at B and C

296b. Korykos, land castle, looking north at the exterior of the circuit between Z and J

296c. Korykos, land castle, looking northwest at the exterior of M

297a. Kum, looking west at the east face of the fort

297b. Mancılık, looking east at the junction of F and G

297c. Meydan, looking south at the south bailey from the north bailey

298a. Meydan, looking at a broken capital south of O

298b. Savranda, looking northwest at A

298c. Savranda, looking southwest at the east flank of K

299a. Meydan, looking northeast at the interior of the wall between A and B (S above)

299b. Sarı Çiçek, looking northeast at the exterior of B

299c. Silifke, looking northwest at the exterior of the outer door of C

300a. Silifke, looking northwest at the exterior of Q, P, and O

300b. Silifke, looking west into the south half of the moat (P and O at right)

300c. Sis, looking east at the south end of the west face of the fortress and outcrop

301a. Sinap (near Çandır), looking east at the exterior of the west wall

301b. Sinap (near Lampron), looking north at the fort

301c. Tumlu, looking northwest at the east flank of the fortress and outcrop

302a. Tece, looking northwest at the exterior of the keep

302b. Tumlu, looking northeast at the exterior of M

302c. Yılan, looking northeast at K and I